CONCRETE MANUAL

A WATER RESOURCES TECHNICAL PUBLICATION

A manual for the control of concrete construction

EIGHTH EDITION, RIVISED REPRINT 1981

REPRINTED 1988

U.S. DEPARTMENT OF THE INTERIOR

BUREAU OF RECLAMATION

UNITED STATES
GOVERNMENT PRINTING OFFICE
WASHINGTON : 1975

For sale by the Superintendent of Documents, U.S. Government Printing Office, Washington DC 20402, and the Water and Power Resources Service, P O Box 25007, Denver CO 80225.

PREFACE TO THE EIGHTH EDITION

This Eighth Edition of the Concrete Manual reflects the Bureau of Reclamation's continuing effort to bring to its construction staff members information on the latest advances in concrete technology which would be useful to them in administering contracts for construction of the Bureau's water resource development projects throughout the western States. Evolving from a set of loose-leaf, blue-printed instructions, the first tentative edition of the manual was published in 1936. Since that time, the Bureau has published seven editions, each recording the many advances in concrete technology developed during the intervening years. The Seventh Edition was published in 1963.

Although fundamental precepts of good concrete practice do not change, continued research and development of the technology bring about significant improvements that keep concrete in the forefront as a versatile, dependable, and economical construction material. This Eighth Edition of the Concrete Manual underscores this progress; it embraces substantial supplemental information relating to many of these improvements in concrete control and technology. Chapter III (Concrete Mixes) now includes compressive strength design criteria for concrete containing water-reducing, set-controlling agents. Chapter VII (Repair and Maintenance of Concrete) has been rewritten to describe in detail techniques introduced in the Seventh Edition for using epoxy in concrete repairs. This latest edition of the manual also includes a brief discussion of concrete-polymer materials, new composites which have considerable potential in construction. Information on the manufacture of concrete pipe has been supplemented and revised. Shotcrete containing coarse aggregate, used for tunnel support, is discussed, and methods for removing stains from concrete surfaces are now described in detail. As in the past, new and helpful suggestions relating to field laboratory sampling and testing equipment are included.

The manual, published primarily for use by the Bureau's construction engineers and inspectors, supplies engineering data and outlines methods and procedures to be followed in administering construction specifications and contracts. References in the manual to "laboratories," "Denver laboratories," "Denver laboratory," and "Denver office" designate the Bureau's Engineering and Research Center at Denver, Colo. Howard J. Cohan, Chief of the Division of General Research, provides overall administrative direction to the many technical research and testing activities, of which preparation of this manual is but one.

Although issued primarily for Bureau of Reclamation staff use, the manual has received widespread acceptance throughout the United States and in many foreign countries. More than 120,000 copies of previous editions, including 40,000 copies of the Seventh Edition, have been distributed throughout the world. Indicative of this world-wide recognition of the technical value of the manual is the fact that it has been translated into Spanish, Italian, and Japanese.

Some procedures in this manual are directly referred to by Bureau of Reclamation specifications. When this is done, these referenced procedures have the full effect of specifications requirements. However, there may be instances where procedures and instructions in the manual are at variance, in some respects, with specifications requirements. In these instances, it must be understood that the specifications take precedence. It is also emphasized that each employee of the Bureau of Reclamation is directly accountable to his supervisor; thus, he should request advice concerning any doubtful course of action from the proper authority.

This edition of the Concrete Manual and earlier editions represent the expertise of individuals too numerous to mention. Their substantial contributions are acknowledged with appreciation, for their efforts provided the foundation on which each succeeding edition has been based.

Engineers in the Concrete and Structural Branch, Division of General Research, prepared the manuscript for the Eighth Edition. Engineer A. B. Crosby, under the direction of E. M. Harboe, then Acting Chief of the Concrete and Structural Branch, coordinated the initial preparation for this edition and was in charge of the major revisions. Substantial contributions were made by the present Chief, Concrete and Structural Branch, J. R. Graham, and engineers H. E. Dickey, N. F. Larkins, L. C. Porter, and J. D. Richards. H. Johns, Applied Sciences Branch, also contributed significantly. The assistance of R. N. Hess for his work on the tables and figures, and his technical review is also acknowledged. Personnel in the Technical Services and Publications Branch, Division of Engineering Support, edited the manuscript; the Publications and Photography Branch in the Commissioner's Office, Washington, D.C., reviewed the manuscript and proofs and arranged for printing. The assistance of these, and many other engineers and technicians, past and present, who contributed in various ways to this publication, is gratefully acknowledged.

This Eighth Edition of the Concrete Manual has had a distribution of approximately 17,000 copies, not including the 16,000 copies of this reprint.

A PERTINENT QUOTATION

Although a concrete manual may fully describe the steps necessary for the accomplishment of first-class work, such an exposition, no matter how perfect, will not in itself insure concrete of good quality. This was recognized by Franklin R. McMillan, member of the Concrete Research Board for Hoover Dam, who, in concluding his "Basic Principles of Concrete Making" published in 1929, stated:

"* * * one further requirement remains. There must be a recognition on the part of someone in authority that uniform concrete of good quality requires intelligent effort and faithfulness to details all along the line—proper materials, proper design, proper mixing and transporting, and special care in placing and protecting. It must be recognized that to obtain the desired results some qualified person must be made responsible for these details, and having been made responsible, must be entrusted with the necessary authority.

"Too often individuals in ultimate authority have the desire for concrete of the proper quality, but fall short of attaining it through failure to delegate the necessary authority and to fix the responsibility for results. It is not uncommon to find a construction superintendent in a position to ignore the recommendations of the engineer where, in his opinion, they impede the progress of the work or increase the cost. If, under such conditions, quality is subordinated to first cost, durable structures cannot be expected.

"It must not be assumed that because it requires well-directed effort to produce uniformly good concrete the cost is necessarily increased. There have been any number of examples in recent years where rigid control of the concreting operations not only has given concrete of the required quality but has shown a distinct saving in first cost as compared with earlier experiences in which only indifferent or unsatisfactory results were obtained. But even if the first cost is increased by the requirements for definite quality, the ultimate cost which must include maintenance and repair charges will be greatly decreased."

CONTENTS

CHAPTER II—INVESTIGATION AND SELECTION OF CONCRETE MATERIALS

A. PROSPECTING FOR AGGREGATE MATERIALS

CHAPTER IV—INSPECTION, FIELD LABORATORY FACILITIES, AND REPORTS

A. INSPECTION

B. FIELD LABORATORY FACILITIES

C. REPORTS AND EVALUATION OF TEST DATA

CHAPTER V—CONCRETE MANUFACTURING

A. MATERIALS

CHAPTER VI—HANDLING, PLACING, FINISHING, AND CURING

A. PREPARATIONS FOR PLACING

CHAPTER VIII—SPECIAL TYPES OF CONCRETE AND MORTAR

A. LIGHTWEIGHT CONCRETE

B. HEAVYWEIGHT CONCRETE

APPENDIX

APPENDIX—Continued

LIST OF FIGURES

LIST OF FIGURES—Continued

LIST OF FIGURES—Continued

LIST OF FIGURES—Continued

LIST OF FIGURES—Continued

LIST OF FIGURES—Continued

LIST OF FIGURES—Continued

LIST OF FIGURES—Continued

LIST OF FIGURES—Continued

LIST OF TABLES

INDEX

Chapter I

CONCRETE AND CONCRETE MATERIALS

A. Introduction

1. Concrete Defined.—Concrete is composed of sand, gravel, crushed rock, or other aggregates held together by a hardened paste of hydraulic cement and water. The thoroughly mixed ingredients, when properly proportioned, make a plastic mass which can be cast or molded into a predetermined size and shape. Upon hydration of the cement by the water, concrete becomes stonelike in strength and hardness and has utility for many purposes.

2. Progress in Concrete.—Concrete has found use in nearly all types of construction—from highways, canal linings, bridges, and dams to the most beautiful and artistic of buildings. With the addition of reinforcement to supply needed tensile strength, advances in structural design, and the use of prestressing and posttensioning, it has become the foremost structural material. The growing popularity of concrete in the United States is attested by the phenomenal growth of the portland cement industry; although it produced less than 2 million tons of cement a year in 1900, it produced at an estimated rate of about 80 million tons of cement per year in 1971.

Concrete technology has progressed and evolved with the times and with new discoveries. In the latter part of the 19th century, concrete was ordinarily placed nearly dry and compacted with heavy tampers. Virtually no reinforcement was used at that time. With the development of reinforced concrete in the early part of this century, very wet mixes became popular and much of the concrete was literally poured into the forms and had neither good strength nor durability. This practice continued until investigations began to emphasize the importance of using scientifically designed mix proportions to produce a uniform concrete of improved workability, durability, and strength. Notable among the early investigations were those of Abrams, who formulated the "water-cement

1

ratio law" and demonstrated the importance of restricting this ratio for a given cement content to the lowest value consistent with the required workability of concrete for the particular work. The development of vibration to consolidate concrete aided materially in the placement of lower slump mixes and eliminated the necessity for sloppy mixes.

The development of special cements, such as high-early-strength cement for use where the concrete is put to early service, low-heat cement for massive construction, sulfate-resisting cement for use in sulfate soils and waters, and the introduction of expansive cement and set-controlled cement have all increased the versatility of concrete. In recent years, the introduction of pozzolanic materials reduced the costs of some concretes. The processing of aggregates to remove undesirable constituents by such methods as heavy media separation, hydraulic jigging, and elastic fractionation, in some instances permits making sound and durable concrete with aggregates which were otherwise unsuitable. Under investigation in the laboratories now is the impregnation of concrete with different monomers followed by polymerization, or hardening, of the monomer. This process increases manyfold the compressive, tensile, and flexural strengths, moduli of elasticity, and other physical properties of the concrete.

About 1938, an outstanding contribution to good concrete was made when it was discovered that small amounts of well-dispersed entrained air not only improved workability of concrete but also multiplied several times its resistance to freezing and thawing. This led to current widespread use of air-entraining agents, both as introduced at the mixer and as incorporated in air-entraining cements. Whereas it was once thought that all desirable properties of concrete depended on securing a maximum of solid substance, it is now recognized that the most dense concrete is not necessarily the most durable. Other admixtures, such as water-reducing, set-controlling agents and nonionic polymeric pumping aids to improve placeability, are now frequently used.

Concrete ingredients were once batched by volume with attendant inaccuracies and nonuniform results. Batching by weight has now superseded this practice, with resulting improvement in the uniformity and economy of concrete. The separation of coarse aggregates into two or more sizes was another improvement in practice, minimizing segregation during handling and bettering concrete quality. Thus, whereas concrete was once considered to be a simple mixture of coarse aggregate, sand cement, and water, mixed and placed in any convenient manner, the modern concept is a carefully proportioned mixture combining admixtures as needed to obtain the optimum quality and economy for any use.

3. Making Good Concrete.—Improved practices and techniques have added greatly to our ability to produce good concrete, and engineers are in close agreement on the practical needs for producing it. They recognize that, in addition to proper ingredients, a modern formula for successful concrete production would include common sense, good judgment, and vigilance.

There is still some concrete which, through carelessness or ignorance in its manufacture and placement, fails to give the service that would otherwise be expected. It is the responsibility of those in charge of construction to make sure that concrete is of uniformly good quality. The extra effort and care required to achieve this objective are small in relation to the benefits. Good engineering dictates acceptance of only the best. This axiom is especially true of concrete, for the best usually costs no more than the mediocre. In fact, good concrete practices result in better quality concrete and often lower costs by reducing placing difficulties. All that is required to achieve the best is an understanding of the basic principles of making good concrete and close attention to proven practices during construction.

B. Important Properties of Concrete

4. General Comments.—The characteristics of concrete discussed in the following sections should be considered on a relative basis and in terms of the degree of quality that is required for any given construction purpose. A concrete that is durable and otherwise satisfactory under conditions which give it protection from the elements might be wholly unsuited in locations of severe exposure to disintegrating influences. Watertightness is essential for a hydraulic structure, but strength and rigidity are obviously the primary structural requisites for an office building. It is apparent that the closest practicable approach to perfection in every property of the concrete would result in poor economy under many conditions and that the most desirable structure is one which meets all reasonable requirements for serviceable life, safety, and appearance. In other words, a structure must be adequately designed and properly constructed of concrete strong enough to carry the design loads and also economical, not merely in first cost but in terms of ultimate service.

The principal properties of good concrete, their interrelationships, and the elements which control these properties are summarized in figure 1.

5. Workability.—Workability has been defined as the ease with which a given set of materials can be mixed into concrete and subsequently handled, transported, and placed with minimum loss of homogeneity. The importance of plasticity and uniformity is emphasized because these essentials to workability have marked influence on the serviceability and appearance of the finished structure.

Workability is dependent on the proportions of the ingredient materials, as well as on their individual characteristics. The degree of work-

OPTIMUM ENTRAINED AIR
LOW WATER-CEMENT RATIO
WITH LOW WATER CONTENT
  WELL-GRADED AGGREGATE
  LOW PERCENTAGE OF SAND
  WELL-ROUNDED AGGREGATE
  REASONABLY FINE GROUND CEMENT
  PLASTIC CONSISTENCY (NOT TOO WET)
  VIBRATION

HOMOGENEOUS CONCRETE
  WORKABLE MIX
  THOROUGH MIXING
  PROPER HANDLING
  VIBRATION

ADEQUATE CURING
  FAVORABLE TEMPERATURE
  MINIMUM LOSS OF MOISTURE

SUITABLE AGGREGATE
  IMPERVIOUS
  STRUCTURALLY STABLE
  LARGE MAXIMUM SIZE

SUITABLE CEMENT
LOW C_3A, MgO, FREE LIME
LOW Na_2O AND K_2O
FREE OF FALSE SET

WATER-TIGHTNESS

LOW VOLUME CHANGE

RESISTANCE TO WEATHERING
TEMPERATURE VARIATIONS
MOISTURE VARIATIONS
FREEZING AND THAWING

RESISTANCE TO ADVERSE
CHEMICAL REACTIONS
  LEACHING (SOLUTION)
  OTHER REACTIONS:
    EXTERNAL IN ORIGIN
    AUTOGENOUS

DURABILITY

RESISTANCE TO WEAR
RUNNING WATER
MECHANICAL ABRASION

LOW WATER-CEMENT RATIO
WITH LOW WATER CONTENT
(SEE ABOVE)
HOMOGENEOUS CONCRETE
(SEE ABOVE)
ADEQUATE CURING
(SEE ABOVE)
INERT AGGREGATE
  STABLE IN CONCRETE ENVIRONMENT
  INCLUDING RESISTANCE TO
  ALKALIES IN CEMENT
SUITABLE CEMENT
(SEE ABOVE)
  RESISTANT TO SALTS IN SOIL
  AND GROUND WATER
SUITABLE POZZOLAN
ENTRAINED AIR

LOW WATER-CEMENT RATIO
WITH LOW WATER CONTENT
(SEE ABOVE)
HIGH STRENGTH
ADEQUATE CURING
(SEE ABOVE)
DENSE, HOMOGENEOUS CONCRETE
(SEE ABOVE)
SPECIAL SURFACE FINISH
  REDUCED FINES IN SAND
  WEAR-RESISTANT AGGREGATE
  MACHINE FINISHING

CONTROLLED QUALITY OF MATERIALS

CONTROLLED PROPORTIONING

GOOD UNIFORM CONCRETE

CONTROLLED HANDLING, PLACING, AND CURING

STRENGTH

ECONOMY

GOOD QUALITY OF PASTE
  LOW WATER-CEMENT RATIO
  ADEQUATE CURING
  APPROPRIATE CEMENT
GOOD QUALITY OF AGGREGATE
  STRUCTURAL SOUNDNESS
  UNIFORM SUITABLE GRADING
  FAVORABLE SHAPE AND TEXTURE
DENSE CONCRETE
  LOW-WATER CONTENT
  PLASTIC, WORKABLE MIX
  EFFICIENT MIXING
  VIBRATION
  LOW AIR CONTENT

EFFECTIVE USE OF MATERIALS
  LARGE MAX. SIZE AGGREGATE
  GOOD GRADING
  POZZOLAN
  MINIMUM WASTE
  MINIMUM SLUMP
  MINIMUM CEMENT
EFFECTIVE OPERATION
  DEPENDABLE EQUIPMENT
  EFFECTIVE METHODS, PLANT
  LAYOUT, AND ORGANIZATION
  AUTOMATIC CONTROL
EASE OF HANDLING
  UNIFORMLY WORKABLE MIX
  HOMOGENEOUS CONCRETE
  VIBRATION
  ENTRAINED AIR

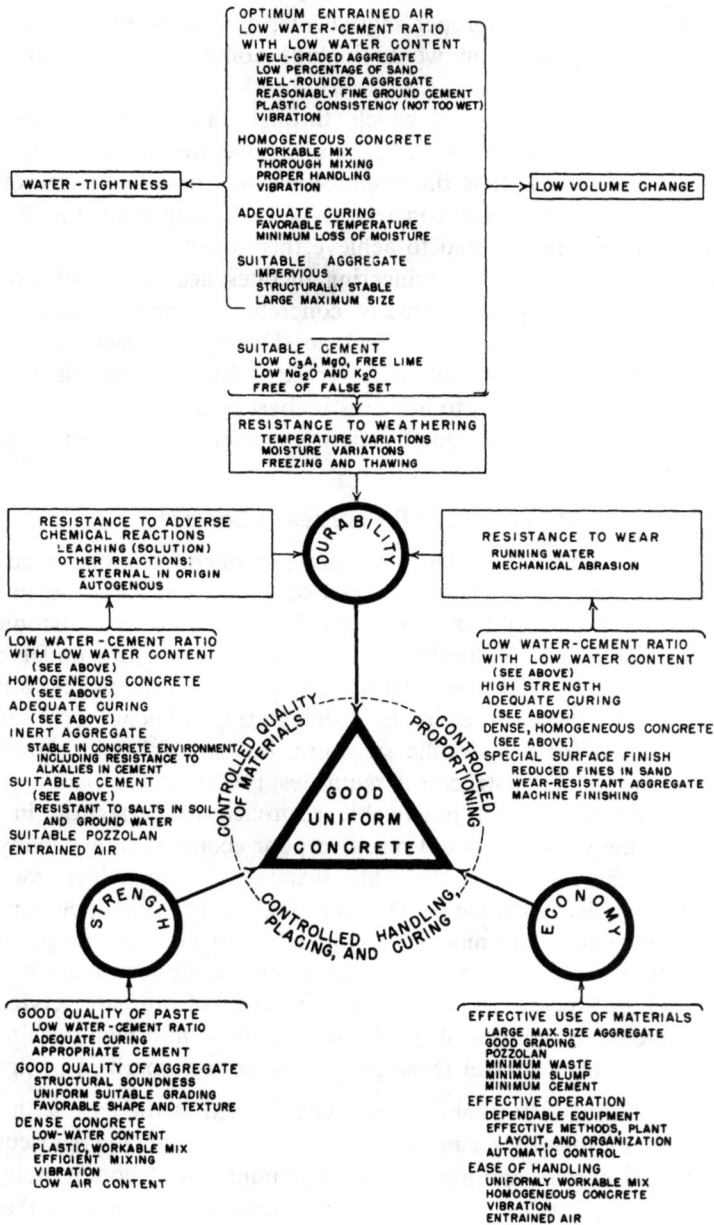

Figure 1.—Chart showing the principal properties of good concrete, their relationship, and the elements which control them. Many factors are involved in the production of good, uniform concrete. 288-D-795.

ability required for proper placement and consolidation of concrete is governed by the dimensions and shape of the structure and by the spacing and size of the reinforcement. For example, concrete having suitable workability for a pavement slab would be difficult to place or would even be unusable in a thin, heavily reinforced section. Over the years many devices for measuring workability of concrete have been developed. However, none of the methods evaluates all of the characteristics involved. These characteristics include ease of placing, finishing qualities, and bleeding or other forms of segregation. The use of entrained air has minimized effects of harshness in a concrete mix, but the determination of workability is still dependent somewhat upon judgment developed by experience.

Consistency or fluidity of concrete is an important component of workability and can be measured with reasonable accuracy by means of the slump test. The standard slump test is used on Bureau of Reclamation work but is conducted in such a manner (see fig. 2 and designation 22 in the appendix) as to provide additional assistance in judging workability of the concrete. Figure 2 shows slump specimens from two mixes having

Figure 2.—Slump test for consistency as performed by the Bureau. By tapping the side of a slump specimen with the tamping rod (see views at right), additional information as to the workability of the concrete is obtained. PX–D–20717.

the same slump. In the two views at the right, the specimens have been tapped with the tamping rod as prescribed in designation 22. The concrete in the upper view is a harsh mix, with a minimum of fines and water. It may be efficient for use in slabs, pavements, or mass concrete where it can readily be consolidated by vibration, but it would be quite unsuitable for a complicated and heavily reinforced placement. The concrete in the lower view is a plastic, cohesive mix; the surplus workability is needed for a difficult placement. However, if it is used where it can be easily placed and vibrated, such a mix would be inefficient because it contains excesses of cement, fines, and water. Thus, it is evident that, while measurement of slump gives a valuable indication of consistency, workability and efficiency of the mix can be judged only by how the concrete goes into place in each part of the structure and how it responds to consolidation by good vibration. Efficient mixes do not have much surplus workability over that needed for good results with thorough vibration.

The influence of temperature on the slump of concrete is indicated in figure 3.

For Bureau of Reclamation work, the maximum permissible slump of concrete, after the concrete has been deposited but before consolidation,

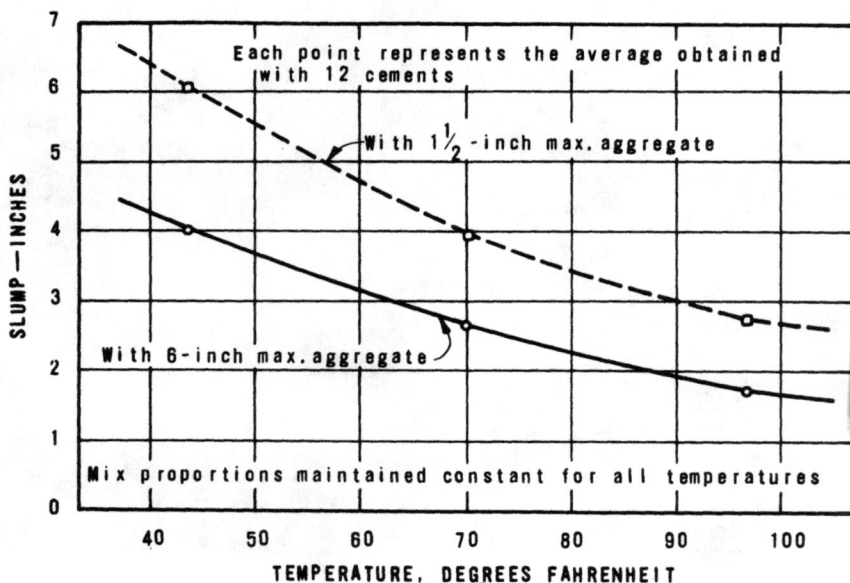

Figure 3.—Relationship between slump and temperature of concrete made with two maximum sizes of aggregates. As the temperature of the ingredients increases, the slump decreases. 288–D–1080.

is restricted by specifications to 2 inches for concrete in tops of walls, piers, parapets, curbs, and slabs that are horizontal or nearly horizontal; 4 inches for concrete in arch and sidewalls of tunnels; and 3 inches for concrete in other parts of structures and in canal linings. The slump of mass concrete is usually restricted to a maximum of 2 inches. If concrete cannot be placed without exceeding specified slump limitations, it may be concluded that the mix proportions are in need of adjustment. The minimum slump that can be used, commensurate with desired workability, requires the least amount of cement and water. In general, the wetter the consistency, the greater the tendency toward bleeding and segregation of coarse aggregate from the mortar.

6. Durability.—A durable concrete is one that will withstand, to a satisfactory degree, the effects of service conditions to which it will be subjected, such as weathering, chemical action, and wear. Numerous laboratory tests have been devised for measurement of durability of concrete, but it is extremely difficult to obtain a direct correlation between service records and laboratory findings.

(a) *Weathering Resistance.*—Disintegration by weathering is caused mainly by the disruptive action of freezing and thawing and by expansion and contraction, under restraint, resulting from temperature variations and alternate wetting and drying. Concrete can be made that will have excellent resistance to the effects of such exposures if careful attention is given to the selection of materials and to all other phases of job control. The purposeful entrainment of small bubbles of air, as discussed in section 14(b), has also helped to improve concrete durability by decreasing the water content and improving placeability characteristics. It is also important that, where practicable, provision be made for adequate drainage of exposed concrete surfaces.

Much has been learned regarding the resistance of air-entrained concrete to frost action, especially with respect to the influence of internal pore structure on durability. Dry concrete, with or without entrained air, sustains no damaging effects from freezing and thawing. Non-air-entrained concrete with high cement content and low water-cement ratio $(0.36\pm)$ develops good resistance to freezing and thawing primarily because of its relatively high density and attendant high impermeability (or watertightness) which reduce the free (or freezable) water available to the capillary system and/or through inflow under pressure. However, within the usual range of water-cement ratio specified for exposed structural concrete (maximum 0.47 to 0.53), greatly increased resistance to freezing and thawing is effected by the purposeful entrainment of air. This entrainment, in the form of multitudinous air bubbles ranging in size from less than 20 micrometers (submicroscopic) to about 3,000

Figure 4.—Typical pattern cracking on the exposed surface of concrete affected by alkali-aggregate action. PX–D–32049.

micrometers (macroscopic), provides relief for pressures developed by free water as it freezes and expands.

(b) *Resistance to Chemical Deterioration.*—Concrete deterioration, attributable in whole or in part to chemical reactions between alkalies in cement and mineral constituents of concrete aggregates, is characterized by the following observable conditions: (1) Cracking, usually of random pattern on a fairly large scale (see fig. 4); (2) excessive internal and overall expansion; (3) cracks that may be very large at the concrete surfaces (openings up to 1½ inches have been observed) but which extend into the concrete only a distance of from 6 to 18 inches; (4) gelatinous exudations and whitish amorphous deposits, on the surface or within the mass of the concrete, especially in voids and adjacent to some affected pieces of aggregate; (5) peripheral zones of reactivity, alteration, or infiltration in the aggregate particles, particularly those particles containing opal and certain types of acid and intermediate volcanic rocks; and (6) lifeless, chalky appearance of the freshly fractured concrete.

Deterioration of concrete also results from contact with certain chemical agents. The chemical action of a number of substances on unprotected concrete is shown in table 1. The table is intended to provide general guidance only, and salts listed as having no action might be aggressive at high concentrations or at high temperatures. Attack may assume one of several forms:

(1) Erosion of concrete results from the formation of soluble products which are removed by leaching. Attack by organic and inorganic acids is in this class. Attack by acids is seldom encountered at sites of Bureau work. This is a fortunate circumstance because no type of portland cement offers resistance to the forms of acid corrosion listed in table 1. Where likelihood of acid corrosion is in-

Table 1.—Effects of various substances on hardened concrete

Substance	Effect on unprotected concrete
Petroleum oils, heavy, light, and volatile	None.
Coal-tar distillates	None, or very slight.
Inorganic acids	Disintegration.
Organic materials:	
Acetic acid	Slow disintegration.
Oxalic and dry carbonic acids	None.
Carbonic acid in water	Slow attack.
Lactic and tannic acids	Do.
Vegetable oils	Slight or very slight attack.
Inorganic salts:	
Sulfates of calcium, sodium, magnesium, potassium, aluminum, iron.	Active attack.
Chlorides of sodium, potassium	None.
Chlorides of magnesium, calcium	Slight attack.
Miscellaneous:	
Milk	Slow attack.
Silage juices	Do.
Molasses, corn syrup, and glucose	Slight attack.
Hot distilled water	Rapid disintegration.

dicated, an appropriate surface covering or treatment should be employed.

When cement and water combine, one of the compounds formed is hydrated lime, which is readily dissolved by water (often made more aggressive by the presence of dissolved carbon dioxide) passing through cracks, along improperly treated construction planes, or through interconnected voids. The removal of this or other solid material by leaching may seriously impair the quality of concrete. The white deposit, or efflorescence, commonly seen on concrete surfaces is the result of leaching and subsequent carbonation and evaporation.

(2) Certain agents combine with cement to form compounds which have a low solubility but which disrupt the concrete because their volume is greater than the volume of the cement paste from which they were formed. Disintegration may be attributed to a combination of chemical and physical forces. In dense concretes this type of attack would be largely superficial. Porous concrete would be affected throughout the mass. Most prominent among aggressive substances which affect Bureau concrete structures are the sulfates of sodium, magnesium, and calcium. These salts which are known as white alkali are frequently encountered in the alkali soils and ground waters cf the western half of the United States.

The stronger the concentration of these salts the more active the corrosion. Sulfate solutions increase in strength in dry seasons when dilution is at a minimum. The sulfates react chemically with the hydrated lime and hydrated calcium aluminate in cement paste to form calcium sulfate and calcium sulfoaluminate, respectively, and

Figure 5.—Disintegration of concrete caused by sulfate attack. PX–D–32050.

Table 2.—Attack on concrete by soils and waters containing various sulfate concentrations

Relative degree of sulfate attack	Percent water-soluble sulfate (as SO_4) in soil samples	mg/l sulfate (as SO_4) in water samples
Negligible	0.00 to 0.10	0 to 150
Positive [1]	0.10 to 0.20	150 to 1,500
Severe [2]	0.20 to 2.00	1,500 to 10,000
Very severe [3]	2.00 or more	10,000 or more

[1] Use type II cement.
[2] Use type V cement, or approved combination of portland cement and pozzolan which has been shown by test to provide comparable sulfate resistance when used in concrete.
[3] Use type V cement plus approved pozzolan which has been determined by tests to improve sulfate resistance when used in concrete with type V cement.

these reactions are accompanied by considerable expansion and disruption of the paste. Figure 5 illustrates the effect of sulfate attack on concrete in a canal lining and a turnout wall. Concrete containing cement with a low content of the vulnerable calcium aluminate is highly resistant to attack by sulfate-laden soils and waters. (See sec. 15(b).) The relative degrees of attack on concrete by sulfates from soils and ground waters are given in table 2.

(3) Where concrete is subjected to alternate wetting and drying, certain salts, such as sodium carbonate, may cause surface disintegration by crystallizing in the pores of the concrete. Such action appears to be purely physical.

(4) In environments such as flash distillation chambers of desalination plants where concrete is exposed to condensing cool-to-hot water vapors or the resulting flowing or dripping of distilled water, the concrete is rapidly attacked by this mineral-free liquid. The liquid rapidly dissolves available lime and other soluble compounds of the cement matrix. Subsequent rapid deterioration and eventual decomposition result. The only palliative known at this time is complete insulation of the concrete from the mineral-free water by coatings or lining materials which are not affected by the water.

(5) Concrete in desalination plants is adversely affected by the feed water, sea water, or brine from wells. At these plants, high-quality concrete has been found unsuitable for use in brine exposures at temperatures of 290° F but suitable at 200° to 250° F provided adequate sacrificial concrete is made available for surface deterioration. Below about 200° F no provision for sacrificial concrete is generally required. Deterioration such as occurs at the higher temperature is a chemical alteration of the peripheral concrete paste which results in extensive microfracturing with resultant reduction of compressive strength, effective cross-sectional area of the member, and

eventual structural integrity. The rate of deterioration has been found to vary directly with temperature. Furthermore, since chemical alteration occurs when the hot sea water brine comes in contact with the concrete, the rate of deterioration could be expected to vary directly with permeability.

(c) *Resistance to Erosion.*—The principal causes of erosion of concrete surfaces are: cavitation, movement of abrasive material by flowing water, abrasion and impact of traffic, wind blasting, and impact of floating ice.

Cavitation is one of the most destructive of these causes and one to which concrete or any other construction material offers very little resistance regardless of its quality. On concrete surfaces subjected to high-velocity flow, an obstruction or abrupt change in surface alinement causes a zone of severe subatmospheric pressure to be formed against the surface immediately downstream from the obstruction or abrupt change. This zone is promptly filled with turbulent water interspersed with small fast-moving bubblelike cavities of water vapor. The cavities of water vapor form at the upstream edge of the zone, pass through it, and then collapse from an increase in pressure within the waterflow at a point just downstream. Water from the boundaries of the cavities rushes toward their centers at high speed when the collapse takes place, thus concentrating a tremendous amount of energy. The entire process, including the formation, movement, and collapse or implosion of these cavities, is known as cavitation.

It may seem surprising that the collapse of a small vapor cavity can create an impact sufficiently severe and concentrated not only to disintegrate concrete but to indent the hardest metals; however, there is abundant evidence to prove that this is possible and of common occurrence. The impact of the collapse has been estimated to produce pressures as high as 100,000 pounds per square inch. Repetition of these high-energy blows eventually forms the pits or holes known as cavitation erosion. Cavitation may occur in clear water flowing at high velocities when the divergence between the natural path of the water and the surface of the channel or conduit is too abrupt, or when there are abrupt projections or depressions on the surface of the channel or conduit, such as might occur on concrete surfaces because of poor formwork or inferior finishing of the concrete. Cavitation may occur on horizontal or sloping surfaces over which water flows or on vertical surfaces past which water flows. Figure 6 is an illustration of cavitation erosion on surfaces on and adjacent to a stilling basin dentate. The collapse of the cavities is accompanied by popping and crackling noises (crepitation).

Data from model studies and from field operation records have enabled designers to eliminate cavitation in most structures, and progress in this direction is still being made.

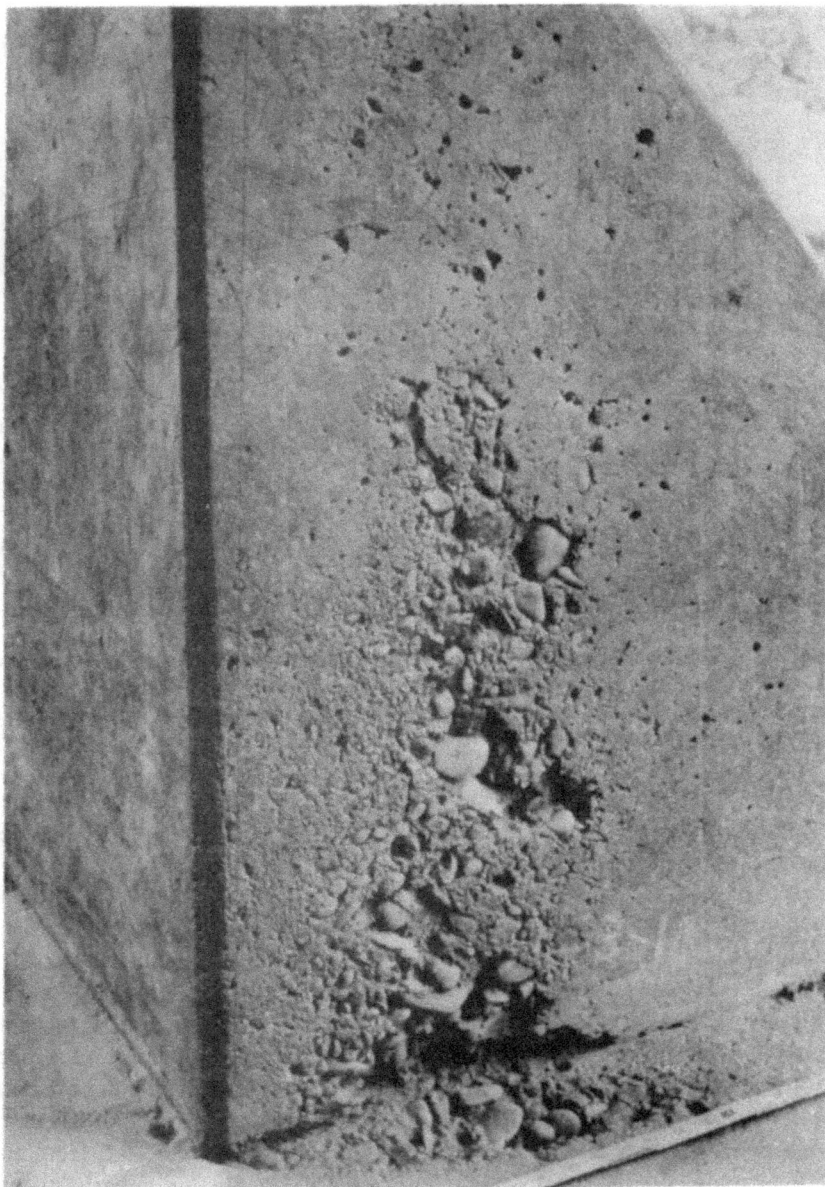

Figure 6.—Cavitation erosion of concrete on and adjacent to a dentate in the Yellowtail Afterbay Dam spillway stilling basin. Fast-moving water during a flood flow caused a pressure phenomenon at the concrete surface which triggered the cavitation damage shown here. P459–D–68902.

Figure 7.—Abrasion erosion of concrete in the dentates, walls, and floor of the Yellowtail Afterbay Dam sluiceway stilling basin. The "ball-mill" action of cobbles, gravel, and sand in turbulent water abraded the concrete, thus destroying the integrity of the structure. P459–D–68905.

Where low pressures cannot be avoided, critical areas are sometimes protected by facing with metal or other appropriate materials which have better resistance to cavitation than concrete. Introduction of air into the streamflow at an upstream point has also been effective in reducing the occurrence of cavitation and diminishing its effects on some structures.

Erosion damage to concrete caused by abrasive materials in water can be as severe as cavitation damage but generally would not cause a catastrophic failure as cavitation can so easily do. The hydraulic jump sections of spillway and sluiceway stilling basins, where turbulent flow conditions occur, are particularly vulnerable to abrasion damage. The water action in these areas tends to sweep cobbles, gravel, and sand from the downstream riverbed back into the concrete-lined stilling basin where the action becomes one of a grinding ball mill. Even the best concrete cannot withstand this severe wearing action. Figure 7 shows the abrasion erosion that occurred to the dentates, walls, and floor areas of the Yellowtail Afterbay Dam sluiceway stilling basin. Characteristic of this type of erosion is the badly worn reinforcing steel and aggregate. Contrast this with cavitation damage (fig. 6) which reflects little or no wearing of the aggregate particles. Although the most severe cases of abrasion damage occur in the areas just described, similar damage could be expected in diversion tunnels, canals, and pipelines carrying wastewater.

Use of concrete of increased strength and wear resistance offers some relief against the forces of erosion brought about by movement of abrasive material in flowing water, abrasion and impact of traffic, sandblasting, and floating ice. However, as is evident with cavitation erosion, the most worthwhile relief from these forces is prevention, elimination, or reduction of the causes by the proper design, construction, and operation of the concrete structures.

7. Watertightness.—Hardened concrete might be completely watertight if it were composed entirely of solid matter. However, it is not practicable to produce concrete in which all spaces between the aggregate particles are filled with solid cementing medium. To obtain workable mixes, more water is used than is required for hydration of the cement. This excess water creates voids or cavities which may be interconnected and form continuous passages. Furthermore, the absolute volume of the products of hydration is less than the sum of the absolute volumes of the original cement and water. Thus, as hydration proceeds, the hardened cement paste cannot occupy the same amount of space as the original fresh paste; consequently, the hardened paste contains additional voids. Purposefully entrained air and entrapped air also produce voids in the concrete, although the former, as will be explained, contributes to the watertightness of the concrete rather than to its permeability.

From the foregoing discussion, it is evident that hardened concrete is inherently somewhat pervious to water which may enter through capillary pores or be forced in by pressure. Nevertheless, permeability may be so controlled that construction of durable, watertight structures is not a serious problem.

The inherent perviousness of concrete can be visualized by considering the internal structure of plastic concrete. Immediately after concrete placement, the solids, including the cement particles, are in unstable equilibrium and settlement forces water upward, thereby commencing the development of a series of water channels, some of which extend to the surface. Gradually the larger pieces of aggregate assume stabilized positions, through point contact or otherwise, and form a skeleton structure within which settlement continues. The mortar settlement forces additional water upward, and part of it comes to rest below the larger pieces of aggregate. Finally, between the sand grains, the cement tends to settle out of the water-cement mixture (a water-cement ratio as low as about 0.30 by weight being required before the cement particles cease to be in suspension) and to leave water voids above the settled cement paste. At the completion of this stage in the mixed concrete, the initial water (the principal contributor to objectionable voids) is no longer homogeneously distributed in the paste but fills (1) relatively large spaces under aggregate particles, (2) the fine interstices among settled cement particles, and (3) a network of threadlike, interconnecting water passages. For air-entrained concrete the internal pore structure is somewhat different because the noncoalescing and separated spheroids of air reduce bleeding considerably and also reduce the water channel structure. As hydration of the cement proceeds (assuming that water is supplied as necessary) gel development reduces the size of the voids and thereby greatly increases the watertightness of the concrete. For this reason, prolonged thorough curing is a significant factor in securing impermeable, watertight concrete.

8. Volume Change.—Excessive volume change is detrimental to concrete. Cracks are formed in restrained concrete as a result of contraction because of temperature drop and drying at early ages before

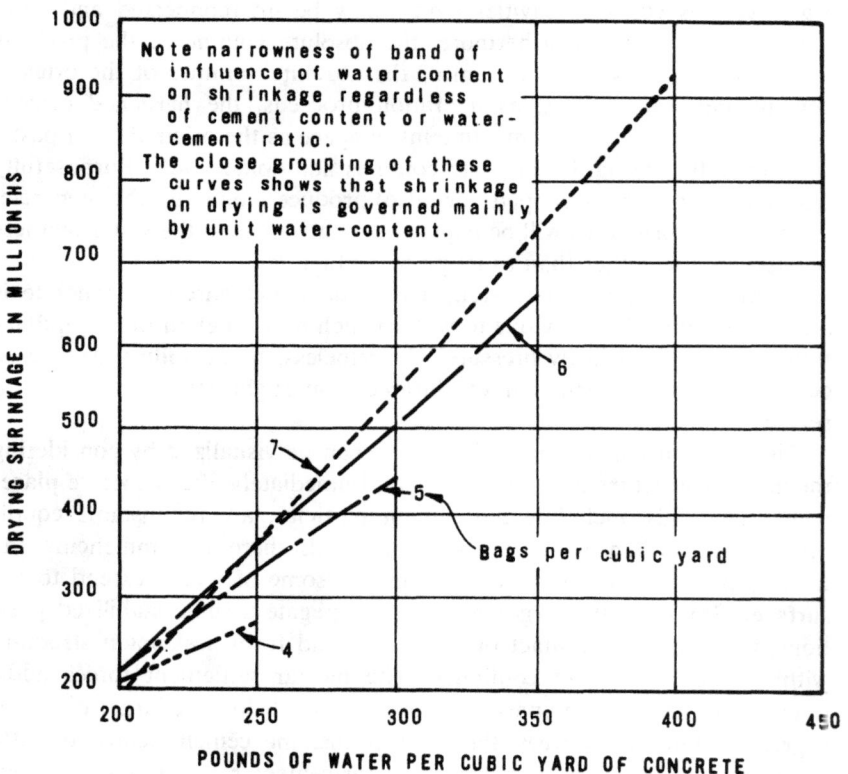

Figure 8.—The interrelation of shrinkage, cement content, and water content. The chart indicates that shrinkage is a direct function of the unit water content of fresh concrete. 288–D–2647.

sufficient tensile strength has developed. Cracking is not only a weakening factor that may affect the ability of concrete to withstand its designed loads, but also may seriously detract from durability and appearance. Durability is adversely affected by ingress of water through cracks and consequent accelerated leaching and corrosion of the reinforcement steel. Further disintegration occurs when cracked concrete is exposed to freezing and thawing. Concrete is also subject to disintegration when it contains alkali-reactive aggregates and high-alkali cement (cement containing in excess of 0.60 percent of equivalent soda) or is subjected to water bearing soluble sulfates. Differential stresses in concrete occasioned by differences in volume change characteristics of ingredients (see sec. 18 (d)) tend to break down the internal structure and the bond between cement paste and aggregate particles and may cause disintegration of the concrete particularly after repeated expansion and contraction. Expansion of concrete, under restraint, may cause excessive compressive stress and spalling at joints.

Drying shrinkage is affected by many factors which include, in order of importance, unit water content, aggregate composition, and duration of initial moist curing (see fig. 8). The principal drying shrinkage of hardened concrete is usually occasioned by the drying and shrinking of the cement gel that is formed by hydration of portland cement. Aggregate size, mix proportions, and richness of mix, among other factors, affect drying shrinkage principally as they influence the total amount of water needed in the mix. Additions of certain pozzolans may increase the drying shrinkage and others may decrease it. This effect is proportional to the pozzolan's relative water requirement. Fly ash typically reduces the drying shrinkage; natural pozzolans are variable in this respect. Initial drying shrinkage, which is somewhat greater than the expansion caused by subsequent rewetting, ranges from less than 200 millionths for dry, lean mixes with good quality aggregates to over 1,000 millionths for rich mortars or some concretes containing poor quality aggregate.

Concrete withstands compressive stress but allowable tensile strength of concrete should seldom exceed 10 percent of the compressive strength. Concrete restrained to the extent that high tensile stresses are produced through shrinkage will invariably crack. Total restraint could theoretically produce tensile stresses ranging between 600 and several thousand pounds per square inch, depending upon the shrinkage characteristics and elastic properties of the particular mix.

Autogenous volume change, although it may occasionally be an expansion, is usually shrinkage and is entirely a result of chemical reaction within the concrete and aging. Furthermore, it is in no way related to volume change resulting from drying or any other external influence. The magnitude of autogenous shrinkage varies widely, ranging from an insignificant 10 millionths, the lowest value observed to date, to somewhat in excess of 150 millionths. Autogenous shrinkage, in contrast to drying shrinkage, is relatively independent of water content but highly dependent upon the characteristics and amount of the total cementing material; it is greater for rich mixes than for lean mixes. Portland cement-pozzolan concretes always produce greater autogenous shrinkage than do similar mixes without pozzolan. Usually the most significant autogenous shrinkage takes place within the first 60 to 90 days after concrete is placed.

The thermal coefficient of expansion is the change (thermal expansion or contraction) in a unit length per degree of temperature change. The thermal coefficient of concrete varies mainly with the type and amount of coarse aggregate and is slightly affected by richness of mix, water content, and other factors. Various mineral aggregates may range in thermal coefficients from below 2 millionths to above 7.5 millionths per degree F. The coefficient for concrete is usually estimated to be the weighted average of the coefficients of the various constituents; thus, the coarse aggregate has the greatest effect.

The neat cement paste (gel) has a minor effect on thermal expansion. The coefficients of neat cement pastes vary from below 6 millionths to above 12 millionths depending upon saturation, age, degree of hydration, and chemical composition. Usual values are between 5 and 8 millionths for well-cured specimens in either dry or saturated condition; however, intermediate moisture contents result in higher thermal expansions.

Normally, concrete aggregates, except crushed materials, are heterogenous mixtures of different rocks and act as an average of the more common materials. Hence, average concrete, for estimating purposes, changes about 5.5 millionths of its length for each degree Fahrenheit of temperature change. Volume changes resulting from temperature variations involve both aggregate and cement paste, and volume changes caused by wetting and drying are usually considered to be principally related to the cement paste. However, volume changes caused by thermal and moisture changes can produce the same disintegrating effect. Deterioration can also be produced by volume changes resulting from chemical reactions between reactive constituents in the aggregate and the alkalies (Na_2O and K_2O) in the cement and also between soluble sulfates occurring in the soil or ground water in contact with a concrete structure and the tricalcium aluminate (C_3A) compound in the cement.

Formation of cracks caused by volume change is largely dependent on the degree to which contraction is resisted by internal and external forces. An example of internal restraint conducive to exterior cracking is a large block of concrete, the surfaces of which are drying or cooling while the interior of the mass is not so affected. Concrete canal lining is a good example of concrete subject to both internal and external restraint. The external restraint varies with the type and condition of subgrade. Unreinforced lining on a subgrade such as sand is not greatly restrained, and cracks resulting from drying shrinkage are relatively far apart and wide. On a rough, tight earth subgrade or on rock, where restraint is high, the cracks in unreinforced lining are more closely spaced and narrower. Reinforcement in the lining, through its bond to the concrete, distributes stresses and thereby reduces the spacing and width of cracks. Difference between moisture contents of the exposed and back faces may produce curling and eventual cracking.

Chemical combination of cement and water (hydration) is accompanied by generation of considerable heat which, under certain conditions, has an important bearing on the volume change of concrete. In small structures heat of hydration is generally of little consequence as it is rapidly dissipated. In massive structures heat of hydration may cause a temperature rise of as much as 50° to 60° F, which may constitute all or a large part of the difference between the maximum and minimum temperatures of the concrete. Much of the heat is generated during the early age of the concrete, when compressive stress developed by restraint of the expansion that accompanies temperature rise is relatively low. Two conditions are responsible for this low stress: at early age the modulus of elasticity is low; and creep, being greater, affords considerable relief of stress.

When heat is dissipated or removed, there is a decrease in the temperature and consequent contraction of the concrete. This volume change occurs at later age, when the modulus of elasticity is greater and stress relief by creep is less. Tensile stress induced when contraction is restrained will cause cracking if the stress exceeds the tensile strength of the concrete.

9. Strength.—Experience on Bureau work has demonstrated that concrete properly placed and cured will usually develop adequate compressive strength when the maximum permissible water-cement ratio has been established on the basis of durability requirement. Where greater strength is required for structural members, it may be necessary to use a lower water-cement ratio.

Tests of drill cores of more than 28 days' age taken from structures almost invariably show greater strengths than those obtained from control cylinders that are standard cured for 28 days. The extent of such excess strength generally varies with the age of the cores and the conditions contributing to continued hydration of the cement. (See table 3.)

Routine compressive strength tests of specimens subjected to standard moist curing give valuable indications of the uniformity and potential quality of the concrete in a structure. Tests of cylinders which have been cured out of doors, exposed to the weather, have no value and may be

CONCRETE MANUAL

Table 3.—Compressive strength of concrete cores and control cylinders

Project and feature	Cement type	Core dia, in	Avg age, mo	Cement, lb/yd³	Pozzolan, lb/yd³	W/(C+P) or W/C	Max size aggregate, in	Slump, in	Core strength, lb/in³	Number of cores tested	Control cylinder, lb/in²*	Strength ratio core to cylinder, percent
Mass Concrete: 22-Inch-Diameter Cores												
Central Valley, Shasta Dam	II	22	6	432	—	0.52	6	2	5,260	3	3,760	140
	II	22	6	376	—	.61	6	2	5,520	3	3,170	174
	II	22	6	320	—	.69	6	2	3,660	3	2,370	155
	II	22	6	263	—	.85	6	2	3,780	3	1,660	228
	II	22	12	432	—	.52	6	2	5,550	3	3,760	148
	II	22	12	376	—	.61	6	2	5,430	3	3,165	172
	II	22	12	320	—	.69	6	2	3,380	3	2,365	143
	II	22	12	263	—	.85	6	1½	3,950	3	1,660	238
	IV	22	24	376	—	.61	6	1½	5,450	3	3,170	172
	IV	22	60	376	—	.61	6	1½	4,920	3	3,170	156
	IV	22	120	376	—	.61	6	2	5,510	3	3,170	174
	IV	22	264	432	—	.52	6	2	6,530	2	3,760	174
	IV	22	264	376	—	.61	6	2	6,470	2	3,170	204
	IV	22	264	320	—	.69	6	2	4,780	2	2,370	202
	IV	22	264	263	—	.85	6	2	5,300	2	1,660	319
Central Valley, Friant Dam	IV	22	6	376	40	.56	6-8	1	5,230	3	4,940	106
	IV	22	6	338	78	.59	6-8	1	4,670	3	4,940	95
	IV	22	6	301	98	.70	6-8	1	4,260	3	3,600	118
	IV	22	6	282	118	.74	6-8	1	4,070	3	3,560	114
	IV	22	6	263	—	.81	6-8	1	3,844	3	3,110	124
	IV	22	12	376	40	.56	6-8	1	5,070	3	4,940	103
	IV	22	12	338	78	.59	6-8	1	4,920	3	4,940	100
	IV	22	12	301	98	.70	6-8	1	4,580	3	3,600	127
	IV	22	12	282	118	.74	6-8	1	4,340	3	3,560	122
	IV	22	24	263	—	.81	6-8	1	4,170	3	3,110	134
	IV	22	24	376	40	.56	6-8	1	5,750	3	4,940	117
	IV	22	24	338	78	.59	6-8	1	5,160	3	4,940	105
	IV	22	24	301	98	.70	6-8	1	4,270	3	3,600	119
	IV	22	24	282	118	.74	6-8	1	4,110	3	3,560	115
	IV	22	60	263	—	.81	6-8	1	4,200	3	3,110	135
	IV	22	60	376	40	.56	6-8	1	5,630	3	4,940	114
	IV	22	60	338	78	.59	6-8	1	4,680	3	4,940	95
	IV	22	60	301	98	.70	6-8	1	4,280	3	3,600	119
	IV	22	60	282	118	.74	6-8	1	4,040	3	3,560	114
	IV	22	60	263	—	.81	6-8	1	3,950	3	3,110	127
	IV	22	120	276	—	.56	6-8	1	5,600	3	4,940	114

Continued

Mass Concrete: 10-Inch-Diameter Cores

IV	22	120	338	40	.59	1	6–8	5,020	3	4,940	102
IV	22	120	301	78	.70	1	6–8	4,580	3	3,600	127
IV	22	120	282	98	.74	1	6–8	4,370	3	3,560	123
IV	22	120	263	118	.81	1	6–8	4,390	3	3,110	141
IV	22	264	376	40	.56	1	6–8	5,930	3	4,940	120
IV	22	264	338	60	0.55–0.59	1	6–8	5,540	3	3,600	112
IV	22	264	301	70	.58–.70	1	6–8	4,320	3	3,560	120
IV	22	264	282	70	.59–.74	1	6–8	4,170	3	3,120	117
IV	22	264	263	79	.62–.81	1	6–8	4,620	3	3,110	149

Hungry Horse, Hungry Horse Dam

II	10	3	188	89	0.51	2	6	3,690	3	2,720	136
II	10	6	188	86	.56	2	6	3,800	24	2,700	141
II	10	9	188	89	.53	2	6	4,040	3	2,720	149
II	10	12	188	89	.54	2	6	4,030	16	2,600	155
II	10	24	188	92	.57	2	6	4,790	8	3,110	154
II	10	36	188	89	.57	2	6	5,040	8	2,530	199
II	10	60	192	81	.62	2	6	5,180	8	2,630	197
II	10	72	188	89	.58	2	6	5,400	5	2,510	175
II	10	84	188	92	.52	2	6	5,230	7	3,090	169
II	10	96	188	89	.57	2	6	5,940	8	2,530	235
II	10	3	282	89	.47	2	6	3,950	4	3,970	99
II	10	6	286		.41	2	6	4,990	4	3,950	126
II	10	6	372	88	.48	2	6	5,900	7	4,480	132
II	10	9	282	89	.47	2	6	5,210	1	3,970	131
II	10	12	282	92	.45	2	6	5,260	14	3,990	132
II	10	24	286		.41	2	6	5,510	8	4,340	121

Central Valley, Shasta Dam

IV	10	60	432		.52	2	6	5,889	3	3,760	156
IV	10	60	376		.61	2	6	5,120	3	3,170	162
IV	10	60	320		.69	2	6	3,840	3	2,370	162
IV	10	60	263		.85	2	6	3,280	3	1,660	196

Solano, Monticello Dam

II	10	7.5	211	69	.58	2	6	3,980	5	3,390	117
II	10	13.4	199	69	.59	2	6	3,800	6	2,710	140
II	10	13.8	214	69	.57	2	6	5,690	5	3,190	178

Central Valley, Friant Dam

IV	10	60	376	40	.56	1	6–8	4,520	3	4,940	92
IV	10	60	338	78	.59	1	6–8	4,780	3	4,940	97
IV	10	60	301	98	.70	1	6–8	3,980	3	3,600	111
IV	10	60	263	118	.74	1	6–8	4,640	3	3,560	130
IV	10	60	376		.81	1	6–8	3,820	3	3,110	123
IV	10	120	338	40	.56	1	6–8	4,400	3	4,940	89
IV	10	120	301	78	.59	1	6–8	4,960	3	4,910	101
IV	10	120	282	98	.70	1	6–8	3,210	3	3,600	89
IV	10	120	263	118	.74	1	6–8	4,340	3	3,560	122
II	10				.81	1¾	6–8	4,210	3	3,110	136

Colorado River Storage, Glen Canyon Dam

II	10	6	214	75	.58	1¾	6	4,200	26	2,720	154
II	10	6	204	77	.59	2	6	4,200	22	2,720	154
II	10	6	199	75	.55	2	6	4,530**	4	2,770**	164
II	10	6	233	94	.51	2	6	5,670	9	3,390	167
II	10	6	528	74	.48			5,920**		—	

Continued

Table 3.—Compressive strength of concrete cores and control cylinders—Continued

Project and feature	Cement type	Core dia, in	Avg age, mo	Cement, lb/yd³	Pozzolan, lb/yd³	W/(C+P) or W/C	Max size aggregate, in	Slump, in	Core strength, lb/in²	Number of cores tested	Control cylinder, lb/in²	Strength ratio core to cylinder, percent
Mass Concrete: 10-Inch-Diameter Cores												
	II	10	12	196	95	.60	6	1¾	4,120	5	2,160	191
	II	10	12	203	80	.59	6	2	4,200	13	2,550	162
	II	10	12	200	74	.55	6	2	5,350**	13	2,990	179
	II	10	12	233	94	.51	6	2	5,520	8	3,390	163
	II	10	12	258	75	0.47	6	2	5,460**	11	—	—
	II	10	24	209	79	.54	6	2	4,000**	14	2,890	138
	II	10	24	187	80	.59	6	2	5,095**	6	2,920**	174
	II	10	24	233	94	.52	6	2	5,060	4	3,430	148
	II	10	60	256	80	.54	6	2	5,008**	14	3,690**	136
	II	10	60	206	78	.56	6	2	4,440	18	2,870	155
	II	10	60	190	76	.52	6	2	5,820**	8	2,930**	199
	II	10	60	239	90	.52	6	2	4,720	12	3,540	133
	II	10	60	256	75		6	2	6,260**		—**	—
Pick-Sloan Missouri Basin Program, Canyon Ferry Dam	II	10	6	244	81	.50	6	2	4,150	6	4,030	103
	II	10	7	177	55	.66	6	2	3,070	5	2,430	126
Colorado River Storage, Flaming Gorge Dam	II	10	6	188	91	.53	6	2	3,370	27	3,120	108
	III	10	12	189	91	.53	6	2	3,130	28	3,170	99
	II	10	36	315	—	.48	6	2	4,570	8	3,625	126
	II	10	36	189	91	.53	6	2	3,210	24	3,280	98
	III	10	60	315	—	.48	6	2	4,640	8	3,280	128
	III	10	60	315	91	.48	6	2	4,700	24	3,625	111
Pick-Sloan Missouri Basin Program, Yellowtail Dam	III	10	6	191	84	.55	6	2	3,800	15	3,230	118
	III	10	6	206	85	.51	6	2	4,100	19	3,530	116
	III	10	6	252	84	.44	3	2	4,170	9	4,180	100
	III	10	6	363	—	.43	3	2	5,690	3	4,950	115
	III	10	12	272	105	.49	3	2	4,260	4	3,990	107
	III	10	12	251	83	.45	6	2	4,220	11	4,190	101
	III	10	12	203	85	.52	6	2	4,230	34	3,470	122
	III	10	60	210	91	.50	6	2	5,480	6	3,740	147
	III	10	60	261	96	.41	6	2	5,120	2	5,080	101
Colorado River Storage, Morrow Point Dam	II	10	2	383	—	.45	4½	2	6,770***	20	5,570***	122
	III	10	6	410	—	.45	4½	2	6,890***	17	5,530***	125
	II	10	12	410	—	.43	4½	2	8,260***	12	5,700***	145

Continued

Mass Concrete: 6-Inch-Diameter Cores

Central Valley, Shasta Dam	IV	6	120	432	—	.52	6	2	4,390	3	3,760	117
	IV	6	120	376	—	.61	6	2	4,560	3	3,170	144
	IV	6	120	320	—	.69	6	2	3,350	3	2,370	142
	IV	6	120	263	—	.85	6	2	3,010	3	1,660	181
Central Valley, Friant Dam	IV	6	120	376	—	0.56	6–8	1	6,030	3	4,940	122
	IV	6	120	338	40	.59	6–8	1	4,040	3	4,940	82
	IV	6	120	301	78	.70	6–8	1	4,000	3	3,600	111
	IV	6	120	282	98	.74	6–8	1	4,270	3	3,560	120
	IV	6	120	263	118	.81	6–8	1	4,240	3	3,110	137
Pick-Sloan Missouri Basin Program, Canyon Ferry Dam	II	6	6	278	92	.54	6	2	3,720	3	3,750	99
	II	6	7	177	55	.67	6	2	3,340	18	2,420	138

Other Than Mass Concrete

Central Valley, Clear Creek Tunnel	I and II	6	23	459	—	.51	2½	4¾	5,530	82	3,690	150
All-American Canal System	II	6	12	519	—	.57	1½	3½	5,080	33	3,740	136
	II	6	18	519	—	.56	1½	3½	5,170	27	3,940	131
	II	6	24	519	—	.56	1½	3¾	5,450	43	3,970	137
	II	6	30	530	—	.56	1½	3	5,530	22	4,030	137
	II	6	36	545	—	.56	1	3¾	5,460	4	3,610	151
	II	6	36	523	—	.55	3	3½	6,380	52	3,670	174
				455	—	.59			4,530	7	3,310	137
Pick-Sloan Missouri Basin Program, Canyon Ferry Dam	II	10	6	188	89	.55	3	2	4,230	3	3,210	132
	II	10	6	248	81	.55	3	2	4,050	2	3,330	122
	II	6	6	248	81	.55			4,100	2	3,820	107

$$W/(C+P) = \frac{water}{cement+pozzolan}$$

$$W/C = \frac{water}{cement}$$

* 6- by 12-inch fog-cured 28-day concrete cylinders
** With WRA 0.37 percent by weight of cement
*** With WRA 0.28 percent by weight of cement

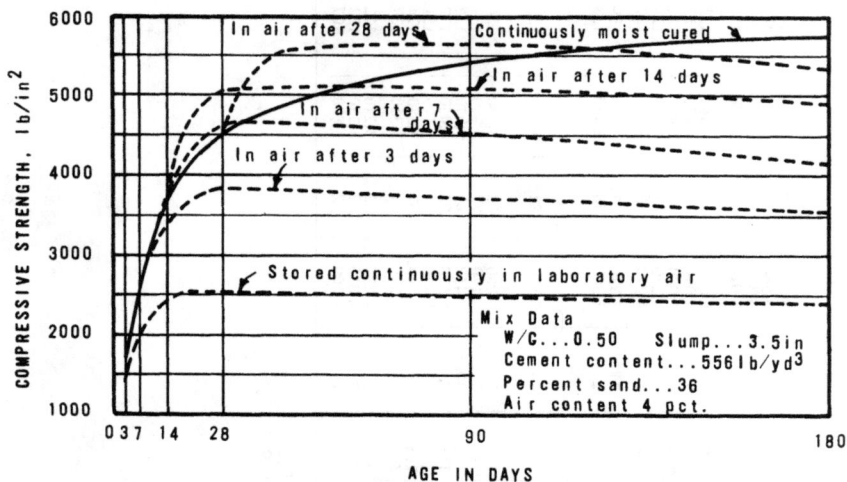

Figure 9.—Compressive strength of concrete dried in laboratory air after preliminary moist curing. 288–D–2644.

entirely misleading. The test results cannot be correlated with those for standard-cured specimens and, because of their high surface-to-volume ratio, the specimens do not simulate conditions in the structure. To determine the adequacy of curing and strength development of concrete representing that in precast pipe or other units, test cylinders are fabricated and cured in a manner similar to that used in the manufacture and cure of the units. In the manufacture of these precast concrete units, steam curing is most generally used to accelerate production.

Figure 9 indicates that development of strength stops at an early age if the concrete specimen is exposed to dry air with no previous curing. Concrete exposed to dry air from the time it is placed is about 50 percent as strong at 6 months' age as concrete moist cured 14 days before being exposed to dry air.

Curing temperatures have a pronounced effect on strength development. Tests indicate that longer periods of moist curing are required at lower temperatures to develop a given strength than are necessary at higher temperatures. Continued curing at higher temperatures for the full 28-day period (see fig. 10) resulted in strength development which varied directly with temperature, the highest strength being developed by the highest temperature at this age. However, at later ages this trend was reversed, the specimens made and cured at lower temperatures developing the higher strengths.

Curves shown in figure 11 represent concrete that was cured at 70° F after the specimens were held at the casting temperature for 2 hours.

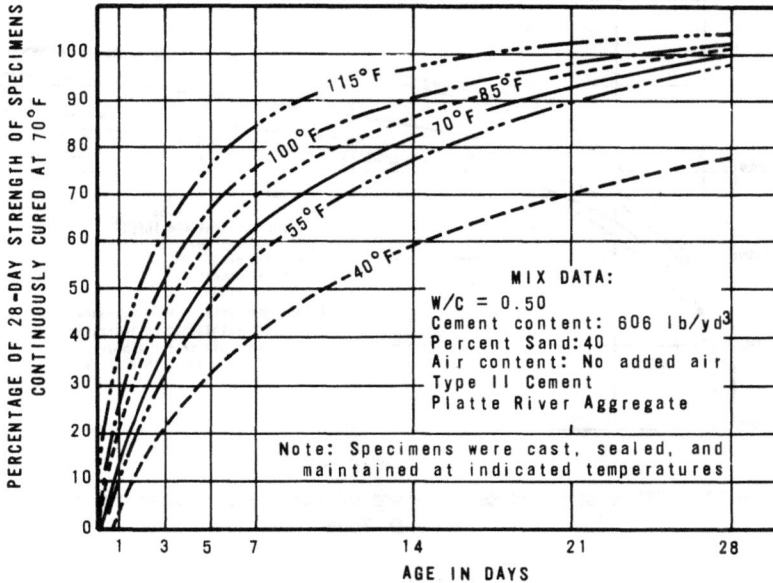

Figure 10.—Effect of curing temperature on compressive strength of concrete.
288–D–2645.

Under such treatment, the specimens made at the lowest temperature attained the highest strengths. These results agree with those obtained on some Bureau projects where the strengths of field control cylinders were higher during the cooler months than during summer months even though all cylinders were moist cured at about 70° F soon after fabrication.

Compressive strength, tensile strength, flexural strength, and shearing strength of concrete are all more or less directly related, and an increase or decrease in one is generally reflected similarly in the others, though not in the same degree. Where flexural strength is an important consideration, as in the construction of road pavement, beam tests are frequently employed for control purposes.

On a few occasions projects have reported significant reductions in concrete compressive strengths at early ages, unexplainable by curing conditions or testing procedures; the lower strengths were the result of change in composition of the cement and/or a decrease in fineness. Lower total amounts of C_3A and C_3S (see sec. 15a) will reduce early strength, but variations in cement fineness cause greater fluctuations than variations in the usual ranges of C_3A and C_3S amounts. These fluctuations are apparent in mill test reports of cube strengths. However, the compressive strength at later ages is usually much closer. (See comparison between types I and III cements with type IV in fig. 23.) Variations in cement

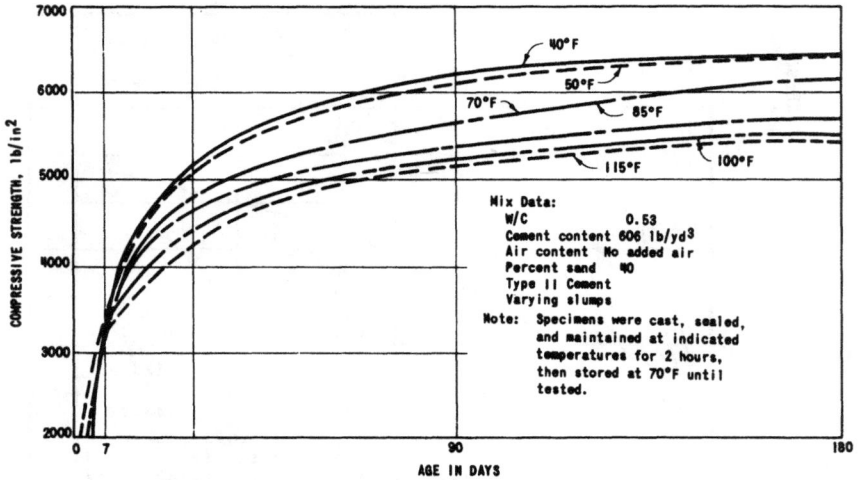

Mix Data:
W/C   0.53
Cement content 606 lb/yd³
Air content No added air
Percent sand   40
Type II Cement
Varying slumps
Note: Specimens were cast, sealed,
       and maintained at indicated
       temperatures for 2 hours,
       then stored at 70°F until
       tested.

Figure 11.—Effect of initial temperature on compressive strength of concrete. 288-D-2646.

fineness occur more frequently, exert more influence on concrete compressive strengths, and affect the uniformity of concrete control since primary control is based on concrete strengths at 28 days' age. In either case, the ultimate strength of the concrete is minimally affected. Where the cement used shows slow strength development, precautions may be necessary to assure adequate strength before subjecting the structure to service loads.

The degree of uniformity of concrete strength is a measure of success or failure in attaining adequate field control. Without adequate quality control of concrete manufacturing operations, wide variations in strength will occur and extra cement will be needed to ensure that the quality of the concrete will meet minimum requirements. Also, for concrete of a given average strength, expectation of wide variations in strength necessitates use of lower working stresses in design. Lack of reasonable uniformity in desirable properties, as indicated by strength variations, can be expected to manifest itself eventually in objectionable variations in durability and higher cost of maintenance.

10. Elasticity.—Concrete is not a truly elastic material, and the graphic stress-strain relationship for continuously increasing loading is generally represented by a curve. For concrete that has hardened thoroughly and has been moderately preloaded, the stress-strain curve is, for all practical purposes, a line of constant slope within the range of usual working

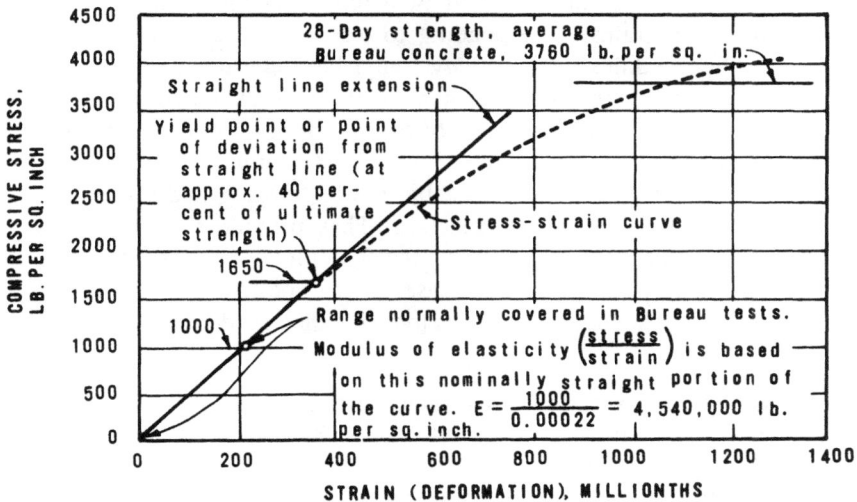

Figure 12.—Typical stress-strain diagram for thoroughly hardened concrete that has been moderately preloaded. The stress-strain curve is very nearly a straight line within the range of usual working stresses, 288–D–799.

stresses. The stress-strain ratio determined from the virtually straight portion of the stress-strain curve is called the "modulus of elasticity." When the loads are increased beyond the working range, the stress-strain curve may deviate considerably from a straight line, indicating that stress and strain are no longer proportional (see fig. 12). However, the stress-strain ratio is fairly uniform for compressive stresses up to 75 percent of the 28-day breaking strength, as indicated in the figure. Usually, concretes of higher strength have higher elastic values, although modulus of elasticity is not directly proportional to strength. The elastic modulus for ordinary concretes at age of 28 days ranges from 2 million to 6 million pounds per square inch.

For most materials, the modulus of elasticity does not vary with age, and the elastic recovery at the time of load removal is equal to the elastic deformation at the time the load was applied regardless of the duration of load application. In concrete, however, the modulus normally increases with age so long as the concrete remains sound; therefore, both initial deformation and subsequent elastic recovery depend on age. The increase in modulus of elasticity as concrete ages accounts for a large part of the tensile stress which develops when concrete that is restrained from expanding and contracting freely is heated at an early age and cooled at a later age.

In addition to the static method of determining stress-strain relationships, in which strains corresponding to test load stresses are measured directly, the modulus of elasticity may be determined by dynamic methods involving either measurement of the natural frequency of vibration of a specimen or measurement of the velocity of sound waves through the material. Dynamic methods are used to determine the extent of deterioration of concrete specimens subjected to freezing and thawing tests or affected by alkali-aggregate reaction. They provide simple and rapid means for frequently determining the modulus of elasticity without damage to the specimen. A decrease in the modulus, measured by a lower natural frequency or wave velocity, indicates deterioration of the concrete.

11. Creep and Extensibility.—When concrete is subjected to a constant sustained load, the deformation produced by the load may be divided into two parts: elastic deformation, which occurs immediately but would entirely disappear on immediate removal of the load; and creep, which develops gradually. In most concrete structures, dead loads that act continuously constitute a large part of the total load; thus, both immediate strain and gradual yielding must be considered when computing deformations of such structures. Gradual yielding also has an important effect on the development of stresses caused by slow temperature changes or drying shrinkage. This behavior has often been called plastic flow, but the term creep is preferred to distinguish it from plastic action of a different sort which may result in stress adjustments when a part of a structure or member is overstressed. Plastic action of concrete, like the plastic flow of metals, is irrecoverable and may be considered to be a type of incipient failure; creep, however, is at least partly recoverable and occurs even at very low stress.

Extensibility is the property of concrete that enables it to withstand tensile deformation without cracking. Extensibility differs from strength in that it involves limiting deformations rather than limiting loads. Elasticity, creep, and extensibility are interrelated properties of considerable importance.

(a) *Creep.*—Under sustained load the creep of concrete continues for an indefinite time. In a long-term test, two concrete specimens under sustained load were still showing deformation after 20 years. However, creep proceeds at a continuously diminishing rate. The Bureau now determines by a computer program the exact relationships of creep variables from values determined in laboratory tests on the same maximum size aggregate as that in the structure.

The following equation can be applied to experimental data from creep tests to obtain an approximate value for the creep function.

$$\epsilon = \frac{1}{E} + f(K) \log_e (t + 1)$$

where:

ϵ = total deformation,

E = instantaneous elastic modulus,

$f(K)$ = a function representing the rate of creep information with time, and

t = time under load in days.

The function $f(K)$ is large when concrete is initially loaded at an early age and small when concrete is loaded later in time. The function $\log_e$ $(t + 1)$ indicates that concrete continues to deform with time at a diminishing rate but with no apparent limit. Although tests made thus far appear to support the view that concrete will creep without limit, it is generally assumed that there is an upper limit to creep deformation.

Figure 13 illustrates the deformation record of a typical laboratory test specimen loaded at the relatively early age of 1 month but removed 6 months later. Because of the increased age of the concrete at the time of unloading, the elastic and creep recoveries are lower than the deformations under load, the result being a nonrecoverable shortening if the load were compression or a nonrecoverable elongation if the load were tensile. The typical curves in figure 14 for 4- by 8-inch cylinders give a general conception of the rate at which creep develops and of the effects of changes in water-cement ratio and intensity of load. The curves show that creep is increased with increasing water-cement ratio and that creep is approximately proportional to load intensity. Most of the factors which increase strength and modulus of elasticity reduce the creep. Generally, concretes made with aggregates of loosely cemented granular structure, such as some sandstones, creep more than those made with dense, compact aggregates such as quartz or limestone. From a 10-year study of the creep properties of five mass concretes, tests indicate that there is a definite relationship between creep and elasticity and that if a creep-strength relationship exists for concrete, it is small and hidden by the effects of type of aggregate, type of cement, cement-aggregate ratio, inclusion of pozzolans, and possibly other conditions.

Creep is often taken into account approximately in design by using a reduced value of the modulus of elasticity. When more exact relationships are needed, such as in the computation of stress from strain measurements in mass concrete, creep is susceptible to mathematical analysis and prediction through the following general properties:

(1) Creep is a delayed elastic deformation involving no changes corresponding to crystalline breakdown or slip and is not the plastic flow of a viscous solid.

(2) At working stress creep is proportional to stress, but when

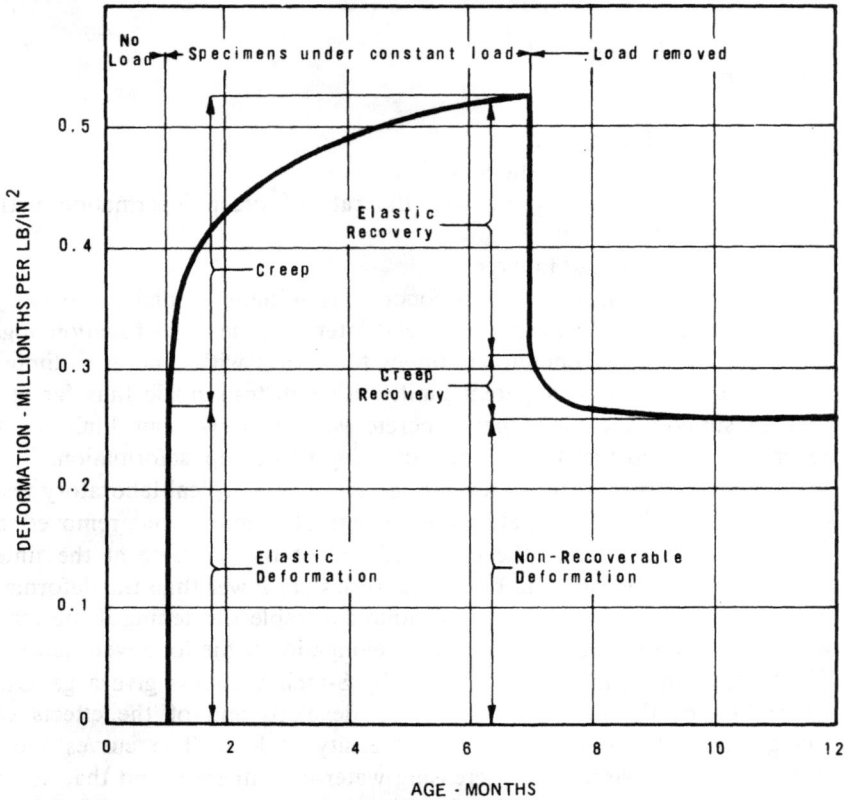

Figure 13.—Elastic and creep deformations of mass concrete under constant load followed by load removal. 288–D–1519.

stress approaches the ultimate strength of concrete, creep increases much more rapidly than stress.

(3) When the effect of age on changing the properties of concrete is taken into account, all creep is recoverable.

(4) Creep is independent of sign; it bears the same proportion to either positive or negative stress.

(5) The principle of superposition applies to creep.

(6) Poisson's ratio is the same for creep strains as for elastic strains.

(b) *Extensibility.*—Measurements have been made of the extensions (strains) on the tension faces of beams which were loaded progressively until the first cracks became visible and of the extensions of direct tension specimens to the point of failure. Extensibility is evidently a function of elasticity, creep, and tensile strength, and its value depends not only on the properties of the concrete but on the rate at which the tensile load is applied. Under fairly rapid loading (too rapid to permit creep), plain concrete beams have been extended about 150 to 190 millionths before

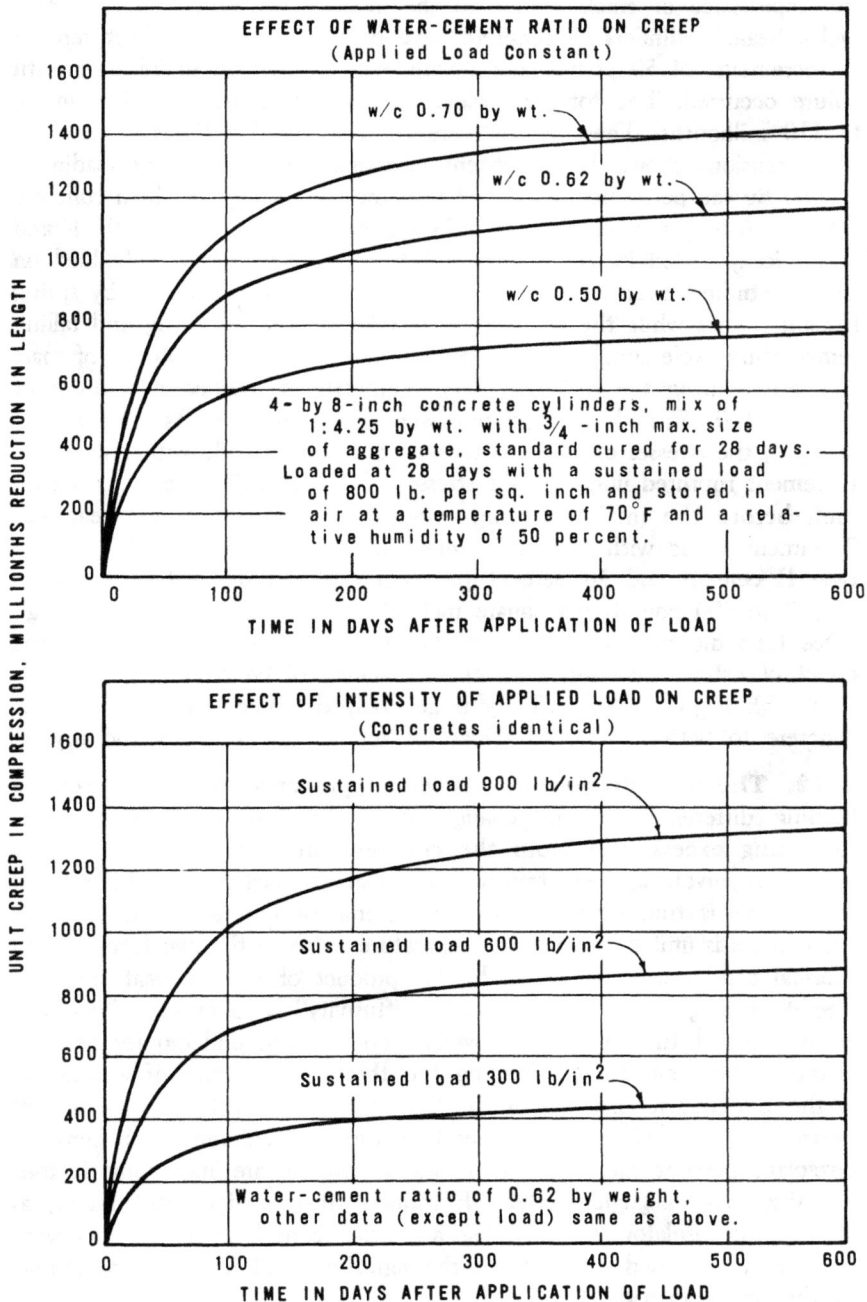

EFFECT OF WATER-CEMENT RATIO ON CREEP
(Applied Load Constant)

w/c 0.70 by wt.

w/c 0.62 by wt.

w/c 0.50 by wt.

4- by 8-inch concrete cylinders, mix of
1:4.25 by wt. with $\frac{3}{4}$ -inch max. size
of aggregate, standard cured for 28 days.
Loaded at 28 days with a sustained load
of 800 lb. per sq. inch and stored in
air at a temperature of 70°F and a relative humidity of 50 percent.

TIME IN DAYS AFTER APPLICATION OF LOAD

EFFECT OF INTENSITY OF APPLIED LOAD ON CREEP
(Concretes identical)

Sustained load 900 lb/in^2

Sustained load 600 lb/in^2

Sustained load 300 lb/in^2

Water-cement ratio of 0.62 by weight,
other data (except load) same as above.

TIME IN DAYS AFTER APPLICATION OF LOAD

UNIT CREEP IN COMPRESSION, MILLIONTHS REDUCTION IN LENGTH

Figure 14.——Rate of creep in concrete as affected by variation in water-cement
ratio and intensity of applied load. 288–D–800.

the appearance of cracks visible to the unaided eye (open about 0.0015 inch). Sealed cylinders of concrete have been subjected to direct tension in increments of 50 pounds per square inch at intervals of 28 days until failure occurred. The total extension at time of failure ranged from 70 to 110 millionths. These values were from 1.2 to 2.5 times as great as the extensions shown by direct-tension specimens under rapid loading.

The Bureau performed a series of tests on extensibility in which concrete cylinders 6 inches in diameter and 24 inches long were cast at 70° F and hermetically sealed in soft copper jackets, with strain gages embedded on the longitudinal axes. The length of cylinders was held constant by spring tension frames while the cylinders were taken through a rising and falling temperature cycle simulating the temperature cycle in the interior of mass concrete. During the first few days, temperatures reached maximums of 100° to 110° F, and the specimens were in compression. As temperatures dropped, the stresses changed to tension. Specimens made with type I and II cement ruptured under tensile stresses of 210 to 225 pounds per square inch before the initial starting temperature of 70° F was reached. Specimens made with type IV cement or a combination of 70 percent type II cement and 30 percent pozzolan ruptured under tensile stresses of 270 to 300 pounds per square inch at approximately 60° F. Although these tests did not permit extensibility measurements, they illustrate the effect of extensibility with respect to cracking of concrete.

A high degree of extensibility is generally desirable, for it permits the concrete to better withstand effects of temperature changes and drying.

12. Thermal Properties.—Thermal properties are significant in keeping differential volume change at a minimum in mass concrete, extracting excess heat from the concrete, and dealing with similar operations involving heat transfer. Thermal conductivity is the rate at which heat is transmitted through a material of unit area and thickness when there is unit difference in temperature between the two faces. When thermal conductivity is divided by the product of specific heat and unit weight, a single coefficient termed "diffusivity" is obtained. Diffusivity is an index of the facility with which concrete will undergo temperature change. The main factor affecting the thermal properties of a concrete is the mineralogic composition of the aggregate, which is a factor not definable in specifications. Specifications requirements for cement, pozzolan, percent sand, and even water content are modifying factors, but they have negligible effect. Entrained air is a significant factor, as it is a good insulator, but economic and other considerations which govern the use of entrained air outweigh the significance of its effect on change in thermal properties.

13. Weight. —The weight of concrete is important in structures that rely on weight for stability, such as gravity dams. Unit weight is increased by the use of aggregate having high specific gravity and by the use of maximum amounts of coarse aggregate well graded to the largest practicable size. Tests of the unit weight of hardened concrete are readily made by displacement when the volume of the specimen cannot be computed accurately.

The unit weight of fresh concrete is employed chiefly as a means for checking the yield of batches, the cement content, and air content, but it is also indicative of the unit weight of the hardened concrete. Average values are shown in table 4.

Table 4.—Observed average weight of fresh concrete

(Pounds per cubic foot)

Maximum size of aggregate, inches	Average values			Unit weight, pounds per cubic foot [1]				
	Air content, percent	Water, pounds per cubic yard	Cement, pounds per cubic yard	Specific gravity of aggregate [2]				
				2.55	2.60	2.65	2.70	2.75
¾	6.0	283	566	137	139	141	143	145
1½	4.5	245	490	141	143	146	148	150
3	3.5	204	408	144	147	149	152	154
6	3.0	164	282	147	149	152	154	157

[1] Weights indicated are for air-entrained concrete with indicated air content.
[2] On saturated surface-dry basis.

C. Effects of Various Factors on the Properties of Concrete

14. Entrained Air Content, Cement Content, and Water Content.—Experience in field and laboratory has conclusively demonstrated that durability and other properties of concrete are materially improved by the purposeful entrainment of 2 to 6 percent air. Purposeful entrainment is accomplished by adding an air-entraining agent to the concrete mix. The use of an agent results in the dispersion throughout the mix of noncoalescing spheroids of air having diameters of from 0.003 to 0.05 inch. The amount of air entrained is a function of the quantity of agent added. Current investigations indicate that various parameters of the air void systems materially affect the properties of the concrete and that the most desirable parameter is that of small, closely spaced air bubbles obtainable with most of the commercial air-entraining agents in common use today.

Since air content has an important effect on water content and also affects cement content to some extent, the effects of these three factors on the properties of concrete are considered together.

(a) *Effects on Workability.*—Entrainment of air greatly improves the workability of concrete and permits the use of aggregates less well graded than required if air is not entrained. This explains why it is possible and usually desirable to reduce the sand content of a mix in an amount approximately equal to the volume of entrained air. Entrained air reduces bleeding and segregation and facilitates the placing and handling of concrete. Reduced bleeding permits finishing of concrete surfaces earlier and usually with less work. Each percent of entrained air permits a reduction in mixing water of 2 to 4 percent, with some improvement in workability and with no loss in slump.

(b) *Effects on Durability.*—Entrainment of 2 to 6 percent air, by use of an air-entraining agent, increases considerably the resistance of concrete to the disintegrating action of freezing and thawing. The entrained air dispersed throughout the concrete in the form of minute, disconnected bubbles provides spaces where forces that would cause disintegration can be dissipated. The effects of different percentages of entrained air on the resistance of concrete to freezing and thawing are indicated in figure 15.

Experience shows that, within the range of water-cement ratios and maximum size aggregates generally used, concretes containing various optimum percentages of entrained air are several times as durable as similar concrete made without entrained air and that low water-cement ratios contribute considerably to the durability of concrete (fig. 16). Entrained air is generally regarded as occurring in the mortar fraction of the concrete; and as mortar is replaced by coarse aggregate with increasing maximum size, the air content is decreased from about 8 percent for concrete containing aggregate graded up to ⅜-inch maximum to about 3 percent for concrete containing aggregate graded up to 6-inch maximum.

Air voids constituting the optimum percentage of entrained air should

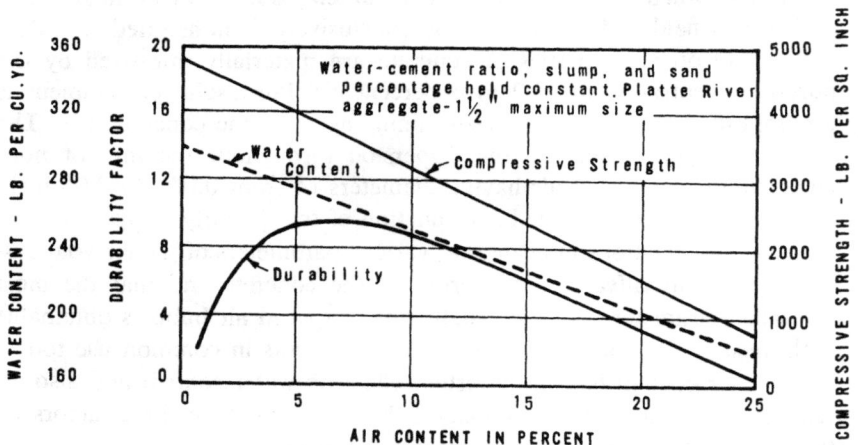

Figure 15.—**Effects of air content on durability, compressive strength, and required water content of concrete. Durability increases rapidly to a maximum and then decreases as the air content is increased. Compressive strength and water content decrease as the air content is increased. 288–D–1520.**

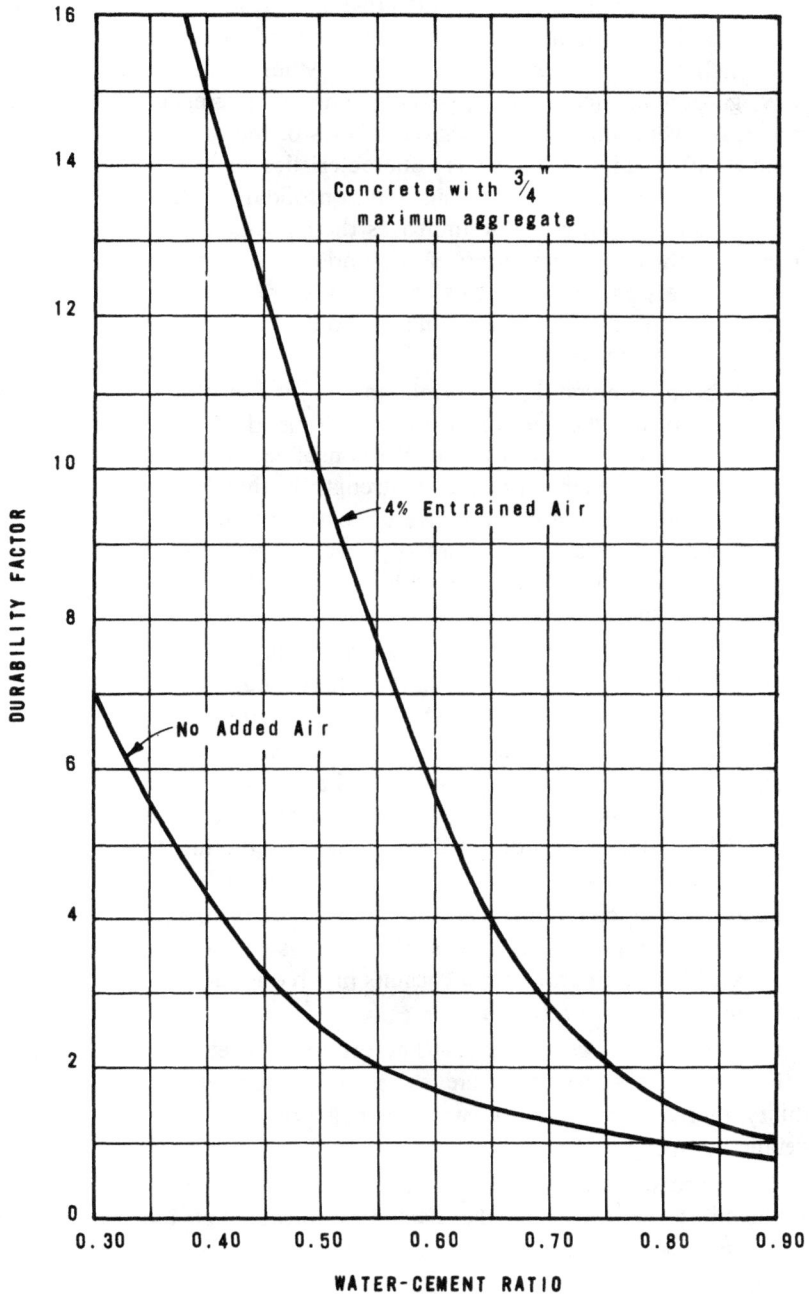

Figure 16.—Relation between durability and water-cement ratio for air-entrained and non-air-entrained concrete. High durability is associated with use of entrained air and low water-cement ratio. 288–D–1521.

be entrained by an approved and effective air-entraining agent and should be dispersed throughout the mortar fraction at an average spacing of 0.007 inch to assure optimum durability. Many factors such as consistency, gradation, sand content, particle shape of aggregate, and type and amount of agent influence the characteristics of the initial air void system formed during mixing. However, characteristics or parameters are little influenced by subsequent handling and consolidation. In fact, consolidation of freshly mixed concrete improves the air void system by decreasing air content through elimination of the undesirable larger air voids; these larger voids are broken up into smaller voids, thus increasing the number but reducing the average size of the air voids with the spacing factor remaining essentially constant.

It has been observed that a normal amount of consolidation or vibration tends to improve the durability of air-entrained concrete even though some of the entrained air is lost in the consolidation process. This reduction in air has a beneficial effect on strength in that it allows recovery of some but not all of the compressive strength lost through initial entrainment of air. An excessive amount of vibration may cause segregation of the mortar and coarse aggregate with detrimental effects on many of the properties of the concrete.

Entrained air further contributes to the durability of concrete because it reduces the water channel structure in hardened concrete by improving workability and reducing bleeding in the fresh concrete.

Reduction in water-cement ratio materially increases the resistance of concrete to sulfate attack. Test results indicate that entrained air, up to 6 percent, slightly increases resistance of concrete to chemical attack. This improved resistance is undoubtedly obtained by the increased watertightness due to the reduction in water channel structure.

Resistance of concrete to erosion is related to compressive strength; therefore, resistance to erosion is increased as the water-cement ratio is decreased. When air entrainment results in a reduction in strength, erosion resistance is likewise reduced.

(c) *Effects on Permeability.*—The pronounced effect of water-cement ratio on permeability of concrete is depicted in figure 17. Note that permeability increases rapidly for water-cement ratios higher than 0.55 by weight.

Water-pressure tests on concrete containing entrained air show that permeability is not appreciably affected by entrained air in the percentages ordinarily used in construction if the water-cement ratio remains unchanged.

Tests of lean mass concretes containing pozzolans indicate increased resistance to the flow of water when finely ground pozzolans are used.

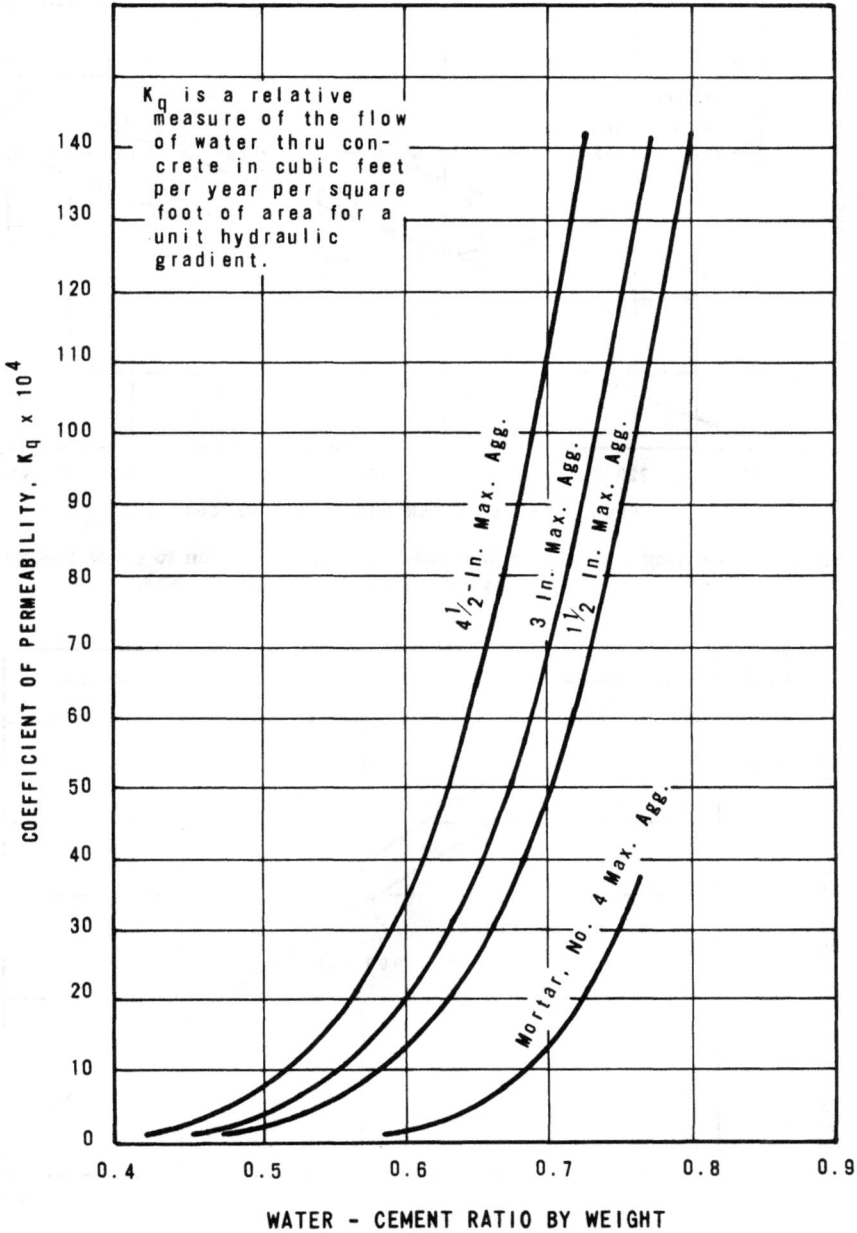

Figure 17.—Relationship between coefficient of permeability and water-cement ratio, for mortar and concrete of three maximum sizes. Relatively low water-cement ratios are essential to impermeability of concrete. 288-D-1522.

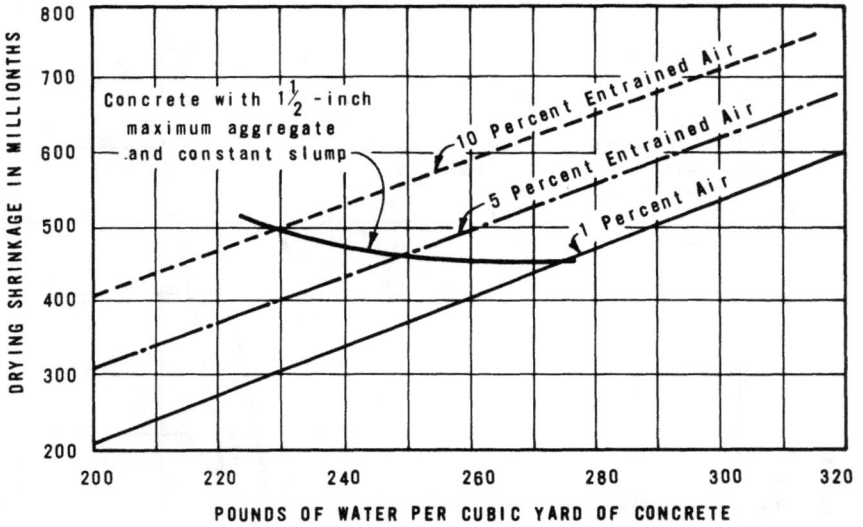

Figure 18.—Drying shrinkage of hardened concrete in relation to water content of fresh concrete, for various air contents. 288–D–1523.

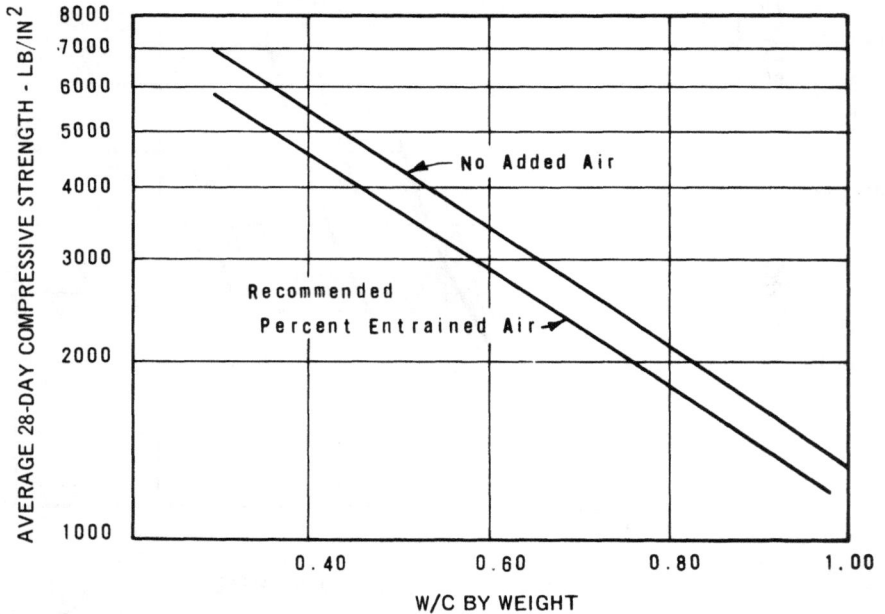

Figure 19.—Strength in relation to water-cement ratio for air-entrained and non-air-entrained concrete. Strength decreases with an increase in water-cement ratio; or with the water-cement ratio held constant, use of air entrainment decreases the strength by about 20 percent. 288–D–1524.

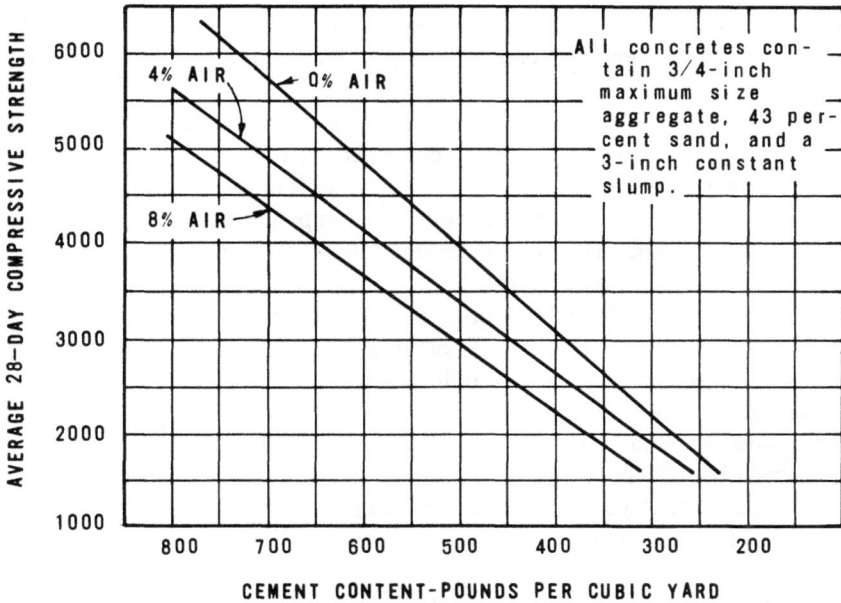

Figure 20.—Strength in relation to cement content for air-entrained and non-air-entrained concrete. 288–D–2654.

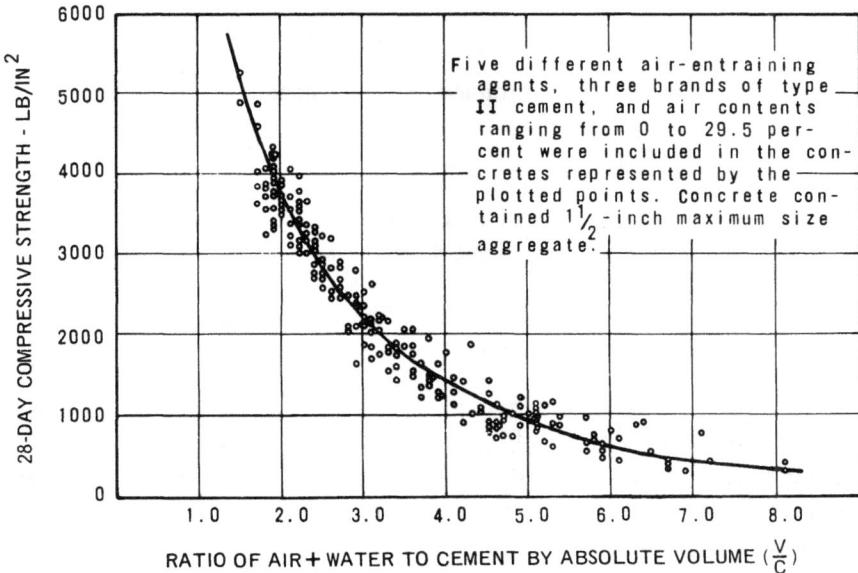

Figure 21.—Compressive strength of concrete in relation to voids-cement ratio. 288–D–1526.

(d) *Effects on Volume Change.*—Drying shrinkage is governed mainly by unit water content. The cement content of a mix has very little effect on the drying shrinkage except as it may influence the water requirement. Ordinarily this effect is small.

Figure 18 reveals that drying shrinkage increases with the water content. This figure also shows that as entrained air content is increased, drying shrinkage increases. However, because entrainment of air permits a reduction in water with no reduction in slump, net shrinkage is not appreciably increased. This fact is demonstrated by the curve for mixes using 1½-inch-maximum size aggregate.

(e) *Effects on Strength.*—Investigations involving thousands of tests and extending over a long period of time have demonstrated conclusively that the most important factor influencing the strength of concrete is the water-cement ratio. Typical graphs in figure 19 show how strength varies with water-cement ratio. For a given water-cement ratio the strength is reduced about 20 percent for air contents recommended in section 20, but when the percentage of air is held constant, as is usually the case for any given maximum size of aggregate, the strength of concrete varies directly with the water-cement ratio.

Entrainment of air in concrete causes a decrease in strength. However, if advantage is taken of the increase in workability through reduction in water, the strength loss will be partially compensated. The curves in figure 20 present the reduction in strength which results from the indicated percentages of air entrained in concrete containing ¾-inch-maximum size aggregate, 43 percent sand, constant cement content, and a constant 3-inch slump.

For a given amount of entrained air, the magnitude of reduction in strength varies with the maximum size of aggregate used. With mixes containing larger size aggregate, the strength reduction becomes less until, with mass concrete containing 6-inch-maximum aggregate and small amounts of cement, the reduction is negligible.

Strength is also a function of the voids-cement ratio, V/C, as shown in figure 21. In this ratio, the term C represents the absolute volume of cement in a unit volume of concrete. The term V represents the total voids in a unit volume of concrete; that is, the combined volume of water and air voids.

(f) *Effects on Elasticity.*—Although modulus of elasticity is not directly proportional to strength, concretes of higher strength usually have higher elastic moduli. Thus, the modulus of elasticity generally increases with a decrease in water-cement ratio or air content.

(g) *Effects on Creep and Extensibility.*—Indications are that creep and extensibility are increased to some extent as air content or water-cement ratio is increased.

(h) *Effects on Thermal Properties.*—The thermal properties are not materially affected by changes in air, cement, or water content within the range of practicable mixes. However, the conductivity of hardened concrete does vary inversely with the air content and directly with the water and cement contents.

(i) *Effects on Unit Weight.*—The unit weight of concrete is reduced in direct proportion to the amount of air entrained. An increase in water content tends to decrease the unit weight. An increase in cement content will increase it.

15. Portland Cement.—Portland cement as a hydrated paste is the binder of concrete. The binder, often called cement gel, governs in large part most of the properties of concrete. Much scientific and technological information is available on the composition of portland cement and on the chemical nature of each of the compounds or phases, as they are sometimes called. In this manual the term compound is used.

(a) *Compound Composition of Cement.*—The composition of the cement clinker may be expressed for practical purposes in amounts of four major compounds, C_3S^1, βC_2S, C_3A, and C_4AF. This simplification is justified as a matter of convenience and, in fact, the deviations from a more exact expression of kind and amount of compounds have no discernible effects on the correlations to be discussed. Generally, no one of the major compounds occurs as a pure compound; glass may also be present as a compound. All four major compounds may contain substantial amounts of extraneous ions, including those of minor constituents, as substituents in the lattice. In the terminology of cement, it is becoming common practice to call "impure" C_3S, alite, and "impure" beta C_2S, belite. The C_3A containing extraneous ions is still generally called C_3A. The C_4AF (in reality the solid solution, $C_6A_2F-C_2F$) generally approaches the composition of C_4AF and is designated C_4AF; the term "ferrite phase" is, however, frequently used. In addition to the compounds described, cement clinker contains minor constituents—some of which, in major part, are substituted in the lattice of one or more of the major compounds.

Portland cement clinker is made by intimately intergrinding selected proportions of an argillaceous material consisting largely of Al_2O_3, Fe_2O_3, and SiO_2, and a calcareous material that supplies the CaO. Small quantities of iron ore or silica may be added to the mix to obtain a desired clinker composition. The mixture as a slurry or dry powder is passed through a rotary kiln at increasingly higher temperatures, the maximum being between about 2,550° F and 2,800° F. At the highest temperatures of burning, about 25 to 30 percent of the solids are present as liquid (melt) which consists largely of CaO, Al_2O_3, and Fe_2O_3, some MgO, and

---

[1] In these and other abbreviated formulas, C=CaO, S=SiO$_2$, A=Al$_2$O$_3$, F=Fe$_2$O$_3$, $\bar{S}$=SO$_3$, N=Na$_2$O, K=K$_2$O, and H=H$_2$O.

a little SiO_2. The CaO in the liquid, the amount of which is continuously replenished, reacts with the SiO_2 to form C_2S. The C_2S in turn reacts in solid state with CaO to form C_3S. The C_4AF and C_3A are crystallization products of the melt as it cools.

The rate of cooling of the product as it discharges from the kiln may markedly affect certain properties of the cement. Clinkers of identically computed compounds may cause marked differences in properties of concrete.

The clinker generally contains small quantities of MgO, Na_2O, K_2O, SO_3, free CaO, TiO_2, P_2O_5, and Mn_2O_3, and trace amounts of many elements. All originate from the raw materials and a small amount from fuel ash, but the manner of burning and cooling the clinker may affect the amounts and distribution of these in the clinker. The K_2O and SO_3 may appear in part as K_2SO_4 and a part of the K_2O may become beta C_2S. The K_2O enters the lattice of beta C_2S but may also be present in part in other compounds. The other minor constituents as well as some of the Al_2O_3 and Fe_2O_3 are also variously distributed in lattices of the major compounds. For example, the major substituents in C_3S are Al_2O_3 and MgO, and the product is called alite.

Portland cement is an intimately interground mixture of cement clinker and predetermined amounts of gypsum (about 2.5 to 8.3 percent) required for controlling set and obtaining optimum strength. The oxide composition of the cement relates to the composition of the raw mix (kiln feed). The compound composition of the clinker relates partly to the thermal history of burning and cooling.

Attempts to correlate compound compositions of cement with properties of cement paste or concrete yield generally only qualitative answers. Three reasons account for lack of precision in the correlations:

(1) Calculated compound composition may differ substantially from the compositions obtained by direct measurements with X-ray diffraction and optical microscopy.

(2) The differences in contribution of pure and impure compounds to the physical properties of concrete could be very substantial. For example, alite gives much higher early strength than C_3S.

(3) The compounds of the cement interact during hydration. The hydrated calcium silicate (the cement gel) may contain in its lattice variable, but up to very substantial, amounts of the Al_2O_3, Fe_2O_3, and SO_3 of the cement.

For practical purposes, approximate compound composition of a cement can be determined from the oxide analysis using the nomographs in figure 22.

Notwithstanding the above-described limitations, the approximate correlations between amounts of calculated compounds and the properties of concrete are of great technological importance. The C_3S (alite) correlates with high early strength; such strength increases are enhanced by larger amounts of C_3A in the cement. These strength-favorable compositions produce more heat per unit during hydration than cements of other compositions. By contrast, beta C_2S (or belite) hydrates more slowly with accompanying lower strength and lower heat of hydration. Larger amounts of C_3A, although increasing early strength, cause lower strengths at later ages. C_3A has other unfavorable properties. Cements with increasing contents (above 5 percent) are much more susceptible to sulfate or saline solution attack. Excluding gypsum recrystallization, C_3A is the principal cause of abnormal setting of cements.

The resistance of concrete to sulfate and saline solutions generally improves with decreasing amounts of C_3A and C_4AF in the cement. C_4AF has a lower heat of hydration than C_3A but is not a significant contributor to strength. When present in large amounts, it may actually decrease strength slightly.

Among the minor compounds of cements, MgO is generally limited to a maximum of 5 percent in specifications. The delayed hydration of MgO in amounts exceeding this limit in moist or intermittently moistened concrete leads to deleterious expansion. Free CaO is a criterion of degree of burning in the manufacture of cement. Federal specifications do not specify limits on permissible amounts, but free CaO is limited to about 1.0 percent maximum by the producers.

The Na_2O and K_2O, if present above certain amounts, when extracted by water in a moist concrete, may cause swelling of certain aggregates, such as opal, and resultant expansion of the concrete. This is known as the alkali-aggregate reaction. The deleterious effects of the reaction can be prevented by using low-alkali cements, aggregate not subject to alkali attack, or certain amounts of selected pozzolans. The limitation on amounts of alkali in Federal specifications is that the summation of percentages of $Na_2O + 0.658$ of percentage of K_2O shall not exceed 0.60 percent. However, there are a few reported instances where this limitation was not sufficient to avoid reaction wnich over a period of years led to significant deterioration.

(b) *Types of Cement.*—The first classification of cements into five types—I, II, III, IV, and V—has persisted in principle. Several modifications, such as the use of air-entraining agents in types I, II, and III to provide air-entraining cement, have not otherwise altered the requirements of the specifications. Federal Specification SS–C–1960/3, classifies the five types according to usage: type I for use in general concrete construction when the special properties specified for

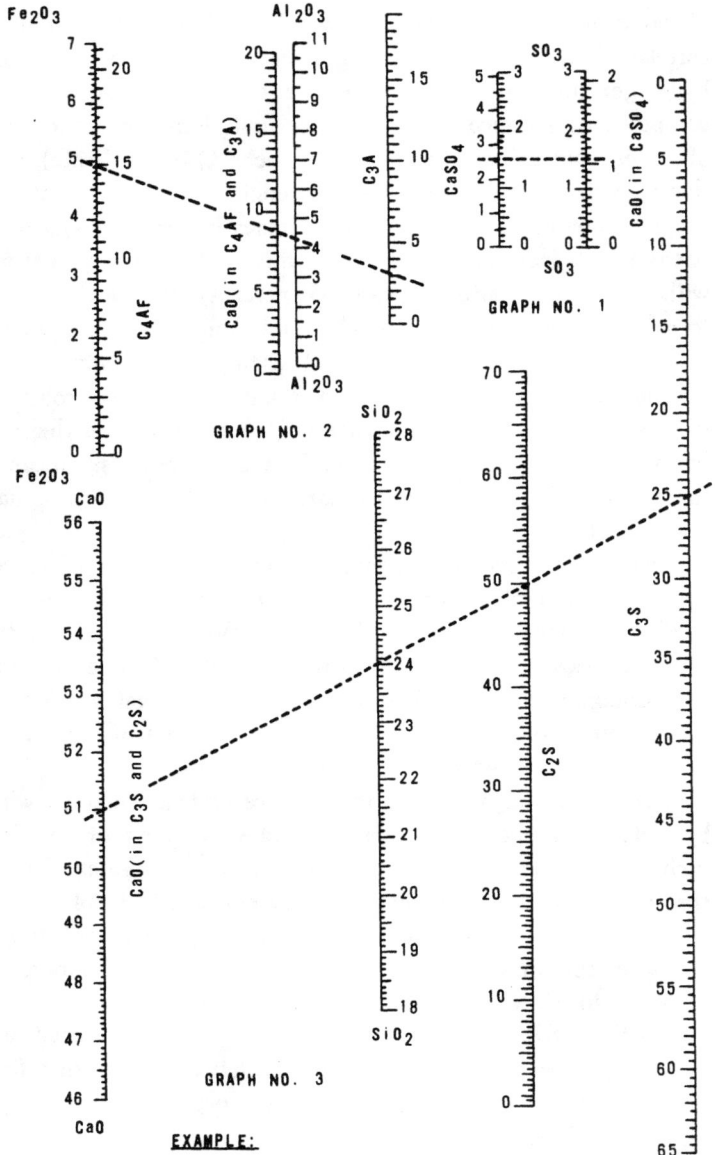

EXAMPLE:

GRAPH NO.	% OXIDES	% CaO	% COMPOUND
1	SO_3 = 1.5	1.0	$CaSO_4$ = 2.5
2	$\begin{cases} Fe_2O_3 = 4.9 \\ Al_2O_3 = 4.3 \end{cases}$	8.8	$\begin{cases} C_4AF = 15.0 \\ C_3A = 3.0 \end{cases}$
	CaO(FREE) 1.0	1.0	CaO(FREE) = 1.0
3	$\begin{cases} CaO = 61.8 - 10.8 = 51.0 \\ SiO_2 = 24.0 \end{cases}$		$\begin{cases} C_2S = 50.0 \\ C_3S = 25.0 \end{cases}$
	TOTAL 96.5		TOTAL 96.5

Figure 22.—Nomographs for determining approximate compound composition of portland cement from its oxide analysis. 288–D–1061.

types II, III, IV, and V are not required; type II for use in general concrete construction and in construction exposed to moderate sulfate attack; type III for use when high early strength is required; type IV for use when low heat of hydration is required; and type V for use when high sulfate resistance is required.

The only major compound limitation on type I and type III is that the amount of C_3A shall not exceed 15 percent. However, the Federal specifications permit limiting the C_3A content for type III to 8 or 5 percent when moderate or high sulfate resistance, respectively, is required. The compound restriction on type II cement is a maximum C_3A content of 8.0 percent and, when needed and specified by the purchaser, a maximum of 58 percent of $C_3A + C_3S$. The type IV cement is limited to a maximum of 7.0 percent of C_3A and 35 percent of C_3S. Type V cement has limits of 5.0 percent C_3A and 20.0 percent of $C_4AF + 2(C_3A)$ as maxima.

The relative proportions of the major compounds in the five types of cement are illustrated in table 5. Figure 23 shows strengths of concretes made and cured under similar conditions for the five types of cements. Heat of hydration and temperature rise for the five types are presented in figure 24.

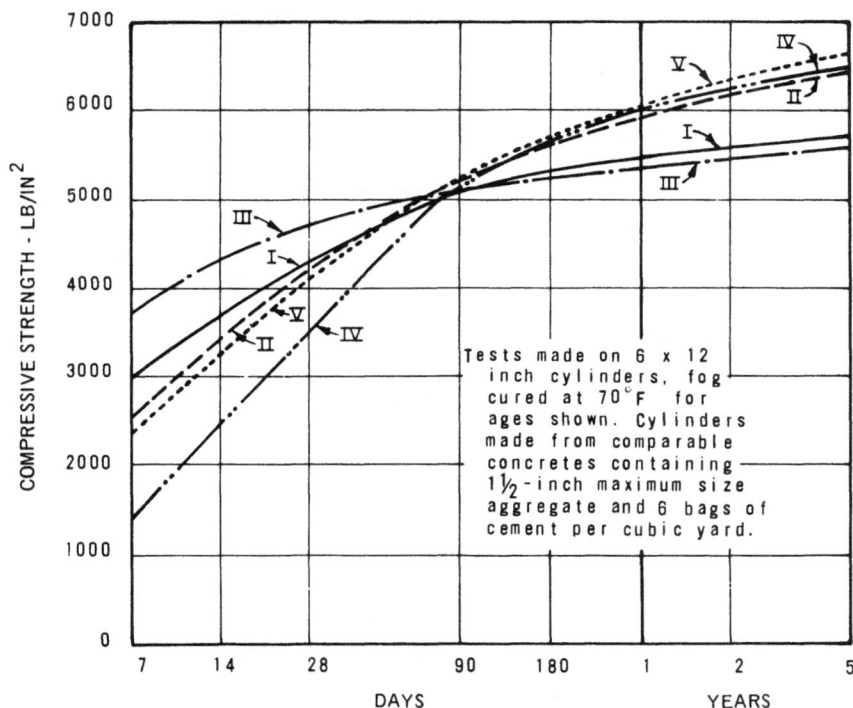

Tests made on 6 x 12 inch cylinders, fog cured at 70°F for ages shown. Cylinders made from comparable concretes containing 1½-inch maximum size aggregate and 6 bags of cement per cubic yard.

Figure 23.—Rates of strength development for concrete made with various types of cement. 288–D–1527.

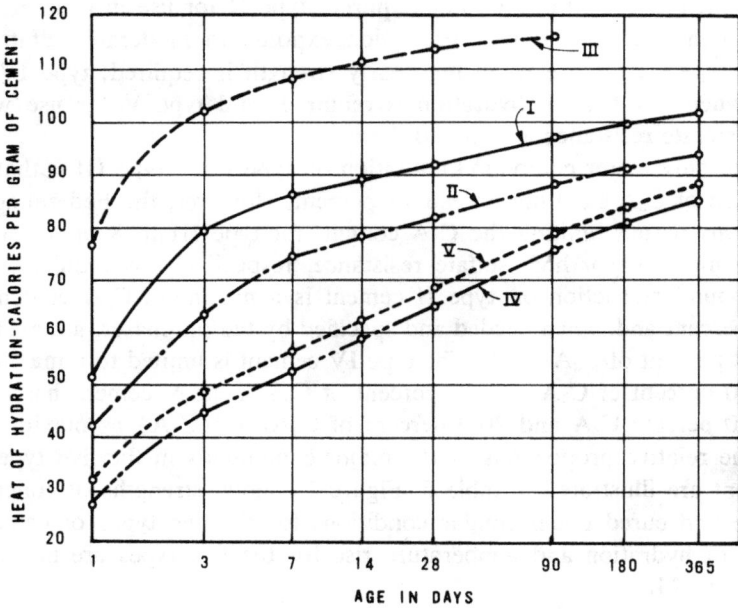

HEAT OF HYDRATION FOR VARIOUS TYPES OF CEMENT

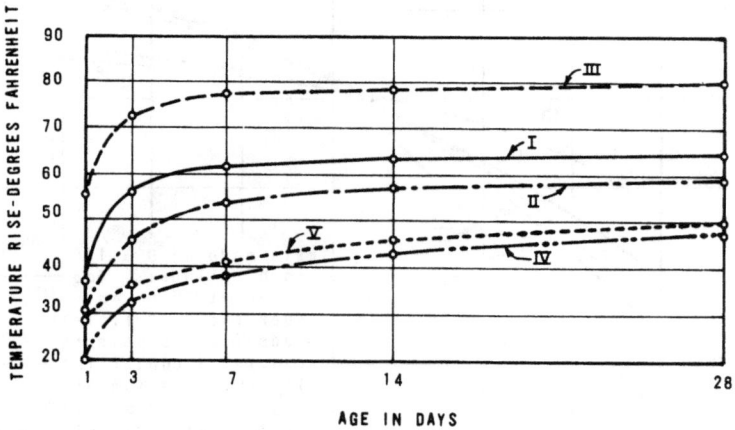

TEMPERATURE RISE OF CONCRETE

Tests of mass concrete with 4½-in. maximum aggregate,
containing 376 pounds of cement per cubic yard in
17-by 17-in. cylinders, sealed and cured in adiabatic
calorimeter rooms

Figure 24.—Heat of hydration and temperature rise for concretes made with various types of cement. 288–D–118.

In addition to the usages of the five types of cement already mentioned, Federal Specification SS–C1960/3 gives further information. Provisions are

Table 5.——Compound composition of portland cements

Type of cement	Compound composition, percentage							
	C_3S	C_2S	C_3A	C_4AF	$CaSO_1$	Free CaO	MgO	Ignition loss
Type I	49	25	12	8	2.9	0.8	2.4	1.2
Type II	46	29	6	12	2.8	0.6	3.0	1.0
Type III	56	15	12	8	3.9	1.3	2.6	1.9
Type IV	30	46	5	13	2.9	0.3	2.7	1.0
Type V	43	36	4	12	2.7	0.4	1.6	1.0

stated under which any of the five types of cement may be required, at the discretion of the purchaser, to meet low-alkali or false set limitations. Additionally, type II or type IV may be required to meet limitations on heat of hydration. For a type II, either a maximum heat of hydration or a maximum limitation on C_3A plus C_3S content, or both, may be specified.

Use of type I cement is generally permitted only in precast or precast-prestressed concrete items not to be in contact with soils or ground water. In such cases, use of this type of cement is an alternative to use of type II or type III. As types I and II are both suitable for use in general construction, use of type II for this purpose in Bureau work is preferable because of the generally moderate sulfate conditions occurring in soils and ground waters in many areas throughout the western part of the United States. Types I and II are normally available at the same cost. Type II cement is also specified for use in mass concrete, and the heat of hydration and C_3A plus C_3S content limitations are required to minimize cracking caused by temperature gradients.

In the past, type IV cement was used in construction of Bureau dams because of lower heat of hyration. This cement has the disadvantage of slow strength development and higher cost. Development of mix designs utilizing pozzolans and water-reducing admixtures to allow decreases in cement content and general improvements in the technology of dam construction made possible the substitution of type II cement for the type IV. These advances have resulted in improved quality and reduction of costs.

Type III cement is used where rapid strength development of concrete is essential, as in emergency construction and repairs and construction of machine bases and gate installations. It is also used in laboratory tests where quick test results are necessary. Where this type of cement is used,

48   CONCRETE   MANUAL

curing and protection of the concrete may be discontinued at earlier ages. Concrete having high early strength may also be produced with an accelerator (see sec. 20). So doing eliminates the need for changing the type of cement and also decreases the cost.

Type V cement is essential where structures such as canal linings, culverts, and siphons will be in contact with soils and ground waters containing sulfates in concentrations that would cause serious deterioration if other types were used. By reference to table 5, it will be seen that the sum of C_3S and C_2S is unusually high in this cement and that the sum of C_3A and C_4AF is less than for any other types. This combination of low C_3A and C_4AF imparts much greater resistance to sulfate attack than is attainable with other cements.

Portland-pozzolan cements, mixtures of portland cement and certain chemically active natural or artificial materials called pozzolans, are covered by Federal Specification SS–C–1960/4 and ASTM Designation C 595. Portland-pozzolan cement are manufactured by intergrinding the pozzolan with the portland cement clinker at the mill.

Expansive cement is a hydraulic cement that expands during early hardening. There is no Federal specification for expansive cements but they are covered by ASTM Designation C 845. Either type I or II portland cement is the major binding material of expansive cements. Various cement formulations have been identified in the industry by letters K, M, and S. The expansive constituents are $C_4A_3\overline{S}$, calcium aluminate cement (CA and $C_{12}A_7$), and C_3A in types K, M, and S cements, respectively. All contain $\overline{S}$ in excess of that normally present in portland cement and range in amount from approximately 4.5 to 6.0 percent.

Expansive cements are used to produce what is known as shrinkage compensating concrete. If the early expansion of the concrete is restrained by reinforcement, formwork or other restraint a compressive stress is developed in the concrete. This compressive stress compensates for later volume change due to drying and prevents or reduces cracking due to drying shrinkage. Most expansive cements are not resistant to sulfate attack and should not be used in concrete that will be exposed to sulfates unless satisfactory resistance has been determined by test.

Laboratory investigations indicate a high sensitivity of expansive cements to aeration and variables of temperature and curing, and careful attention must be given to these factors during any concrete mixing and placing operation. Laboratory investigations also indicate this high sensitivity of expansive cements to variables would generally make questionable their effective use in combination with water-reducing, set-controlling agents and pozzolans.

Set-controlled cement, a type applicable to work of an unusual requirement for hardening rate, is now on the market. White cement, a readily available product, can be used for special architectural and esthetic treatments of structures which would otherwise contain ordinary gray-colored portland cement.

(c) *Fineness of Cement.*—Higher fineness increases the rate at which cement hydrates, causing greater early strength and more rapid generation of heat. Although total heat generation and strengths at later ages are somewhat greater for the finer cements, the effects of higher fineness are manifested principally during the early period of hydration.

Because of their extremely small size, the finer cement particles are not susceptible to separation into size fractions by means of screens, and special methods have been devised for making quantitative approximations of size distribution. Instruments known as turbidimeters and air-permeability apparatus are in common use for this purpose. The measure of fineness is known as specific surface and is the summation of the surface area, in square centimeters, of all the particles in 1 gram of cement, the particles being considered as spheres. The Wagner turbidimeter method for determining the specific surface of cement, the accepted standard for many years, has been replaced in Federal specifications by the Blaine air-permeability method.

As determined by the Blaine method, specific surface of most modern cements ranges from 2,600 to 5,000 square centimeters per gram. Federal Specification SS-C-1960/3 stipulates that for all types of cement, except type III on which there is no fineness requirement, the average specific surface determined with samples representing a bin of cement shall be not less than 2,800. Although there is no definite ratio between the surface areas of cement as determined by the Blaine and Wagner methods, and approximation of the Wagner values may be made by dividing the Blaine specification requirements by 1.8.

Cements having a specific surface less than about 2,800 (Blaine) may produce concrete with poor workability and excessive bleeding (water gain at the top of concrete caused by settlement of solids prior to initial set). Bleeding often causes unsightly sand streaking on concrete surfaces. Within the normal fineness range, decreased fineness increases water requirements. Greater fineness improves early strength development. However, tests indicate that resistance to freezing and thawing is slightly lower when finely ground cement is used in concrete cured under conditions similar to those used in the field.

Evidence of differences in strength, heat of hydration, production of laitance, bleeding tendency, and durability has been observed in comparing cements otherwise considered to be similar on the basis of fineness tests and chemical analyses. Causes of these differences are not fully understood, but it is suspected that dissimilarities in raw materials and manufacturing processes are responsible. Differences in inherent air-entraining characteristics of the cements may be contributing factors. Some attempts to analyze and regulate these more obscure effects on concrete quality have been made, and the matter is receiving increasing attention.

16. Abnormal Set of Portland Cement.—Abnormal set, or premature stiffening, of cement impedes or prevents proper placing and consolidation of concrete. A normal setting concrete may be defined as one that retains its workability for a sufficient period of time to permit proper placing and consolidation. The period of time required between completion of mixing and completion of consolidation may be as short as 10 minutes or may extend up to 2 hours. The loss of workability during the interval is called slump loss, measurable either by the slump test or Proctor Needle Penetration Test (ASTM C 403 Standard). In the laboratory, abnormal setting is determinable as decrease of penetration of a 1-centimeter-diameter, 400-gram Vicat needle in a mortar, following the method of ASTM C 359 Standard.

Abnormal set may be due to one or more causes, and different types of set are known (or designated) as false, delayed false, quick, delayed quick, and thixotropic. In the following definitions, paste, mortar, and concrete are interchangeable words. According to ASTM C 359 Standard, "False set is the rapid development of rigidity in a mixed portland cement paste (without evolution of much heat) which rigidity can be dispelled and plasticity regained by further mixing without addition of water." False set as described is often caused by recrystallization of gypsum (which was dehydrated during grinding) in the immediate postmixing period. The corrective for this type of false set is the maintenance of sufficient amount of gypsum in the cement during manufacture to cause total precipitation of dehydrated gypsum during the mixing of concrete. False set is also occasionally caused by continuation of ettringite precipitation for several minutes in the postmixing period. Ettringite $(C_3A \cdot 3CS \cdot H_{32})$ is formed by the reaction of the C_3A, gypsum, and water. In a normal setting cement, ettringite precipitates as a slightly previous coating over the exposed surfaces of C_3A crystals and stops temporarily the fast hydration of C_3A. This is the generally accepted theory explaining gypsum as a set retarder.

Delayed false set is phenomenologically and chemically the same as false set except that the recrystallization of gypsum (and infrequently ettringite precipitation) occurs after the remixing at 11 minutes in ASTM

C 359 Method. Both false set or delayed false set can be dispelled by further mixing.

According to ASTM C 359 Standard, "Quick set is the rapid development of rigidity of a cement paste (usually with the evolution of considerable heat) which rigidity cannot be dispelled nor can plasticity be regained by further mixing without addition of water." Quick set is caused by rapid and uninterrupted precipitation of ettringite. Quick set has not been encountered in Bureau work for several years. Delayed quick set occurs when the ettringite reaction has temporarily stopped during mixing but is reactivated during remixing at 11 minutes or shortly thereafter. Pastes or mortars exhibiting delayed ettringite precipitation continue to set; therefore, this set is not dispelled by further mixing. The dispelling or nondispelling of delayed sets is the criterion for calling one delayed false set and the other delayed quick set.

Thixotropic set may be defined as a very rapid and pronounced development of rigidity of a cement paste immediately upon cessation of mixing. This rigidity is dispelled without recurrence by additional mixing up to 2 minutes, but infrequently longer mixing may be required. This type of set was determined in the Bureau laboratories to be caused by interaction of opposite electrostatic surface charges on different compounds in ground cement clinker. Such charges, detected in a few cements obtained from different projects, were probably induced by aeration. It has been found that electrostatic charges can be caused by aeration of ground clinker or cement at 50 percent relative humidity. An instrument called a thixometer (adapted from a Stormer paint viscometer) has been developed to measure the relative strengths of bonds between particles in a cement-benzene slurry. The difference in the total load to shear the set slurry and the load to maintain free flow after set is broken divided by the total load provides an index ratio to express thixotropic set.

17. Classification and Use of Pozzolans.—Pozzolans are siliceous or siliceous and aluminous materials which in themselves possess little or no cementitious value but will chemically react, in finely divided form and in the presence of moisture, with calcium hydroxide at ordinary temperatures to form compounds possessing cementitious properties.

All pozzolans owe their chemical activity to one or more of five kinds of substances: (1) siliceous and aluminous, artificial or natural glass; (2) opal; (3) calcined clay minerals; (4) certain zeolites; and (5) hydrated oxides and hydroxides of aluminum. They can be classified petrographically as follows:

    (1) Clays and shales (must be calcined to activate).

        Kaolinite type

        Montmorillonite type

        Illite type

(2) Opaline materials (calcination may or may not be required).
    Diatomaceous earth
    Opaline cherts and shales
(3) Volcanic tuffs and pumicites (calcination may or may not be required).
    Rhyolitic types
    Andesitic types
    Phonolitic types
(4) Industrial byproducts.
    Ground brick
    Fly ash
    Silica fume

Except for rare occurrences, natural pozzolans must be ground before use. The clayey pozzolanic material, including altered volcanic ashes and tuffs as well as shales, must be calcined at temperatures between 1,200° and 1,800° F to activate the clay constituent.

Pozzolans are normally not specified for concrete unless advantages in their use outweigh the disadvantages of storing and handling an extra material. Pozzolans may be used to improve the workability and quality of concrete, to effect economy, or to protect against disruptive expansion caused by the reaction between certain aggregates and the alkalies in cement. Most good quality pozzolans also increase the resistance of concrete to deterioration in exposure to soluble sulfates in soil or ground water. Fly ash is more effective and consistent for this purpose than the natural types. In addition to improving workability of concrete, most pozzolans will reduce heat generation, thermal volume change, bleeding, and permeability of concrete. Some pozzolans, particularly calcined clay and shales, require more water than portland cements. When additional water is required, additional cement is also required to maintain a specified water-cement ratio and to assure that the concrete will meet design strengths. The additional cement increases the cost of the concrete, and the additional water increases drying shrinkage, which may result in increased cracking. Also, investigations demonstrate that concrete containing pozzolan must be thoroughly cured; otherwise resistance to freezing and thawing will be reduced.

The following pozzolans are known to control alkali-aggregate reaction effectively, even when reactive aggregate and high-alkali cement are used:

(1) Highly opaline material, such as diatomaceous earth and opaline chert.
(2) Certain volcanic glasses.
(3) Certain calcined clays.
(4) Fly ash.

All the materials listed here reduce expansion caused by alkali-aggregate reaction, with fly ash being generally the least effective. However, the effectiveness of these pozzolans in controlling disruptive alkali-aggregate expansion is generally diminished if calcium chloride is added to the mix.

Pozzolans that will control alkali-aggregate reaction can be divided into two groups: (1) certain amorphous siliceous and aluminous substances; and (2) certain calcined montmorillonite-type clays. Materials of the first group include opal and highly opaline rocks of any type; kaolin clays calcined in the range 1,200° to 1,800° F; diatomaceous earth; some rhyolitic pumicites; and some artificial siliceous glasses. Fly ashes as a group are moderately effective in reduction of reactive expansion when compared to the better materials of groups (1) and (2); however, some fly ashes significantly reduce expansion. Calcined clays of the montmorillonite group often are effective in controlling alkali-aggregate reaction. Calcination at 1,600° F or higher is necessary for these materials to avoid causing excessive water requirement, shrinkage cracking, or abnormal stiffening of a concrete mix.

Some pozzolans appreciably increase the water requirements of concrete when used in sufficient quantity to control alkali-aggregate reaction. These materials include diatomaceous earth, several industrial byproducts composed of amorphous hydrous silica, and some of the clayey pozzolans. The increase in water requirement with pozzolans results from their high absorption, low specific gravity, and in some cases high fineness. Fly ash, of low carbon content, generally decreases the water requirement.

Caution must be exercised in the selection and use of pozzolans, as their properties vary widely and some may introduce adverse qualities into the concrete, such as excessive drying shrinkage and reduced strength and durability. Moreover, when used in insufficient proportions with some chemically reactive aggregate-cement combinations, certain pozzolans have increased the expansion in mortars. Before accepting a pozzolan for a specific job, it is advisable to test it in combination with the cement and aggregate to be used, so as to determine accurately the advantages or disadvantages of the pozzolan with respect to quality and economy of the concrete. Any pozzolan proposed for use in Bureau construction must meet the requirements of Federal Specification SS–C–1960/5, Pozzolan For Use in Portland Cement Concrete.

18. Quality and Gradation of Aggregates.—Concrete aggregate usually consists of natural sand and gravel, crushed rock, or mixtures of these materials. Natural sands and gravels are by far the most common and are used whenever they are of satisfactory quality and can be obtained economically in sufficient quantity. Crushed rock is widely used for coarse aggregate and occasionally for sand when suitable materials from natural deposits are not economically available, although production of workable concrete from sharp, angular, crushed fragments usually requires more vibration and cement than that of concrete made with well-rounded sand and gravel. However, through the extra workability imparted by entrained air, the difficulty of making workable concrete with crushed aggregate has been greatly reduced. The shape of the particles of crushed rock depends largely on the type of rock and the method of crushing.

Artificial aggregates in common use in certain localities consist mainly of crushed, air-cooled blast-furnace slags and specially burned clays. Slags are economically available only in the vicinity of blast furnaces. Lightweight aggregate, manufactured by vitrifying and expanding clays in kilns, is used by the Bureau principally for insulation, fireproofing, and lightweight floor and roof slabs. (For further discussion of lightweight aggregates, see secs. 140 through 143.)

Deterioration of concrete has been traced in many instances to the use of unsuitable aggregate. Suitable aggregate is composed essentially of clean, uncoated, properly shaped particles of strong, durable materials. When incorporated in concrete, it should satisfactorily resist chemical or physical changes such as cracking, swelling, softening, leaching, or chemical alteration and should not contain contaminating substances which might contribute to deterioration or unsightly appearance of the concrete. The elements contributing to unsoundness through physical and chemical changes or through deleterious contamination are mentioned in the following subsections and discussed in detail in chapter II.

The choice in selecting aggregate, for economic reasons, is usually limited to local deposits. Good judgment in making this choice involves an appreciation of the desirable and undesirable characteristics that determine aggregate quality and of the practicability of improving available materials by suitable processing.

(a) *Contaminating Substances.*—Aggregate is commonly contaminated by silt, clay, mica, coal, humus, wood fragments, other organic matter, chemical salts, and surface coatings and encrustations. Such contaminating substances in concrete act in a variety of ways to cause unsoundness, decreased strength and durability, and unsightly appearance; their presence complicates processing and mixing operations. They may increase the water requirement, may cause the concrete to be physically weak or susceptible to breakdown by weathering, may inhibit the development of maximum bond between the hydrated cement and aggregate, may hinder the normal hydration of cement, or may react chemically with cement constituents. One or more of these substances contaminate most aggregates but the amounts that are allowable depend on a number of factors, which vary in individual cases. Permissible percentages, by weight, are commonly stipulated by specification. Fortunately, excesses of contaminating substances may frequently be removed by simple treatment. Silt, clay, powdery coatings, soluble chemical salts, and certain lightweight materials are usually removable by washing. Special and more complicated processing may be necessary for other, less amenable substances such as clay lumps, or their removal may not be possible by methods which are economically practicable. Deleterious substances such as tree roots and driftwood are discussed in section 69.

(b) *Soundness.*—An aggregate is considered to be physically sound if it is adequately strong and is capable of resisting the influences of weathering without disruption or decomposition. Mineral or rock particles that are physically weak, extremely absorptive, easily cleavable, or swell when saturated are susceptible to breakdown through exposure to natural weathering processes. The use of such materials in concrete reduces strength or leads to premature deterioration by promoting weak bond between aggregate and cement or by inducing cracking, spalling, or popouts. Shales, friable sandstones, some micaceous rocks, clayey rocks, some very coarsely crystalline rocks, and various cherts are examples of physically unsound aggregate materials; these may be inherently weak or may deteriorate through saturation, alternate wetting and drying, freezing, temperature changes, or by the disruptive forces developed as a result of crystal growth in the cleavage planes or pores.

The most important properties affecting physical soundness of aggregate are the size, abundance, and continuity of pores and channelways within the particles. These pore characteristics influence freezing and thawing durability, strength, elasticity, abrasion resistance, specific gravity, bond with cement, and rate of chemical alteration. Aggregate particles that contain an abundance of internal channelways of very small size (particularly those less than 0.004 millimeter in diameter) contribute most toward reduced freezing and thawing durability of concrete. Such particles readily absorb water and tend to retain a high degree of saturation when enclosed in concrete; consequently, with progressive freezing, drainage of excess water from the freezing zone may not be accomplished before high internal hydrostatic pressure causes failure of portions of the concrete.

Chemical soundness of an aggregate is also important. In many instances, excessive expansion causing premature deterioration of concrete has been associated with chemical reaction between reactive aggregate and the alkalies in cement. Known reactive substances are the silica minerals, opal, chalcedony, tridymite, and cristobalite; zeolite, heulandite (and probably ptilolite); glassy to cryptocrystalline rhyolites, dacites and andesites and their tuffs; and certain phyllites. Any rock containing a significant proportion of reactive substances will be deleteriously reactive; thus, although pure limestones and dolomites are not deleteriously reactive, limestones and dolomites that contain opal and chalcedony are related to deterioration of concrete as a result of alkali-aggregate reactivity. Similarly, normally innocuous sandstone, shales, granites, basalts, or other rocks can be deleteriously reactive if they are impregnated or coated with opal, chalcedony, or other reactive substances.

Other types of chemical alteration, such as oxidation, solution, or hydration, may decrease the physical soundness of susceptible aggregate particles after their incorporation in concrete, or may produce unsightly exudations or stains.

(c) *Strength and Resistance to Abrasion.*—An aggregate should and usually does have sufficient strength to develop the full strength of the cementing matrix. When wear resistance is important, aggregate particles should be hard and tough. Quartz, quartzite, and many dense volcanic and siliceous rocks are well qualified for making wear-resistant concrete.

(d) *Volume Change.*—Volume change in aggregate resulting from wetting or drying is a common source of injury to concrete. Shales, clays, and some rock nodules are examples of materials which expand when they absorb water and shrink as they dry. Thermal coefficients of expansion vary widely in different minerals (see sec. 8), and it has been suggested that damaging internal stresses may also develop when the change in volume of aggregate particles caused by temperature variations is substantially different from that of cement paste or when there are large differences in the coefficients of expansion among the aggregate particles. Instances of cracking and spalling have been ascribed to this cause. However, aggregates are usually heterogeneous masses, and even when such variations could theoretically cause failure, the proof of such failure is infrequent and doubtful.

The coefficient of thermal expansion of a material is the rate at which thermal volume change takes place. Coefficients of expansion of individual rock specimens may vary widely. (Limestones range from 2 to 6.5 millionths.) However, the following are given as average coefficients for some common rocks frequently found in concrete aggregates:

Rock	Coefficient of expansion
Basalts and gabbros	3.0
Marbles	3.9
Limestones	4.4
Granites and rhyolites	4.4
Sandstones	5.6
Quartzites	6.1

Some crystalline rocks are anisotropic; in other words, they have different coefficients along each of the various crystalline axes. For example, the coefficient of feldspar is about 0.5 millionths on one axis, and 9 millionths on another axis.

The aggregate in concrete, making up from 70 to well over 80 percent of the total solid volume, will essentially control the coefficient of expansion of the concrete when estimated by the usual method of using the weighted averages of the coefficients of the different components. Since natural stream gravels are usually heterogeneous mixtures, concretes made from such aggregate will be about average, with a coefficient of about 5.5 millionths. Mineral aggregates vary from below 2 millionths to above 7 millionths in thermal coefficient of expansion. Cements and frequently sands exhibit somewhat higher average expansion, and hence mortars without coarse aggregate should be estimated separately (see sec. 8).

At a temperature of 1,063° F, which is commonly reached in a burning building, quartz changes state and suddenly expands 0.85 percent, usually producing a disruptive effect at the surface of concrete. This sudden change of 0.85 percent represents a linear change of 8,500 millionths and is equivalent to several hundred degrees temperature change. Expansion which accompanies chemical reactions between certain aggregates and alkalies in cement has been discussed previously in this section.

(e) *Particle Shape.*—The chief objection to flat or elongated particles of aggregate is the detrimental effect on workability and the resulting necessity for more highly sanded mixes and consequent use of more cement and water. A moderate percentage (on the order of 25 percent of any size) of flat or elongated fragments in the coarse aggregate has no important effect on the workability or cost of concrete.

(f) *Specific Gravity.*—Specific gravity of aggregate is of direct importance only when design or structural considerations require that the concrete have minimum or maximum weight. When lightness is desired, artificially prepared aggregates of low unit weight are frequently used in place of natural rock.

Specific gravity is a useful, quick indicator of suitability of an aggregate. Low specific gravity frequently indicates porous, weak, and absorptive material, and high specific gravity often indicates good quality; however, such indications are undependable if not confirmed by other means.

(g) *Gradation.*—The particle size distribution of aggregate as determined by separation with standard screens is known as its gradation. Sieve analysis, screen analysis, and mechanical analysis are terms used synonymously in referring to gradation of aggregate. For the sake of uniformity, the term "screen" has been adopted for general use in this manual. In Bureau work, gradation of sand is now expressed in terms of the individual percentages retained on United States standard screens designated by the numbers 4, 8, 16, 30, 50, and 100. Gradation of coarse aggregate is determined by means of screens having openings according to the specifications or special requirements for the job, as described hereinafter.

From the percentages of sand and total coarse aggregate to be used (dependent on maximum size, character, and grading of the material) the combined grading of aggregate may be computed. A grading chart is useful for depicting the size distribution of the aggregate particles. Figure 25 is such a chart, illustrating grading curves for sand, gravel, and combined sand and gravel. The fineness modulus (F.M.) shown in the table for sand is an index of coarseness or fineness of the material but gives no idea of grading. (See appendix, designation 4.)

Test results shown in tables 6 and 7 indicate that changes in sand grading over an extreme range have no material effect on compressive strength of mortar and concrete specimens when water-cement ratio and slump are held constant. However, such changes in sand grading under the conditions mentioned do cause the cement content to vary inversely with the fineness modulus of the sand. Although effect on cement content is relatively small (see fig. 26), grading of sand has a marked influence on workability and finishing quality of concrete. The effect on workability is somewhat intensified as a result of holding the percentage of sand constant.

Experience has demonstrated that either very fine or very coarse sand, or coarse aggregate having a large deficiency or excess of any size fraction, is usually undesirable, although aggregates with a discontinuous or gap grading have sometimes been used with no apparent disadvantages. Aggre-

Table 6.—Effects of sand grading on mortar

Mix	Cement content, pounds per cubic yard [1]	Unit compressive strength at 28 days [1]	F. M. of sand	Sand grading (individual percentages retained)					
				No. 8	No. 16	No. 30	No. 50	No. 100	Pan
1	846	5,620	3.29	30	23	17	13	10	7
2	831	5,460	3.20	21	23	26	18	9	3
3	850	5,350	3.17	24	20	21	21	12	2
4	850	5,330	3.03	22	21	20	17	15	5
5	876	5,300	2.94	15	20	30	19	11	5
6	876	5,390	2.91	21	20	19	17	15	8
7	895	5,420	2.79	20	19	18	17	15	11
8	884	5,510	2.75	12	20	24	24	15	5
9	891	5,230	2.71	6	16	36	29	11	2
10	910	5,170	2.70	17	11	15	42	12	3
11	929	5,210	2.56	4	4	46	38	6	2
12	921	5,570	2.54	17	17	17	17	16	16

[1] Each value represents the average of tests made with Hoover, Grand Coulee, and Friant Dam sands at a constant W/C of 0.50 and slump of 2¼ inches. Strength values were obtained from three 2- by 4-inch cylinders made with each of 3 sands.

SCREEN SIZE	% RETAINED		COMBINED % RET.	
	INDI-VIDUAL	CUMU-LATIVE	INDI-VIDUAL	CUMU-LATIVE
6 Inch	0	0	0	0
3 Inch	28	28	21	21
1½ Inch	26	54	20	41
3/4 Inch	22	76	16	57
3/8 Inch	16	92	12	69
No. 4	8	100	6	75
No. 4	0	0	0	
No. 8	12	12	3	78
No. 16	20	32	5	83
No. 30	24	56	6	89
No. 50	24	80	6	95
No. 100	16	96	4	99
Pan	4	100	1	100

FINENESS MODULUS = 2.76
PERCENT SAND (clean separation) = 25
(Screen sizes are based on square openings)

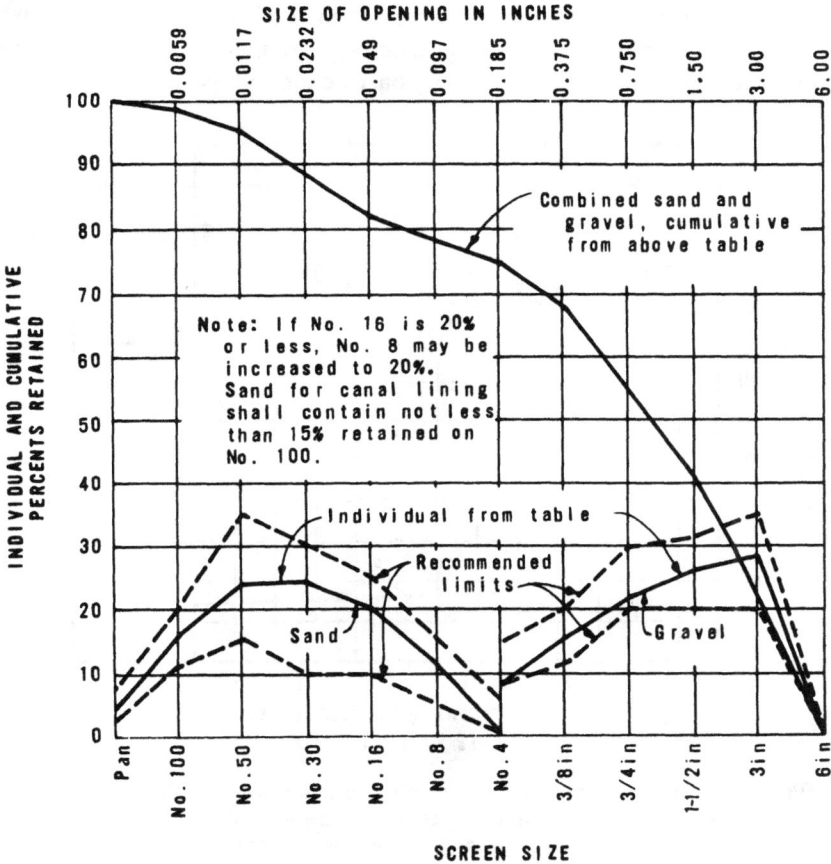

Figure 25.—Typical size distribution of suitably graded natural aggregate. 288-D-803.

Table 7.—Effects of sand grading on concrete

Mix [1]	Cement content, pounds per cubic yard	Unit compressive strength at 7 days	F. M. of sand	Sand grading (individual percentages retained)					
				No. 8	No. 16	No. 30	No. 50	No. 100	Pan
1	478	2,760	3.10	27	20	17	15	14	7
2	481	2,690	2.93	21	16	22	22	14	5
3	489	2,840	2.91	21	20	19	17	15	8
4	493	2,700	2.89	16	12	30	31	9	2
5	481	2,810	2.74	15	15	25	24	16	5
6	504	2,730	2.70	10	10	34	34	10	2

[1] The following were common to all mixes: $W/C=0.57$; slump $=4$ inches; sand $=37$ percent; grading of coarse aggregate $=20$ percent of No. 4 to ⅜-inch, 30 percent of ⅜- to ¾-inch, and 50 percent of ¾- to 1½-inch material. Strength values were obtained from 6- by 12-inch cylinders.

gate grading is important principally because of its effect on water-cement ratio and paste-aggregate ratio, which affect economy and place-ability of concrete. As far as practicable, grading occurring in natural deposits should be used in Bureau construction unless it has been demon-

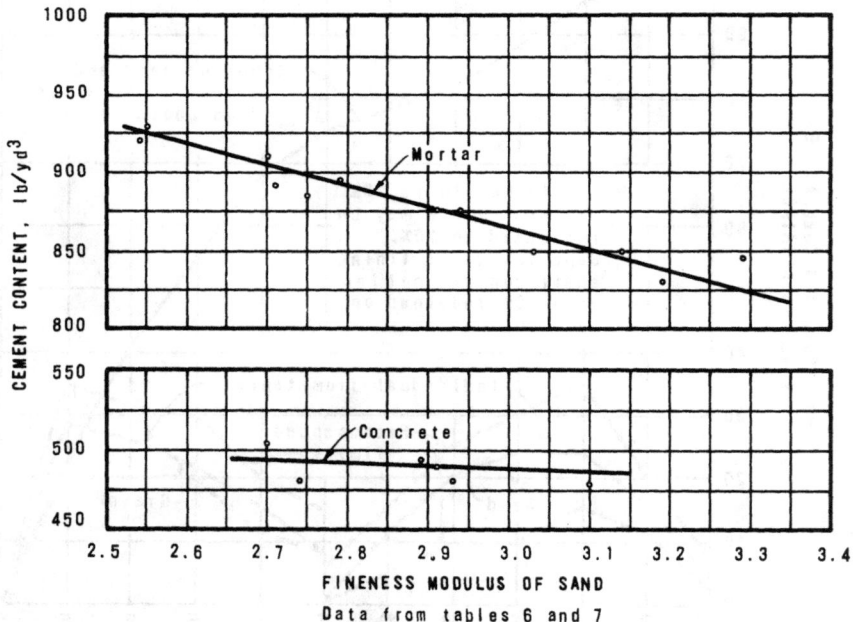

Figure 26.—Cement content in relation to fineness modulus of sand. With mortars having the same water-cement ratio and slump, more cement per cubic yard is required when sand of lower fineness modulus is used. 288–D–120.

strated through experience or laboratory investigations that corrections in gradings would be advantageous.

Allowable grading limits for sand depend to some extent on shape and surface characteristics of the particles. A sand composed of smooth, rounded particles may give satisfactory results with coarser grading than would be permissible for a sand made up of sharp, angular particles with rough surfaces. It is not hard to visualize the interlocking positions taken by angular particles in close contact, nor the contrast between such particles and smooth, rounded particles, with respect to freedom of movement in fresh concrete. It is also evident that roughness of the surfaces of the grains increases internal friction.

Sand having a smooth grading curve of regular shape cannot always be obtained economically. However, if the results of screen analyses fall within certain limits and if variation in fineness modulus is properly restricted, the sand will almost invariably be satisfactory with respect to grading. Bureau specifications usually provide that screen analyses be within the following limits:

Screen size	*Percentage retained (individual)*
No. 4	0 to 5.
No. 8	5 to 15.
No. 16	10 to 25.[1]
No. 30	10 to 30.
No. 50	15 to 35.
No. 100	12 to 20.[2]
Pan	3 to 7.

[1] If the individual percentage retained on the No. 16 screen is 20 percent or less, the maximum limit for the percentage retained on the No. 8 screen may be increased to 20 percent.

[2] Sand for concrete canal lining shall contain not less than 15 percent of material passing the No. 50 screen and retained on the No. 100 screen.

For large jobs it is desirable that grading of the sand be controlled so that the fineness moduli (appendix, designation 4) of at least 9 out of 10 consecutive test samples of finished sand, when samples are taken hourly, will not vary more than 0.20 from the average fineness modulus of the 10 test samples.

Correction of sand grading by classifying, screening, and recombining is uneconomical on small jobs, but such processing of coarse aggregate can readily be accomplished. Methods for correction of sand gradings are described in section 64. Table 8 shows approximate practicable ranges in grading of coarse aggregates. Bureau specifications usually restrict the

Table 8.—Approximate ranges in grading of natural coarse aggregates for various concretes

| Maximum size aggregate in concrete, inches | Percentage of coarse aggregate fractions (clean separation) | | | | | |
| | Cobbles, 3 to 6 inches | Coarse gravel, 1½ to 3 inches | Medium gravel, ¾ to 1½ inches | Fine gravel | | |
				3/16 (No. 4) to ¾ inch	⅜ to¾ inch	3/16 (No. 4) to ⅜ inch
¾	0	0	0	100	55 to 73	27 to 45
1½	0	0	40 to 55	45 to 60	30 to 35	15 to 25 [1]
3	0	20 to 40	20 to 40	25 to 40	15 to 25	10 to 15
6	20 to 35	20 to 32	20 to 30	20 to 35	12 to 20	8 to 15

[1] In concrete for canal lining, the percentage of 3/16- to ⅜-inch fraction is reduced to about 5 percent of the total aggregate (see sec. 108).

maximum nominal size of aggregate to 6 inches. Use of cobbles larger than 6 inches generally accomplishes little or no saving in the cost of concrete or improvement in mix characteristics. The larger cobbles increase grinding action in the mixer, segregate easily, and make placing more difficult. Under certain conditions, however, the inclusion of larger cobbles is advantageous. For some Bureau mass concrete dams where use of 3- to 6-inch material was desirable but the pit-run material was deficient in 3- to 6-inch size coarse aggregate, the usual maximum limit for oversize was extended to 8 inches. The use of such large-size aggregate may increase concrete mixer maintenance and handling and placing difficulties which could offset any savings in materials costs.

Although size separations of coarse aggregate at $\frac{3}{16}$, ¾, 1½, and sometimes 3 and 6 inches are in general use in Bureau work, there are instances in which it is advantageous to use other separations. An outstanding example is the division of the $\frac{3}{16}$- to ¾-inch size, which has a size range ratio of 4 to 1, into two fractions, $\frac{3}{16}$ to ⅜ and ⅜ to ¾. This procedure results in a reduction of segregation during handling and enables control of the amount of $\frac{3}{16}$- to ⅜-inch material, which often has a critical effect on concrete workability. Pit-run aggregate often contains an excess of one or more sizes, which must be eliminated to produce satisfactory gradation. Sometimes this may be accomplished, without incidental loss of desirable sizes, by establishing size separations that closely bracket the objectionable excess.

Concrete containing 1½-inch-maximum size aggregate is occasionally specified for tunnel linings less than 12 inches in thickness and with double reinforcement curtains. Tunnel linings greater than 12 inches in thickness with no reinforcement or only a single row of reinforcement, or cutoff walls and other structures may often be constructed using aggregate with a maximum size of 2½ inches, thereby effecting a saving in cement. Where 2½-inch-maximum size aggregate is to be used in tunnel lining this size should also be used in the massive portions of other work in lieu of 3-inch material which could otherwise be used. This practice eliminates the need to produce and use two different aggregate sizes in the larger size range. The actual maximum size of aggregate selected and used is dependent on quantity involved, thickness of section, and amount and spacing of reinforcement steel. This procedure might involve corresponding adjustment of other size fractions. Concrete containing coarse aggregate separated at $\frac{3}{16}$, ½, 1¼, and 2½ inches can readily be pumped through 8-inch pipe. Aggregate larger than 2½ inches may cause difficulty in pumping.

Because perfect screening of aggregates on the job cannot be done at reasonable cost, each sized product contains some undersize material. Oversize is also frequently present because of screen wear or the use of screens with effective openings somewhat larger than the openings specified. The amount of undersize is increased by breakage and attrition during handling operations. However, there are considerable portions of undersize and oversize that are only slightly smaller or larger, respectively, than the specified limits of an aggregate fraction, and these portions are not sources of trouble. The significant, or objectionable, portion of the undersize may be considered as the relatively small material that will pass a test screen having openings five-sixths of the specified minimum size of the aggregate fraction.

To control screening effectiveness and improve concrete uniformity, Bureau specifications require that the aggregates as batched will be within specified limits for significant undersize when tested on screens having openings five-sixths of the nominal minimum size of each separation. The allowable percentage will vary somewhat depending on job conditions. When final screening is done at the batching plant, it is practicable to restrict undersize in each size fraction to 2 percent. No significant oversize is permitted; that is, no material is to be retained on the designated test screens that have openings approximately seven-sixths of the nominal size of the material. Sizes of openings in screens for determining significant undersize and oversize for coarse aggregate are shown in table 9. The nominal separation points include those commonly used. In any coarse aggregate size fraction, Bureau specifications usually require a certain percentage of material to be retained on an intermediate "index screen" to assure inclusion of sufficient larger size material to provide uniform size distribution of aggregate particles. Index screens for various nominal coarse aggregate sizes and typical minimum percentages to be retained on them are shown in table 9.

The undersize in fine gravel is usually composed largely of material retained on a No. 8 screen. An objectionable amount of pea gravel and

Table 9.—Sizes of square openings in test screens for various nominal sizes of coarse aggregate (inches)

Designation of size Nominal size range [1]	3/8 3/16 to 3/8	3/4 3/8 to 3/4	3/4 3/16 to 3/4	1½ 3/4 to 1½	3 1½ to 3	6 3 to 6
Significant oversize	7/16	7/8	7/8	1¾	3½	7
Nominal oversize	3/8	3/4	3/4	1½	3	6
Nominal undersize	3/16 (No. 4)[2]	3/8	3/16 (No. 4)[2]	3/4	1½	3
Significant undersize	No. 5 [2]	5/16 [2]	No. 5 [2]	5/8	1¼	2½
Next nominal screen	No. 8 [2]	No. 4 [2]	No. 8 [2]	3/8	¾	1½
Index screen	—	—	3/8	1¼	2½	5
Minimum percent retained on index screen	—	—	50	25	25	varies

[1] For other nominal separation points the test screen openings would bear the same 7/6 and 5/6 relationship to the maximum and minimum nominal sizes, respectively.
[2] U.S. standard screen.

undersize No. 8 material in a concrete batch is generally occasioned by breakage and segregation in all sizes of coarse aggregate during handling and stockpiling operations, rather than by ineffective processing; it is difficult to compensate satisfactorily for excessive fluctuations in pea gravel content of the various sizes by continual changing of the mix. The obvious and practicable way of minimizing such erratic grading is to improve handling methods, to divide the fine gravel into two fractions using a 3/8-inch screen, or to finish screen the corase aggregate at the batching plant as it is used and waste the minus 3/16-inch undersize material.

A sudden increase in pea gravel brings about an increase in the voids between aggregate particles which, if not corrected by changing the mix, may result in a serious decrease in workability. This probably occurs because insufficient mortar is present to fill the excessive void space. Adjustment of the mix by increasing mortar content will restore the lost workability. This expedient, which involves abnormally high cement content and water content, should be necessary only on infrequent occasions when it is impracticable to maintain a reasonably uniform pea gravel content.

In general, crushed aggregate, as compared with gravel, requires more sand to compensate for the sharp, angular shape of the particles to obtain a mix comparable in workability to one in which no crushed material is used. About 27 percent natural sand was used with the 6-inch-maximum crushed limestone in much of the concrete for Angostura Dam, but only about 22 percent was required with gravel in mixes at Hungry Horse and Canyon Ferry Dams, which contain natural aggregate.

Figures 27 and 28 portray the significant degree of benefit derived from using concrete containing aggregate graded to the largest maximum size and show the decrease that occurs in water and cement contents with an increase in maximum size of aggregate. The latter is the primary factor in reducing drying shrinkage, as illustrated in figure 18. From figure 27 the appreciable economy of such concrete is clearly evident in the reduction in cement content that is possible as the maximum size of aggregate is increased, particularly in the range of sizes smaller than 3 inches. With the larger maximum sizes the reduction in cement content is not so pronounced.

These reductions in water and cement content with larger aggregate are possible because coarse aggregate contains fewer voids as its range of sizes is increased, and less mortar is required to make workable concrete. The amount of cement (fig. 29) required to produce maximum compressive strength at a given age with a given aggregate will vary with each maximum size of aggregate involved. Greater strengths can be obtained at higher cement contents for all sizes of aggregates until a maximum strength is reached beyond which the addition of cement produces no increase in strength. The compressive strength at which the addition of ce-

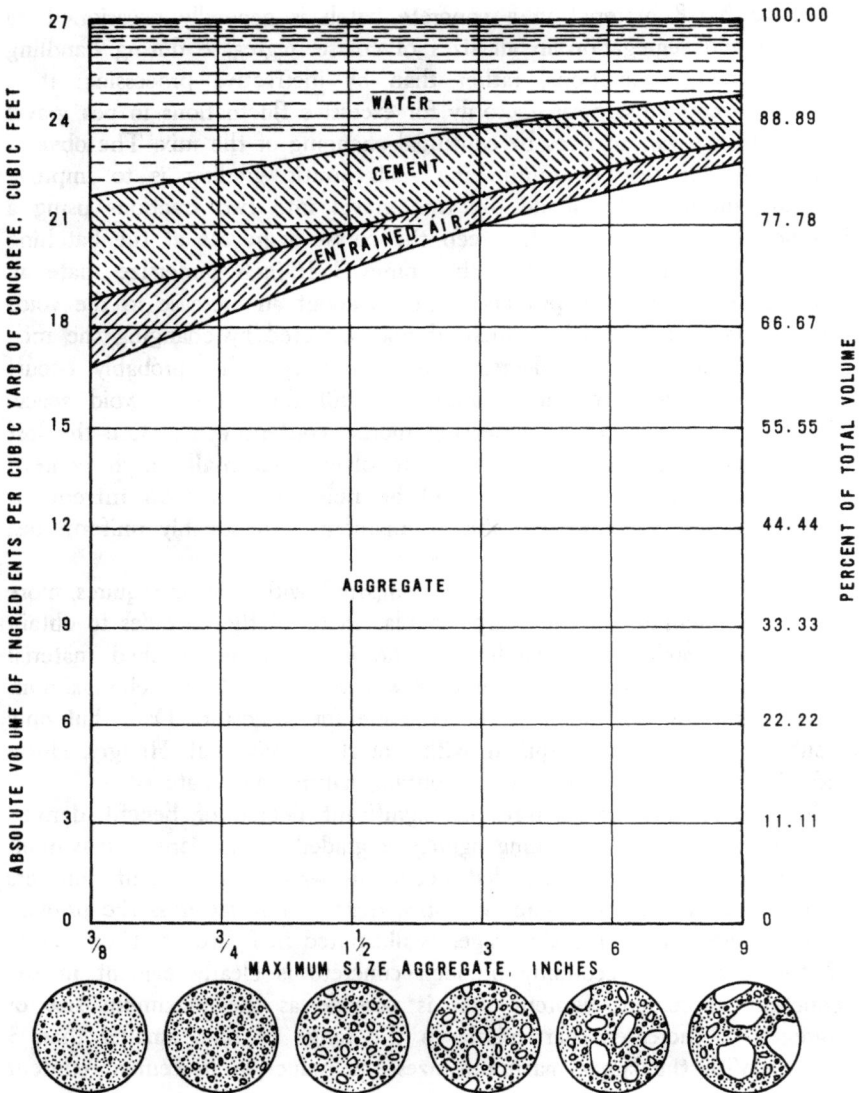

Chart based on natural aggregates of average grading in
mixes having a w/c of 0.54 by weight, 3-inch slump,
and recommended air contents.

Figure 27.—Absolute volume of water, cement, and entrained air for various
maximum sizes of aggregate. Mixes having larger coarse aggregate require
less water and less cement per cubic yard than do mixes with small coarse
aggregate. 288–D–805.

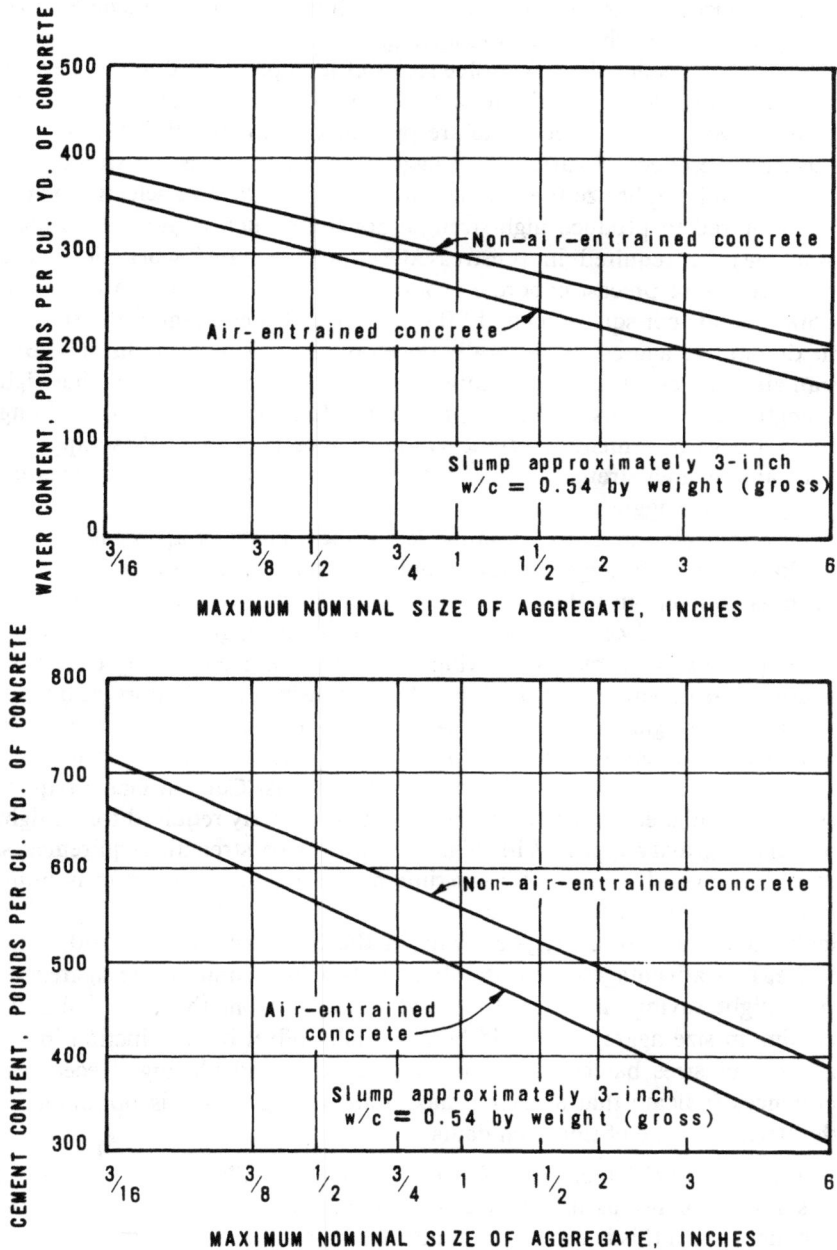

Figure 28.—Cement and water contents in relation to maximum size of aggregates, for air-entrained and non-air-entrained concrete. Less cement and water are required in mixes having large coarse aggregate. 288–D–1528.

ment produces no further increase in strength is higher for the smaller size aggregates than for the larger size aggregates.

Significant results of an extensive series of tests performed in the Denver laboratories to determine the influence of maximum size of aggregate on compressive strength of concrete are presented in figure 29. These curves illustrate the effect of various size aggregates on compressive strength of concrete and emphasize the narrow limits of aggregate size selection when it is required to produce high-strength concrete. Such high-strength concrete may be required in prestressed or posttensioned work or in the manufacture of precast concrete products. At strength levels in excess of 4,500 pounds per square inch at 90 days, concretes containing the smaller maximum size aggregates generally develop the greater strengths. The Bureau annually constructs many miles of concrete pressure pipe having high strength requirements in which producers often use concrete containing up to 8 bags of cement per cubic yard of concrete. It is in these applications and at high strength levels that the concretes containing the smaller maximum size aggregates are most effective.

Also indicated in figure 29 is the less critical effect of aggregate size in the lower strength ranges such as would be encountered in mass concrete. In mass concrete mix design, studies are conducted to establish the optimum grading and cement content for a particular aggregate source, since a few pounds per cubic yard saving in cement represents many dollars in a large mass concrete dam. It should be emphasized that important overall economies are gained by use of 3- to 6-inch-maximum size aggregate in lean mass concrete for the interior of heavy arch and gravity structures.

Thin arch dams such as Morrow Point and East Canyon Dams require concrete of higher compressive strength than formerly required for straight or curved gravity types. With higher compressive strength requirements, the advantage of using 6-inch-maximum size aggregate concrete is minimized. Since the curves (fig. 29) are relatively flat between the 3- and 6-inch-maximum size aggregate limits in the 3,500-lb/in² to 4,500-lb/in² range, it is advantageous to use about a 4-inch-maximum size aggregate. Any slight savings in cement content accruing from the use of 6-inch-maximum size aggregate would be essentially offset by a reduction in the number of sizes batched and the easier mixing and placing. Recent experiences indicate that a 4-inch-maximum size aggregate is optimum for the strength range of thin arch dams.

19. Quality of Mixing and Curing Water.—Usually, any potable water is suitable for use as mixing water for concrete. However, there may be instances when this is not true and also where water not suitable for drinking is satisfactory for use in concrete. Under certain conditions, acceptable concrete has even been made with sea water. Two criteria should be considered in evaluating suitability of water for mixing concrete. One is whether the impurities will affect the concrete quality and the other is the degree of permissible impurity. When the water quality is questionable, it should be analyzed chemically. Also, its effect on compressive strength should be determined and compared to that of a similar control concrete made with water of known purity. This determination should be made at

Each point represents an average of four 18-by 36-inch and two 24- by 48-inch concrete cylinders tested at 90 days for both Clear Creek and Grand Coulee aggregates. Mixes had a constant slump of 2"± l" for each maximum size aggregate.

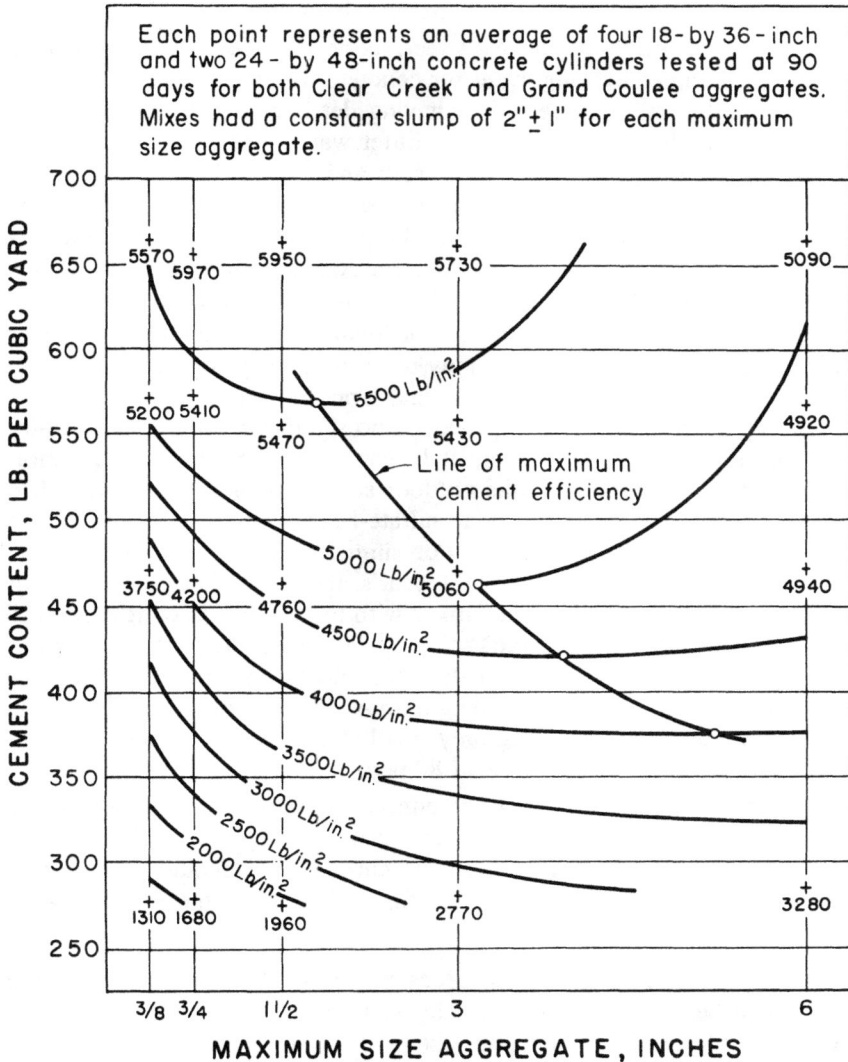

Figure 29.—Variation of cement content with maximum size of aggregate for various compressive strengths. Chart shows that compressive strength varies inversely with maximum size of aggregate for minimum cement content. 288–D–2656.

different ages, as detrimental effects often do not become apparent until later ages.

Certainly, a mixing water should not contain an excessive amount of silt or suspended solids. As a guide, a turbidity limit of 2,000 p/m is a reasonable maximum, although water containing five times this amount has been used to produce good concrete.

If clear water does not have a sweet, saline, or brackish taste, it may be used as mixing and curing water for concrete without further testing. Proposed mixing water suspected of having detrimental amounts of sulfate should be analyzed. Hard and very bitter waters are apt to contain high sulfate concentrations. Water from wells and streams in the arid Western States often contains dissolved mineral salts, chlorides, and sulfates and should be regarded with suspicion. The purest available water should be used for mixing and curing. However, a concentration of 3,000 p/m of dissolved sulfates has no detrimental effect when used for mixing or curing.

Researchers have found that use of mixing water containing considerable amounts of soluble sulfate may result in a delayed reduction of compressive strength of the concrete. In one series of tests, 1.0 percent (10,000 p/m) sulfate in mixing water produced no significant reduction in 28-day compressive strength. However, at 1 year's age (1 month standard moist curing followed by 11 months outdoor storage) compressive strength of concrete made with 10,000 p/m sulfate-bearing mix water and type V cement declined by 10 percent. For similar concrete made with type I cement, compressive strength at 1 year's age was 15 percent below the strength of the control concrete mixed with tap water. Concrete made with sodium chloride (common salt) solutions showed significant reductions in strengths at ages greater than 7 days for all concentrations. Five percent of ordinary salt reduced the strength about 30 percent. A highly carbonated mineral water containing only small quantities of sulfates and chlorides gave a strength ratio as low as 80 percent.

There are little test data on the effect of impurities in curing water on the quality of concrete. A concentration of 3,000 p/m dissolved sulfates should not be harmful. However, the curing water should be free from organic matter or other impurities that might stain the surface of the concrete.

20. Use of Admixtures.—(a) *Accelerators.*—The early strength of concrete can be materially increased by inclusion of an accelerator such as calcium chloride in the concrete mix, as illustrated in figure 30. Increased early strength during cold weather affords better protection against damage from freezing at the end of the specified protection period. Also, high early strengths may be desirable for expediting form removal or to permit early loading of anchor devices. The amount of calcium chloride used is restricted to that necessary to produce the desired results and should never

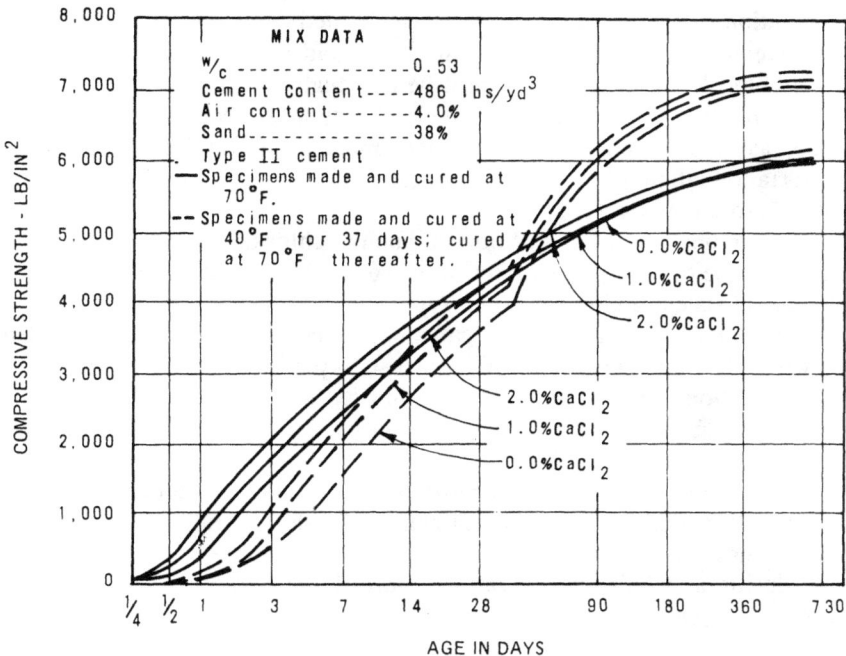

Figure 30.—Rate of compressive strength development of concrete made with addition of calcium chloride, for two different curing conditions. Addition of calcium chloride increases the compressive strength. 288–D–1529.

exceed 2 percent, by weight, of the cement. One percent will usually result in sufficiently increased early strength to meet requirements.

The use of this most common of accelerators should not be undertaken lightly; it should be used only when other readily available means will not suffice. This compound has become universally associated with winter concrete protection to such an extent that it is considered by some to be an antifreeze solution which will protect concrete from freezing and thus eliminate the need for heat, insulation, and other forms of protection. This is not so and creates a false sense of security. Calcium chloride, in the amounts permissible as a concrete admixture, has so little effect upon the freezing temperatures of the mix as to be totally insignificant. Thus, use is never justification for reducing the amount of protective cover, heat, or other winter protection normally used. The reduction in quality, as explained in the following paragraphs, is of sufficient importance that care should be exercised to ascertain which of the properties of the concrete will be adversely affected when calcium chloride is used.

Use of accelerators requires special precautions in handling and placing of concrete to avoid delay, as slump loss and stiffening of the concrete—

in addition to strength gain—will be accelerated. A reduction in thickness of placement layers may be necessary to avoid cold joints. In using calcium chloride it is important that it be thoroughly dissolved in the mixing water and that the solution be evenly distributed throughout the batch. Also, as calcium chloride precipitates most air-entraining agents, it is important that they be kept in separate solutions and introduced separately into the mixer.

Tests indicate that the addition of calcium chloride to concrete reduces its resistance to attack by sulfates in the soil. Bureau specifications prohibit the use of calcium chloride when type V cement is required and also in concrete in which aluminum or galvanized metalwork is to be embedded or when the concrete will be in contact with prestressed steel. Where sulfate conditions are encountered requiring type V cement and concreting operations must be continued during cold weather, additional cement— about one bag per cubic yard of concrete—will result in high early strengths approximately equivalent to those obtainable with calcium chloride, as illustrated in figure 31. Calcium chloride generally should not be used in hot weather as the time for concrete set will be accelerated, making placing and finishing operations more difficult and possibly affecting adversely strength development.

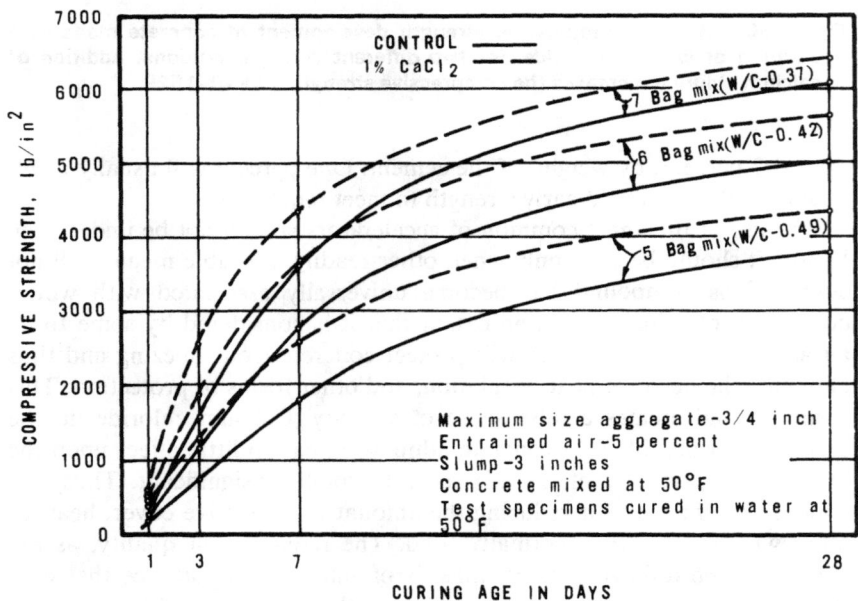

Figure 31.—The effects of calcium chloride on the strength of concrete of different cement contents and at different ages with type II cement. 288-D-1530.

Other findings indicate that the effectiveness of calcium chloride in producing high early strength of concrete containing pozzolan is proportional to the amount of portland cement in the mix. The action between high-alkali cement and reactive aggregates is increased by the admixture.

Commercial accelerators, principally of sodium and aluminum, were recently introduced in the United States. These accelerators are used to produce early strength in shotcrete, especially when it is used in lieu of steel sets for tunnel support. Their use is described in detail in sections 173 through 180.

(b) *Air-Entraining Agents*.—Use of air entrainment in concrete is a general requirement for Bureau construction. It is also required that the agent be added to the batch in solution in a portion of the mixing water and, except for very small jobs, that this solution be batched by a mechanical batcher. Bureau specifications require that air-entraining agents conform to "Specifications for Air-Entraining Admixtures for Concrete" (ASTM Designation C 260), except that the limitation and test on bleeding by concrete containing the agent and requirement for time of set do not apply. A number of commercially available agents have been tested by the Bureau and approved for use. These agents are sold under various trade names in powder form or as solutions. Commercially available air-entraining agents from companies that have demonstrated abilitiy to supply such materials in accordance with specifications requirements are accepted on manufacturers' certification that the materials meet Bureau specification requirements. However, this does not preclude sampling and testing after delivery if the materials appear to be abnormal or difficulties are encountered in their use. Agents not previously tested or so demonstrated in use should be tested and approved for use by the Chief, Division of General Research. Approval of air-entraining agents is contingent on laboratory test data indicating that the agent conforms to ASTM requirements. The agent should be uniform in consistency within each batch and uniform in quality between batches and between shipments.

As most of the air entrained is contained in the mortar, it follows that the percentage of air required in concrete made with aggregate having a small maximum size (high mortar content) is larger than that for concrete in which large aggregate is used (low mortar content). Desirable air contents, for concrete at the mixer, are as follows:

Coarse aggregate, maximum size in inches	Total air, percent
¾	6.0±1
1½	4.5±1
3	3.5±1
6	3.0±1

To minimize the damaging effects of ice-removal salts on concrete surfaces such as bridge decks, pavements, etc., the recommended air content should be increased 1 percent. For concrete not subject to severe freezing, air contents may be reduced as much as one-fourth if strength is essential and sufficient workability can still be maintained. Air contents after placing and vibration are decreased about one-fifth from the values listed. The maximum air content for concrete for precast concrete pipe should be less than 2½ percent.

Among the factors that influence the amount of air entrained for a given amount of air-entraining agent are: grading and particle shape of aggregate, richness of mix, mixing time, slump, and temperature of concrete. Organic material in aggregates and in pozzolans can also influence the amount of air entrained for a given amount of agent. Air content increases with increase in slump and decreases with increase in fineness of portland cement or pozzolan, temperature of the concrete, or mixing time.

(c) *Water-Reducing, Set-Controlling Admixtures (WRA)*.—A water-reducing, set-controlling admixture, or WRA, is included in a concrete mix designed primarily as a means of reducing the water requirement. By function, these materials are subdivided into (1) water-reducing, (2) water-reducing, set-retarding, and (3) water-reducing, set-accelerating admixtures. Admixtures of this category are also divided into four recognized classes according to chemical composition. Class I admixtures are the lignosulfonic acids and their salts; class II materials are derivatives of class I admixtures modified to perform a slightly different function. Class III admixtures are the hydroxylated carboxylic acids and their salts. Similarly, modifications or derivatives of hydroxylated carboxylic acids and their salts fall into the class IV category.

Generally, class I and III admixtures reduce the quantity of mix water and retard the initial and final set of the concrete compared to an air-entrained concrete without the admixture.

The class II and IV materials are the class I and III admixtures modified so that they will reduce the unit water content necessary for a given consistency and that they will either not affect the time to reach initial set or will accelerate the occurrence of this event depending on the degree of modification.

Extensive laboratory and field tests have shown that nominal dosages of class I and III materials at normal temperatures will, at the same air content, allow an average water reduction for structural concrete of 6 percent and 4 percent, respectively. These water reductions may range from 0 to 25 percent depending on the dosage. (Nominal dosages for these materials are 0.25 to 0.35 percent by weight of cement and rarely exceed 0.40 percent—solid basis class I and fluid basis class III.) Generally, class I materials bring about greater reductions than class III materials. The average water reduction produced by optimum amounts of class I admixture in mass concretes is about 8 percent.

Figure 32 shows the reduction in unit water content resulting from the addition of various amounts of three class I admixtures and one class III admixture with two type I and two type II cements. It is evident that, as the dosage of admixture was increased, the unit water requirement decreased. The decrease in water requirement with increasing dosage was much greater for class I than for class III admixtures, which is caused by the increased air entrained by the larger doses of class I admixtures. (See fig. 33.) The class III admixture entrained very little air and, in all cases, was supplemented with a special air-entraining agent, while most of the class I materials entrained slightly more air than required at 0.3 percent dosage. Still larger doses of class I admixtures entrained undesirably large quantities of air.

Nominal dosages of class I and class III admixtures usually extend the average time to reach vibration limit, as determined by the Proctor needle test (ASTM Designation C 403), by about 25 and 40 percent, respectively. However, the range of time intervals to reach vibration limit

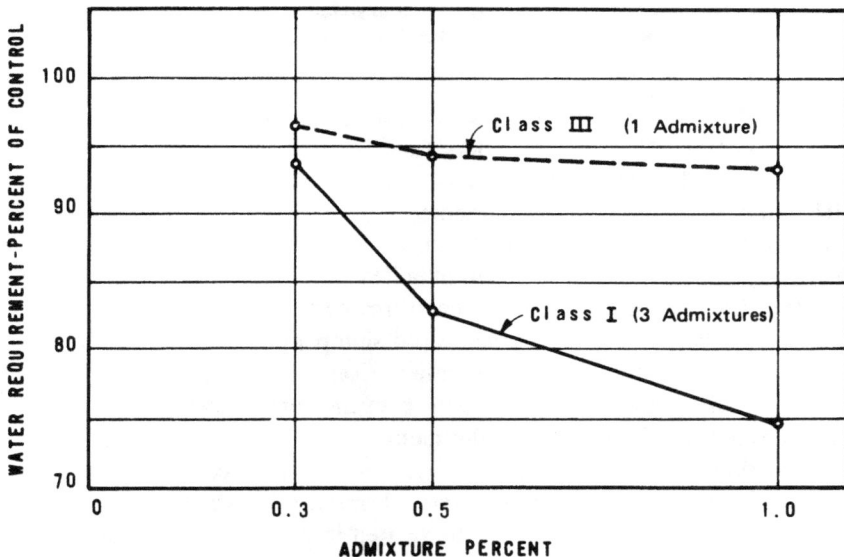

Figure 32.—Influence of lignin and hydroxylated carboxylic acid admixtures on water requirement of structural concrete. 288–D–2640.

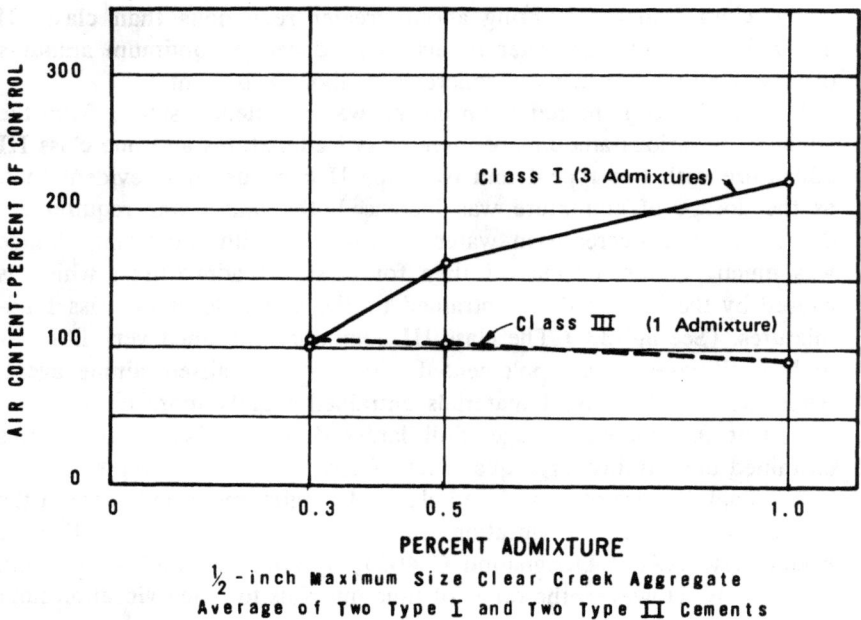

Figure 33.—Influence of lignin and hydroxylated carboxylic acid admixtures
on air content of structural concrete. 288-D-2641.

produced by various admixtures with different cements is quite broad,
extending from 4½ to 12¾ hours for concretes with class I agents and
from 5¾ to 17 hours for concretes with the class III admixtures. Class
III admixtures may be better adapted for some concrete placements requiring extended retardation, because they do not induce premature
stiffening which has been noted in a few instances when greater than
normal doses of various class I admixtures were used. Even normal doses
of type I admixture may produce rapid slump loss when used with some
cements, particularly in warm weather. Concretes with class III admixtures remain plastic until the setting cycle commences and thereafter
undergo a normal strength development.

Occasionally, rapid slump loss in concrete has occurred with either
class I or class III admixtures. Some laboratory investigations have indicated that this phenomenon may be related in part to the moisture condition of cement grains induced by aeration during manufacturing, transporting, or batching. (Refer to discussion in sec. 16.—Abnormal Set
of Portland Cement.) Aeration by dry air appears to be the primary
source of the difficulty. Laboratory studies have indicated that by introduction of a small amount of moisture in vapor form, abnormal behavior
of the cement can in most cases be eliminated. Complete elimination

appears to have been achieved in laboratory tests by slight prehydration of the cement and use of a WRA.

Care must be taken to see that dispensing equipment is reliable and accurate, since an overdose of any retarding agent may produce objectionably long setting periods.

Considering compressive strength tests of 6- by 12-inch cylinders at all ages, 95 percent of the specimens containing recommended dosages of the admixtures developed greater compressive strength than companion control specimens; the agents increased the overall average compressive strength by 18 percent. Overdosage decreased to 66 percent the number of specimens developing compressive strength greater than control specimens, the overall average compressive strength increase equaling 12 percent. Severe overdosage of 1 percent caused 72 percent of the specimens to develop less compressive strength than the companion control specimens and accounted for a decrease of 25 percent in average compressive strength. This emphasizes the need for dependable dispensers, properly calibrated and maintained to deliver the correct dosage of admixtures to each batch of concrete.

Compressive strengths developed at various ages by 18- by 36-inch cylinders containing mass concretes and lignin agents are significantly higher at all ages than strengths of companion cylinders without admixture. This is true for concrete proportioned with 2½, 3, and 3½ bags of total cementitious materials and with 0, 20, 33, and 43 percent pozzolan replacements.

Drying shrinkage is not significantly affected by the use of class I or class III admixtures. Generally, concretes with admixtures have shown slightly less shrinkage than concretes without admixtures.

As a result of lower cement requirements attendant to use of water-reducing admixtures, the temperature rise and the tendency of mass concrete to crack are lessened. Some savings in cooling of mass concrete may also be realized through wider spacing of cooling coils.

Resistance to freezing and thawing of concrete containing an admixture may be improved only when the use of the admixture produces a more watertight concrete with the air void system increased. The ability of a water-reducing agent to provide factors contributing to improved resistance to freezing and thawing depends on its chemical characteristics as well as the properties of other materials used in the concrete. Consequently, each specific agent proposed should be investigated in combination with all other materials in the mix, including pozzolans, to determine its ability to produce durable concrete.

All other qualities of concrete, including autogenous volume change, sulfate attack, and tensile and shearing strength, are either improved or at the least not adversely affected by such admixtures.

Laboratory tests should be performed as outlined in designation 40 before using these admixtures, to ascertain suitability with the particular cement and materials to be used in the proposed construction and under proposed construction conditions.

21. Field Control.—After materials have been selected and relative proportions determined, use should be controlled to best advantage. This is the purpose of field control which involves correct procedures of proportioning, mixing, handling, placing, and curing. These subjects are discussed in detail in chapters II, III, IV, V, and VI. Field control governs quality, uniformity, and ultimate economy of the structure. Much potential value of first-class materials and optimum proportioning may be lost through ineffective control of these procedures. The poorer the quality of the ingredients, the greater the need for rigid control to attain satisfactory durability and strength and therefore maximum serviceable life.

Production and handling of materials for concrete, particularly the aggregate, must first be controlled if efforts to produce good, uniform concrete are to be successful. Concrete of uniformly high quality cannot be obtained economically unless materials having uniform properties are supplied, batch after batch, to the mixer. The cost of processing materials to obtain better quality and uniformity is usually justified by improvement in quality of concrete and by economies resulting from better workability and lower cement content.

Concrete mixes for Bureau work are proportioned on the basis of a close approach to maximum net water-cement ratios stated in the specifications. With water-cement ratio, mix proportions, and consistency predetermined, the next step is to control accurately the quantities of ingredients entering successive batches. Adequate and uniform mixing will then produce uniform concrete. Proper and careful handling and placing, whereby the concrete will be consolidated in final position without loss of uniformity and quality, need only be supplemented by suitable finishing and adequate curing to complete the cycle of satisfactory construction operations.

Segregation of ingredients is one of the most detrimental characteristics of concrete. It is directly or indirectly related to substantially all the imperfections, visible and invisible, to which concrete is heir, including rock pockets, weak and porous layers, lack of bond at construction joints, surface scaling, crazing, pitting, and sand streaks. It is much easier and far safer to guard against segregation than to repair its damaging effects and, fortunately, practical means are available whereby it may be controlled within a tolerable limit.

Effective control of segregation enhances economy as it obviates use of excess cement otherwise required to overcome the effects of lack of uniformity and ultimately results in lower maintenance cost.

The last opportunity for controlling quality of concrete while in the plastic state is afforded during consolidation in the forms. Thorough consolidation is necessary if maximum durability, strength, economy, and uniformity are to be attained. Developments in vibration equipment have made possible the use of stiffer mixes containing less cement and water and have resulted in an appreciable higher quality of concrete.

The condition of the forms may influence not only appearance of the structure but also quality. Use of good form materials and proper form construction and maintenance is important in field control. The molding surfaces of forms must be clean and sufficiently tight to prevent leakage of mortar and leakage of excessive amounts of water where sand streaking is objectionable. Form surfaces should be suitably oiled.

In finishing unformed surfaces, control must be exercised in every operation involved. Floating and troweling must be regulated as to time and manner of performance so that the surface quality is not impaired by overworking and by bringing excessive fines and water to the surface.

Proper hydration of the cement particles to form hard and durable concrete requires that the concrete be maintained in a moist condition for a suitable period. Within reasonable limits, the desirable properties of concrete are improved as effectiveness and duration of curing are increased. This fact is well demonstrated by the appearance, in figure 34, of concrete test panels subjected to an abrasion test after different degrees of moist curing. The trowel-finished surfaces of these panels were held before the nozzle of a grit blast so that they would be struck in successive spots for 60 seconds each. The panel with no fog curing was so inferior in resistance that only one spot was blasted.

Special precautions are necessary to ensure proper curing during hot or cold weather.

The effect of curing temperature on strength of concrete is primarily a matter of rate of hydration, an increase in curing temperature being attended by more rapid hydration and correspondingly accelerated gain in strength. At later ages concrete cured at lower temperatures will attain higher strengths.

Further discussion on curing of concrete is included in sections 124, 125, and 126.

22. Control of Heat Generation and Cracking in Concrete.—In addition to quality control regularly exercised, control of concrete construction methods may contribute largely to the quality of the completed structure. Foremost among control measures are means for limiting the temperature rise of mass concrete and thus reducing undesirable thermal stresses and cracking.

Under rapid rates of mass concrete placement made possible by

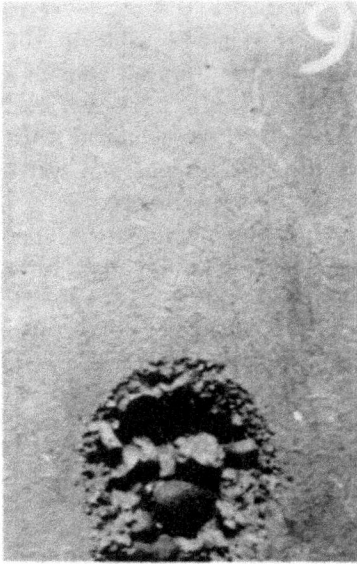

No Fog Curing
28 Days Dry

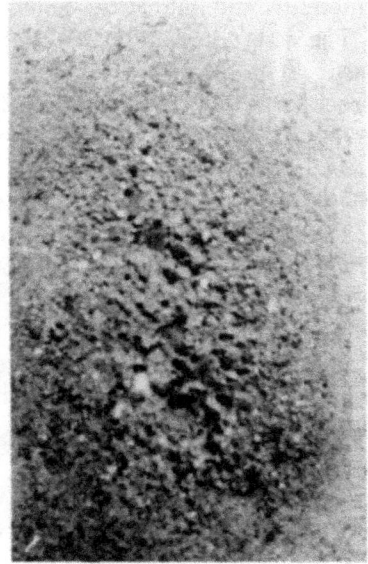

7 Days Fog
21 Days Dry

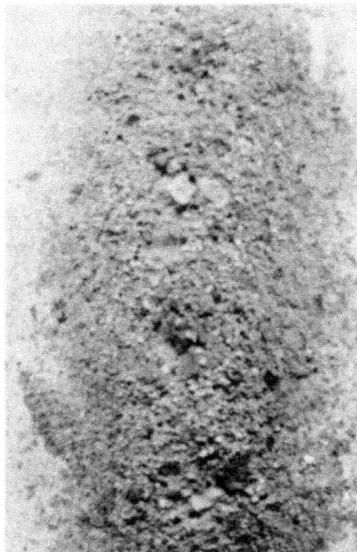

14 Days Fog
14 Days Dry

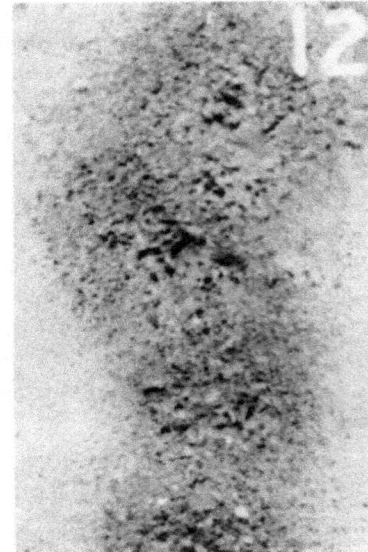

21 Days Fog
7 Days Dry

Figure 34.——Concrete test panels subjected to abrasion test after various degrees of moist curing. Good curing enhances the resistance of concrete to abrasion. PX–D–20715.

modern construction methods, the heat of hydration of the cement cannot escape from massive structures as fast as it is generated during early ages of the concrete, and the temperature rises above the placing temperature. The maximum value reached depends on initial concrete temperature, amount of cement used, heat-generating characteristics of the cement, dimensions of the mass and rate of placing, surrounding temperature conditions, thermal properties of the concrete, and amount of heat extracted by artificial cooling.

Reduction in temperature to an ultimate stable condition, by gradual dissipation of the stored-up heat or its removal by artificial means, is accompanied by contraction, which in mass concrete necessitates use of extensive and costly contraction joints to prevent development of uncontrolled cracks. These joints are grouted after the temperature of the concrete has attained a predetermined value which is usually lower than the ultimate mean value if design conditions require the structure to function as a monolith. Successful grouting of contraction joints requires an opening of about 0.02 inch or more. Consequently, if joints are used, design and construction should be coordinated to provide openings that can be grouted effectively and will prevent objectionable intermediate cracking.

Reduction of the difference between maximum and ultimate temperatures of concrete will decrease the tendency for uncontrolled cracking or will permit wider spacings between contraction joints. Several methods or combinations of methods have been used for reducing the temperature rise in mass concrete. These include (a) use of low-heat-developing cement; (b) reduction in cement content; (c) use of pozzolanic material; (d) limitation of the rate of placement so that a greater part of the heat of hydration is lost from the top surface of the lift during construction; (e) placement of concrete during cool weather so that the heat of hydration will raise the temperature to, or only slightly above, the ultimate temperature; (f) precooling concrete ingredients to reduce placing temperature; (g) introduction of ice into the mix; (h) early removal of forms and use of steel forms to facilitate loss of excess heat from the surfaces; and (i) artificial cooling, begun at the time or soon after the concrete is placed, which not only reduces maximum temperature rise but also cools concrete to any desired temperature within a short time, permitting grouting of contraction joints within a reasonable time after concrete placement.

Low-heat cements are characterized not only by a lower total heat generation but also by a lower rate of generation. Artificial cooling is most effective in reducing the maximum temperature for a given rate of construction when cement having low-heat characteristics is used.

Precooling of concrete may be accomplished in several ways. Coarse aggregates may be sprayed with water, cooled with refrigerated air, vacuum cooled, or inundated in river or refrigerated water. Sand and cement may be cooled in hollow-flight screw-type equipment (a heat exchanger in which cold water is circulated through the interior and around the periphery of a screw device conveying the cement and sand). Refrigerated water and ice may be used in the mix. Precooling of concrete permits use of lower water and cement contents, or if these are held constant it will produce concrete of greater strength and durability at later ages.

Another construction procedure that has an important effect on the final condition of the concrete from the standpoint of cracking is the protection of closed water carriers, such as tunnels, conduits, siphons, and pipelines, from dryout during the interval between completion of the specified curing and the operation of the structures. The procedure has particular application to structures that are in mild climates and are to be operated without extended periods of shutdown. The protection may be provided at relatively small cost by installing tight bulkheads at the ends of the structures and maintaining a pond of water on the inside. While tunnel-lining operations are in progress, excessive drying of sections already placed and cured may be avoided by placing a fan in the bulkheaded portal where lining was commenced so that moist air, from the section being water cured, will pass through the finished portion. The benefit of protection from dryout of tunnel lining on the Colorado River Aqueduct is illustrated in figure 35.

Concrete structures to be backfilled may often be protected from drying and given the benefit of prolonged curing by completing the backfill as soon as the strength of the concrete is sufficient to support the load.

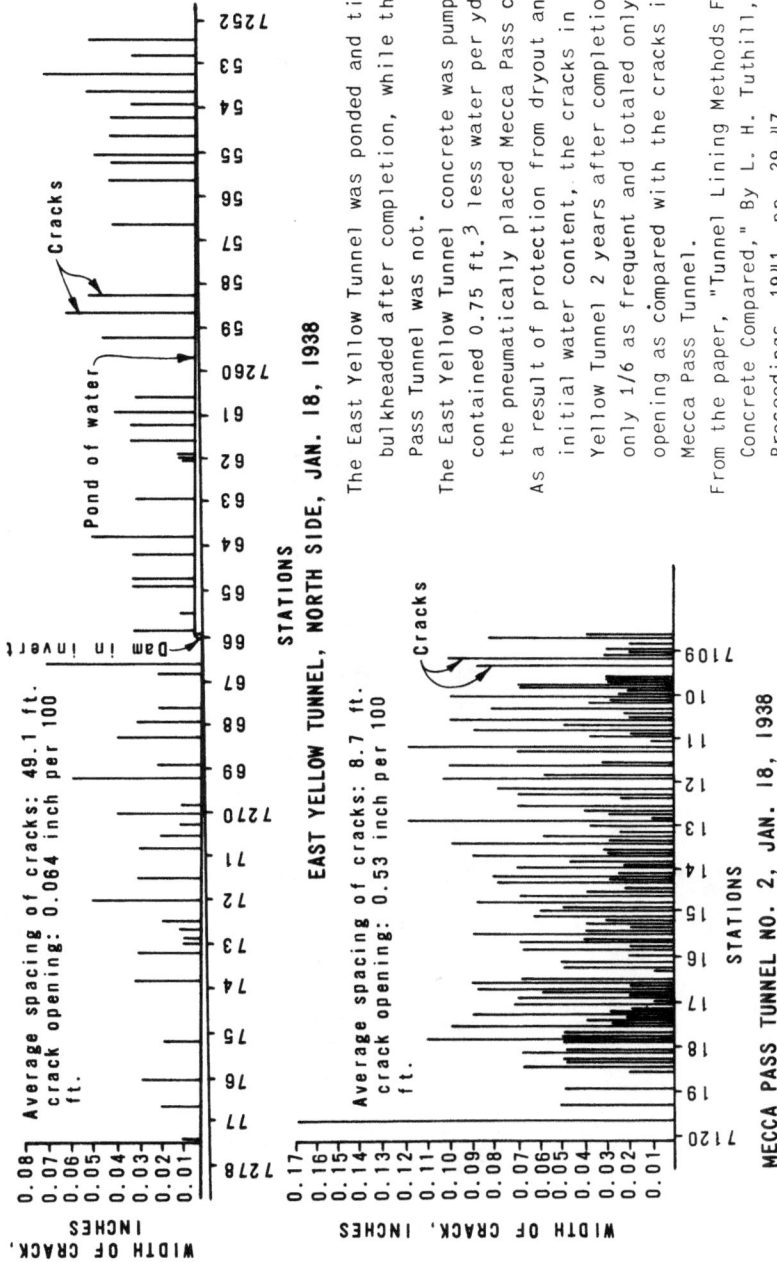

The East Yellow Tunnel was ponded and tightly bulkheaded after completion, while the Mecca Pass Tunnel was not.

The East Yellow Tunnel concrete was pumped and contained 0.75 ft.^{3} less water per yd.^{3} than the pneumatically placed Mecca Pass concrete.

As a result of protection from dryout and lower initial water content, the cracks in the East Yellow Tunnel 2 years after completion were only 1/6 as frequent and totaled only 1/8 the opening as compared with the cracks in the Mecca Pass Tunnel.

From the paper, "Tunnel Lining Methods For Concrete Compared," By L. H. Tuthill, A.C.I. Proceedings, 1941, pp. 29–47.

Figure 35.—Effect of ponding and bulkheading in concrete-lined tunnels. Such protection from dryout measurably reduces cracks. 288–D–809.

Chapter II

INVESTIGATION AND SELECTION OF CONCRETE MATERIALS

A. Prospecting for Aggregate Materials

23. General Comments.—Field investigations for concrete materials prior to construction are confined chiefly to prospecting for aggregate and to exploration and sampling of available deposits. Whenever practicable, the engineer in charge is informed regarding the approximate quantity of aggregate required, the maximum size to be used, and the general nature of the proposed construction. Those to whom prospecting work is assigned should be familiar with the effects on properties of concrete of grading, physical characteristics, and composition of aggregates. Judgment and thoroughness in conducting preliminary field investigations are usually reflected in durability and economy of the completed structures.

24. Maps and Materials Information.—The Denver laboratories maintain test reports and, on individual State maps, records showing locations of sources of concrete aggregate and riprap tested by the Bureau. Maps are maintained for the 17 Western States in which the Bureau functions. New sources are constantly being added; and thus the maps are not available for general distribution. However, information regarding the location of tested sources of material may be obtained by request from the Denver office.

Other more detailed maps useful in location of sources of concrete aggregates or in layout of aggregate processing plants or other structures are topographic maps, aerial photographs, river survey maps, and in some instances geologic maps. Aerial photographs are commonly used as the basis for topographic mapping of aggregate sources as shown in figure 36. Before undertaking map making, a thorough search should be made for existing maps. The U.S. Geological Survey should be contacted for information on the availability of maps. This organization has made many standard topographic maps of the United States and Puerto Rico.

Locations and true geodetic positions of triangulation stations and permanent benchmarks are recorded on the maps.

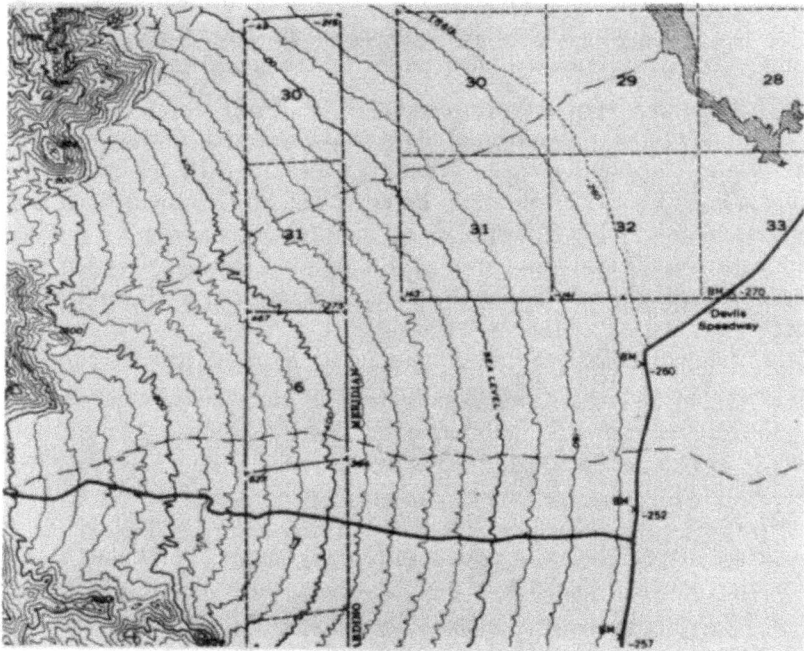

Figure 36.—Aerial view and topography of an alluvial fan, a potential source of sand and gravel. (Courtesy U.S. Geological Survey.) PX–D–16262.

Requests and inquiries on published maps and on the availability of maps, manuscripts, or other information should be directed to U.S. Geological Survey, Denver Federal Center, Denver, Colo., or Washington, D.C.

25. Geological and Related Characteristics of Aggregates and Aggregate Deposits.—Most factors pertaining to suitability of aggregate deposits are related to the geological history of the region. The geologic processes by which a deposit was formed or by which it was subsequently modified are responsible for many of the characteristics that may influence decision as to utilization. Among these are size, shape, and location of the deposit; thickness and character of the overburden; types and condition of the rocks; grading, rounding, and degree of uniformity of the aggregate particles; and ground-water level.

(a) *Types of Deposits.*—Aggregate may be obtained from deposits of natural sand and gravel or from quarries in areas of bedrock outcroppings. In the West, natural sands and gravels are prevalent and are usually the most economical source of aggregate. They are commonly obtained from stream deposits, glacial deposits, and alluvial fans. Talus accumulations may sometimes be processed for use. Fine blending sand may sometimes be obtained from windblown deposits.

Stream deposits are the most common and generally most desirable because (1) individual pieces are usually rounded, (2) streams exercise a sorting action which may improve grading, and (3) abrasion caused by stream transportation and deposition leads to a partial elimination of weaker materials. Extensive deposits of sand and gravel frequently occur along the borders of a stream or in its channel, but often the search must be extended to include terrace deposits at higher elevations. For example, while the Colorado River in Texas is flowing on (or near) bedrock, extensive gravel deposits are found only in adjacent areas of higher topography.

Glacial deposits are restricted to northerly latitudes or high elevations. They occur abundantly in the Northwest and as far south as Colorado but are rare or lacking in southern California, New Mexico, Arizona, and Texas. Glacial deposits are of two types—true glacial and fluvial glacial—and each has very different characteristics. True glacial deposits have been transported by glacial ice and have not been subjected to the abrasive or sorting actions of river transportation. Therefore, such deposits will usually contain material having heterogeneous shapes and sizes and ranging widely in quality, the weaker constituents not having experienced the abrasive disintegration associated with stream action. Fluvial-glacial deposits consist of glacial materials that have been subjected to stream action. True glacial deposits usually occur as hummocky hills and ridges (moraines); fluvial-glacial deposits occur mainly in stream channels or on

Table 10.—General classification of rocks commonly encountered

IGNEOUS
(Solidified from a molten state)

COARSE GRAINED CRYSTALLINE	FINE GRAINED CRYSTALLINE (OR CRYSTALS AND GLASS)	FRAGMENTAL (CRYSTALLINE OR GLASSY)
ORIGIN: Deep intrusion slowly cooled	ORIGIN: Quickly cooled volcanic or shallow intrusive	ORIGIN: Explosive volcanic fragments deposited as sediments
Granite Diorite Gabbro Note: Rock names are based on mineral content (see glossary). Color may be used as a rough index as noted above. *Increasing quartz and light minerals*	*Increasing dark minerals* Rhyolite Andesite Basalt Essentially glass (suddenly chilled, few or no crystals) Obsidian, Pitchstone, Etc.	Ash and pumice (Volcanic dust or cinders) Tuff (consolidated ash) Agglomerate (coarse and fine volcanic debris)

SEDIMENTARY
(Sediments transported by water, air, ice, gravity)

MECHANICALLY DEPOSITED	CHEMICALLY OR BIO-CHEMICALLY DEPOSITED
A-UNCONSOLIDATED: Clay Silt Sand } According to Particle size Gravel Cobbles B-CONSOLIDATED: Shale (Consolidated clay) Siltstone (Consolidated silt) Sandstone (Consolidated sand) Conglomerate (Consolidated gravel or cobbles-rounded) Breccia (Angular fragments)	A-CALCAREOUS: Limestone ($CaCO_3$) Dolomite ($CaCO_3 \cdot MgCO_3$) Marl (Calcareous shale) Caliche (Calcareous soil) Coquina (Shell limestone) B-SILICEOUS: Chert Flint Agate } Spring deposit, Vein or cavity filling Opal Chalcedony C-OTHERS: Coal, Phosphate, Salines, Etc.

METAMORPHIC
(Igneous or sedimentary rocks changed by heat, pressure)

A-FOLIATED
    Slate: Dense, dark, splits into thin plates (Metamorphosed shale)
    Schist: Predominantly micaceous, semi-parallel lamellae
    Gneiss: Granular, banded, subordinately micaceous

B-MASSIVE
    Marble: Coarsely crystalline, calcareous (Metamorphosed limestone)
    Quartzite: Dense, very hard, quartzose (Metamorphosed sandstone)

outwash plains, downstream from moraines. True glacial deposits, being uninfluenced by fluvial action, are usually too heterogeneous to be suitable as aggregate and, at best, are usable only after elaborate processing. Fluvial-glacial deposits frequently yield satisfactory aggregate materials.

An alluvial fan is a gently sloping, semiconically shaped mass of detrital material deposited at the mouth of a ravine. Alluvial fans are characteristic of semiarid and arid regions and are formed by repeated torrential floods. Where the stream leaves the mountains and enters an adjacent valley, the abrupt flattening in gradient causes deposition of the greater part of the load. Sands and gravels laid down under such conditions are very different from those of normal stream deposition: the particles are angular, and the material is poorly stratified and graded. Alluvial fan deposits are frequently used as sources of aggregate, but they commonly require more than usual processing.

Talus accumulations form at the bottoms of sharp topographic elevations by the sliding and falling of loosened rock. There is no grading action, very little rounding, and no segregation of different materials. Normally, however, there is little variety in rock type. In some cases, talus accumulations may be crushed and otherwise processed to form suitable aggregate.

Windblown material is confined to the fine-sand sizes and is useful as blending sand. It is normally very well rounded and composed predominantly of quartz because the intense attrition produced by the wind effectively removes the less durable constituents.

Natural sand and gravel are not always available, and it is sometimes necessary to produce concrete aggregates by quarrying and processing rock. Quarrying normally is done only where other materials of adequate quality and size cannot be obtained economically. Some geological considerations are mentioned in section 27 (b) in regard to instructions for sampling bedrock outcrops in quarry investigations.

(b) *Classification and Characteristics of Rocks.*—In table 10 rocks are classified, on the basis of their origin, into three main groups. This classification, much abbreviated, includes only the most important types. Table 11, showing the principal mineral constituents of common igneous rocks, includes some rocks not mentioned in table 10. Although these intermediate types are less common than the granites, diorites, and gabbros and their fine-grained equivalents, they are frequently mentioned in reports on field and laboratory investigations of aggregates, and their inclusion in table 11 will probably be helpful to engineers who read the reports. Accompanying tables 10 and 11 is a glossary of mineralogical and lithological terms.

Most igneous rocks are excellent aggregate materials; they are normally

Table 11.—Principal mineral constituents of common igneous rocks

Coarsely crystalline rocks	Principal constituent minerals [1]	Finely crystalline or porphyritic rocks
Granite.....................	Q+ O +(P)+A..............	Rhyolite.
Syenite.....................	O +(P)+A..............	Trachyte.
Quartz monzonite............	Q+ O + P +A..............	Dellenite.
Monzonite...................	O + P +A..............	Latite.
Quartz diorite...............	Q+(O)+ P +A or B.........	Dacite.
Diorite.....................	(O)+ P +A or B.........	Andesite.
Gabbro.....................	P +B..............	Basalt.

[1] Mineral symbols:
　　Q—quartz (hard, shiny, conchoidal fracture).
　　O—orthoclase feldspar (commonly pinkish, unstriated, regular cleavage faces).
　　P—plagioclase feldspar (commonly white or nearly so, good cleavage faces which are often striated).
　　A—amphibole and/or biotite.
　　B—pyroxene.
　　( )—minerals in parentheses are subordinate in amount.

NOTES

1. Minerals other than those listed in table 11 usually occur as accessory or minor constituents of igneous rocks. Such accessory minerals may sometimes be abundant as exemplified by the occasional abundance of muscovite ("white" mica) in granite (called muscovite granite). Certain specific mineral combinations are given separate rock names.

2. Amphibole, biotite (black mica) and pyroxene are dark colored (greenish black to black) and their increase in quantity from granite (rhyolite) to gabbro (basalt) is responsible for the color differences which are noted in table 10.

3. For more complete rock descriptions and more comprehensive classification, consult a textbook on petrography (e.g., Grout, "Kemp's Handbook of Rocks" or Pirsson and Knopf, "Rocks and Rock Minerals").

GLOSSARY OF MINERALOGICAL AND LITHOLOGICAL TERMS

Agglomerate: A mass of unsorted, volcanic fragments, which may be either loose or consolidated by interstitial fine material, the fragments being angular or rounded by volcanic action, but showing essentially no effects of running water.

Amorphous: Noncrystalline (e.g., opal, volcanic glass).

Amygdule: A vesicle filled with secondary minerals. (Adj. Amygdaloidal.)

Aphanitic: An igneous texture having no grains visible to the unaided eye (compare phaneritic).

Amphibole: Group name for related minerals containing magnesium, iron, calcium (sometimes sodium and aluminum) in various combinations (e.g., hornblende).

Arenaceous: Sandy.

Argillaceous: Clayey.

Arkose: Sandstone containing essential amounts of feldspar.

Ash (Volcanic): Loose, fragmental material, the particles being largely less than 4 millimeters in diameter, blown from a volcano; includes volcanic dust (compare tuff, breccia, cinder, and agglomerate).

Bentonite: Earth materials which contain 75 percent or more of clay minerals of the montmorillonite group.

Breccia: Consolidated, angular fragments, the fragments being largely more than 4 millimeters in diameter, which originated through (1) weathering and erosion, (2) mechanical crushing along a fault zone, or (3) volcanic explosions.

Carbonaceous: Containing organic matter.

Chalcedony: Microcrystalline silica mineral, usually containing combined water.

Chert: Usually dense (sometimes porous), microcrystalline, hard rock composed of chalcedony, quartz, and occasionally opal.

Chlorite: A hydrous aluminum silicate containing iron and magnesium, having micaceous structure, commonly formed through weathering of amphibole, pyroxene, or biotite but also formed through metamorphism.

Cinder (Volcanic): Loose agglomerate composed largely of highly vesicular (scoriaceous), glassy fragments of lava, 4 to 32 millimeters in size; also the individual fragments of this description as they occur in ash, tuff, agglomerate, or breccia.

Clay: A naturally occurring, fine-grained material containing as essential constituents the usually platy, crystalline minerals of the kaolinite, montmorillonite (bentonite), illite (hydromica), or related groups, which are primarily hydrous aluminum silicates, formed largely as a result of chemical alteration of other rocks and minerals.

Cleavage: The property of minerals to separate easily along regular planes determined by the crystal structure; the property of rocks (e.g., slates) to split into thin sheets along planes of weakness produced by earth stresses.

Conchoidal Fracture: Fracture, resulting in curved surfaces, as in broken glass (conchoidal: literally, "shell-like").

Concretion: A nodular, lenticular, or irregular structure formed by localized deposition of mineral matter in sediments.

Conglomerate: A sedimentary rock composed largely of rounded particles (gravel and boulders) more than 4 millimeters in diameter and consolidated by cementation or compaction.

Cryptocrystalline: A rock texture in which the crystals are so small as to be irresolvable even with the petrographic microscope.

Diatomite: Consolidated sedimentary rock composed primarily of diatom skeletons (diatomaceous earth).

Equigranular: Igneous texture characterized by grains of essentially equal size (compare porphyritic).

Facies: A geologic or petrographic term designating a portion of a deposit or rock formation differing lithologically in a significant manner from other portions, as a shale facies interbedded with or grading into sandstone; a facies of diorite enclosed in a body of granite.

Feldspar: Name for a group of minerals of igneous origin with similar physical characteristics and composition. Varities are—orthoclase and microcline, potassium-aluminum-silicate; albite, sodium-aluminum-silicate; anorthite, calcium-aluminum-silicate; and others.

Ferromagnesian: Group of generally dark-colored silicate minerals (pyroxene, amphibole, olivine, biotite) containing iron and magnesium.

Ferruginous: Iron-bearing, as reddish-brown ferruginous stain.

Foliation: The banding or lamination of metamorphic rocks arising from parallelism of mineral grains (distinct from stratification in sedimentary rocks).

Glass, volcanic: Igneous rock material (formerly liquid) which has been prevented from crystallizing by sudden cooling.

Gneiss: A metamorphic rock composed of alternate bands of granular minerals and bands of predominantly platy or prismatic minerals.

Hardness: Mineralogically, a measure of the ability of a particle to scratch another, or be scratched; designated quantitatively by Mohs' scale, ranging from 1 (talc) to 10 (diamond).

Kaolinite: A clay mineral (hydrous aluminum silicate) commonly formed through decomposition of feldspar or feldspathic igneous rocks.

Luster: The appearance of the surface of a substance as influenced by its light reflecting qualities—e.g., metallic, vitreous, resinous.

Megascopic (macroscopic): A general term applied to observations made on minerals and rocks, and to the characters observed by means of the unaided eye or pocket-lens, but not with a microscope (compare microscopic).

Meta-: Prefix to rock name (e.g., meta-andesite) indicating the rock has been changed by metamorphic processes.

Microcrystalline: A rock texture in which the crystals are so small as to be observable only with the aid of a microscope.

Microscopic: A general term applied to observations made on minerals and rocks, and to the characters observed, by means of the microscope.

Montmorillonite: Commonest and most widespread member of the bentonitic clay minerals, usually called the montmorillonite group; other members are beidellite, saponite, and nontronite. These clays typically swell with wetting and become unusually soft and slick.

Opal: Colorless to pale gray or brown, amorphous, hydrous silica sometimes varicolored due to impurities).

Pegmatite: Extremely coarse or giant-grained igneous rock commonly composed of quartz, feldspar, mica; often containing rare minerals.

Phaneritic: An igneous texture with grains visible to the unaided eye (compare aphanitic).

Phenocryst: See Porphyritic.

Phyllite: Fine-grained metamorphic rock intermediate in grain size between slate and schist.

Porphyritic: An igneous rock texture in which larger crystals (phenorcrysts) are set in a groundmass of smaller crystals or glass. (Porphyry: rock with porphyritic texture.)

Pozzolan: A siliceous or siliceous and aluminous material, which in itself possesses little or no cementitious value but will, in finely divided form and in the presence of moisture, chemically react with calcium hydroxide at ordinary temperatures to form compounds possessing cementitious properties.

Pumice: A very highly porous and vesicular lava composed largely of glass drawn into approximately parallel or loosely entwined fibers, which themselves contain sealed vesicles (commonest in viscous, silica-rich lavas, like rhyolite and dacite). Compare scoria.

Pumicite: Naturally occurring, very fine-grained, pumiceous, volcanic ash, usually rhyolitic in composition.

Pyroclastic: Refers to fragmental rocks produced by volcanic explosions (tuff, volcanic ash, etc.).

Pyroxene: Group name for related minerals containing magnesium, iron, calcium (sometimes manganese, lithium, and aluminum) in various combinations (e.g., augite).

Quartz: Most abundant crystalline form of silica (SiO_2). Hard, shiny, conchoidal fracture, no true cleavage, usually colorless but sometimes pink, gray, or purple.

Quartzite: Extremely hard, quartzose sandstone, firmly cemented by intergranular, secondary quartz.

Schist: A fine- to coarse-grained planar metamorphic rock predominantly composed of platy or prismatic minerals.

Scoria: A very highly vesicular lava in which the vesicles typically are rounded or elliptical in cross section, the interstitial glass occurring as thin films (commonest in fluid, basic lavas like basalt). Adjective: scoriaceous.   Compare pumice.

Serpentine: Generally greenish rock high in magnesium silicate, usually of secondary origin.

Shale: A sedimentary rock composed primarily of clay or silt and possessing marked fissility parallel to the stratification.

Slate: A very fine-grained metamorphic rock possessing marked fissility which is typically not parallel to the stratification.

Tufa: Generally porous deposits of calcium carbonate, formed around mineral springs (travertine is a variety).

Tuff: Consolidated volcanic ash.

Vermiculite: A group of hydrated mica-like ferro-magnesian silicate minerals that expand perpendicular to the cleavage with rapid heating or treatment with strong hydrogen peroxide; intermediate between biotite or clorite and certain clay minerals.

Vesicular: Refers to volcanic rocks containing "bubble" holes called vesicles (compare amygdaloidal).

Weathering: Rock alteration (decomposition and disintegration) produced by atmospheric agents.

hard, tough, and dense. Tuffs and certain lavas which have been rendered extremely porous by the inclusion of gas bubbles may be exceptions. These are usually unsuitable for concrete aggregates except in lightweight concrete manufacture because of their low strength, light weight, and high absorption.

The sedimentary rocks range from hard to soft, heavy to light, and dense to porous; suitability as aggregate varies correspondingly. Sandstones and limestones, when hard and dense, are suitable as aggregate. But sandstones are frequently friable or excessively porous because of imperfect cementation of the constituent grains. Either sandstone or limestone may contain clay, which renders the rock friable, soft, and absorptive; with increased clay content, these rocks grade into sandy or limey shales. Shales are generally poor aggregate materials, as they are soft, light, weak, and absorptive. Moreover, because they were originally thin bedded, shales are prone to assume flat and slabby shapes when reduced to sand and gravel. Conglomerates may not be suitable as aggregate because of the tendency to break down progressively to smaller sizes during handling and processing. Cherts and flints are widely used as aggregate throughout the country, but so many cherts have exhibited unsatisfactory service histories that suitability of each must be judged individually, preferably on the basis of service records and tests of concrete. Lacking service records, the suitability of cherts that are comparatively low in specific gravity or high in absorption, and which comprise a significant proportion of the aggregate, may be questioned. It has been established that resistance of many cherts to disintegration by freezing and thawing varies with degree of saturation. Cherts that disintegrate readily when saturated may be quite sound when dry. Moreover, when once dried they do not quickly or readily become equally saturated again. The same considerations may apply to other similarly absorptive rocks containing small pores. The presence of particles characterized by extremely small pores, and consequently the need for these special considerations, can be established only by laboratory tests.

Characteristics of metamorphic rocks also vary widely. Marbles and quartzites are usually massive, dense, and adequately tough and strong. Gneisses are usually very durable and tough but may have the undesirable characteristics of schists. Schists frequently are thinly laminated and thus tend to assume slabby shapes; they usually contain large amounts of soft, micaceous minerals and often lack the strength desirable in concrete aggregate. On the other hand, some schists are entirely suitable as aggregate. Slates characteristically possess a thin lamination, which is undesirable.

The characteristics of fresh, unaltered rocks are usually more or less modified by secondary processes, such as weathering, leading to chemical decomposition and physical disintegration; even the strongest rocks may ultimately be reduced to incoherence by these processes. Other secondary processes, such as the action of ground water, may also modify the original character of rock materials by deposition of coatings or cementing substances. Such substances may be deleterious in themselves or may be objectionable because they make processing more difficult. Conversely, secondary processes may occasionally render a rock stronger or less porous by addition of deposited material and, thus, rarely ameliorate initial unsatisfactory characteristics. As explained later, disadvantageous features of a rock type or of a deposit may be susceptible to improvement by washing, selective quarrying, or other treatment.

(c) *Chemical Suitability of Aggregates*.—Some aggregate materials undergo chemical changes that may possibly be beneficial but are often definitely injurious. Such reactions may be of different kinds, including reaction between aggregate material and the constituents of cement, solution of soluble materials, oxidation by weathering, and complicated processes that impede the normal hydration of cement. Volume changes of clays by absorption and dehydration are physical changes which may be considered under this topic because of their relation to the crystallographic structure and chemical composition of different clay minerals.

Reaction between certain aggregate materials and the alkalies in cement is associated with expansion, cracking, and deterioration of concrete. Small amounts of opal, rhyolites, and certain other rocks and minerals in aggregates that are otherwise unobjectionable have caused excessive expansion and rapid deterioration. Opal (amorphous, hydrous silica) is the most reactive constituent in aggregates, but the acidic and intermediate volcanic rocks are the most significant because they are most numerous. Opaline silica may occur as a minor constituent in many rock types or may form coatings or encrustations on sand or gravel particles.

Rocks and minerals known to react deleteriously with cement alkalies are volcanic rocks of medium to high silica content; silicate glasses (artificial or natural excluding the basic type such as basaltic glass); opaline and chalcedonic rocks (including most cherts and flints); some phyllites, tridymite, and certain zeolites. In general, aggregates petrographically similar to known reactive types, or which on the basis of service history or laboratory experiment are suspected of reactive tendencies, should be used only with cement that is low in alkalies. Such reactions are reduced in intensity, and probably eliminated in some instances, by the limitation of the alkalies (Na_2O plus 0.658 K_2O) to 0.5 to 0.6 percent of the cement and/or the use of an effective pozzolan. Zeolites and montmorillonite-type clay minerals can augment the supply of alkalies by cation exchange reactions.

A reaction, similar in effect to that just described, is the alkali-carbonate reaction that occurs when certain dolomitic limestones are used as coarse aggregate in conjunction with a high-alkali cement. There is some disagreement on the mechanism of expansion. One well-supported hypothesis is that excessive expansion is caused by moisture uptake of clay constituents enclosed in dolomite crystals. Finely divided carbonate material, particularly calcite, also may contribute to expansion through uptake of moisture.

Certain sulfide minerals, such as the iron sulfides, pyrite, and marcasite, readily oxidize through atmospheric attack, resulting in unsightly rust stains and loss of strength and coherence of the affected particle. Such reactions may also generate acidic compounds injurious to the surrounding concrete matrix and cause associate reactions that result in volume increase conducive to popouts. Coal is objectionable because of its low strength and undesirable appearance on concrete surfaces and because it decreases the resistance of concrete to freezing and thawing. Other organic substances, such as certain vegetable matter and humus, contain organic acids that inhibit the hydration of cement. Sands producing a color darker than the standard in the colorimetric test for organic impurities may be rejected, although such results are usually interpreted as meaning that additional testing is desirable to determine the type of organic matter present and specific effect on concrete. Clays are subject to swelling and shrinking by absorption and dehydration, and when they occur as constituents of rocks, such as limestones, this absorptive characteristic greatly increases the susceptibility of the rock to disruption by weathering. Chemical salts, such as sulfates, chlorides, carbonates, and phosphates, may occur in aggregates in numerous forms. Some of these substances react chemically to modify or impede the normal setting processes of cement; others are undesirable because of low strength or because they tend to dissolve. Such contaminations may also contribute to exudation and effloresence, and if powdery and fine-grained, they may augment the undesirable silty fractions of an aggregate.

The extremely fine fractions of aggregate materials are commonly classed as silt or silt and clay and should not be permitted in large amounts because of their tendency to increase water requirements of a mix and thus contribute to unsoundness or to decreased strength or durability. Rarely, natural inclusions of siliceous material, which might be classed in the field as silt or clay, may be beneficial to concrete. On the other hand, montmorillonite-type clays which constitute a part of the sand particles have produced detrimental effects by causing excessive loss in slump. Mica is a common contaminating substance in aggregates and is undesirable because its soft, laminated and absorptive character and susceptibility to disintegration along cleavage planes contribute to reduced strength and durability.

Chemically deleterious or contaminating substances may occur in coarse or fine aggregate as coatings on the aggregate particles, as encrustations cementing them together, or as distinct layers in aggregate deposits. These contaminations are more prevalent in arid or semiarid regions. When powdery or pulverulent, such material is sometimes loosened by handling so that the silt fraction is augmented. Some clays and some chemically active minerals are formed through weathering of rocks that are normally hard, sound, and chemically suitable. For this reason, the degree and kind of weathering of aggregates are important from a chemical standpoint as well as from that of physical soundness.

Simple washing will usually remove silt, clay coatings, some fine free mica, easily soluble salts, and light organic matter. Clay lumps are removable only with difficulty. Coal can be removed from aggregates by washing, if the particles are not too coarse, or by reverse flow processing or heavy-media separation, if relatively large particles are present. Hard and adherent coatings and encrustations require vigorous abrasion processing, such as tumbling, in order to loosen them so that they may be removed by screening and washing. Some coatings cannot be removed at reasonable cost.

Soluble substances in aggregates may dissolve and contaminate the mixing water if they are not removed by processing. Surface coatings and encrustations, especially if loose and powdery, may be partly removed in the mixer; but any resulting improvement in bond between cement and aggregates may be offset by the tendency of the loosened material to increase the water requirement and to promote formation of laitance.

26. Prospecting.—When searching for suitable aggregate, it is important to bear in mind that ideal materials are seldom found. Deficiencies or excesses of one or more sizes are very common; objectionable rock types, coated and cemented particles, or particles of flat or slabby shape may occur in excessive amounts; clay, silt, or organic matter may contaminate the deposit; or weathering may have seriously reduced the strength of the particles.

It is most important to obtain a reasonable interpretation of the materials through proper sampling procedures. Moreover, depth of ground water or excessive overburden may seriously impede operation at a deposit. Unfortunately, the strata within the body of the deposit cannot be directly observed at the surface. However, interpretations based on surface observations are greatly aided by an understanding of the geological processes that have acted on the material. Frequently such an understanding will permit a distinction to be made between conditions that are merely superficial and those that may be expected at some depth. Final conclusions on these matters will usually require thorough exploration, but as much pertinent information as practicable should be obtained during the reconnaissance and preliminary exploration.

Many objectionable features of sand and gravel deposits are remediable by proper processing. Crushing may alleviate deficiencies in fine gravel or even in sand sizes, or blending sand may be available. Washing may serve to remove deleterious clay, silt, or organic matter. Selective excavation may be a satisfactory means of avoiding the use of objectionable parts of the deposit. Whether these or other methods of processing are justifiable will usually depend on the magnitude of the project and availability of satisfactory materials from other sources. Such considerations must influence preliminary explorations. Accessibility, proximity to the job, and workability of a deposit are essential considerations in evaluating suitability.

The quantity of aggregate that a deposit may yield should be roughly estimated and compared with the probable requirements. Areas may be estimated roughly by pacing the dimensions. Depth and grading of material may be judged by examining banks of channels or other exposures. Except for an estimated deduction for waste, which may run from 20 to 50 percent based on appearance of the material, it may generally be assumed that a cubic yard of material in place will produce aggregate for a cubic yard of concrete.

27. Preliminary Sampling of Prospective Aggregate Sources and Reporting of Related Information.—Methods used in obtaining preliminary samples for tests in the Denver laboratories are largely dependent on such factors as type and design of structure and characteristics of the aggregate deposit with respect to uniformity, size and shape, overburden, ground-water conditions, etc. Establishment of standardized instructions applicable to all prospective sources of aggregates is impracticable. Sources considered to be most feasible and economical should be sampled and the samples sent to the Chief, Division of Research, Attention: Code 1510, Bureau of Reclamation, P. O. Box 25007, Denver, Colorado 80225. Sizes of samples are as follows: 600 pounds of pit-run sand and gravel; or, if screened, 200 pounds of sand, 200 pounds of No. 4 to ¾-inch size, and 100 pounds of each of other sizes produced. The size of sample of quarry rock proposed for crushed aggregate is 600 pounds. A letter of transmittal containing information requested in subsections (a) and (b) should be sent for each shipment of samples. The letter should indicate the purpose for which the samples are submitted, such as for concrete aggregate, filter material, or pervious embankment. A copy of the letter should be enclosed in a sample sack.

(a) *Sand and Gravel Deposits*.—Samples of representative pit-run materials from noncommercial deposits may be obtained from exposed faces or from trenches or pits excavated at appropriate locations. In sampling commercial sources or other sources where screening facilities are available, it is desirable to obtain individual samples of separate sizes.

The following information relative to each promising noncommercial deposit investigated will assist in selection or approval of the source of aggregates and in preparation of specifications:

(1) Ownership of the deposit.

(2) Location of deposit, indicated on a map, with reference to section, township, and range.

(3) Type of deposit, character of topography, and description of vegetation.

(4) Roughly estimated volume and average depth of deposit and average overburden. Also, the ground-water level with comments relative to fluctuation.

(5) Approximate percentage of material larger than maximum size included in the samples.

(6) Roads affording access from highways.

(7) Service history of concrete made with the aggregate, if any, or of concrete made with similar aggregates in the locality.

(8) Photographs and any other information that may be useful or necessary.

For commercial sand and gravel deposits and rock quarries that are equipped for operation, the following information is desirable:

(1) Name and address of operator. If deposit is not active, a statement relative to ownership or control.

(2) Location of deposit and plant.

(3) Age of plant and, if inactive, the approximate date when operations ceased.

(4) Transportation facilities and difficulties.

(5) Extent of deposit.

(6) Capacity of plant and stockpiles.

(7) Description of plant, including type and condition of equipment for excavating, transporting, crushing, screening, washing, classifying, and loading.

(8) Approximate percentages of the various sizes of materials produced by the plant.

(9) Location of scales on which materials would be weighed for rail shipment.

(10) Approximate prices of materials f.o.b. carriers at the plant.

(11) Principal users of plant output.

(12) Service history of concrete made with the aggregate, including type and size of structure, mix proportions, type of cement used, and quality of concrete.

(13) Any other pertinent information.

(b) *Prospective Rock Quarries for Concrete Aggregate.*—It is often necessary to sample quarries or undeveloped rock formations. For preliminary aggregate investigations, requirements for sampling operating quarries, or inactive quarries where finished materials are in storage, are similar to those for commercial sand and gravel deposits.

Samples from undeveloped rock formations must be taken very carefully so that the material selected will be, to the greatest possible extent, typical of the deposit and inclusive of any significant variations of rock type. Representative samples may be difficult to obtain. Overburden may restrict the area from which material can be taken and obscure the true character of a large part of the deposit. Moreover, surface outcrops will frequently be more weathered than the interior of the deposit. Samples obtained from loose pieces on the ground or collected from the weathered outer surfaces of outcrops are rarely representative. Fresher material may be obtained by breaking away the outer surfaces or, if necessary, by trenching, blasting, or core drilling.

In sampling undeveloped bedrock formations, certain geological considerations are pertinent. In stratified deposits, such as sandstones or limestones, uniformity in a vertical direction must be evaluated because successive strata are often very different. The dip of stratified formations must also be considered because inclination of the strata with respect to surface slope will bring different strata to the surface in different parts of the area, and excavation may be uneconomical because of excessive overburden. Attention must be directed to the possibility of zones or layers of undesirable material. Clay or shale layers or seams may be so large or prevalent as to necessitate selective quarrying, excessive wasting, or special processing.

The information needed in a report on the investigation of a prospective quarry is similar to that previously described in this section. A complete discussion of procedures for investigation of rock quarries for riprap production is given in designation E–39 of the Second Edition of the Bureau's Earth Manual. Descriptions of any observable conditions relating to the accessibility or workability of the deposit, such as thickness and uniformity of overburden, ground-water conditions, and area available for operations, are desirable. If further field investigation is necessary, it will be specifically requested by the Denver laboratories.

B. Exploration of Natural Aggregate Deposits

28. General Procedure.—If the results of preliminary investigations restrict the choice of aggregate source to one of several undeveloped deposits of sand and gravel, the most promising deposit (or deposits, if there is question as to relative merits or whether it is advisable to use more than one source) is explored and sampled by means of steel-cased test holes, open test pits, or trenches. Where deposits are exposed in highway or railroad cuts or along gulches, little excavation is required. The methods used depend on local topography; area, shape, and depth of deposit; ground-water conditions; prevalence of large rocks; and considerations affecting economy. Test excavations should be distributed at intervals in keeping with uniformity and extent of the deposit and held to the minimum effective number. The principal objectives are to obtain sufficient representative samples to permit accurate estimation of quality and quantity of materials available; to enable reliable prediction of what processing operations will be required and what concrete mixes are best suited for the work; and to supply information for use of bidders and of field forces during construction.

29. Exploratory Excavations.—(a) *Steel-Cased Test Holes.*—The mechanically driven, steel-cased test hole as now developed is a very accurate method of sampling extensive aggregate deposits. It is also the most economical method presently in use. When the ground to be prospected is reasonably free from oversize rocks, a deposit may be readily sampled by driving steel pipe or casing, through which samples are removed with a post-hole auger or other suitable device. The casing also prevents sand and gravel from caving and running into the bottom of the hole. This method should be used wherever practicable, unless more economical means can be devised. Casing sections should be short enough to permit ready handling, and the joints should be smooth enough on the outside to facilitate driving. Screwed or locked joints are necessary if the casing is to be pulled after sampling is finished. If the casing must be driven with a sledge because a weighted platform is insufficient for sinking it, a heavy driving ring will prevent the casing from being injured. A three-pronged fishing tool is essential for removing rocks that cannot be handled by the auger.

At Parker Dam a deposit of fine blending sand was explored by a 6-inch pipe casing through which samples were removed with a 4-inch auger. On the Colorado River project in Texas, 51 cased holes 3 to 9 feet deep were sunk in a very short time during the exploration of a deposit of fine blending sand proposed for use in construction of Buchanan and Inks Dams. The systematic procedure for handling and treating samples from cased holes is the same as described in section 29(c) for samples from test pits, except that cased hole samples consist of the full amount of material excavated.

During exploration for aggregate for Shasta Dam, when water was encountered at the bottoms of test pits in the Cottonwood Creek deposit, the full depth of material (containing about 50 percent sand and few rocks larger than 4½ inches) was explored by means of cased holes driven in the bottoms of test pits by equipment similar to a well drill. A churn drill was used at the start and in hard spots, but otherwise the casing was

lowered by a driving weight. Samples of the lower material were taken by
a bailer equipped with a suction piston and a check valve at the lower end.

A truck-mounted rotary boring machine, designed primarily for drilling
holes 18 to 60 inches in diameter in soft ground was used in the investigation of a deposit proposed as a source of aggregate for Glen Canyon Dam.
Use of such equipment was occasioned by the unstable condition of materials below ground-water elevation. The rig is shown in figure 37. The

Figure 37.—Rotary digger used in exploring aggregate deposits. In the foreground
is 20-inch casing, inside of which an 18-inch bucket was used for boring below
ground-water level. PX–D–32042.

holes were bored to ground water with a 30-inch bucket that had two
radial openings and excavating blades extending below the openings. The
equipment handled rocks up to 5 inches in diameter. Below ground-water
elevation an 18-inch bucket was used to operate inside a 20-inch casing.
The smaller bucket had one opening that would handle rocks up to 8
inches in diameter. Both buckets had a thick rubber flap over each opening which allowed material to pass inside but sealed the opening when a
loaded bucket was being hoisted. The casing was forced down by turning
under the weight of a "Kelly" bar. The operations of digging with the
bucket and setting the casing were alternated, the casing being kept as

near the bottom of the hole as possible. Three-foot sections of casing were added as digging progressed. A hole can be dug with this equipment until the pressure against the side of the casing exceeds the turning capacity of the machine or until too hard a layer of material is encountered.

During explorations at Canyon Ferry Dam, a rather ingenious core shovel was used to secure aggregate samples (fine sand, gravel, and large cobbles) from cased test holes that could not be dewatered. The device illustrated in figure 38 is secured to a Kelly bar and controlled by a cable attached to the shovel. The equipment is lowered into the casing, and, as the shovel takes its full bite, the cable releases, tilting it back, thus obtaining a representative sample of material from the test hole.

Figure 38.—"Core shovel" used to secure representative aggregate samples from cased test holes that cannot be dewatered. PX–D–32043.

Exceptionally well-designed, efficient sampling equipment (fig. 39) was used for aggregate exploration and investigation of precious metal content of existing gravel bars along the North Fork of the American River during Auburn Dam investigations. A 36-inch-diameter, flush-coupled casing of ⅝-inch wall thickness was used. The length of the driving shoe was 5 feet. The other sections were 4 feet long. A string of cable-operated digging tools excavated the sample within the casing. The casing was driven sufficiently ahead of the excavation to prevent run-in. Hydraulic rams, mounted on the clamshell rig tower and capable of exerting a thrust in excess of 50 tons, were operated to remove the casing. A detachable heavy ring was attached to the casing to prevent distortion while being driven. The excavation within the casing was accomplished through the operation of a 3,500-pound string of digging tools. A 365 ft^3/min compressor supplied air to a ram contained within and which operated the clamshell bucket. The bucket opened a maximum of 36 inches. Figure 40 is a view of the clamshell bucket extended. The digging string was suspended by cable, and a paralleling air hose was synchronized with the cable reel. Outriggers on the truck were extended to level it and to provide stability during operation. A pan mounted on a front-end loader received the excavated materials from the clamshell bucket and transported them to the screening operations.

(b) *Uncased Test Holes.*—Where the soil is suitable and an abundant water supply is available, it may be occasionally possible to explore an aggregate deposit below the ground-water table by use of the reverse-flow method of drilling. This method (fig. 41), using a 30-inch-diameter churn bit, was selected by the contractor to determine the depth and percentage of cobbles available in the deposit proposed for use in Glen Canyon Dam.

The reverse-flow method was developed to overcome deficiencies of the

Figure 39.—Clamshell excavator and backhoe unit with reverse shovel exploring the gravel bars at Auburn Dam, California. P1859–245–5533.

Figure 40.—Closeup of clamshell bucket in extended open position; Auburn Dam aggregate investigations. P859–245–5490.

Figure 41.—Reverse-circulation rotary drilling machine. Dike built with bulldozer forms reservoir of water required for operation. P557–420–254.

direct-flow method, in which water is forced down the inside of the drill rod and is lost or returns to ground surface outside of the drill rod. In the direct-flow method if churn bits are used, casing is generally required for the full depth of the hole to prevent caving. In the reverse-flow method, caving is prevented by a hydrostatic head maintained, often in a casing at the top of the hole, 8 to 10 feet above the water table. This provides a column of water around the outside of the drill rod which, with material from the hole, mixes with air injected near the bottom of the drill rod and is pumped up through the center of the drill rod. Except for some specific conditions at the top of the hole, casing is not usually required. With the reverse-flow method, test holes can be drilled at a faster rate than is normally possible in a cased hole or by the direct-flow method, particularly if the drilling is in stable fine aggregate and sand.

At Glen Canyon, using a caliper to measure the diameters of the drill holes so that volumes of the holes could be computed, fairly uniform estimates of the percentage of cobbles in the pit were obtained. A 6-inch inside-diameter drill rod and a 6-inch-diameter discharge pipe were used. The discharge was collected in two vats, each 150 cubic feet in volume; these were inadequate since excessive amounts of sand and silt were lost in discharge splashing. Larger catch basins which would have retained the sand and fines would have permitted a complete grading analysis of the aggregate. Material too large to pass through the drill rod and discharge pipe was recovered with an orange-peel bucket operated with the same drill rig and was added to the sample. This operation slowed the drilling rate. Some caving occurred in layers of windblown sand and added to the difficulty of maintaining a uniform hole diameter. Nonuniformity of the hole diameter decreases the accuracy of this method of estimating the percentage of cobbles in an aggregate deposit. However, the facility with which the test holes can generally be drilled with the reverse-flow method, permitting rapid exploration of several locations in a deposit without the necessity of casing the holes, compensates in large measure for the above deficiencies of the method.

(c) *Test Pits.*—Aggregate deposits sometimes contain large rocks that prevent use of cased test holes for sampling. Unless accurate sampling can be accomplished by easier means (that is, mechanical equipment), it is necessary to hand dig test pits. Machine digging of test pits has frequently saved considerable time and money. Bulldozers, clamshells, backhoes, or draglines have been used to good advantage. Usually such methods are not suitable for depths greater than 15 to 30 feet. Although machine-dug pits do not permit such precise sampling as hand-dug and shored test pits, they do give a good general idea of the material in a deposit and indicate the advisability of excavating one or more shored test pits for evaluation of the deposit.

Hand-dug test pits may be about 16 square feet in bottom area and will require shoring if deeper than 5 feet or in unstable material. A 3- by 5-foot hand-dug hole is easier to work than a hole 4-foot square. The total depth of a pit cannot be determined in advance and is left to the judgment of the engineer in charge. Precautions should be taken to prevent contamination of the samples by material that has fallen from the side of the pit. The bottom of the hole should be kept fairly level and of full size while excavation is in progress so that material removed in each lift may represent the corresponding portion of the deposit in both quantity and quality. As noted in figure 42, the depth of each lift may be 5 feet where there is no marked change of materials with depth; otherwise, the depths of the lifts should correspond with differences in grading or in type of material, as suggested by the data in figure 43.

Test pits over 5 feet deep must be shored in accordance with the minimum requirements for trench shoring in the Bureau of Reclamation publication, "Safety and Health Regulations for Construction," (based on 29

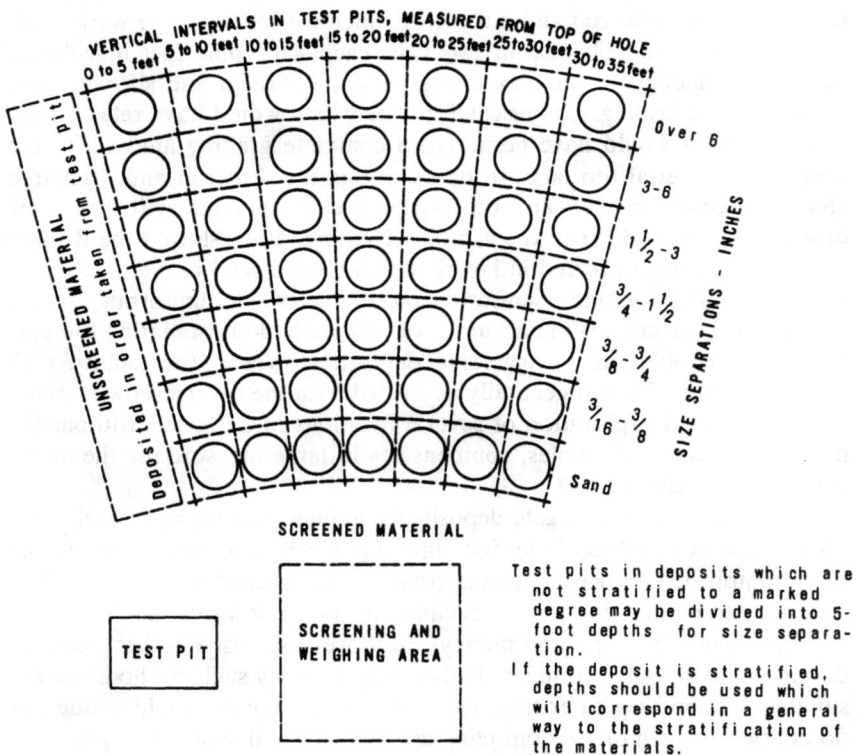

Figure 42.—Sand and gravel from a test pit stored in systematically arranged piles. Sampling and inspection are facilitated by such an arrangement. 288–D–125.

7-1324

UNITED STATES
DEPARTMENT OF THE INTERIOR
BUREAU OF RECLAMATION

GRADING OF AGGREGATE FROM TEST PIT

PROJECT__ _Colorado River Storage_ __FEATURE__ _Glen Canyon Unit_ _____

DEPOSIT_ _ _Wahweap_ _ _ _DESCRIPTION_ _Probably remnants of terrace deposits accumulated within Wahweap Creek_ _ _ _ _ _

LOCATION OF DEPOSIT__Wahweap Creek_ _

TEST PIT NO._ _ _ _109_ _ _ _ _ _SIZE_ _ _Dragline excavation 20' by 50' at surface_
Sta 12+50 0/s Line "b"

MATERIAL REPRESENTED	SCREEN OPENINGS	SCREEN ANALYSIS, PERCENT RETAINED (EACH SIZE) DEPTH OF TEST PIT (FEET)				
		0 to 4.5	4.5 to 8.0	8.0 to 13.0	13.0 to 18.0	18.0 to 22.0
Wt. of Sample, lb.			1319	1985	1662	
Aggregate	6-inch	No	0	0	0	No
	3-inch	sample,	10	7	6	Sample
	1½-inch	fine	26	18	21	
	¾-inch	sand	27	29	21	
	3/8-inch	and	20	27	30	
	3/16-inch	gravel	17	19	22	
	Percent		76	62	69	

MATERIAL REPRESENTED	SCREEN OPENINGS	4.5 to 8.0		8.0 to 13.0		13.0 to 18.0		18.0 to 22.0	
Sand	Sample No.	1	2	1	2	1	2		
	No. 8	28	29	21	21	25	32		
	No. 16	19	21	11	11	13	22		
	No. 30	10	10	9	9	7	7		
	No. 50	15	15	23	23	19	16		
	No. 100	20	20	28	30	23	16		
	Pan	8	5	8	6	13	7		
	Percent	24		38		31			

ALL TEST SCREEN OPENINGS TO BE SQUARE. IF OTHER SIZE SEPARATIONS ARE USED,
INDICATE THE SIZES OF TEST SCREEN OPENINGS.
INDICATE THE PRESENCE OF ORGANIC MATERIAL, SOFT STONE, SHALE, CLAY, MUD BALLS,
BOULDERS, COATED MATERIALS, CONGLOMERATE, MICA IN THE SAND, OR OTHER DELET-
ERIOUS MATERIALS. INDICATE GROUND WATER LEVEL AND GENERAL MOISTURE COND-
ITION OF THE MATERIAL, ALSO ANY OTHER PERTINENT INFORMATION. USE BACK OF
SHEET IF NECESSARY.

_ _ _Water table 8.0 ft. Pumps utilized and an attempt made to sample below_ _ _
_ _ _ _ _ _water table._ _
_ _ _Sampled by dragline bucket_ _
_ _ Depth of hole 22.0 ft._ _
_ _ Sand sample No.1 unwashed. No.2 washed_ _ _ _ _ _ _ _ _ _ _ _ _ _ _ _ _ _ _ _
_ _ Organic test OK, Silt 3% (Designation 14)_ _ _ _ _ _ _ _ _ _ _ _ _ _ _ _ _ _ _ _
_ _ _Large sandstone boulders at 22 ft._ _
_ F.M. unwashed sand 2.96, 2.50, 2.59, _ _ Washed sand 3.09, 2.52, 3.17 _ _ _ _ _ _ _
Date_ _ 8 -20 _ _19 56 _ _ _ _ _ _ _ _ _ _ _ INSPECTOR _ L.T.L. - C.A.C. - M.B._ _ _ _ _ _ _ _ _

Figure 43.—A systematic and comprehensive form for recording test-pit exploration data. 288–D–2633.

CFR 1926.652). To ensure contamination-free sampling, closed-wall cribbing is preferred to skeleton shoring. Details of cribbing which will meet the safety standards for depths up to about 20 feet (assuming non-saturated loading) and provide adequate protection from sample contamination are shown in figure 44.

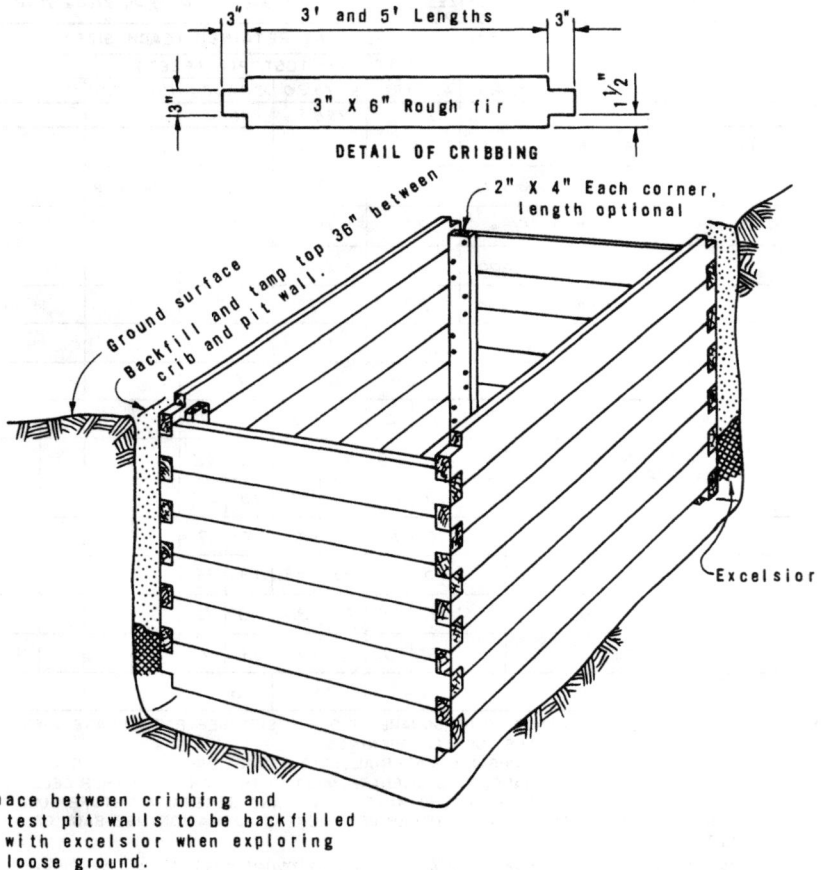

Figure 44.—Test-pit cribbing. 288–D–2624.

Except for relatively shallow test pits, the plank for cribbing should have a nominal thickness of not less than 3 inches. Thicker plank should be used wherever necessary to ensure safety. Six inches is a convenient width. The pieces of cribbing should be kept level to ensure a vertical pit. In loose materials it is advisable to (1) keep a minimum space between the pit walls and the cribbing, (2) pack this space with hay or excelsior, and (3) keep the bottom round of cribbing close to the bottom of the pit. Cribbing details are shown in figure 44.

When water is encountered, an efficient pumping system is necessary. A number of small, readily portable, gasoline-powered, self-priming, centrifugal pump units are manufactured. It is desirable that the suction hose be one-half inch larger than the discharge opening of the pump and not more than 15 feet long. This length limitation requires setting of the pump in the pit, on a frame attached to the cribbing, at intervals of about 12 feet. With the gasoline engine in the pit, it is necessary to pipe the exhaust gases well away from the pit.

Figure 45 shows a test-pit operation on the Hungry Horse project. Of particular interest in figure 45 are the pit collar, which prevents loose rocks from being kicked down into the pit; the safety belts and safety ropes; the hard hats and the signal code board. All these are essential to safe test-pit operation. Figure 46 shows construction details of a screening hopper which facilitates hand screening of materials from test pits and is also used for tests of coarse aggregates in field laboratories. The apparatus may be used to particular advantage at the test pit if powered by a small gasoline engine. An electric motor is commonly used for operation in the field laboratory. (See fig. 47.) The power drive has also made the screening apparatus generally useful for preparation of small quantities of special aggregate. The sides of the frame in figure 46 may be built to support additional sets of rollers, which will permit independent use of two or more superimposed screens. Such an arrangement reduces the number of handling operations. During cold weather on the Colorado-Big Thompson project, a three-sided portable canvas shelter containing a stove protected test-pit excavation and screening operations so that it was possible to continue the work throughout the winter.

It is not necessary that all the material from a hand-dug test pit be screened to determine the grading if the following sampling procedure is carefully practiced. The objective is to secure for screening all the material from a continuous column, within the pit, having a diameter of approximately 2 feet. When the bottom of the pit is cleared to a new level, the sample material is taken from a 2-foot-diameter hole as deep as clean excavation permits. The remainder of the pit bottom is then excavated to the level of the bottom of the sample hole, and the procedure is then repeated.

Where the sides of machine-dug pits are less than 5 feet deep in stable material, samples for pit screen analysis should be obtained by taking all the material from a vertical channel of adequate size in the face of the standing material. Where pits are deeper than 5 feet or are in unstable material, they shall be shored, laid back to stable slopes, or otherwise protected from caving so that a person can safely obtain representative samples.

The sample material from each 5-foot lift or from differing strata should be separated carefully into the sizes contemplated for use. If the material is too moist to permit reasonably clean size separation, it should be al-

Figure 45.—View of test-pit collar setup. PX–D–32045.

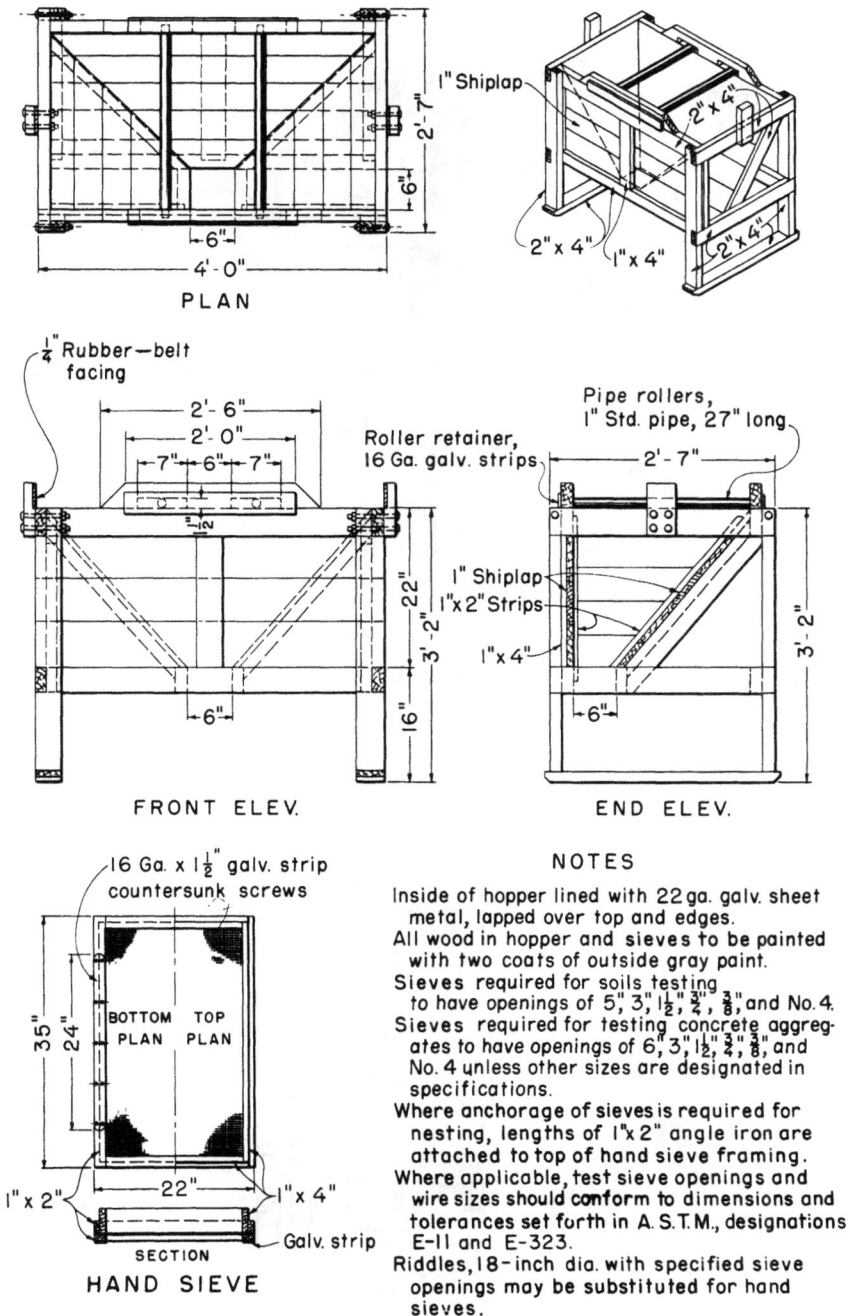

PLAN

$\frac{1}{4}$" Rubber—belt facing

2'-6"
2'-0"
7"-6"-7"
$\frac{3}{12}$"

1" Shiplap

2"x 4"
2"x 4"
1"x 4"
2"x 4"

FRONT ELEV.

Pipe rollers, 1" Std. pipe, 27" long

Roller retainer, 16 Ga. galv. strips

2'-7"

1" Shiplap
1"x 2" Strips
1"x 4"

3'-2"

6"

END ELEV.

16 Ga. x 1$\frac{1}{2}$" galv. strip countersunk screws

35"
24"

BOTTOM TOP
PLAN PLAN

1" x 2"
22"
1"x 4"
Galv. strip
SECTION
HAND SIEVE

NOTES

Inside of hopper lined with 22 ga. galv. sheet metal, lapped over top and edges.

All wood in hopper and sieves to be painted with two coats of outside gray paint.

Sieves required for soils testing to have openings of 5", 3", 1$\frac{1}{2}$", $\frac{3}{4}$", $\frac{3}{8}$", and No. 4.

Sieves required for testing concrete aggregates to have openings of 6", 3", 1$\frac{1}{2}$", $\frac{3}{4}$", $\frac{3}{8}$", and No. 4 unless other sizes are designated in specifications.

Where anchorage of sieves is required for nesting, lengths of 1"x 2" angle iron are attached to top of hand sieve framing.

Where applicable, test sieve openings and wire sizes should conform to dimensions and tolerances set forth in A.S.T.M., designations E-11 and E-323.

Riddles, 18-inch dia. with specified sieve openings may be substituted for hand sieves.

Figure 46.—Screening hopper and hand screens for gradation tests of coarse aggregate and earth materials. 101-D-112.

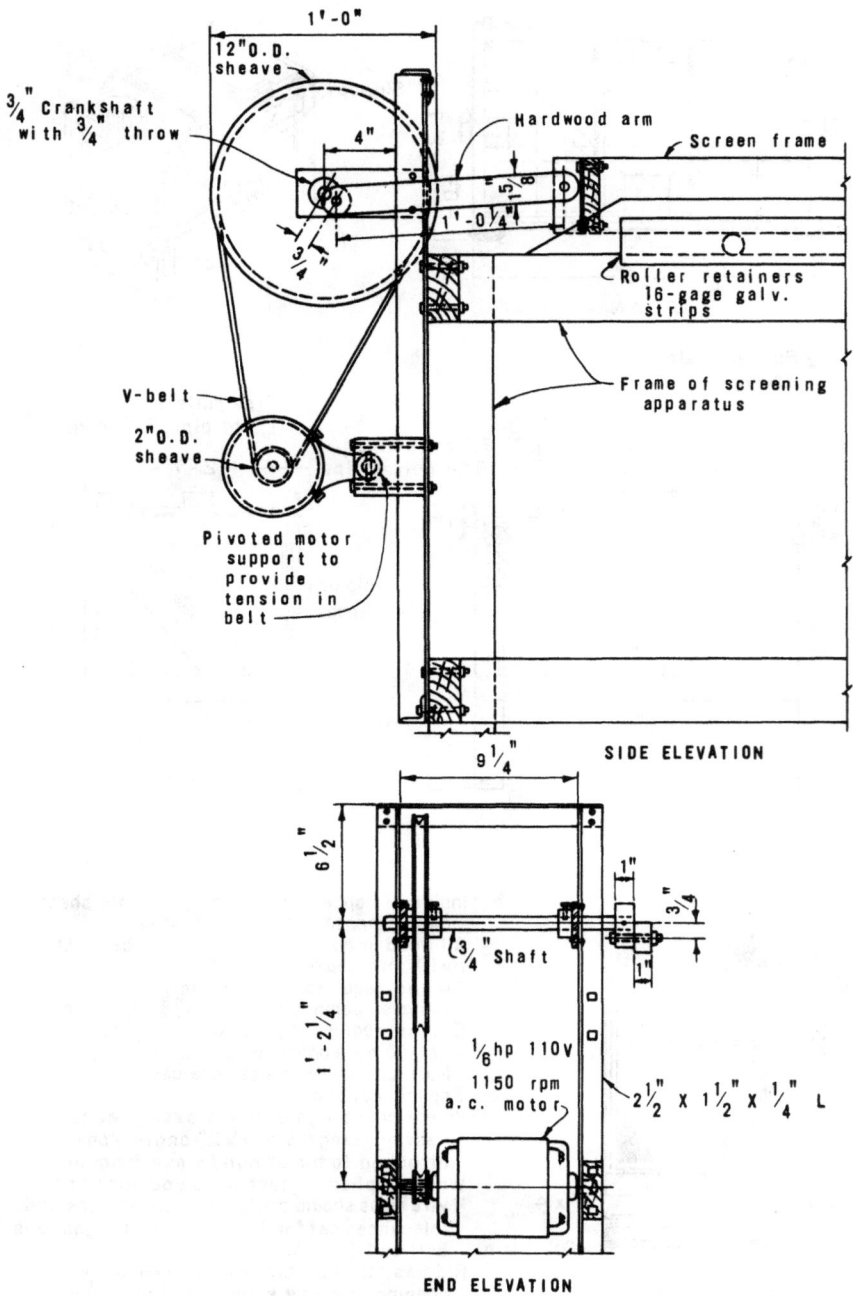

Figure 47.—A power drive for the screening apparatus illustrated in figure 46. Such a drive is usually used in field laboratories and will increase the output considerably. 288–D–810.

lowed to dry until screened properly. After each size from each depth interval is weighed carefully, the weights should be recorded. (See fig. 43.)

At least two representative samples should be taken from the sand from each 5-foot lift or separate stratum and complete screen analyses (appendix designation 4) should be made and recorded as shown in figure 43. If the sand samples are not dry, they should be dried by spreading on a canvas in the sun or by a suitable heating apparatus and the moisture content determined by weighing the sample before and after drying. If gradings of sand samples are not in close agreement, additional tests should be made and the average of all tests reported. Any other data that will help in obtaining an accurate knowledge of the material in the deposit should also be recorded.

All screened excavated materials should be placed in stockpiles arranged in concentric rows adjacent to the pit, each pile consisting of one size-fraction from one lift. Figure 42 illustrates a satisfactory arrangement. The piles should be placed on ground from which loose material has been removed and spaced so as to prevent intermixing. An effective method for labeling the stockpiles is to build them around marked stakes previously driven into the ground and long enough to project above the tops of the finished piles. It should be kept in mind that the purpose of piling materials in this manner is to facilitate future check tests and inspection. The unscreened excavated materials should be deposited in a long pile (fig. 42), conveniently located, so that the material from each depth will remain visible for inspection.

After each test pit is opened, it should be covered with solid timber not less than 2 inches thick, and a guard railing or fence should be erected around the pit.

(d) *Trenches.*—It is usually impracticable to excavate trenches of sufficient depth to expose the entire vertical section of a deposit and thus enable procurement of a complete sequence of accurate samples such as may be obtained from test pits. However, in certain topographic situations, trenching is a practical means of procuring accurate samples at much lower cost.

Detail exploration by trenching is most feasible for deposits in which deep erosion has exposed vertical sections on steeply sloping walls of gulleys. Trenches extending from top to bottom of such slopes may traverse the entire thickness of the deposit or the greater part of it. The slopes are usually covered with considerable material that has rolled down from the top and that which must be penetrated so that representative samples may be obtained. This difficulty was overcome in the aggregate explorations for Davis Dam by digging deep trenches with a bulldozer. Progress was very rapid and the method efficient.

Trenches were used to explore the Henry's Fork deposit during aggregate investigations on the Flaming Gorge unit of the Colorado River Stor-

Figure 48.—Trench in Henrys Fork aggregate deposit, Flaming Gorge unit, Utah. P591–421–1228.

age project to supplement information derived from test pits. Operators using a backhoe dug through trenches normal to the apparent direction of deposition to expose as long a face as possible to aid identification and tentative evaluation of various types of materials present. Figure 48 shows the type of trench, and figure 49 shows graded aggregate representing ma-

Figure 49.—Graded aggregate piles representing sampled material from trench excavation. P591–421–1217.

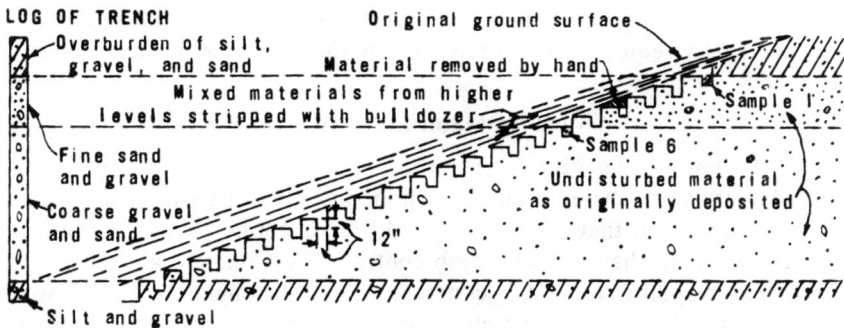

LOG OF TRENCH   Original ground surface
Overburden of silt,
gravel, and sand   Material removed by hand
        Mixed materials from higher
        levels stripped with bulldozer   Sample 1
Fine sand
and gravel   Sample 6
Coarse gravel   Undisturbed material
and sand   as originally deposited
        12"
Silt and gravel

Figure 50.—Exploration of aggregate deposits by trenching. Where this method is practicable, reliable samples may be secured at a much lower cost than from test pits. 288–D–2623.

terial from the deep cut of the trench. To more fully determine the depth of suitable materials, test holes at selected locations were extended below the trench bottom.

Trenches must be finished by hand, care being taken to remove sloughed material. Samples are then taken at appropriate vertical intervals, the procedures being governed by the same considerations as pertain to sampling in test pits. (See fig. 50.) Contamination will usually be minimized if the final cleanup and the sampling are done together, proceeding from top to bottom.

30. Designation of Deposits and of Test Holes and Test Pits.—Test areas or deposits, usually designated by names, should be referenced by quarter or half section, township, and range, and test excavations should be referenced by numbers or combinations of numbers and letters. Designations for holes and pits are marked on nearby stakes and shown on the map of the deposit.

31. Reports and Samples Required.—Bureau administrative instructions require that reports of explorations be submitted to the Chief, Division of General Research, at the conclusion of aggregate investigations. When explorations extend over several months, progress reports should be submitted each month. Reports should describe field activities in detail and should be accompanied by test-pit grading data on form 7–1324, illustrated in figure 43. Photographs, maps, and other drawings are helpful and desirable for the record of investigations. Indication of powerlines, rights-of-way, fences, structures, and other important surface features enhances the usefulness of maps. Figures 51, 52, 53, and 54 indicate the nature of information needed for preparation of specification drawings for large jobs when the aggregate source is specified.

Samples for preliminary investigations should be submitted as provided in section 27. Samples for mix investigations and other special studies will be specifically requested when required by the Chief, Division of General Research.

C. Facilities for Materials Testing at Denver

32. Laboratory Facilities.—The functions of the laboratories in Denver include service to the Bureau for conducting investigations and tests relating to concrete, thereby relieving field laboratories of most laboratory work other than that of routine job control. The Concrete and Structural Branch is responsible for the concrete quality throughout Bureau work and thus monitors field concrete control activities. The laboratories are equipped to conduct physical, chemical, and petrographic investigations of all concrete materials and to study the behavior of these materials under a variety of conditions. The facilities permit fabrication and testing of concrete specimens for strength, elasticity, permeability, volume change, temperature rise, thermal characteristics, and durability. There are also facil-

ESTIMATED QUANTITIES OF GRAVEL IN DEPOSIT AND OF QUANTITIES REQUIRED FOR CONSTRUCTION

AGGREGATE (CLEAN SEPARATION)	ESTIMATED TONS OF SEPARATED AGGREGATE					TOTAL ESTIMATED TONS SEPARATED AGGREGATE	ESTIMATED TONS REQUIRED
	BETWEEN ELEVATIONS 3240 & 3200	BETWEEN ELEVATIONS 3200 & 3160	BETWEEN ELEVATIONS 3160 & 3120	BETWEEN ELEVATIONS 3120 & 3100			
SAND	135,000	596,000	2,240,000	981,000		3,952,000	1,500,000
COARSE AGGREGATE $\frac{3}{4}$" TO $\frac{3}{8}$"	231,000	1,081,000	1,728,000	451,000		3,491,000	1,100,000
COARSE AGGREGATE $\frac{3}{4}$" TO 1$\frac{1}{2}$"	202,000	816,000	1,367,000	408,000		2,793,000	1,100,000
COARSE AGGREGATE 1$\frac{1}{2}$" TO 3"	168,000	732,000	1,192,000	396,000		2,488,000	1,100,000
COARSE AGGREGATE 3" TO 6"	103,000	552,000	681,000	174,000		1,510,000	800,000

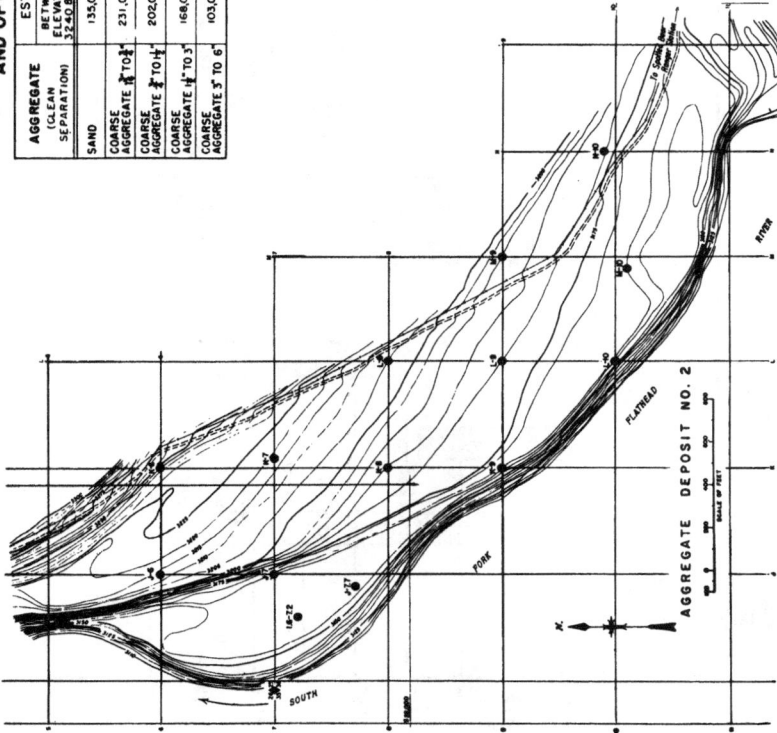

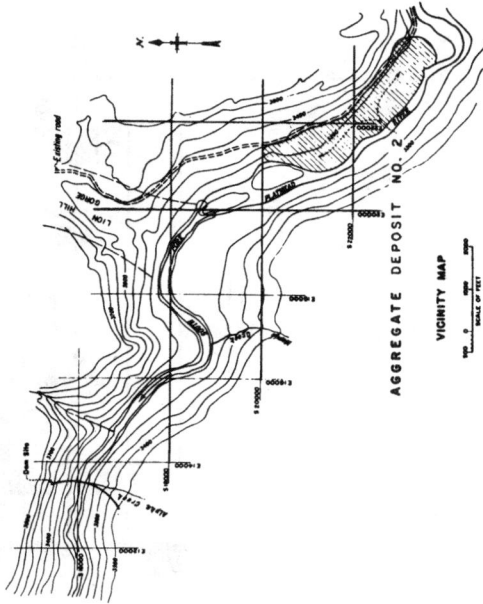

Figure 51.—A plan of an aggregate deposit, estimated quantities, and a vicinity map as prepared for the specifications. 288-D-3277.

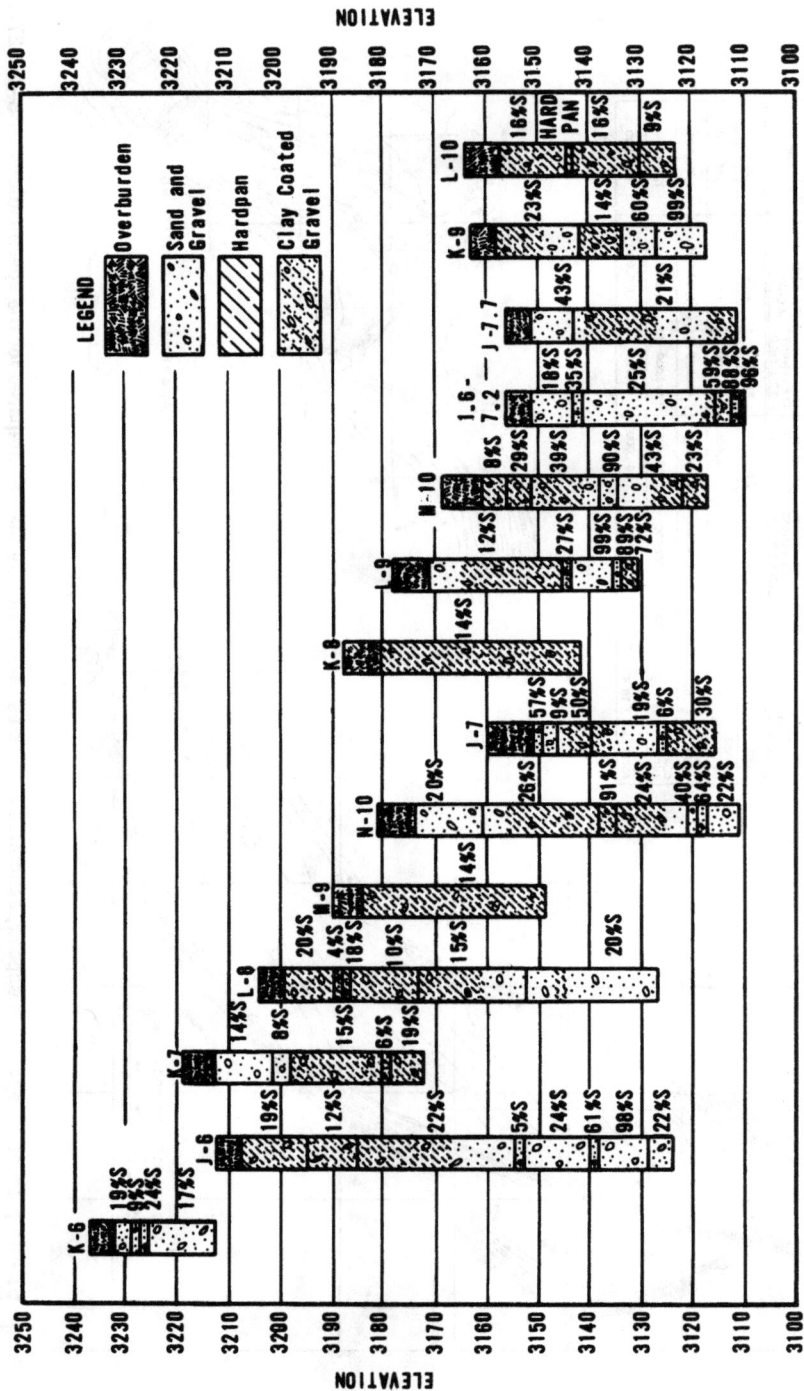

Figure 52.—A log of test pits as included in the specifications. 288–D–1532.

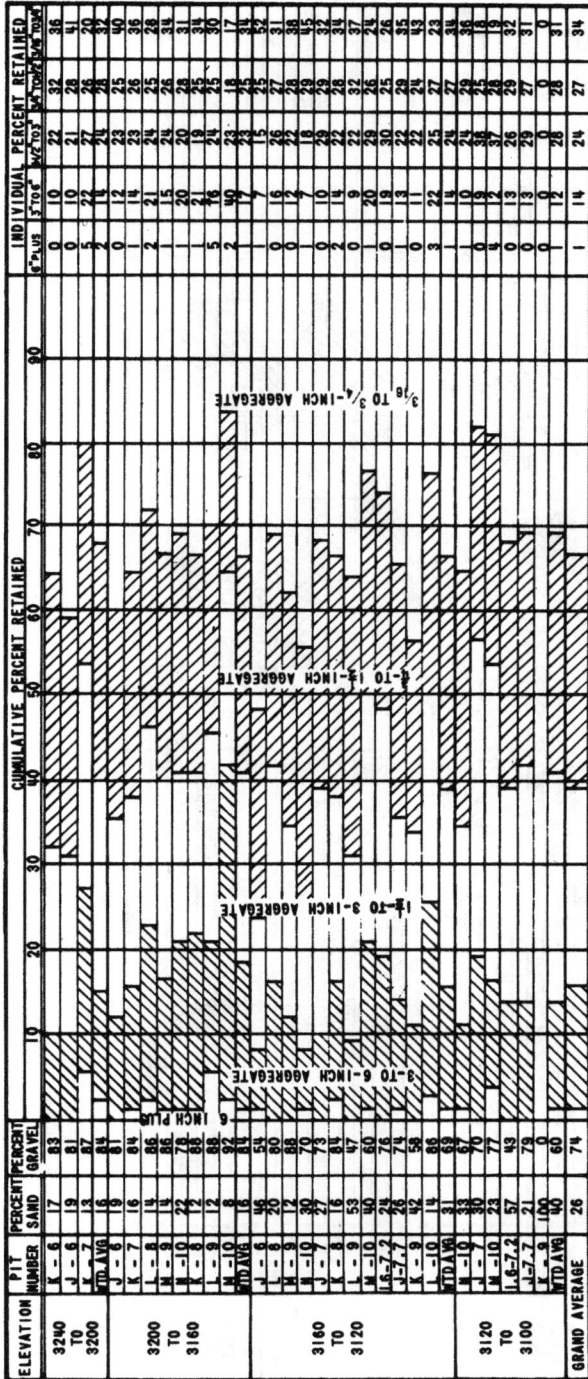

Figure 53.—Aggregate depths and grading data as indicated by test pits. These are included in the specifications to aid the bidders and the contractor. 288–D–1533.

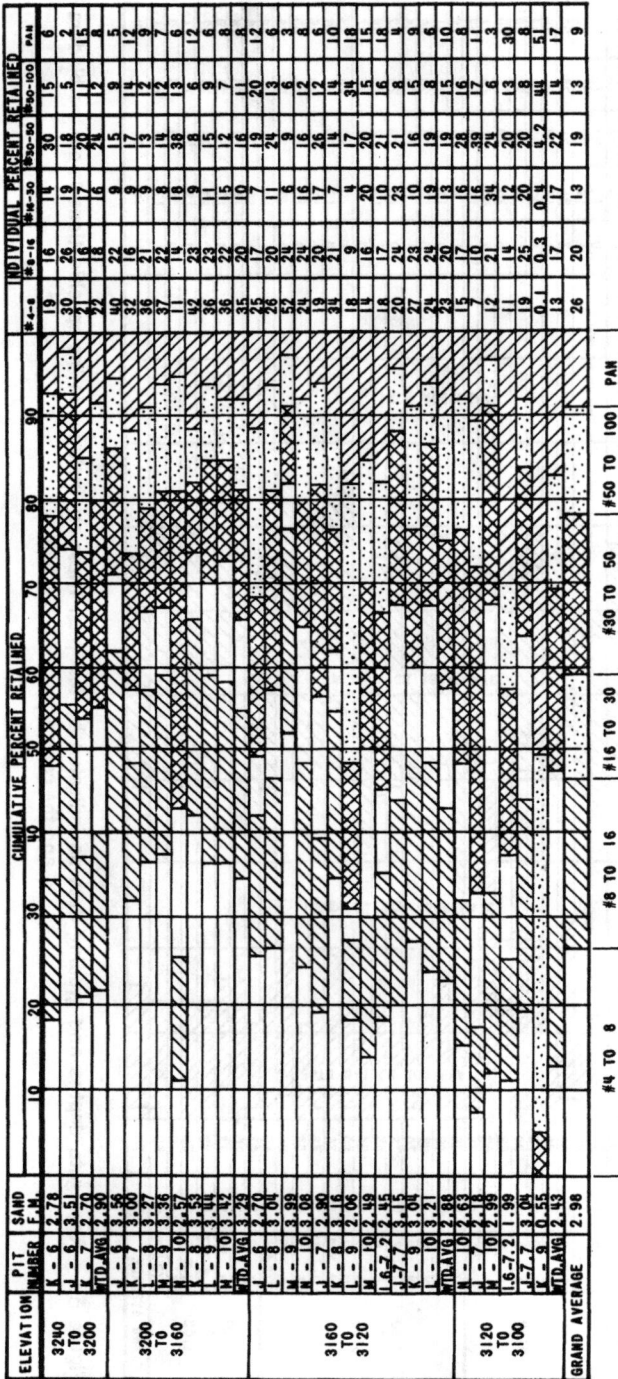

Figure 54.—Screen analyses of test-pit material. These are graphically shown in the specifications as a further aid to bidders and the contractor. 288–D–1534.

ities for testing metals and fabricating special instruments and apparatus for solution of special problems that constantly arise in concrete design and contruction.

D. Denver Tests and Selection of Aggregates

33. Tests of Aggregates.—The procedure followed in handling concrete aggregate samples shipped to Denver and the tests used to establish quality are shown graphically in figure 55. Physical properties are determined and samples are analyzed petrographically. Aggregates are tested for soundness by the sodium sulfate soundness test, for toughness and abrasion resistance in the Los Angeles abrasion machine, and for potential reactivity with alkalies in cement by the mortar bar test. Durability of concrete made with aggregates is determined by freezing and thawing tests. Tests performed under the heading of "Physical Properties" include specific gravity and absorption tests, colorimetric test for organic impurities in sand, determination of percentage of material that will pass the No. 200 screen, and gradings of sand and coarse aggregate.

Specific gravity and absorption tests are made as routine investigations, principally because of their importance in concrete mix design. Aggregates of higher specific gravity are usually more satisfactory with respect to soundness and strength. A low specific gravity does not necessarily mean rejection of an aggregate but serves as a warning that additional tests are

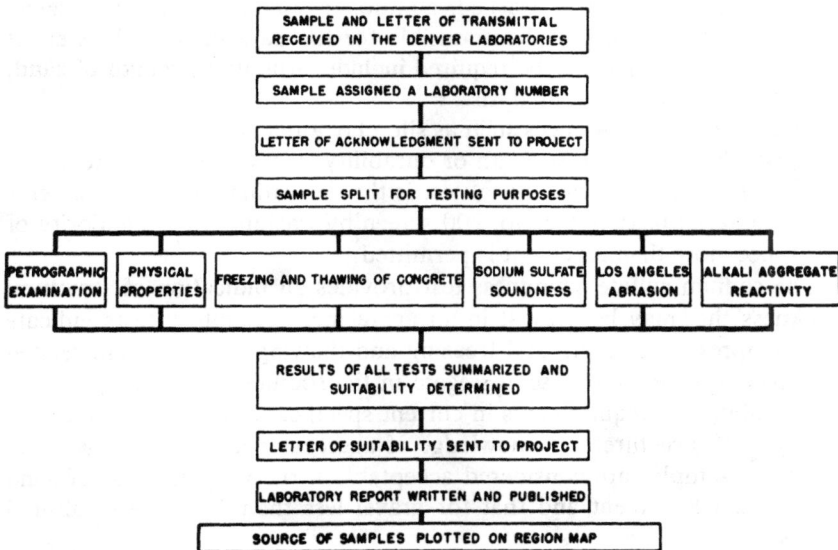

Figure 55.—Flow chart for concrete aggregate suitability determination.
288–D–3278.

required before the aggregate can be considered acceptable. Unit weight of concrete is dependent, to a large extent, on the specific gravity of the aggregate. Specific gravity of sand and gravel is frequently limited by specification to a minimum value of 2.60. On some projects (Glen Canyon and Flaming Gorge Dams, for example) special processing to remove light-weight particles has been required.

The absorption test determines the amount of water that an aggregate will absorb when submerged during a predetermined period of time, usu-ally 30 minutes for mix design purposes and 24 hours for aggregate eval-uation. Since the water-cement ratio of a concrete is based on use of saturated surface-dry aggregate, it is necessary to determine the absorption value of the aggregate. An absorption value much over 1 percent indicates that the aggregate may be of poor quality but does not necessarily mean that it should be rejected. For example, lightweight aggregates usually have high absorption yet are used successfully for structural concrete. Maximum limits on absorption are not specified by the Bureau, as limitations on other physical properties will generally serve for rejection of aggregates having unusually high absorption characteristics. Natural aggregates of high absorption have occasionally been used. One example occurred on the Southern Nevada water project, yet concrete of adequate quality was obtained.

The colorimetric test on sand is useful to indicate the presence of harm-ful amounts of organic material. If a color darker than standard is obtained with a sand that has been washed, additional tests are required to deter-mine the nature of material responsible for the dark color and its effect on mortar. Tests that may be required include structural strength of sand, time of set, and chemical analysis.

Contaminating substances such as silt, clay, organic matter, and soluble salts, which may reduce strength or durability, can very often be removed by washing. The simplest control test is the determination of the percent-age of material passing the No. 200 screen by washing. In the majority of cases, not more than 3 percent is permitted.

The sodium sulfate soundness test provides an indication of structural weakness that may be present in an aggregate. Laboratory tests indicate that compressive strength and freezing and thawing durability are related to percentage loss of coarse aggregate in the sodium sulfate test. The so-dium sulfate test requirements in current specifications vary with location and type of structure and knowledge of aggregates available. In general, aggregate samples are considered acceptable if the weighted loss of sand is less than 8 percent and that for gravel less than 10 percent, after 5 cycles.

The Los Angeles abrasion test provides valuable information regarding hardness and toughness of an aggregate and gives an indication of the breakdown expected to occur with a given material during stockpiling, handling, and transporting. There is a definite relationship between the strength of concrete and the quality of coarse aggregate as measured by the Los Angeles abrasion test. General practice dictates that coarse aggregate should lose not more than 10 percent after 100 revolutions nor more than 40 percent after 500 revolutions.

Petrographic examination aids the interpretation of physical and chemical tests of aggregates and may disclose weaknesses not discovered by standard physical tests. Aggregate is visually examined and identified according to mineralogical and chemical differences. The extent to which the particles are coated, the nature of the coating substance, and particle shape are determined. Potential deleterious reactivity of aggregates with cement alkalies may be detected, and, when sufficient time is available, the reactivity of the aggregates may be checked by a chemical test devised in the Denver laboratories.

Freezing and thawing durability tests are made when it is deemed necessary to obtain additional information to determine the quality of a certain aggregate. Strength-producing characteristics are evaluated by comparing the compressive strength with the average compressive strength of more than 400 concretes (using different aggregates) made for previous freezing and thawing durability tests. Also, the resistance of concrete containing the test aggregate to cycles of freezing and thawing is evaluated. Concrete specimens are made with standard laboratory blended cement, standard aggregate grading, 0.51 ± 0.01 water-cement ratio, 4 to 6 percent entrained air, and 2 to 4 inches slump. The specimens are cured 28 days in the fogroom and are then subjected to alternate cycles of freezing and thawing until 25 percent of the original weight is lost or until 1,000 cycles of freezing and thawing are obtained. Generally, if the concrete withstands 500 cycles of freezing and thawing without failure, the aggregate is considered satisfactory.

Following petrographic examination, if the mortar bar test for potential alkali-aggregate reactivity is recommended, concrete aggregates are tested in 1- by 1- by 11¼-inch mortar bars made with Bureau standard high- and low-alkali cements. To establish the existence of the pessimum condition in which smaller percentages of reactive aggregates may produce increased expansion, the aggregates, including both sand and coarse aggregate crushed to sand sizes, are tested in amounts of 25, 50, and 100 percent with high-alkali cements only. Neutral quartz also crushed to sand sizes in amounts of 75, 50, and 0 percent, respectively, constitutes the remainder of the aggregate. If any of these combinations results in expansions equal to or greater than those described immediately hereafter, the aggregate is considered to be reactive and the recommended precautions should be taken in its use. The low-alkali cement is used in some bars in lieu of high-alkali cement to establish whether expansions other than those caused by alkali-aggregate reactions will occur. Reactivity of the aggregate can thus be definitely established, and effectiveness of using low-alkali cement can be determined.

Results of the mortar bar tests have been correlated with rate and magnitude of deterioration caused by alkali-aggregate reaction in field structures. It has been established that any combination of aggregate and standard high-alkali cement causing a mortar expansion in excess of 0.20 percent linearly in 1 year will produce readily recognizable expansive deterioration of concrete through alkali-aggregate reaction. Aggregates that cause an expansion of less than 0.10 percent in 1 year when used with standard high-alkali cement have been found to be innocuous so far as alkali-aggregate reactivity is concerned. Aggregates causing an expansion in the range of 0.10 and 0.20 within 1 year when used with high-alkali cement do not show clear-cut evidence of this reaction; but several aggregates in this category are associated with concrete deterioration, the cause of which is conjectural. Consequently, these aggregates should be used only with low-alkali cement or a suitable combination of portland cement and pozzolan.

34. Analysis of Field and Laboratory Data.—Field reports on aggregate investigations are reviewed, and the grading and other data submitted are reduced to average values for the full usable depth of each test pit. When exploration is sufficiently extensive, weighted average data are compiled for the whole area of the deposit or for the probable working area. Results of the Denver laboratory tests are tabulated and arranged for easy comparison and consideration with the field data.

After analysis of the compiled field and laboratory data relating to different aggregate deposits under consideration, usually one deposit is tentatively selected and additional samples are tested in Denver to determine the properties of concrete made with that material. The data are also used in the preparation of specification drawings and tabulations similar to those shown in figures 51, 52, 53, and 54.

35. Quantity of Aggregate.—An important element in the analysis of data relative to aggregate supply is the quantity of aggregate available in relation to quantity required. As indicated in section 26, very rough methods of estimating may be employed during prospecting, but after one or more selected areas have been explored, reasonably accurate estimates must be made of the different sizes of aggregate required and available. The quantity of aggregate required is determined from the type and volume of concrete in the projected work. The quantity of suitable material in a deposit is computed from test-pit data, screen analyses, and areas represented by test pits.

The following example illustrates a simple method of estimating with sufficient accuracy the quantities of aggregate required.

Suppose the job calls for concrete made with 1½-inch-maximum size aggregate containing 34 percent sand, a water content of 245 pounds per cubic yard, a cement content of 490 pounds per cubic yard, and an air content of 4½ percent. The average unit weight of such concrete is about 146 pounds per cubic foot (see table 4) or $27 \times 146 = 3,942$ pounds per cubic yard. The net weights of sand and gravel required per cubic yard of concrete are:

Total aggregate, $3,942 - (245 + 490) = 3,207$ lb. per cu. yd.

Sand, $0.34 \times 3,207 = 1,090$ lb., or 0.545 ton

Gravel, $3,207 - 1,090 = 2,117$ lb., or 1.058 tons.

Coarse aggregate may be further divided in accordance with the size-separation proportions selected or those the supply is expected to provide. By multiplying the derived weights by the number of cubic yards of concrete and making allowance for waste and overrun, the total quantities of the several sizes of aggregate required for the work are obtained.

For reasons stated in section 78, aggregate quantities are much better expressed in tons than in cubic yards. Moreover, transportation costs are based on tonnage. However, as there are times when it is necessary to convert from weight to volume, or vice versa, the approximate data in table 12 are given.

Table 12.—Weight of aggregate in tons per cubic yard

Kind of aggregate	Compacted	Loose
Sand, dry	1.40 to 1.55...	1.30 to 1.45
Sand, moist	1.20 to 1.45...	1.05 to 1.35
Coarse aggregate (separated):		
³⁄₁₆ to ¾ inch	1.35 to 1.45...	1.25 to 1.35
¾ to 1½ inches	1.30 to 1.40...	1.25 to 1.35
1½ to 3 inches	1.25 to 1.40...	1.20 to 1.35
3 to 6 inches	1.20 to 1.35...	1.15 to 1.30
Coarse aggregate (combined):		
³⁄₁₆ to 1½ inches	1.35 to 1.55...	1.30 to 1.45
³⁄₁₆ to 3 inches	1.40 to 1.60...	1.30 to 1.55
³⁄₁₆ to 6 inches	1.45 to 1.70...	1.35 to 1.60
Sand and gravel combined, dry.	1.60 to 1.85...	1.50 to 1.75

In estimating aggregate requirements for jobs where mixes of different maximum sizes of aggregate are to be used, the computations are similar to the foregoing but more extended.

36. The Selected Aggregate.—Several factors must be considered in making final selection of aggregate when more than one source is available. The relative quality of material in the several sources is the most important consideration and should receive greatest weight in making a choice. Records of previous use of aggregate from a particular source and examinations of concrete made with such aggregate provide valuable information concerning its quality. Such indications should be evaluated along with the characteristics of aggregate which affect the concrete as discussed in chapter I.

Economy will dictate a choice of aggregate sources if quality fails to differentiate materially among them. The study will include appraisal of location and amount of processing that each source may require. The aggregate that can be delivered to the mixing plant at lowest cost may not be the most economical one. It may require a cement content that exceeds that of another aggregate from a more costly source. Also, very often the cost of some processing, such as correction of grading, may be fully recovered when the processing accomplishes a reduction in cement content. In general, the aggregate that will produce the desired quality with least overall expense should be selected.

E. Prospecting for Pozzolanic Materials

37. Geologic Occurrences of Pozzolan.—Natural pozzolanic materials originate as volcanic tuffs and ashes or as shales and clays, all of which accumulate in stratified deposits. The formations may be thick or thin; they may be highly variable in character and composition, or they may be uniform over large areas. For example, the Monterey formation is widely distributed and is fairly uniform in general lithological characteristics over a distance of about 350 miles through western California. However, within itself the Monterey formation is composed of highly lenticular strata of siliceous shale, cherty shale, clay shale, sandstone, siltstone, and diatomaceous earth, any of which may or may not be present in a particular section and which typically vary considerably in composition and petrographic character from place to place. Fly ash is an artificial pozzolan obtained from the stacks of powerplants in which pulverized coal is used for fuel. Because of possible variability, prospective sources of pozzolan must be subjected to thorough investigation, sampling, and testing to establish the extent and tonnage of usable materials.

A comprehensive survey of natural pozzolan occurrence within the United States was conducted by the Bureau of Mines. The report of the investigation including qualitative tests is published in U.S. Bureau of Mines Circular 8421 "Pozzolanic Raw Material Resources in the Central and Western United States".

38. Samples and Information Required.—Testing pozzolan samples in the Denver laboratories is a part of the preliminary investigation of construction materials. Instructions pertaining to sampling and transmittal of pozzolan samples are similar to those for concrete aggregates described in sections 27 and 31. Samples should be approximately 50 pounds each. Samples of promising materials from undeveloped deposits may be obtained from any natural exposure or excavation or, if necessary, from test pits or trenches.

F. Denver Tests and Investigation of Pozzolanic Materials

39. Tests and Analyses of Pozzolanic Materials.—Samples of pozzolanic material submitted for preliminary investigation are subjected to petrographic analysis and chemical and physical tests. These determinations permit elimination of materials of inferior quality and those for which processing would entail excessive cost. Promising materials that appear to fulfill the requirements of the job and are economically available are subsequently tested exhaustively in concrete to establish quantitatively their effects on the concrete and to determine the mixes that will take full advantage of potential benefits and minimize any adverse qualities of the pozzolan.

Properties of pozzolanic materials that control quality include grindability, need for calcination or other treatment, specific gravity, fineness, water requirement, strength development with portland cement, effect on alkali-aggregate reaction, heat generation, etc. For example, grindability is reflected in processing costs. Specific gravity controls the weight-volume relationship between cement and pozzolan. Increased water requirement increases drying shrinkage and decreases freezing and thawing durability, but use of an air-entraining agent will diminish, to some degree, these effects.

G. Denver Investigations of Other Materials

40. Cement Investigations.—Consideration is given to the choice of type of cement for each project. For large concrete dams, cement having low heat of hydration is usually selected if low heat generation is not assured by use of pozzolan. Concrete investigations for these structures generally include complete tests under conditions which simulate those occurring within massive concrete by curing hermetically sealed test specimens under the temperature cycles that will apply to the dam concrete. These tests have been useful in defining the properties of special cements and aiding the selection of the proper type for use with materials to be used in construction.

Selection of cement and the solution of problems incident to temperatures encountered in large dams also involve determination of thermal properties of the concrete. As the flow of heat in different concretes varies considerably, and the thermal characteristics of concrete are influenced markedly by the nature of the aggregate, tests for determining specific heat and thermal conductivity are made after the aggregate and approximate mix proportions have been chosen.

There are occasions when difficulty from false set or other abnormal behavior of cement is encountered in construction. These experiences should be reported in detail, and a bag of the cement should be furnished for analysis and study. Data on the particular shipment involved, the mill bin number, and age of the cement should be included in the field report.

41. Investigations of Admixtures and Curing and Bonding Compounds.—Admixtures of different kinds are tested on occasion, as circumstances warrant. In view of the significant benefits contributed to concrete by entrainment of a small percentage of air, investigations for evaluation of air-entraining agents have been extensive. These include comprehensive study of the properties of air-entrained concrete.

Admixtures, such as water-reducing agents and set-controlling agents, are specified when sufficient quantities of concrete are involved to make their use economical. Presently, they are usually specified where more than 2,000 cubic yards of concrete will be placed. These agents do not react the same with cements from different sources nor always the same with cement from the same source. It is usually advisable to test the proposed agent in combination with the cement and aggregate being used before specific recommendations are made. Tests to determine suitability of these agents are performed in conformance with designation 40 in the appendix. When a contractor selects the agent and cement source, specifications require that samples be submitted to the Chief, Division of General Research, attention Code 1510, Building 56, Bureau of Reclamation, Denver Federal Center, Denver, Colorado 80225.

Chlorinated rubber base, clear resin base, acrylic-modified chlorinated rubber base solutions, and pigmented resin base curing and bonding compounds for curing concrete are being tested for compliance with specification requirements, for collecting additional information on behavior under various methods and conditions of use, and for exploring possibilities for development of more versatile curing compounds.

42. Sampling and Analysis of Water and Soil.—Steps should be taken in the field to ascertain whether the water in the area meets specification requirements applying to use in mixing and curing of concrete and in aggregate washing operations. Similar studies should be made to determine whether the soil or water with which the concrete will be in contact contains harmful sulfate concentrations. These requirements are based on considerations relating to durability discussed in sections 6 and 19.

If the available water is of questionable quality for the intended use, a sample should be sent to the Denver laboratories, together with a complete sampling report, for determination of quality. If the surface water, ground water, or soil is known or suspected to contain alkalies or soluble sulfates, samples of the soil or ground water should be submitted in accordance with designation 3 of the appendix to determine the protective measures to be employed. This should be done in the early phases of investigation to assure sufficient time for testing. These samples should be submitted in such a manner that final concentrations of salts (usually sulfates) to which the structure may be exposed can be determined. Sulfate salts will concentrate in low-drainage regions and may cause rapid disintegration of thin-wall concrete structures. Particular attention should be given to the soil where structures such as canal linings, retaining walls, and concrete pipe are constructed. Procedures for sampling water and soils are given in the appendix, designation 3.

Chapter III

CONCRETE MIXES

43. Introduction.—Proportions of ingredients for concrete should be selected to make the most economical use of available materials that will produce concrete of the required placeability, durability, and strength. Established basic relationships and laboratory tests provide guides for optimum combinations. However, final proportions should be established by actual trial and adjustment in the field.

Concrete is composed essentially of water, cement, and aggregate. In some cases an admixture is added—most often to entrain air, but sometimes for other reasons. Water-reducing, set-controlling admixtures (WRA) are used extensively in Bureau of Reclamation concrete. Types of aggregate and cement have a marked effect on strength and durability and on the amount of mixing water required. When sources of ingredients remain the same, the quantity of cement, grading and maximum size of aggregates, and consistency (or slump) of concrete can be varied without materially affecting strength, provided the quality of the cement paste, as determined by the water-cement ratio, is maintained constant.

When sources of ingredients vary, as in the case of aggregates from different deposits and cements from different mills, concrete strength and durability may differ appreciably even though the water-cement ratio is held constant. Therefore, laboratory tests are desirable prior to major construction on the project to determine the properties of the hardened concrete. The tests are usually performed in the Denver laboratories prior to establishment of a field control laboratory on the project. For small jobs, where Denver laboratory tests of the concrete are not practicable, a reasonably good combination of ingredients can be developed from a knowledge of the characteristics of aggregates and application of established empirical relationships. However, regardless of the procedure followed in selection of the initial proportions, the mix will usually require adjustment in the field.

Laboratory data used in estimating concrete mix proportions from established relationships include screen analysis, specific gravity (saturated surface-dry basis), absorption of both fine and coarse aggregate, and dry-rodded unit weight of the coarse aggregate. The actual specific gravity of the cement also should be known; but if it is not known and cannot be readily determined, the value of 3.15 for specific gravity may be used with assurance of sufficient accuracy in mix design calculations. In addition, the moisture content of the aggregate must be known to compute batch weights for field use.

As discussed in section 14, purposeful entrainment of air greatly improves workability and resistance of concrete to weathering. Entrainment will sometimes reduce the strength of concrete; but if the cement content is not changed and advantage is taken of lower water requirements and thus the lower water-cement ratio, the reduction in compressive strength is not great and becomes apparent only in the range of richer mixes. For lean mixes, strengths are generally increased by entrainment in proper amounts.

44. Selection of Proportions.—Mix proportions should be selected to produce concrete with:

(1) The stiffest consistency (lowest slump) that can be placed and consolidated efficiently with vibration to provide a homogeneous mass.

(2) The maximum size of aggregate economically available and consistent with job requirements.

(3) Adequate durability to withstand satisfactorily the weather and other destructive influences to which it may be exposed.

(4) Sufficient strength to withstand the loads to be imposed.

45. Estimate of Water Requirements.—For best strength, durability, and other desirable properties, concrete should be placed with the minimum quantity of mixing water consistent with proper handling.

The quantity of water per unit volume of concrete required to produce a mix of desired consistency is influenced by the maximum size, particle shape, and grading of the aggregate and by the amount of entrained air.

Within the normal range of mixes, the water requirement is relatively unaffected by the quantity of cement. Table 13 gives recommended limitations for slump. Overwet concrete should always be avoided as it is difficult to place without segregation and is almost certain to be weak and lacking in durability.

The quantities of water given in table 14 are of sufficient accuracy for preliminary estimates of proportions; they are the averages that should be expected for various maximum sizes of fairly well-shaped aggregates graded within limits of usual Bureau specifications. If aggregates otherwise suitable have higher water requirements than those given in table 14, they probably have less favorable shape and grading than may

Table 13.—Recommended maximum slumps for various types of concrete construction

Type of construction	Maximum slump in inches [1]
Heavy mass construction	2
Canal lining thickness of 3 or more inches [2]	3
Slabs and tunnel inverts	2
Tops of walls, piers, parapets, and curbs	2
Sidewalls and arch in tunnel lining	4
Other structures	3

[1] These maximum slumps are for concrete after it has been deposited, but before it has been consolidated, and are for mixes having air contents as indicated in table 14.
[2] On machine-placed canal and lateral lining, less than 3 inches thick, the slump should be increased to 3½ inches.

Table 14.—Approximate air and water contents per cubic yard of concrete and the proportions of fine and coarse aggregate

(For concrete containing natural sand with an F.M. of 2.75 and average coarse aggregate, and having a slump of 3 to 4 inches at the mixer)

Max. size of coarse aggregate, inches	Recommended air content, percent	* Sand, percent of total aggregate by solid volume	Percent dry-rodded unit weight of coarse aggregate per unit volume of concrete	Air entrained concrete average water content, lb/yd [3]	Air entrained concrete with WRA – average water content, lb/yd [3]
⅜	8	60	41	320	300
½	7	50	52	305	285
¾	6	42	62	280	265
1	5	37	67	265	250
1½	4.5	34	73	245	230
2	4	30	76	230	215
3	3.5	28	81	205	190
6	3	24	87	165	155

* When WRA is used in concrete the sand content should be increased 1 or 2 percent to allow for the loss in mortar volume due to water reduction.

ADJUSTMENT OF VALUES FOR OTHER CONDITIONS

Changes in materials or proportions	Effect on other proportions		
	Water content percent	Percent sand	Percent of dry-rodded coarse aggregate
Each 0.1 increase or decrease in F.M. of sand	—	±0.5	∓1
Each 1-inch increase or decrease in slump	±3	—	—
Each 1-percent increase or decrease in air content	∓3	∓0.5 to 1.0	—
Each 0.05-increase or decrease in water-cement ratio	—	±1	—
Each 1-percent increase or decrease in sand content	±1	—	∓2

If aggregates are proportioned by percent sand method, use the first and second columns; if by dry-rodded coarse aggregate method, use the first and third columns.

normally be expected. Proportions of such aggregates should be adjusted to maintain the desired workability.

Some materials may require less water than indicated in table 14. Accordingly, laboratory tests for strength and durability may suggest appropriate adjustments in cement content. However, a rounded gravel and an angular coarse aggregate, both similarly graded and of good quality, usually will produce concrete of about the same compressive strength for the same cement content, despite differences in water-cement ratio. Also, for the same proportions, different cement may produce concretes having strengths that differ appreciably.

The weight of water is usually assumed as 62.3 pounds per cubic foot. For water at temperatures below 60° F, the weight is taken as 62.4 pounds per cubic foot.

46. Estimate of Cement Requirements.—Concrete quality is measured in terms of workability, durability, and strength. For a given water requirement, quality of concrete is proportional to cement content. While durability and strength of concrete are influenced by many variables, proportions should be selected to provide cement pastes of adequate quality to withstand expected exposures and ensure adequate strength. Suitable control of other factors will then ensure strong, durable concrete. The estimated cement requirement for a given concrete is, therefore, computed from the water requirement and the water-cement ratio. Table 15 will serve as a guide in selecting maximum permissible water-cement ratios for different severities of exposure when proper use is made of air entrainment.

The maximum water-cement ratio or minimum cement content to produce the required strength will usually be determined by laboratory tests, in which are used only those materials that will be used in the project. Table 16 shows an approximation of the minimum average strengths to be expected for both air-entrained concrete and concrete containing WRA for different water-cement ratios. This table, which lists some values depicted in figure 19 of section 14, is conservative and can be used in estimating the strength of concrete until verified by laboratory tests.

The cement content is calculated using the maximum permissible water-cement ratio selected from table 15 or table 16 (whichever value is lower) and the water requirement from table 14. The calculation is made by dividing the water content by the water-cement ratio. If a minimum cement content is specified, the corresponding water-cement ratio for estimating strength can be computed by dividing water content by cement content.

47. Estimate of Aggregate Requirements.—To be expected, the minimum amount of mixing water and the maximum strength will result for

Table 15.—Net water-cement ratios for concrete

Type or location of concrete or structure and degree of exposure	Water cement ratio by weight	
	Severe climate, wide range of temperature, long periods of freezing, or frequent freezing and thawing	Mild climate, rainy or arid, rarely snow or frost
A. Concrete in portions of structures subject to exposure of extreme severity, such as the top 2 feet of walls, boxes, piers, and parapets; all of curbs, sills, ledges, copings, corners, and cornices; and concrete in the range of fluctuating water levels or spray. These are parts of dams, spillways, wasteways, blowoff boxes, tunnel inlets and outlets, tailrace walls, valve houses, canal structures, and other concrete work.	0.45±0.02..	0.55±0.02
B. Concrete in exposed structures and parts of structures where exposure is less severe than in A, such as portions of tunnel linings and siphons subject to freezing, the exterior of mass concrete, and the other exposed parts of structures not covered by A.	0.50±0.02..	0.55±0.02
C. Concrete in structures or parts of structures to be covered with backfill, or to be continually submerged or otherwise protected from the weather, such as cutoff walls, foundations, and parts of substructures, dams, trashracks, gate chambers, outlet works, and control houses. (If severe exposure during construction appears likely to last several seasons, reduce W/C for parts most exposed by 0.05.)	0.58±0.02..	0.58±0.02
D. Concrete that will be subject to attack by sulfate alkalies in soil and ground waters and will be placed during moderate weather.	....	0.50±0.02
E. Concrete that will be subject to attack by sulfate alkalies in soil and ground water but will be placed during freezing weather when calcium chloride would normally be used in mix. Do not employ $CaCl_2$, but decrease W/C to the value shown.	0.45±0.02..	....
F. Concrete deposited by tremie or pump in water	0.45±0.02..	0.45±0.02
G. Canal lining	0.53±0.02..	0.58±0.02
H. Concrete for the interior of dams	The properties of this concrete will be governed by the strength, thermal properties, and volume change requirements that will be established for each structure.	

Table 16.—Probable minimum average compressive strength of concrete for various water-cement ratios, pounds per square inch

Water-cement ratio by weight	Compressive strength at 28 days	
	Air-entrained concrete	Air-entrained concrete with WRA
0.40	5,700	6,500
0.45	4,900	5,600
0.50	4,200	4,800
0.55	3,600	4,200
0.60	3,100	3,600
0.65	2,600	3,100
0.70	2,200	2,700

given aggregates and cement when the largest quantity of coarse aggregate is used consistent with job requirements. The quantity of coarse aggregate that can be used increases with the maximum size of aggregate. This quantity can be determined most effectively from laboratory investigations of the materials. However, a good estimate of the best proportions can be made from established empirical relationships shown in table 14.

Table 17 gives recommended limitations for maximum size of aggregate. Within the limits of economy and consistent with job requirements, the largest possible aggregate size should be used because this permits a reduction in water and cement requirements, as discussed in section 18. When all or most of the concrete must be placed through a curtain of steel, it is necessary that the maximum size generally not exceed two-thirds of the minimum clear distance between reinforcement bars. However, if there is space to place concrete between or in front of the steel curtains and the concrete is effectively vibrated, experience has shown that mixes do not need to contain maximum size aggregate smaller than bar spacing or form clearance to ensure good filling consolidation. That

Table 17.—Maximum sizes of aggregate recommended for various types of construction

Minimum dimension of section, inches	Maximum size of aggregate,[1] in inches, for—		
	Reinforced walls, beams, and columns	Heavily reinforced slabs	Lightly reinforced or unreinforced slabs
5 or less		¾ to 1½	¾ to 1½
6 to 11	¾ to 1½	1½	1½ to 3
12 to 29	1½ to 3	3	3 to 6
30 or more	1½ to 3	3	6

[1] Based on square screen openings.

portion of the mix that is molded by vibration around bars, and between bars and forms, is not inferior to that which would have filled those parts had a smaller maximum size aggregate been used. The remainder of the mix in the interior of the structure, as long as it is properly consolidated, is superior because of its reduced mortar and water content.

Concrete of comparable workability can be expected with aggregates of comparable size and grading, provided the volume of mortar and air content remains constant. The solid volume of cement, water, and sand may be interchanged to maintain a constant mortar content. The optimum quantity of coarse aggregate depends to some extent on the grading of the sand. This relationship is reflected in the adjustment for fineness modulus of sand, as indicated under "Adjustment of values for other conditions" in table 14.

The percentage of sand in the concrete mix has been used extensively as a means of identifying the proportions of sand and coarse aggregate. Recommended percentages of sand for each maximum size of coarse aggregate are listed in table 14. In the following section it is demonstrated that aggregates can be proportioned by estimating the quantity of coarse aggregate or by computing the total solid volume of sand and coarse aggregate in the concrete mix and multiplying by the recommended percentage of sand. Either method is satisfactory and will produce approximately the same proportions under average conditions.

48. Computations of Proportions.—Computing proportions for concrete mixes can best be explained by means of specific examples. Calculations are based on saturated surface-dry aggregates. For these examples, the following materials will be used:

(1) Type II cement with a specific gravity of 3.15; suitable pozzolan (as selected for example 2) with a specific gravity of 2.50.

(2) Coarse aggregate with a specific gravity of 2.68.

(3) Sand with a specific gravity of 2.63 and a fineness modulus of 2.75.

(4) Dry-rodded unit weight of coarse aggregate of 105 pounds per cubic foot.

(5) Sufficient air-entraining agent to entrain the air contents shown in table 14.

(6) A lignin-type water-reducing set-controlling admixture (example 1) and a hydroxylated-carboxylic acid type (example 2).

(a) *Example 1.*—The first example involves a reinforced retaining wall, having a minimum thickness of 11 inches. Tables 13 and 17 indicate that a 3-inch slump (under "Other structures") and 1½-inch-maximum size aggregate will be satisfactory. The concrete will be subjected to rather severe climatic exposure and will therefore fall in class

Computation of trial mix (example 1):

Mix ingredients	Weight, pounds per cubic yard	Conversion of weight to volume	Solid volume, cubic feet per cubic yard
Water: Estimated value from table 14	245	$\dfrac{245}{62.3}$	3.93
Cement: W/C for durability, class B concrete (table 15) = 0.50			
W/C for strength (table 16) = 0.56			
Durability controls, use 0.50			
Cement = $\dfrac{\text{water}}{\text{W/C}}$ =	490	$\dfrac{490}{3.15 \times 62.3}$	2.50
Air: From table 14, 4.5 percent; 0.045 x 27 =	....	....	1.21
Sand: From table 14, 34 percent of volume of total aggregate; 27.00 − (3.93 + 2.50 + 1.21) = 19.36 cubic feet of total aggregate			
Volume of sand = 19.36 x 0.34 = 6.58 ft.³ Weight of sand = 6.58 x 62.3 x 2.63 =	1,078	....	6.58
Coarse aggregate: Volume of coarse aggregate = 27.00 − (3.93 + 2.50 + 1.21 + 6.58) = 12.78 ft.³ Weight = 12.78 x 62.3 x 2.68 =	2,134	....	12.78
Total	3,947	....	27.00

B of table 15. The wall has been designed on the basis of 80 percent of the standard 6- by 12-inch cylinders having compressive strength greater than 3,000 pounds per square inch in 28 days. Accordingly, average Bureau control, which is considered as having a coefficient of variation of 15 percent, requires an average strength of 3,460 pounds per square inch at 28 days, as discussed in section 62.

When the aggregate is proportioned by the dry-rodded unit weight of coarse aggregate method, the amount of coarse aggregate is calculated as follows: From table 14, the amount of dry-rodded unit weight of coarse aggregate per unit volume of concrete is 73 percent and the weight of coarse aggregate per cubic foot of concrete is 0.73 × 105 = 76.6 pounds. The weight of coarse aggregate per cubic yard is 76.6 × 27 = 2,068 pounds. The volume of coarse aggregate and sand can then be calculated as shown above.

Suppose a ligno-sulfonate type of water-reducing, set-controlling admixture (WRA) is to be used. This type of admixture is most often in the form of a suspension of lignin solids in water and is batched much like an air-entraining admixture. The dosage is based on the amount of cement or cement plus pozzolan in the mix. In this example, assume the dosage desired is 0.25 percent lignin solids, by weight of cement. It contains 40 percent solids; that is, each milliliter of the solution contains 0.40 gram of lignin solids. If WRA is to be used, the initial estimate of the water requirement would have been 230 pounds per cubic yard rather than 245 pounds (table 14), and the weight of cement would have been 460 pounds. The volume of WRA solution per cubic yard of concrete is calculated as follows:

$$\text{Milliliters of WRA} = \frac{460 \times 0.0025 \times 453.6}{0.40 \text{ g/ml}} = 1,304 \text{ ml/yd}^3.$$

(b) *Example 2.*—The second example requires a concrete mix for a powerhouse foundation which will not be exposed to freezing and thawing. This condition will permit the use of class C concrete, from table 15. The designers have specified a design strength of 2,500 pounds per square inch at 28 days; therefore, an average strength of 2,880 pounds per square inch will be required for 80 percent of the tests to fall above 2,500. Tables 13 and 17 indicate that a 3-inch slump (again under "Other structures") and 3-inch-maximum size aggregate will be satisfactory. A suitable pozzolan is available at a cost lower than cement, and laboratory investigations have indicated some danger of alkali-aggregate reaction. The cementing materials will, therefore, be composed of 30 percent pozzolan and 70 percent portland cement.

Assume a hydroxylated carboxylic acid WRA is to be used in example 2 at a dosage of 0.30 percent, by weight of cement plus pozzolan. As this

Computation of trial mixes (example 2):

Mix ingredients	Weight, pounds per cubic yard	Conversion of weight to volume	Solid volume, cubic feet per cubic yard
Water: Estimated value from table 14	205	$\dfrac{205}{62.3}$	3.29
Cement and pozzolan: W/C for durability, class C concrete (table 15) = 0.58			
W/C for strength (table 16) = 0.62			
Durability controls, use 0.58			
Cementing materials: $\dfrac{\text{Water content}}{W/(C+P)} = \dfrac{205}{0.58} = 353$ pounds			
Portland cement = 353 x 0.70 =	247	$\dfrac{247}{3.15 \times 62.3}$	1.26
Pozzolan = 353 x 0.30 =	107	$\dfrac{107}{2.50 \times 62.3}$	0.69
Air: From table 14 = 3.5 percent 0.0035 x 27 =			0.95
All ingredients except aggregates	559		6.19
Aggregate: Volume = 27 − 6.19 =			20.81
Percent sand (table 14) = 28 percent			
Volume of sand = 0.28 x 20.81 =			5.83
Volume of coarse aggregate = 20.81 − 5.83 =			14.98
Weight of sand = 5.83 x 2.63 x 62.3 =	955		
Weight of coarse aggregate = 14.98 x 2.68 x 62.3 =	2,502		
Total	4,016		27.00

type of WRA is a liquid, rather than a suspension of solids, the calculation of amount per cubic yard is somewhat different than in example 1. As a WRA is being used, the initial estimate of the water requirement would have been 190 pounds per cubic yard rather than 205 pounds (table 14), and weight of cement plus pozzolan would then be 328 pounds. The specific gravity of this type is usually about 1.17 and the procedure is as follows:

$$\text{Milliliters of WRA} = \frac{328 \times 0.0030 \times 453.6}{1.17} = 381 \text{ ml/yd}^3.$$

49. Batch Weights for Field Use.—The preceding trial-mix computations provide batch quantities for a cubic yard of concrete. It is seldom possible to mix concrete in exactly 1-cubic-yard batches. Therefore, these quantities must be converted in proportion to the size of batch to be used. This conversion can be accomplished by multiplying the 1-cubic-yard quantity of each ingredient by the volume of the new batch in cubic yards. This volume can readily be computed from proportions of any one of the ingredients in the 1-cubic-yard batch and the new batch. For example, assume that a three-bag mixer is available and that the trial mix in example 1 is used. Volume of the new batch is three times the weight of a bag of cement, divided by the weight of cement for the 1-cubic-yard batch, or $\dfrac{3 \times 94}{490} = \dfrac{282}{490} = 0.575$ cubic yard. The field batch proportions will be:

Water	$0.575 \times 245 = 141$ pounds
Cement	$0.575 \times 490 = 282$ pounds (3 bags)
Sand	$0.575 \times 1,078 = 620$ pounds
Coarse aggregate	$0.575 \times 2,134 = 1,227$ pounds

Aggregates were assumed to be in a saturated surface-dry condition. Under field conditions they will generally be moist, and quantities to be batched must be adjusted accordingly. Assume that tests show the sand to contain 5 percent and the coarse aggregate 1 percent free moisture. As the quantity of saturated surface-dry sand required was 620 pounds, the amount of moist sand to be weighed is 651 pounds (620×1.05). Similarly, the weight of moist coarse aggregate is 1,239 pounds ($1,227 \times 1.01$). Coarse aggregate is sometimes drier than saturated surface-dry. Assuming an absorption of 1 percent, the amount of dry aggregate required would be $1,227 \times 0.99 = 1,215$ pounds.

Free water in the aggregate must be considered as part of the mixing water, whereas in the case of dry aggregate, water must be added to allow for absorption. In the example, free water (mixing water) in the sand is 31 pounds ($651 - 620$), and in the coarse aggregate is 12 pounds

(1,239 − 1,227). If the coarse aggregate were dry, 12 pounds of water (1,227 − 1,215) must be added to the mixing water to allow for absorption.

50. Adjustment of Trial Mix.—Because of the large number of variables introduced into a concrete mix through air entrainment, two or more trials may be necessary to establish definitely the water requirement and air-entraining agent requirement. Figure 56 illustrates a convenient form for recording trial mix data. The data presented conform to trial mixes of the preceding examples with an adjustment of the mix for example 1.

Assume the first trial batch has a 2-inch slump and 4 percent air instead of the desired 3-inch slump and 4.5 percent air. Table 14 indicates that to increase the slump the desired 1 inch an increase in water content of 3 percent is required. Similarly, to increase the air content by 0.5 percent an adjustment in water content must be made, but for this it must be decreased by 1.5 percent. As these two adjustments are occurring simultaneously, a net change of plus 1.5 percent in water content is the result.

To find the amount of agent required to produce 4.5 percent air, a linear adjustment is made. As 12 ounces of agent produced 4 percent air

7-1302
BUREAU OF RECLAMATION

CONCRETE MIX COMPUTATION SHEET

SERIES CM DATE CAST 6-30-71
SAND GRADING % PAN 5, #100 15, #50 25, #30 25, #16 15, #8 15 F.M. 2.75
C. AGG. GRADING % #4-3/8" 22, 3/8-3/4" 33, 3/4-1½" 45, 1½-3" , 3-6"
AIR METER NO. 2 FACTOR
UNIT WT. CAN NO. 2 FACTOR 8.03 lb(factor = 4.037)
DRY RODDED UNIT WEIGHT LB/CU FT 105.0

INGREDIENTS	NAME AND LAB NO	SP G	DENSITY LBS./CU FT	% ABS. % MOIST.
WATER	Tap	1.0	62.3	
CEMENT	Laboratory Blend	3.15	196.2	
POZZOLAN		2.50	155.8	
SAND	Laboratory Reference	2.63	163.8	5.0
C. AGG.	Laboratory Reference	2.68	167.0	1.0
AEA	NVX 12 % SOLIDS			

ADMIXTURE

TRIAL MIX COMPUTATIONS / BATCH COMPUTATIONS

MIX NO.	W/C+P BY WT	% SAND OR % GRAV	ESTIMATED QUANTITIES PER CUBIC YARD: AGENT		AIR	WATER	CEMENT	POZZ	SAND	C. AGG.	TOTAL	VOL OF BATCH IN CU YD	QUANTITIES PER BATCH: AGENT	WATER	CEMENT	POZZ	SAT. SUR. DRY: SAND	C. AGG.	TOTAL	ADJUSTMENT FOR MOISTURE: ADJUSTED WEIGHT: SAND	C. AGG.	MIXING WATER: IN SAND	IN C. AGG.	ADDED	TOTAL
Example			12.0	LB	4.5%	245	490	-	1,078	2,134	3,947		6.9		Batch 1										
1	0.50	34	oz.	CU FT	1.21	3.93	2.50	-	6.58	12.78	27.00	0.575		141	282	-	620	1227	2270	651	1239	31	12	98	141
			13.5	LB	4.5%	249	498	-	1,071	2,123	3,941		7.8		Batch 2										
1	0.50	34	oz.	CU FT	1.21	4.00	2.54	-	6.54	12.71	27.00	0.575		143	286	-	616	1221	2266	647	1233	31	12	100	143
				LB																					
				CU FT																					
Example			10.0	LB	3.5%	205	247	107	955	2,502	4,016														
2	0.58	28	oz.	CU FT	0.95	3.29	1.26	0.69	5.83	14.98	27.00														

RESULTS OF TESTS ON FRESH CONCRETE

MIX NO.	TEMP °F	SLUMP IN	WEIGHT OF CONCRETE LBS	UNIT WEIGHT LBS/CU FT	VOL BATCH CU FT/CU YD*	ACTUAL QUANTITIES PER CUBIC YARD: AGENT		WATER	CEMENT	POZZ	SAND	C. AGG.	TOTAL	% AIR: GRAV	PRESS	W/C+P BY WT.	% SAND OR % GRAV	SPECIMENS CAST	SPECIMEN NUMBERS
Batch 1					*		LB	247	495	-	1087	2152	3981						
Example 1		2.0	36.53	147.47	1.754	12.1	CU FT	3.96	2.52	-	6.64	12.89	26.01	3.7	4.0	0.50	34	None	Waste-low slump
							LB												
							CU FT												
Batch 2					*		LB	249	498	-	1072	2126	3945						
Example 1		3.0	36.20	146.14	1.741	13.6	CU FT	4.00	2.54	-	6.54	12.73	25.81	4.4	4.6	0.50	34	4 6x12	1-4
							LB												
							CU FT												

*Batches per cubic yard MIX NOTES

* OR BATCHES PER CU YD

Figure 56.—Typical trial computations for concrete mix. 288–D–2634.

and 4.5 percent is desired, then $\dfrac{4.5}{4.0}\times 12$, or 13.5 ounces will be used in the second trial mix.

The percentage of entrained air in the mix can be measured directly with an air meter or obtained by computing the difference between the calculated, or theoretical, volume and the measured volume, in accordance with designations 24 or 23, respectively, of the appendix. It is advantageous to record both air contents because any marked difference indicates an error and may lead to discovery of mistakes in mix design, trial mix computations, or test methods. A difference in the indicated air contents of as much as 0.3 percent is considered normal. The actual unit weight is necessary for determination of actual water, cement, and aggregate contents regardless of the method used for air determination.

The job mix should not be adjusted for minor fluctuations in water-cement ratio. A difference in water-cement ratio of plus or minus 0.02 by weight, resulting from maintaining a constant slump, is considered normal and should be considered in selecting the water-cement ratio so that with the usual variation the specified maximum is not exceeded.

51. Concrete Mix Tests.—Values of strength and other design factors listed in tables 14 and 16 may be used for establishing a trial mix, as discussed in section 48. However, these values are based on averages obtained from tests on a large number of aggregates and do not necessarily apply exactly to materials being used on a particular job. Therefore, if facilities are available, it is preferable to make a series of mix tests to establish the relationships needed for selection of appropriate proportions based on particular materials to be used. For concrete used on Bureau projects, these mix tests are usually made in the Denver laboratories. This procedure not only establishes the properties of concrete for each project but permits a comparison of these properties with those of concrete used on other Bureau projects.

An example of a minimum series of concrete mix tests is illustrated in table 18. The first mix of the series was a computed trial mix, obtained as previously discussed. The second mix was adjusted to increase the slump but appeared to be oversanded and to contain too little coarse aggregate. In the next mix the amount of coarse aggregate was increased to an estimated maximum amount that would still produce a mix of satisfactory workability. When the proper amount of coarse aggregate was determined, three additional mixes were made in which the water-cement ratio was varied over a range of 0.45 to 0.60. From these mixes the relationships among water-cement ratio, cement content, and strength were established for materials to be used on the job, and it was unnecessary to use the empirical values established for average conditions. Field mixes could be interpolated directly from table 18.

Table 18.—Typical minimum series of concrete mix tests

Mix No.	Net water-cement ratio	Quantities of ingredients, pounds per cubic yard				Air content, percent	Strength at 28-day age		Slump, inches	Finish	General appearance
		Water	Cement	Sand	Coarse aggregate		Compressive	Flexure			
1..	0.50	248	496	1,255	1,940	4.5	3,500	550	1	Fair	Low on slump.
2..	.50	264	528	1,194	1,948	4.1	3,610	560	3.3	Very good ...	Over-sanded.
3..	.50	259	519	1,116	2,023	4.7	3,570	555	2.5	Good	Good.
4..	.45	262	582	1,076	2,039	3.9	4,120	600	3	Very good ...	Very good.
5..	.55	261	474	1,160	2,031	4.3	3,200	525	2.7	Good	Good.
6..	.60	260	433	1,191	2,027	4.5	2,650	450	3	Good	Good.

In laboratory tests desired adjustments will seldom develop as smoothly as indicated in table 18, even with experienced operators. Furthermore, field results cannot be expected to check exactly with laboratory results. An adjustment of the selected trial mix on the job is usually necessary. Closer agreement between results obtained in the laboratory and in the field will be assured if machine mixing is employed in the laboratory. This is especially desirable when air-entraining agents are used, as the type of mixer influences the amount of air entrained. Before mixing the first batch, the laboratory mixer should be primed with a small batch of sand, cement, and water, as a clean mixer retains a percentage of mortar. Similarly, any processing of materials in the laboratory should simulate as closely as practicable the corresponding treatment in the field.

The minimum series of tests illustrated in table 18 may be expanded as the size and special requirements of the work warrant. Alternative aggregate sources and different aggregate grading, various types and brands of cement, different admixtures, different maximum sizes of aggregate, and considerations of concrete durability, volume change, temperature rise, and thermal properties are variables that may require a more extensive program.

52. Mixes for Small Jobs.—For small jobs where time and personnel are not available to determine proportions in accordance with recommended procedure, mixes in table 19 will provide concrete that is amply strong and durable if the amount of water added at the mixer is never large enough to make the concrete overwet. These mixes have been predetermined in conformance with recommended procedures by assuming conditions applicable to the average small job and for aggregate of average specific gravity. Three mixes are given for each maximum size of coarse aggregate. Table 17 may be used as a guide in selecting an appropriate maximum size of aggregate. Mix B for each size of coarse aggregate is intended for use as a starting mix in table 19. If this mix is undersanded, change to mix A, or, if it is oversanded, change to mix C. Note that the mixes listed in the table apply where the sand is dry. If the sand is moist or very wet, make the corrections in batch weight prescribed in the note.

The approximate cement content in bags per cubic yard of concrete listed in the table will be helpful in estimating cement requirements for the job. These requirements are based on concrete that has just enough water in it to permit ready working into forms without objectionable separation. Concrete should slide, not run, off a shovel.

Table 19.—Concrete mixes for small jobs [1]

(May be used without adjustment)

Maximum size of aggregate, inches	Mix designation	Approximate bags cement per cubic yard of concrete	Pounds of aggregate per 1-bag batch		
			Sand [2]		Gravel or crushed stone
			Air-entrained concrete [3]	Concrete without air	
½	A	7.0	235	245	170
	B	6.9	225	235	190
	C	6.8	225	235	205
¾	A	6.6	225	235	225
	B	6.4	225	235	245
	C	6.3	215	225	265
1	A	6.4	225	235	245
	B	6.2	215	225	275
	C	6.1	205	215	290
1½	A	6.0	225	235	290
	B	5.8	215	225	320
	C	5.7	205	215	345
2	A	5.7	225	235	330
	B	5.6	215	225	360
	C	5.4	205	215	380

[1] Procedure: Select the proper maximum size of aggregate. Then, using mix B, add just enough water to produce a sufficiently workable consistency. If the concrete appears to be undersanded, use mix A; and if it appears to be oversanded, use mix C.

[2] Weights are for dry sand. If damp sand is used, increase the weight of sand 10 pounds for a 1-bag batch, and if very wet sand is used, add 20 pounds for a 1-bag batch.

[3] Air-entrained concrete is specified for all Bureau of Reclamation work. In general, air-entrained concrete should be used in all structures that will be exposed to alternate cycles of freezing and thawing.

Chapter IV

INSPECTION, FIELD LABORATORY FACILITIES, AND REPORTS

A. Inspection

53. Concrete Control.—Many factors enter into concrete control. It is a combination of (1) testing and inspection of various materials selected for use, (2) proper proportioning and adequate mixing of materials, (3) proper handling, placing, and consolidating procedures, and (4) proper curing. With these factors closely controlled through testing and inspection, the objective of concrete control—which is the construction, at minimum practicable cost, of structures in which quality is uniform and sufficient to assure satisfactory service throughout the intended operating life—will be realized. The best of materials and design practices will not be sufficiently effective unless the construction practices and procedures are properly performed. Competent inspection and testing staffs functioning under general supervision are required for proper coordination of services and for the class of construction involved.

Many phases of inspection are involved in concrete control. To make any inspection effective, such phases should be considered in an orderly manner. With this objective, a summary of inspection items is presented following the designations in the appendix; detailed discussion of these subjects is found in appropriate sections of the manual text. The summary of inspection items is included as a general aid in carrying out inspection assignments; it is not intended for daily use, as it obviously must include many items not applicable to a specific assignment. An inspector should make his own list for daily use on his particular assignment.

54. Administrative Instructions.—The source of administrative instructions governing inspection on Bureau construction is Reclamation Instruction Series 170.

55. The Inspector.—A competent inspector is thoroughly conscious of the importance and scope of his work. He is observant, alert, and properly trained; he knows both how as well as why the work is required to be done in a certain way. As designs and specifications change from job to job, it is important that the inspector be thoroughly acquainted with the specifications for the particular work with which he is involved. Armed with this knowledge and with judgment gained from experience, he will not only detect faulty construction, but he will also be in a position to forestall it by recognizing causes in advance and preventing use of improper procedures.

Although the inspector will require special instructions or advice from his supervisor concerning unusual problems or controversial matters, his initiative is continually brought into play. The inspector should not delay the contractor unnecessarily nor interfere with the contractor's methods, unless it is evident that acceptable work will not otherwise be produced. Fairness, courtesy, and cooperativeness, along with practicality, firmness, and a businesslike demeanor, will engender respect and cooperation. Avoidance of needless requirements and restrictions will facilitate the accomplishment of the primary purpose of inspection, which is fulfillment of the specification requirements, and also will enable the contractor to perform his work in the most advantageous and profitable manner.

The inspector should follow the Bureau's manual "Safety and Health Regulations for Construction." That manual officially establishes the health and safety requirements for construction by contract; compliance with the safety standards contained in the manual constitutes a contractual obligation on the part of the contractor to the Government. The inspector and supervisor should have a thorough knowledge of this manual and should always be alert to deviations from outlined safety requirements and take proper action to initiate corrective measures.

Through his supervisor's continual training, guidance, and support, the inspector will perform his work with increasing confidence and ability—he knows when decisions should be referred to his supervisor, what contacts he should have with the contractor's men, and what special arrangements or agreements pertinent to his work have been made with the contractor. Occasional joint meetings of the inspectors and their supervisors are of great benefit to all concerned and particularly so if some meetings can be arranged before the construction begins. Also, preconstruction meetings of inspectors, supervisors, and contractors' representatives are especially beneficial in resolving questions and interpretations of specifications requirements. In such meetings the "whys" of the specifications requirements can be explained, leading to better understanding and cooperation of those involved in the construction effort.

56. Daily Inspection Reports.—A most important part of an inspector's responsibility is preparation of a good inspection report. Some purposes of an inspection report are to acquaint supervisors with current progress of the construction; to inform supervisors and inspectors who work on other shifts; to accumulate information for a technical report of construction; to provide a record of engineering data; and, often very important, to substantiate or refute a contractor's claim.

The report should be a written narrative but can follow a printed form. The report should be neat, legible, clearly stated, and concise, yet fully describe activities on the shift. Mere chronological listing of activities should be avoided; the wording, however, should be kept to a minimum. It is advantageous to use lettering rather than to write the report in longhand. Well-labeled sketches to indicate locations and dimensions, as well as photographs, are very useful in explaining a condition to be recorded.

It is important to remember that the inspector's daily report often constitutes the only permanent record of any activities related to that particular phase of work. Consequently, that record should be complete and intelligible, not only when the matter is fresh but also years later when details have been forgotten. As such, complete recording of all happenings is necessary. It is sometimes necessary to definitely reestablish, without question, events of a situation to prepare findings of fact. Specific information recorded in the inspector's daily reports must include statements concerning the condition and progress of the work, the important factors affecting such condition and progress, and any instructions given to the contractor or his representatives. Also included in a report are data on tests made by the inspector, samples taken by him, concrete mixes used, and concrete cylinders made.

Daily inspection reports may be grouped into several categories, such as materials inspection, concrete batching and mixing, or general inspection. Forms should be developed to aid in efficiently and completely recording the information so that it becomes a permanent part of project engineering data.

In some instances, as in large dam construction, inspection at an aggregate plant is warranted, although acceptance tests of aggregate are made at the batching and mixing plant. Such inspection serves to provide data of aggregate processing and as an aid to the contractor. A report of these activities should include notes of the area where the equipment is working, accurately referenced to the pit survey control points. A complete record of the rate at which the pit run material is excavated and processed would include rate of plant feed, quantities of finished product obtained each day, sizes, gradations, and waste quantities. A description of the equipment, modifications, screen capacities, plant efficiency, and down time should be noted. Occasional photographs of the plant and working area, including aerial photographs, are sometimes very helpful.

The extensiveness of batching and mixing plant reports varies with the quantities of concrete placed on a project, the extent of records needed, and the contractor's method of operation. Since aggregates must meet speci-

CONCRETE MANUAL

BUREAU OF RECLAMATION
FRYINGPAN - ARKANSAS PROJECT

BATCH PLANT INSPECTION REPORT

STRUCTURE _Plug Area - Block No 3_ DATE _12-30-70_ SHIFT _Day_ TIME _12:25 PM_
STATION _4+35 to 4+92.5_ TEST BY _RB, RK & DC_ CHECKED BY _RB_
MIX NO. _455-2 (Interior)_

SCREEN SIZE	CUMULATIVE WT RET.	% RET.	IND. % RET.	SPEC. LIMIT
#4	0.5	1.8	1.8	0-5
#8	42.2	8.7	6.9	5-15
#16	97.3	20.1	11.4	10-25
#30	235.0	48.6	28.5	10-30
#50	386.2	79.8	31.2	15-35
#100	464.9	96.1	16.3	12-20
PAN	483.8	100.0	3.9	3-7
TOTAL	483.8		100.0	

F.M. = 2.55

BATCH WEIGHTS (LBS./C.Y.)

REMARKS:

*S.O - SIGNIFICANT OVERSIZE   ** 0 - % OVERSIZE
S.U - SIGNIFICANT UNDERSIZE   U - % UNDERSIZE
M.O - MARGINAL OVERSIZE
M.U - MARGINAL UNDERSIZE

BUREAU OF RECLAMATION
FRYINGPAN - ARKANSAS PROJECT

INSPECTOR'S CONCRETE BATCHING AND MIXING REPORT

CONTRACTOR	SPEC. NO.	DATE AND SHIFT	PLANT	NUMBER OF MIXERS
Drevo Corporation	DC-6820	12-30-70 Day	Drevo	1 - 1 cu. yd

INSPECTOR _Rex D Berger_
CHECKED BY _DC, RK & RD_

Figure 57.—Typical daily summary report for batching and mixing plant inspection. 288-D-3271.

fications requirements as batched, concrete batching and mixing plant daily reports should include aggregate test data. Specific gravities, gradation, and absorption or free moisture can be conveniently combined with concrete mix design adjustments and scale settings in tabular report form, as shown at the top of figure 57. Other forms are needed as work sheets for calculating concrete test batch data and to summarize these data for project construction records. A very useful form is shown in figure 69 of section 61, outlining procedures for completing the calculations involved in its preparation. The bottom form shown in figure 57 is a typical daily summary of quantities of concrete manufactured and ingredients used. In such a report, each class of concrete should be grouped with the referenced specifications item number, total quantities produced, concrete yield, record of waste, and unusual happenings during the shift.

The project records should contain a daily report of control tests performed on fresh concrete and test specimens fabricated. A typical form report of daily concrete testing is shown at the top of figure 58.

Since the amount of waste concrete is often a disputed item, it is very helpful if the amount (agreed to by project and contractor's representatives) can be established for each shift and recorded at the time when an accurate determination can be made. A form such as shown at the bottom of figure 58 can be completed in duplicate, each retaining a copy for reference.

Figure 59 is a general form that can be used for daily concrete placing or equipment installation inspection. Figure 60 is a form developed in the Denver laboratories for recording results of tests on sand and coarse aggregate.

In addition to daily inspection reports, the inspector may be required to keep an official diary containing a summary of all instructions issued relative to the work and a brief record of all important conversations with the contractor or his representatives. Also, the inspector may be required to maintain a record of the day and hour of beginning and of completing each section of the work, the time lost, and the cause of each delay.

57. The Inspection Supervisor.—Area-of-work assignments are made by the inspection supervisor. The competent inspection supervisor provides, in advance, the necessary instructions for carrying out the assignment. He regularly visits the worksite to discuss problems and questions that arise as the work progresses. He carries on a continual on-the-job training in these discussions supported by broad experience. He informs the inspector as to what decisions should be referred to him. He makes certain that inspectors have access to and read all necessary instructions pertaining to the work and information concerning special problems and conditions apt to arise. He also informs the inspector of special arrangements or agreements that have been made with the contractor which affect

BUREAU OF RECLAMATION
FRYINGPAN-ARKANSAS PROJECT
FEATURE _____ *Pueblo Dam* _____

INSPECTOR'S DAILY CONCRETE TESTING REPORT

STRUCTURE: *Plug Area Block No. 3*   DATE: *12-30-70*

STATION: *4+35 to 4+92.5*   ELEV. *4677.6 to 4680.0*

INSPECTOR Burger, Kehler & Carstens		CONTRACTOR Dravo Construction Company						SPECIFICATION NUMBER DC-6820				SHIFT Day	
MIX NO.	LOAD NO.	TEST TIME	CONCRETE TEMP.	SLUMP	UNIT WEIGHT	AIR CONTENT METER	AIR CONTENT GRAV.	GAL. OF WATER	OZ. OF AEA	TEST BATCH	STARTED MIXER (TIME)	LOAD OUT (TIME)
7545-4	2	10:35 AM	54	2.0	139.02	7.0	7.4	91	24			
455-2	5	10:50 AM	52	3.0	143.62	5.5	6.4	72	17			
7545-4	9	11:00 AM	55	4.0	137.01	8.4	8.3	93	20			
455-2	14	11:45 AM	49	2.0	146.13	4.2	4.8	67	10			
455-2	19	12:25 PM	49	2.25	146.83	4.0	4.3	69	9	x		
455-2	36	1:40 PM	50	1.75	146.83	4.1	4.3	64	9			
455-2	46	2:35 PM	49	1.75	147.23	3.9	4.0	64	9	x		

CONCRETE WASTE _____ *12.00 cu yds. 3/4" mix* _____ REASON FOR WASTE *Too old-batching delayed because pozz silo was not operating properly*

CONCRETE WASTE _____ *2.00 cu yds. 4" mix* _____ REASON FOR WASTE *Batch #41 time 2:00 PM Bucket partially open - dumped on ground*

REMARKS _____

BUREAU OF RECLAMATION
FRYINGPAN - ARKANSAS PROJECT

CONCRETE WASTE RECORD

FEATURE _____ *Pueblo Dam* _____ SPEC. NO. _____ *DC-6820* _____ DATE *12-30-70*

STRUCTURE _____ *Plug Area Block #3* _____

STATION _____ *4+35* _____ TO _____ *4+92.5* _____

ELEVATION _____ *4677.6* _____ TO _____ *4680.0* _____

CUBIC YARDS WASTED: *12 cu. yds. 3/4" mix*   BATCH PLANT   PLACEMENT
                        2 cu yds. 4" mix   *14 cu. yds.*

NAME: _____ *J. B. Steward* _____
CONTRACTOR'S REPRESENTATIVE

NAME: _____ *Rex D. Burger* _____
INSPECTOR

**Figure 58.—Summary of daily concrete control tests and waste record.
288-D-3272.**

his work assignment, particularly in regard to concessions or special interpretations of the specifications. He makes the inspector aware of his full support in all proper execution of inspection work. He obtains opinions of inspectors and engineers close to the job before making decisions relative to requests and representations made by the contractor. He encourages informal meetings between inspectors and contractors' representatives to promote efficiency and cooperation.

UNITED STATES DEPARTMENT OF THE INTERIOR, BUREAU OF RECLAMATION
INSPECTORS DAILY REPORT

PROJECT_____, DIVISION_____

SHIFT_____

DATE_____, 19____

SPEC. NO._____,FEATURE_____

LOCATION OF WORK_____

TYPE OF WORK_____

_____

CONTRACTOR'S REPRESENTATIVE_____

I. ORDERS RECEIVED_____

_____

_____

_____

2. UNUSUAL OR UNSATISFACTORY CONDITIONS_____

_____

_____

_____

_____

3. INSTRUCTIONS TO CONTRACTOR_____

_____

_____

_____

_____

4. RECOMMENDATIONS AND GENERAL COMMENTS_____

_____

_____

_____

(USE REVERSE SIDE OF SHEET
FOR EXTENDED REMARKS)   INSPECTOR_____

**Figure 59.—Inspector's daily report—general form for any class of inspection.
288-D-824.**

B. Field Laboratory Facilities

58. The Field Laboratory.—The primary purpose of the field laboratory is to perform routine testing of concrete and concrete aggregate. The

CONCRETE MANUAL

LABORATORY WORK SHEET FOR AGGREGATE TESTS

EL-255
(3-59)

FEATURE _____ PROJECT _____ DATE _____

SAMPLE _____ LABORATORY NO. _____

TESTED BY _____ DATE _____

CHECKED BY _____ DATE _____

GRAVEL-RIPRAP

SIZE	GRADING WEIGHT RET. LB.	IND. % RET.	CUM. % RET.	A SAT. WT. IN AIR	B WT. OF AGG. IN WATER	A-B WEIGHT WATER DISPL.	C OVEN DRY WEIGHT	A-C WEIGHT ABS. WATER	% ABS.	SPECIFIC GRAVITY
6" +										
3" TO 6"										
1½" TO 3"										
¾" TO 1½"										
⅜" TO ¾"										
NO. 4 TO ⅜"										
TOTAL GVL.										
F.M.				Average Specific Gravity (A/A-B) _____						
SAND										
% SAND				Average Absorption (A-C/A) _____						

LOS ANGELES ABRASION-GRADING

REVOLUTIONS	DESCRIPTION	WT. RET. No.12	WT. OF LOSS	PERCENT LOSS
100				
500				

SAND

SIZE	SAMPLE No.1 WEIGHT RET. GRAMS	CUM. % RET.	IND. % RET.	SAMPLE No.2 WEIGHT RET. GRAMS	CUM. % RET.	IND. % RET.	AVERAGE IND. %	CUM. %
#4*								
#8								
#16								
#30								
#50								
#100								
PAN								
F.M.								
*Oversize								

PHYSICAL PROPERTIES (24 HR)

SPECIFIC GRAVITY
Jar No. _____
Temp. of H₂O _____
Sample Wt. (A) _____
Wt. Jar filled with H₂O _____
Total _____
Wt. Jar, Sample, H₂O _____
Displacement _____
Specific Gravity (A/B) _____

ORGANIC IMPURITIES
Color _____

ABSORPTION
SSD Wt. _____
Dry Wt. _____
Wt. Abs. _____
% Abs. _____

AGGREGATE
PASSING #200
Dry Wt. _____
Washed Wt. _____
Wt. Silt _____
% Silt _____

SOUNDNESS OF AGGREGATE

BOWL No.	SIEVE SIZE	TYPICAL GRADING %	SAMPLE WT. BEFORE TEST	SAMPLE WT. AFTER TEST	ACTUAL % LOSS	WEIGHTED % LOSS
SAND	#8	20				
	#16	20				
	#30	30				
	#50	30				
		100				
GRAVEL RIPRAP	¾"	50				
	⅜"	30				
	#4	20				
		100				

REMARKS:

Figure 60.—Form developed in the Denver laboratories for recording results of tests on sand and coarse aggregate. (Front of Form EL–255)

SAND GRADING (WASHED)								
	SAMPLE NO.1			SAMPLE NO 2			AVERAGE	
SIZE	WEIGHT RET. GRAMS	CUM. % RET.	IND % RET.	WEIGHT RET. GRAMS	CUM. % RET.	IND % RET.	IND % RET.	CUM. % RET.
#4*								
#8								
#16								
#30								
#50								
#100								
PAN								
F.M.								

*Oversize

PHYSICAL PROPERTIES (24-HR WASHED)

SPECIFIC GRAVITY________
Jar No.________
Temp. of H_2O________
Sample Wt. (A)________
Wt. Jar filled with H_2O________
Total________
Wt. Jar, sample, H_2O________
Displacement________
Sp. Gr. $\left(\frac{A}{B}\right)$________

ABSORPTION
SSD Wt.________
Dry Wt.________
Wt. Abs.________
% Abs.________

ORGANIC IMPURITIES
Color________

30-MINUTE SPECIFIC GRAVITIES AND ABSORPTIONS FOR LABORATORY MIXES

GRAVEL							
SIZE	A SAT. WT. IN AIR	B WT. OF AGG. IN WATER	A-B WEIGHT WATER DISPL.	C OVEN DRY WEIGHT	A-C WEIGHT ABS. WATER	% ABS.	% GR.
6" +							
3" TO 6"							
1½" TO 3"							
¾" TO 1½"							
⅜" TO ¾"							
#4" TO ⅜"							

Average Specific Gravity $\left(\frac{A}{A-B}\right)$________

Average Absorption $\left(\frac{A-C}{A}\right)$________

SAND WASHED – AS REC'V'D

SPECIFIC GRAVITY________
Jar No.________
Temp. of H_2O________
Sample Wt. (A)________
Wt. Jar filled with H_2O________
Total________
Wt. Jar, Sample, H_2O________
Displacement________
Sp. Gr. $\left(\frac{A}{B}\right)$________

ABSORPTION
SSD Wt.________
Dry Wt.________
Wt. Abs.________
% Abs.________

SPECIFICATION GRADING 1/ SPEC. NO.

SIEVE SIZE \ SAMPLE SIZE	CUM. WT. RETAINED	CUM. % RET.	CUM. WT. RETAINED	CUM. % RET.	CUM WT. RETAINED	CUM. % RET.
7 IN.						
6 IN.						
3½ IN						
3 IN						
2½ IN						
2 IN.						
1¾ IN						
1½ IN						
1¼ IN						
1 IN.						
⅞ IN						
¾ IN						
⅝ IN						
½ IN						
7/16 IN						
5/12 IN						
⅜ IN						
5/16 IN						
NO. 4						
NO. 5						
TOTAL WT.						
% OVER						
% UNDER						

1/ *Significant Oversize Sieve
**Significant Undersize Sieve

CLAY LUMPS IN AGGREGATE				
SIZE				
Sample Wt.				
Washed Wt.				
% Clay				

% LIGHTW'T MATERIAL SP. GR.				
SIZE				
Sample Wt.				
Wt. Light				
% Light				

REMARKS:

Figure 60.—Form developed in the Denver laboratories for recording results of tests on sand and coarse aggregate. (Back of Form EL-255)

test data serve as a basis for determining and ensuring compliance with the specifications, for adjusting concrete-mix proportions, for securing the maximum value from materials being used, and for providing a complete record of concrete placed in every part of the work.

Laboratory investigations generally are made in the Denver laboratories, but tests for solution of construction problems often can be performed in the field laboratory. Typical examples of the latter are the studies of concrete-mixer performance conducted at Grand Coulee Dam; the effect of fly ash on dosage of air-entraining agent and on calcium chloride at Hungry Horse Dam; the investigation of aggregate segregation at Hoover Dam; heavy media beneficiation and water-reducing, set-controlling agent studies at Glen Canyon Dam; and heavy media aggregate beneficiation tests at Flaming Gorge Dam. Early submittal of progress reports and complete final reports of nonroutine investigations to the Chief, Division of Research, is desirable, as such information can often be used to advantage on other projects. Part C of this chapter contains information relative to presentation of test data.

The size and type of laboratory depends upon the job. When it is necessary to construct a laboratory building, the requirements will usually be met by one of the three plans shown in figure 61 or a suitable modification. Laboratory C is appropriate for major centralized concrete construction requiring more than 500,000 cubic yards of concrete, such as Yellowtail, Glen Canyon, and Flaming Gorge Dams. Laboratory B is applicable to noncentralized concrete work, such as that on the Columbia Basin irrigation division; to projects where facilities are required for both earth and concrete testing, such as those embracing Davis, Enders, Angostura, Boysen, and Trinity Dams; and to concrete dams requiring from 25,000 to 500,000 cubic yards of concrete, such as Bartlett, Kortes, and Canyon Ferry Dams. Laboratory A is a small laboratory suitable for isolated projects, for divisions of large noncentralized work, and for concrete dams requiring less than 25,000 cubic yards of concrete, such as small diversion dams.

In figure 61, laboratory B is equipped with facilities for both earth and concrete testing, as generally used on projects where both types of construction are involved. Other examples of floor plans of combined earth-work and concrete laboratories can be found in the Bureau's Earth Manual, second edition. The other two laboratories may also require equipment for earth-control operations. As the equipment for laboratory A, and in some instances for laboratory B, does not include a compression testing machine, it is necessary that concrete test cylinders be shipped to the Denver laboratory or to a nearby project for testing. Procedures for curing, packing, and shipping the cylinders are contained in the appendix, designation 31. In recent years, trailer-housed, mobile-satellite laboratories have been used to supplement a central laboratory where construc-

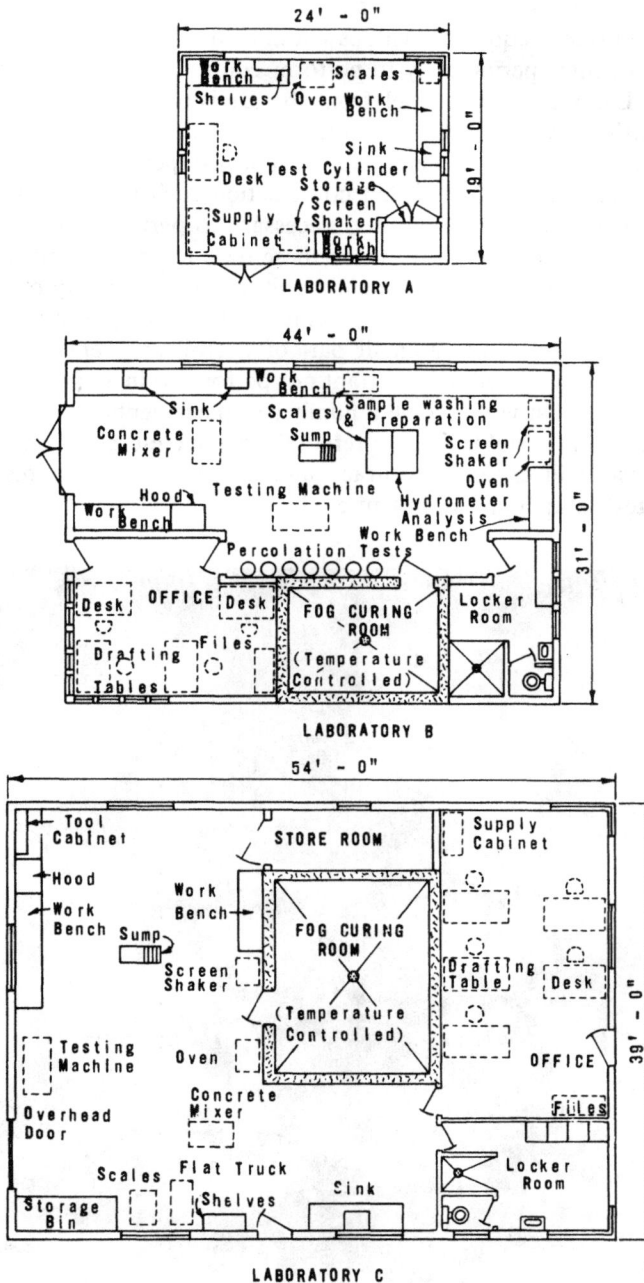

Figure 61.—Three typical plans for Bureau field laboratories. Facilities should be provided which are commensurate with the requirements of the work. 288–D–1536.

tion activities are widespread. It is occasionally necessary that laboratory personnel in trucks equipped with essential field control testing equipment visit the worksites periodically to perform control tests and fabricate test cylinders. Laboratory personnel from the Division of General Research make occasional visits to construction laboratories to calibrate equipment and discuss concrete control activities with project personnel.

On smaller features at isolated construction sites some distance from semipermanent facilities, it is often necessary to perform control tests of fresh concrete. The manner in which these tests are performed influences uniformity of results and consequently uniformity and quality of concrete.

Although tests for entrained air, slump, and unit weight involve simple procedures and constitute a small part of overall concrete control, they should not be considered unimportant nor performed in a careless manner. For a structure that will involve a number of placements, the provision of a test area with a stable base, preferably a concrete slab, on which to perform tests will enhance accuracy and acceptance of test results. An excellent test setup is shown in figure 62.

Figure 62.—A concrete platform provides a stable test area on the Wichita project. P835-526-682.

59. Lists of Laboratory Equipment.—Equipment for concrete control appropriate for a well-equipped laboratory of each type is listed in table 20. Quantities listed are sufficient for the work during early stages of construction; additional items should be obtained as required. The need for a compression-testing machine in smaller laboratories depends, to a large extent, on whether the project is conveniently located with respect to Denver or to other projects where 6- by 12-inch cylinders can be tested. If a laboratory has no compression-testing machine, the need for capping equipment, with accessory apparatus and tools, is of course eliminated.

Table 20.—Field laboratory equipment

Item No.	Laboratory A	B	C	Description
1	1	1	1	Aggregate screening hopper, coarse (fig. 46)
2	1	1	1	Airmeter (Pressure type) ¼ cubic foot
3	....	....	1	Airmeter (Pressure type) 1 cubic foot
4	....	....	1	Analytical balance
5	1	1	1	Aspirator with necessary pipe fittings for attaching to water supply
6	1	1	1	Balance, trip, with agate bearings and brass weights, capacity 2 kilograms, 1/10-gram gradations
7	....	....	1	Blaine fineness meter
8	3	12	12	Bottles, 12-ounce graduated prescription
9	2	4	4	Bottles, 32-ounce graduated prescription
10	....	1	1	Calculator
11	1	1	1	Concrete Manual and job specifications
12	1	3	6	Cones, concrete slump (fig. 219)
13	....	2	2	Glass graduates, 50, 100, 500, and 1,000 milliliter
14	1	1	1	Hose, ³⁄₁₀-inch inside diameter, 25 feet, plastic or stiff rubber for attaching to aspirator
15	1	1	1	Hydrometer
16	....	1	1	Machine, compression testing, 300,000 pounds
17	....	....	1	Microscope
18	....	1	1	Mixer, portable, 2½ to 3½ cubic feet capacity
19	100	100	100	Molds, can, cylinder, 6- by 12-inch
20	....	6	24	Molds, standard cylinder, 6- by 12-inch, with base plates (fig. 223)
21	....	1	1	Molds, capping and alining jig (fig. 225)
22	....	2	2	Plates, cast iron, 10-inch diameter by ½-inch thick, machined for capping, with plain surface
23	1	1	1	Pressure gage, capacity 30 pounds per square inch, with necessary pipe fittings for attaching to waterline
24	2	3	3	Scale, platform counter, 1/100-pound gradations. 240-pound capacity
25	1	1	2	Sampler, tube, for sand
26	1	1	1	Screen, 18- by 18-inch, No. 4 opening, for mixer efficiency test
27	1	2	2	Screen shaker and timer, automatic
28	1	1	1	Screens, set of coarse aggregate, including oversize and undersize screens (tables 9 and 31)
29	1	1	2	Screens, set of standard 8-inch brass, consisting of numbers 200, 100, 50, 30, 16, 8 and 4, with pan and cover

Table 20.—Field laboratory equipment—Continued

Item No.	Laboratory A	Laboratory B	Laboratory C	Description
30	1	1	2	Splitter, sample, 10- by 18-inch, complete with pans
31	....	1	1	Sulfur melting pot, electrically heated, thermostatically controlled
32	1	3	6	Tamping rods, ⅝-inch diameter by 24 inches long, bullet nosed
33	2	2	3	Thermometers, 0° to 220° F, armored
34	1	2	2	Thermometers, maximum and minimum registering
35	2	6	6	Thermometers, 0° to 220° F, pocket, in metal carrying case with clip
36	1	1	1	Vacuum gage, capacity 30 inches of mercury, with pipe fittings for attaching to vacuum line
37	1	1	1	Vibrator, laboratory model, immersion type
38	....	....	1	Vicat apparatus for false set test
39	1	1	1	Wire basket, 8-inch diameter by 2½ inches high for specific gravity (appendix, designation 10)
40	1	1	1	Wire basket, No. 8 mesh, with sufficient capacity to weigh 18 pounds of aggregate in water, for mixer efficiency tests
				Miscellaneous Apparatus and Supplies
41	2	6	6	Brush, brass, for cleaning sand screens
42	....	2	2	Buckets, aggregate, 1 cubic foot
43	1	2	2	Canvas, 12 feet square, 12-ounce duck, double flat seams, hemmed edges
44	1	2	3	Dial gages, operating range 1 inch, graduated in 0.0001-inch divisions
45	1	1	1	Electric drying oven and hotplate
46	1	2	2	Electric fans
47	1	1	1	Fire clay, standard mill, 100-pound sack
48	1	1	1	First-aid table
49	1	1	3	Funnels, 4-inch galvanized
50	1	1	3	Funnels, 8-inch galvanized
51	1	1	1	Hydraulic jack
52	12	12	12	Jars, 1-quart, fruit, ground edge with fitted glass disk lids
53	....	1	1	Ladle, 3-inch diameter, steel
54	2	6	12	Pails, 12-quart galvanized, reinforced
55	2	2	4	Pans, cement, 24- by 24- by 2-inch
56	....	1	1	Pans, concrete batch
57	1	3	6	Pans, concrete slump
58	6	6	6	Pans, deep pudding, 3-quart
59	12	12	12	Pans, drip, 10- by 14-inch
60	....	1	1	Sacks, paper, No. 6, bundle of 500
61	25	50	50	Sacks, sample, cloth
62	1	3	6	Scoops, large hand
63	1	3	6	Scoops, small hand
64	....	1	2	Scoops, shovel, short-handled
65	....	3	6	Shovels, round-point, long-handled
66	1	3	6	Shovels, square-point, short-handled
67	1	1	1	Sodium hydroxide, 1-pound bottle
68	1	2	2	Stopwatch
69	1	1	....	Stove, small wood or coal cook, 4-hole with pipe
70	....	1	3	Sulfur, commercial powdered, 100-pound sack
71	....	1	1	Truck, rubber-tired, flat, 2 by 4 feet
72	....	2	4	Weights, standard, 50-pound
73	....	1	2	Wheelbarrows, rubber-tired

In addition to the items listed, handtools and supplies should be provided as needed.

60. Facilities for Curing Concrete Test Specimens.—The kind and capacity of laboratory curing equipment depend on project requirements. In general, laboratories for the small jobs (A, fig. 61) use water-curing tanks and damp-sand pits. The larger laboratories (B and C, fig. 61) are usually equipped with insulated, temperature-controlled fog-curing rooms. Detailed procedures pertaining to field laboratory curing of test cylinders are included in the appendix, designation 31.

(a) *Water Tanks.*—Figure 63 shows a satisfactory double-deck tank having a capacity of 168 6- by 12-inch cylinders. Standard galvanized stock-watering tanks are satisfactory for use on many of the smaller jobs and are much less expensive than the tanks illustrated in figure 63. When the laboratory is not equipped with approximately constant air temperature control, uniform curing water temperature can be maintained at $73.4° \pm 3°$ F by electric heaters thermostatically controlled. Simple electric coils and thermostats similar to the GE, HS 11040, Soil Heating Set, consisting of 60 feet of 400-watt, lead-sheathed cable and a HSC-7 thermostat assembly have been used successfully. The cable should be placed in a 3- by 9-foot tank and covered with 1 inch of fine gravel for protection. For extremely cold weather and for laboratories that will be in service for a long time, it is advisable to construct the tank with double walls filled with insulating material and to provide a suitable cover. In hot weather, the water temperature can be kept down by circulating colder water through the tank or by using ice.

Wiring and schematic diagrams for another curing-water temperature control system used successfully at some Bureau projects are shown in figure 64. The curing-water temperature is sensed and the system controlled by two thermoregulators mounted as shown in figure 65. One regulator is set at 72.5° F and the other at 74.5° F. A small water-circulating pump is mounted on the end of the tank. The system is designed so that nothing operates while the water temperature ranges between the maximum and minimum limits.

The plumbing is arranged so that valves are available to regulate the amount of hot or cold water added and also the amount of water recirculated. The hot and cold water is supplied from the building water system. If such a system is not available, a small electric water heater of 3- to 5-gallon capacity and a small water cooler of 1-gallon capacity could be substituted. The mixed water is introduced at the upper water surface on one side of the tank through a ¾-inch horizontal pipe which has $\frac{1}{16}$-inch-diameter holes drilled on 6-inch centers along the entire length. The end is capped. The water is withdrawn from the tank through a similarly fabricated pipe near the bottom of the tank on the opposite side and is circulated through the valve hookup where hot and cold water is added as

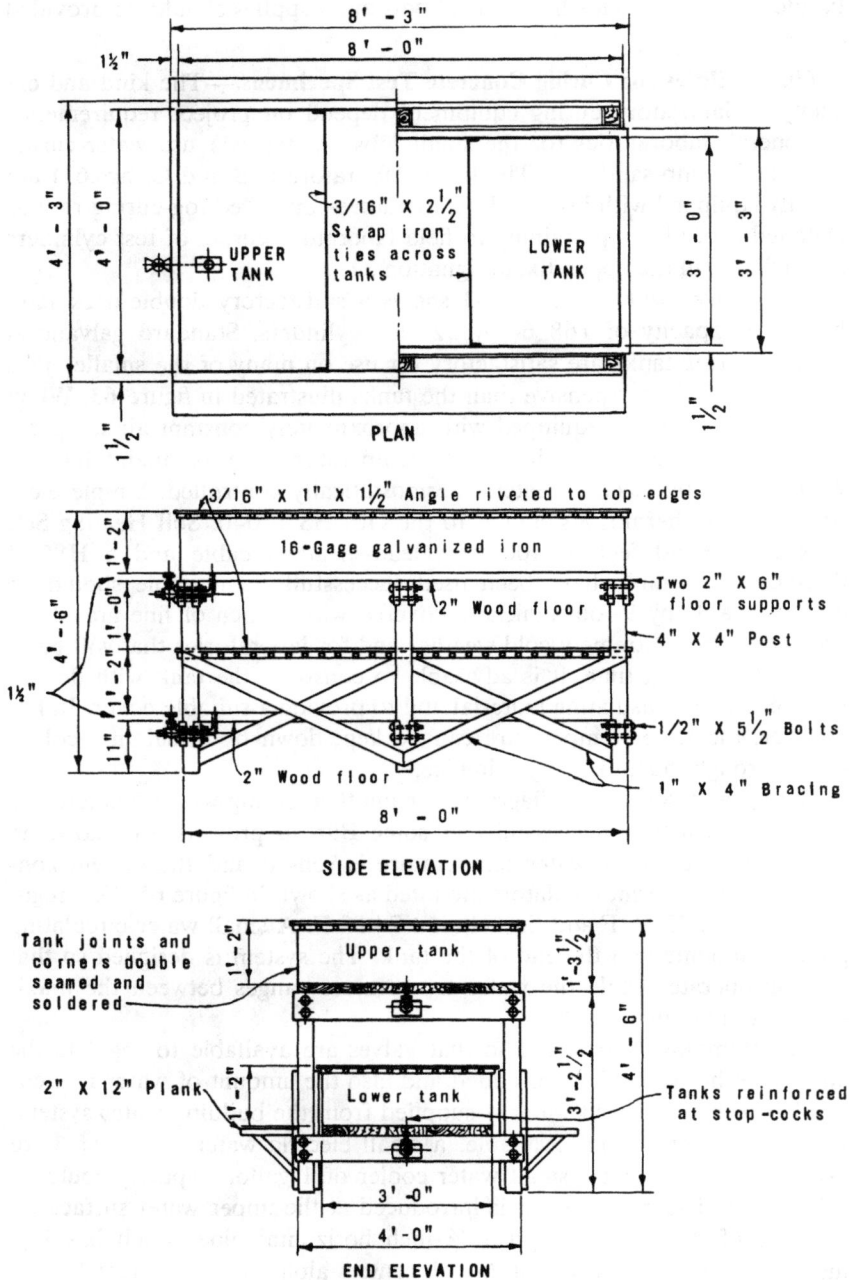

8' - 3"
8' - 0"
1½"

4' - 3"
4' - 0"

UPPER TANK

3/16" X 2½" Strap iron ties across tanks

LOWER TANK

3' - 0"
3' - 3"

1½"
1½"

PLAN

3/16" X 1" X 1½" Angle riveted to top edges

16-Gage galvanized iron

2" Wood floor

Two 2" X 6" floor supports

4' - 6"
1' - 2"
1' - 0"
1' - 2"
1' - 0"
11"

4" X 4" Post

1½"

1/2" X 5½" Bolts

2" Wood floor

1" X 4" Bracing

8' - 0"

SIDE ELEVATION

Tank joints and corners double seamed and soldered

1' - 2"

Upper tank

1'-3½"

2" X 12" Plank

1' - 2"

Lower tank

3'-2½"

4' - 6"

Tanks reinforced at stop-cocks

3' -0"

4' - 0"

END ELEVATION

Figure 63.—Construction details of framework and tanks for use in laboratory curing room. 288–D–186.

72.5° F Thermoreg'r
74.5° F Thermoreg'r

Hot water solenoid

Water-circ.-pump mtr.
Water-circ.-pump mtr.

Cold water solenoid
72.5° F Thermoreg'r
74.5° F Thermoreg'r

Water-circ.-pump mtr.
Cold water solenoid
Hot water solenoid

120 V.a.c. return
120 V.a.c.
Ground

SYSTEM
ON–OFF
ON

NEMA Type I enclosure

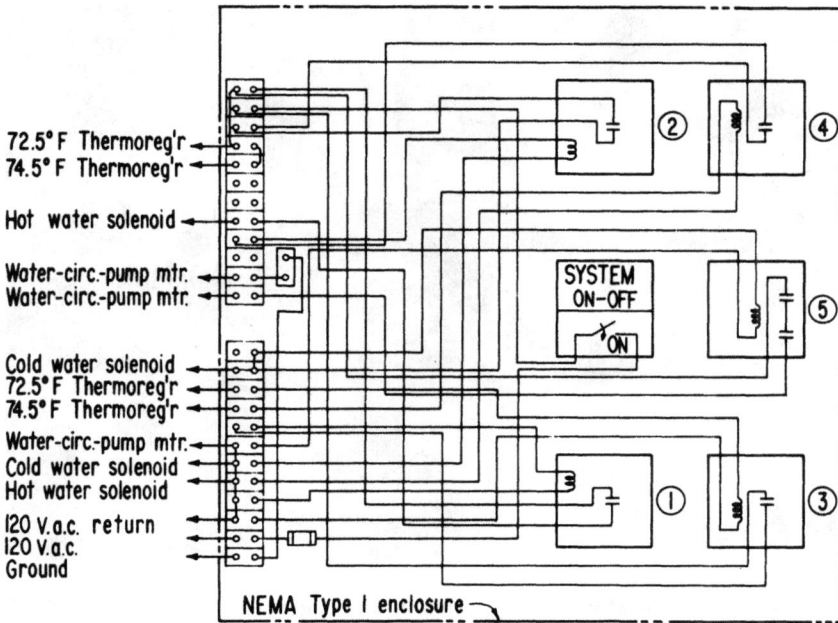

WIRING DIAGRAM

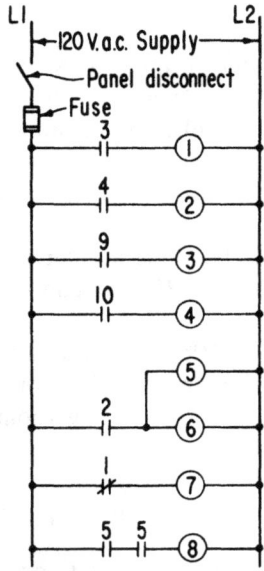

L1 L2

120 V.a.c. Supply

Panel disconnect

Fuse

SCHEMATIC DIAGRAM

1. Relay – Scientific Supplies Co., No. 61841-1* N.C. contact.
2. Relay – Scientific Supplies Co., No. 61841* N.O. contact.
3. Relay – Scientific Supplies Co., sensitive relay No. 61841-2.*
4. Relay – same as 3.
5. Relay – General purpose type.
6. Cold water solenoid.
7. Hot water solenoid.
8. Water-circulating-pump motor – from Proven Pumps.*
9. Thermoregulator – Scientific Supplies Co., No. 61840* closed above 72.5° F
10. Same as 9, except closed above 74.5° F
* Not to be construed as the only supplier.

Figure 64.—Wiring and schematic diagrams for a water temperature control system for concrete test specimen curing tanks. 288–D–3285.

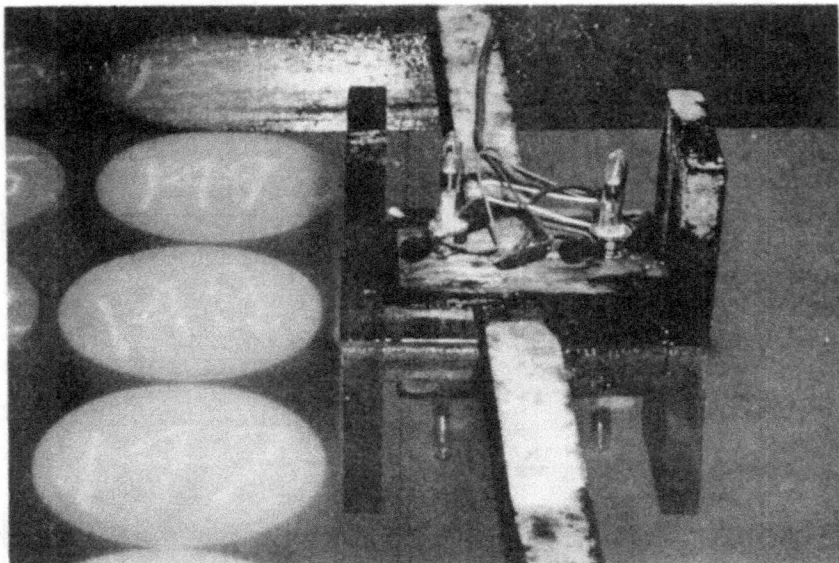

Figure 65.—A view of thermoregulators in a curing water tank. P126–D–75009.

necessary and then introduced into the curing tank. The mixing faucet of the circulating system, like the type used on automatic washing machines, serves to control the amount of hot or cold water introduced. An overflow leading to a drain is necessary to dispose of excess water. Figure 64 is a view of the arrangement of the automatic controls, recirculating pump, and overflow. The tank may be of a size selected sufficient for the expected storage capacity and should be fitted with a metal cover for protection.

(b) *Storage in Moist Sand.*—If tests are to be made at later ages, the concrete may be removed from the water tank after 28 days and stored in moist sand maintained at proper temperatures.

(c) *Fogrooms.*—A fogroom maintained at 100 percent relative humidity and $73.4° \pm 3°$ F is ideal for moist curing and permits more effective use of available space than do other methods. A curing room constructed and operated at Grand Coulee Dam maintained the humidity and temperature well within permissible limits. The room had a capacity of one-thousand 6- by 12-inch cylinders and was built and equipped as follows:

The room was 13 feet wide, 16 feet long, and 10 feet high. The walls were made of 2- by 8-inch wood studs covered with shiplap on the outside and cement plaster on metal lath on the inside. They were insulated with rock-wool filling between studs, and the cement plaster was waterproofed with emulsified asphalt paint. Ceiling construction was similar to that of the walls. The floor was of concrete and provided with a drain. Specimen racks were built of welded steel angles supporting eight shelves spaced at 11 inches. Each shelf was 19 inches wide and constructed of 1- by 1- by ⅛-inch angles set in a position to drain. The racks were placed along each side of the room and a double row along part of the central portion. Wood grating floor strips were provided for walkways.

Fog was supplied by water sprays above each specimen rack. Sprays were constructed by inserting small gasoline-lamp generator tips into tapped holes in water-pipes suspended from the ceiling. Streams of water from the tips were atomized by striking the flattened ends of wires secured to the pipes. Metal gutters, suspended under the spray equipment, discharged drip water through downspouts to the floor. Figure 66 shows three types of fog sprays.

During the coldest weather, heat was supplied to the room by 500-watt strip heaters. During moderately cold weather, the ceiling lights were left burning to supply the necessary heat. A fan was used to circulate the air and maintain uniform temperature in the room.

In moderately warm weather, when the water was cooler than 73.4° F, the sprays kept the temperature down. In hot weather, when the spray water and the outside air were both above 73.4° F, the gates of two cooling ducts near the top of opposite walls were opened as required. A circulating fan directly in front of one of the ducts drew warm air of very low relative humidity from the outside, and evaporation of water by this dry, incoming air cooled the curing room to the desired point. The saturated air escaped through the opposite duct. The temperature was controlled by manual operation of the duct openings.

C. Reports and Evaluation of Test Data

61. Reports.—Reclamation Instruction Series 170 requires that a summary of concrete construction data be submitted each month to the Director of Design and Construction as part of the Construction Progress Report. These reports serve a fourfold purpose:

(1) They enable the project and Denver offices to keep informed regularly as to construction activities and effectiveness of concrete control.

(2) They are valuable in the development of specifications for construction.

(3) They add to the fund of information gained from practical experience on a broad variety of Bureau projects, which information, supplemented by technical knowledge acquired through laboratory tests and scientific research, serves as a basis for continued advancement in concrete construction practices.

(4) They constitute a permanent record for use, if required, in settlement of contracts.

1½" ROUND BRASS STOCK

100-MESH SCREEN SOLDERED TO NIPPLE

SOLDER

4"

SEE NOTE "B"

SEE NOTE "A"

SEE NOTE "C"

18 GAGE COPPER PLATE

¼" CLOSE NIPPLE

THUMB SCREW

No. 9 GALV. WIRE

SECTION A-A

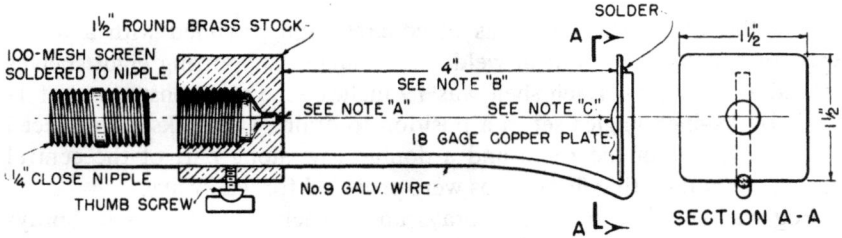

NOTE A: Start hole with ⅛-inch drill and complete with No. 80 drill for water aperture.

NOTE B: Distance from nozzle to plate is about 4 inches. This distance will vary with differences in water pressure.

NOTE C: The raised surface is formed by making a depression on the back of the plate with hammer and punch.

NOTE D: Sprays are located 20 inches down from ceiling.

½" REDUCTION COUPLING

RUBBER GASKET

¼" GLASS TUBING

BRASS SCREW

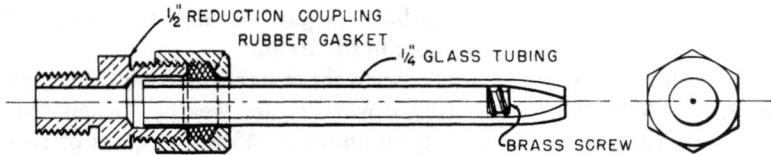

NOTE: The nozzle is made by heating a piece of glass tubing and stretching it to obtain a water aperture approximately .025 of an inch. The tubing is cut off and smoothed with a file.

Two threads from a brass screw are placed inside the glass tube to cause the water to whirl, making a spray as it emerges from the opening.

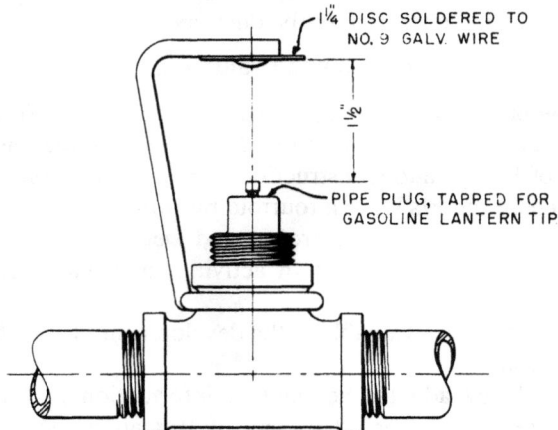

1¼" DISC SOLDERED TO NO. 9 GALV. WIRE

1½"

PIPE PLUG, TAPPED FOR GASOLINE LANTERN TIP

NOTE: Spray tips are placed on 3-foot centers along pipe and pipe is carried above each cylinder rack close to ceiling.

Plugging of tips with sediment can usually be eliminated by a trap in the water line.

NOTE: THESE TYPES OF SPRAYS OPERATE UNDER WATER PRESSURES OF AT LEAST 20 Lb/in.^{2}

Figure 66.—Fog-spray devices for curing room. 288–D–1537.

The value of these reports to engineers in the Denver office has been demonstrated through many years of use. They are reviewed by those who must keep in close touch with concrete construction control.

The general nature of the reports and the forms on which summarized data are to be recorded are hereinafter prescribed as set forth in Reclamation Instruction Series 170.

(a) *Narrative Portion.*—The narrative summary of concrete construction data should be restricted to matters of special importance or interest relating to technical control exercised in concrete construction. The text should be clear, concise, and free from copious repetition in the individual report or from month to month. Where appropriate and practicable, photographs and drawings should be included. Of special interest and value are descriptions of such features, conditions, and experiences as:

(1) *Concrete ingredients.*—Important or novel changes in arrangement or equipment of aggregate plants or finish-screening facilities at the batching plant; difficulties encountered in meeting specifications requirements pertaining to aggregates, and methods for correction; and interesting facts concerning the handling of concrete ingredients. On large projects, information should be included on delivery of bulk and bagged cement, pozzolans, blending operations, and processing of cement used in grouting contraction joints.

(2) *Batching and mixing.*—Procedures or changes that result in improvements; means used for heating or cooling concrete materials and concrete; tests for mixer performance; and dates when scale checks were made and any corrective measures taken, but not detailed data.

(3) *Transporting concrete.*—Difficulties encountered, such as loss of slump, segregation, false set, etc., and reasons for such difficulties and methods of correction.

(4) *Placing concrete.*—Problems encountered in preparation for placing and in difficult placements; bleeding, harshness, excessive surface voids, etc.; remedies employed and improved practices developed.

(5) *Finishing concrete.*—Problems encountered and solutions, removal of stains, and experience with special equipment.

(6) *Protection and curing of concrete.*—Unusual difficulties encountered and remedies therefor; novel procedures or precautions used in protection and curing, and surface damage, if any, resulting from these operations. Include a general statement concerning extreme weather conditions, but do not report copious weather statistics.

(7) *Precast concrete.*—Items of special interest pertaining to concrete pipe or other precast products, such as defects and repairs or rejections; also, outstanding accomplishments or qualities.

(8) *Field laboratory activities.*—Special tests or investigations; condition of laboratory equipment, etc.

(9) *Administrative matters affecting concrete construction.*— Discussions with contractor and protests by contractor; job training of inspectors, etc.

(10) *Suggestions.*—Means for improving concrete control, including improvements in designs and specifications.

(b) *Summarized Tabulations.*—Two forms for summarizing concrete construction data have been used extensively and have proved to be practicable. Figure 67 shows the form for reporting summarized data of aggregate tests. Because specifications requirements for grading are for aggregates as batched, the data shown will generally be the results of tests on samples from the batching plant. However, the form may be used for reporting results of tests on samples taken from the aggregate processing plant or from stockpiles, in which case appropriate comment should be made on the form.

Figure 68 shows the form for reporting data on concrete mixes. Data should be shown for each set of tests made during the month and may be entered on the form from day to day as data become available. No field summarizing is needed. The form has been developed so that the original may be retained at the project office and prints submitted with the monthly concrete construction data reports.

The heading of this form will indicate whether it is the "first," "second," or "final" report for the particular mixes. The first report carries the name of the month during which the reported mixes and cylinders were made and results of 7- and 28-day cylinder strengths available at the end of the month. The second report, submitted a month later, shows the same month and data as the first with the addition of all 28-day cylinder strengths. The final report, submitted 2 months after the second, shows the same month and data recorded in previous reports supplemented with a record of 90-day strengths.

On jobs where cylinders are made for breaking at ages later than 90 days, the report giving 90-day strengths should be marked "third" instead of "final" to indicate that cylinders have been made for breaking at later ages. Strengths of these cylinders should be reported when available and the cylinders identified with respect to mixes they represent. The mixes should also be grouped according to the maximum size of aggregate and separated according to interior, exterior, backfill, or structural concrete when listed on form 7–1317.

An appropriate worksheet similar to that in figure 69 is quite helpful in computing the concrete-mix data shown in figure 68. Worksheets should not be included in the concrete construction data reports but may be retained by the project office for future reference. The following steps explain the use of the form in figure 69 (complete instructions for use of this

7-1316 (4-73)
Bureau of Reclamation

CONCRETE CONSTRUCTION DATA
AGGREGATE TESTS

Project ... CENTRAL UTAH

Feature ... Water Hollow Tunnel and Channel No. 2 (Tunnel arch - Item 27)

For the month of December, 1970, Sheet 1 of 1   Spec No. DC-6575

TESTS OF SAND FOR 16 SAMPLES						TESTS OF COARSE AGGREGATE		
	PERCENT RETAINED					PERCENT IN EACH SEPARATION		
TEST SCREEN	COARSEST		FINEST		AVERAGE		NUMBER OF SAMPLES	16	13
	INDIV.	CUM.	INDIV.	CUM.	INDIV.	CUM.	NOMINAL SIZE RANGE	#4 TO 3/4 TO	3/4 TO 1-1/2 TO
NO. 4	1.7	1.7	2.7	2.7	2.6	2.6	SIGNIFICANT OVERSIZE	7/8	1-3/4
NO. 8	12.4	14.1	10.6	13.3	11.2	13.8	AVERAGE	0.0	0.0
NO. 16	20.7	34.8	16.3	29.6	18.7	32.5	MAXIMUM	0.0	0.0
NO. 30	23.7	58.5	20.5	50.1	22.0	54.5	MINIMUM	0.0	0.0
NO. 50	26.9	85.4	28.6	78.7	27.9	82.4	SIGNIFICANT UNDERSIZE	-#5	-5/8
NO. 100	11.5	96.9	16.7	95.4	13.7	96.1	AVERAGE	1.4	1.5
NO. 200	2.3	99.2	3.2	98.6	2.7	98.8	MAXIMUM	2.0	3.1
PAN	0.8	100.0	1.4	100.0	1.2	100.0	MINIMUM	0.9	0.4
FM		2.91		2.70		2.82	MARGINAL OVERSIZE	3/4	1-1/2
MOISTURE CONTENT						sand	AVERAGE	3.7	9.1
AVERAGE						+3.7	MAXIMUM	7.4	14.8
MAXIMUM						+6.0	MINIMUM	1.5	2.2
MINIMUM						+2.7	MARGINAL UNDERSIZE	#5	5/8
NUMBER OF SAMPLES						17	AVERAGE	2.1	4.0
SPECIFIC GRAVITY						2.60	MAXIMUM	3.5	7.2

COMMENTS ON SAND Average absorption on sand was 1.2%. Sand continues to run within specified limits after blending about 50% sand from Borrow Area "E" with about 50% sand from Wallace and Hobusch pit. One sample was a little low on the No. 100 screen.

MINIMUM	1.4	0.0
PASSING NEXT NOMINAL SCREEN		
RETAINED ON 3/8 SCREEN	69.1	
RETAINED ON 1-1/2 SCREEN		30.8
RETAINED ON 2-1/2 SCREEN		
MOISTURE CONTENT	3/4"	1-1/2"
AVERAGE	+3.2	+0.9
MAXIMUM	+4.4	+1.9
MINIMUM	+1.9	-0.8
NUMBER OF SAMPLES	16	14
SPECIFIC GRAVITY	2.42	2.42

COMMENTS ON COARSE AGGREGATE: Average absorption on 3/4-in. aggregate was 2.5%, and 1-1/2-in. aggregate was 1.3%. Significant undersize was high on one sample of 1-1/2-in. aggregate due to plugged screen on shaker.

SOURCE OF SAND: Same as aggregate.

SOURCE OF COARSE AGGREGATE: Borrow Area "E" near U.S. Highway No. 40 in NW1/4 sec. 15,

REMARKS: T 35, R 11 W. Sand and aggregate tested and approved by Denver laboratory, Report No. C-1031, dated July 20, 1962.

Continued blending sand from Borrow Area "E" with sand from Wallace and Hobusch tracts, north of South Street, Salt Lake City, Utah. Sec. 5, T 3 S., R 1 E., SL B&M sand tested and approved by Denver teletype dated January 14, 1969.

High moistures on aggregate due to using hot water spray bars on vibrating shaker.

NOTES: Insert nominal size separations of coarse aggregate. See Concrete Manual for size of screen openings for tests for significant undersize and oversize

Figure 67.—Typical report of aggregate tests. 288–D–2637.

7-1817 (Special Rev. for 8th Ed. Conc. Man. Bureau of Reclamation)						CONCRETE CONSTRUCTION DATA																		SHEET 1 OF 1
PROJECT		Central Utah																INSTRUCTIONS: 1 Class: 1=Exterior, 2=Interior, 3=Structural
FEATURE		Water Hollow Tunnel and Channel No. 2												2 Shift: 1=Graveyard, 2=Day, 3=Swing
						3rd REPORT OF MIXES USED DURING THE MONTH OF September, 19 70									3 See Designations 22, 23, 24, and Fig. 68 of Concrete Manual, 8th Edition.
SPECIFICATIONS NO. DC-6575 (1-4)															4 Right adjust data in appropriate spaces.

CLASS	DAY	MONTH	SHIFT	YEAR	SPEC ITEM NO.	SAND	#4 TO 3/4	3/4 TO 3/2	3/2 TO 6	CU YD OF CONC	WATER	CEMENT	POZZO-LAN	SAND	GRAVEL	AEA	WRA	TEMP °F	SLUMP	UNIT WT LB PER CU FT	C+P BY DY	GRAVI-METRIC	AIR% PRESS METER	7 DAY	28 DAY	90 DAY	180 DAY	1 YEAR
												1½-INCH MSA TUNNEL INVERT MIX																
	DESIGN					34	33	33			254	540		1037	1874				3.00	143.7	.47	4.5	4.5					
3	28	8	2	0	27	36	31	31	2	149	256	538		1061	1845	96		59	3.00	137.0	.48	4.6	5.5	-	4345 / 4505			
3	1	9	2	0	27	36	33	28	3	153	250	536		1036	1866	95		55	3.75	136.6	.47	5.0	5.5	-	4400 / 4240			
3	3	9	2	0	27	36	31	31	2	120	240	539		1043	1877	96		55	3.75	137.0	.45	5.1	5.4	-	4435 / 4950			
3	8	9	2	0	27	36	31	30	3	144	265	539		1026	1853	96		48	3.25	136.4	.49	4.7	5.6	-	3350 / 3400			
3	10	9	2	0	27	36	32	30	2	135	259	543		1065	1864	97		46	2.50	138.2	.48	3.8	4.6	2635	5090 / 5290			
3	16	9	2	0	27	36	33	29	2	163	259	543		1059	1866	96		42	2.75	138.0	.48	3.9	4.9	2545	4950 / 4610	5250		
3	22	9	2	0	27	36	32	28	4	147	268	544		1051	1869	111		60	3.50	138.2	.49	3.7	4.9	3025	5090 / 5090			
3	23	9	2	0	27	37	33	29	1	156	262	539		1067	1843	192		55	2.50	137.5	.49	4.1	5.2	-	4610 / 4345			
3	25	9	2	0	27	35	33	27	5	161	254	537		1029	1882	191		59	3.50	137.1	.47	4.6	5.4	-	4080 / 3990			

CEMENT:	BRAND Ideal		TYPE V low alkali	SOURCE Devils Slide, Utah
POZZOLAN:	TRADE NAME None		SP. G.	SOURCE
AEA:	TRADE NAME Protex		SP. G.	METHOD OF CURING CYLINDERS Water Tank-Thermostat Control
WRA:	TRADE NAME None	None	SP. G.	CURING TEMPERATURE RANGE 70° to 78° F.
CALCIUM CHLORIDE: PERCENT BY WT. None				

Figure 68.—Typical monthly report of concrete mixes. 288–D–2635.

form are printed on its reverse side).

Step 1—Insert in column 1 sizes of aggregate.

Step 2—Insert in column 2 actual batch weights.

Step 3—Record percentages of saturated-surface-dry moisture in column 3A from moisture tests performed on each fraction of aggregate.

Step 4—In column 3C record the saturated-surface-dry weight computed for each fraction of aggregate $= \dfrac{\text{wet weight} \times 100}{100 + \text{percent moisture}}$.

Step 5—Subtract the saturated-surface-dry weight in column 3C from the batch weight in column 2 and record in column 3B. The difference, if plus, is water being contributed to the mix by the aggregate. If minus, it is weight of water that will be absorbed by the aggregate to obtain a saturated-surface-dry condition. The algebraic sum of these weights for all fractions of aggregate in column 3B is entered at the bottom of column 3B on the line opposite "water."

Step 6—Insert size of test screens in column 4A.

Step 7—In column 4B, record from screen analysis tests percentages retained and passing the nominal screens for each fraction of aggregate.

Step 8—Multiply the percentages in column 4B by the weights in column 3C and record in column 4C.

7-1557
(7-64)
BUREAU OF RECLAMATION

CONCRETE MIX DATA WORK SHEET
(For use with Form No. 7-1317)

PROJECT: _ _ Fryingpan-Arkansas_ _ _ _ _ _ _ _ FEATURE:* _ _Pueblo Dam_ _ _ SPEC. NO.:* DC-6820 ITEM NO.:* _ _46_ _

MONTH AND YEAR:* December 1970 DAY OF MONTH & SHIFT:* _ _30th Day_ _ _ _ _ _ TYPE OF CONCRETE: 3-inch MSA-Interior

CEMENT: PORTLAND CEMENT BRAND _ _Ideal II L. A._ _ _ _ _ _ _ _ POZZOLAN _ Fly Ash_ _ _ _ _ ADMIXTURE WRA PDA-25 _ _ _ _

INSPECTOR:_ _ _ _ _ _ _ _ _ _ Rex D. Burger_ _ _ _ _ _ _ _ _ _ _ _ _ _ _ _ _ CYLINDER NUMBERS: I-325 through I-330 _ _ _

1. INGREDIENT	2. BATCH WEIGHTS FROM MIXING PLANT	3. CORRECTION FOR MOISTURE IN AGGREGATE			4. CORRECTION FOR OVERSIZE AND UNDERSIZE IN AGGREGATE				5. CORRECTED BATCH WEIGHTS IN POUNDS	QUANTITIES PER C.Y.		8. PERCENT GRADING OF AGG. BASED ON CLEAN SEPARATION
		A. PERCENT OF WATER IN AGG.	B. WT. OF WATER IN AGG.	C. S.S.D. WEIGHTS OF AGG.	A. SCREEN SIZE	AGGREGATE IN EACH SIZE				6. CONTENTS PER CUBIC YARD	7. SOLID VOLUME OF EACH INGREDIENT	
						B. PERCENT	C. WT. IN POUNDS	D. CORRECTED WT. IN POUNDS				
AIR-ENTRAINING AGENT	—	—	—	—					—	—	—	—
CEMENT	254	—	—	—					254	253	1.28	—
POZZOLAN	85	—	—	—					85	85	0.53	—
SAND	997	4.2	40	957	PASSING 64	98.2	940	940	952	947	5.80	27.4
					RETAINED ON 64	1.8	17	12				
COARSE AGGREGATE 64 TO 3/4	852	0.6	5	847	PASSING 64	1.4	12	17	859	855	5.26	24.8
					#4 TO 3/4	96.0	813	813				
					RETAINED ON 3/4	2.6	22	29				
COARSE AGGREGATE 3/4 TO 1-1/2	961	-0.5	-5	966	PASSING 3/4	3.0	29	22	930	925	5.67	26.8
					3/4 TO 1-1/2	88.6	856	856				
					RETAINED ON 1-1/2	8.4	81	52				
COARSE AGGREGATE 1-1/2 TO 3	696	-0.8	-6	702	PASSING 1-1/2	7.4	52	81	731	727	4.46	21.0
					1-1/2 TO 3	92.6	650	650				
					RETAINED ON 3	0	0	—				
COARSE AGGREGATE TO					PASSING TO							
					RETAINED ON							
AGGREGATE PLUS												
WATER	152	—	34	—					186	185	2.97	
TOTALS	3997	—	—	3472		—	—	—	3472	3997	3997	25.97	100.00

VOLUME OF CONCRETE REPRESENTED BY TEST:

NUMBER OF BATCHES X VOLUME OF BATCH = _ _188_ _ _ _ C.Y.*

TOTAL COARSE AGGREGATE: 2507 _

AIR CONTENT BY AIR METER:* _ _ _ _ _ _3.2% _ _ _ _ _ _ _ _ _

SLUMP:* 1.75' _ TEMP OF CONCRETE:* _ _49°F_ _ UNIT WEIGHT OF CONCRETE:* _ _ _147.23_ _ _ _ _ _ _ _ _ _ _ _ _ _ _ _

BRAND OF AIR-ENTRAINING AGENT:* _ _ _ _ _ _ _Protex_ $\frac{w}{c+p}$ .185/(253 + 85) = 0.55

$\frac{AIR\ CONTENT}{(BY\ GRAVIMETRIC\ METHOD)} = \frac{(27 - 25.97)}{27} = 3.8\ %$

VOLUME OF BATCH = $\frac{3997}{147.23(27)}$ = 1.005 CU. YD.

*Data and information to be submitted on Form No. 7-1317

Figure 69.—Typical work sheet for concrete mix data. 288-D-2636.

Step 9—Transfer to column 4D the undersize and oversize weights, as indicated by the arrows, to the fraction of aggregate to which they belong for clean separation.

Step 10—Add the combined weights of each size fraction of aggregate in column 4D and record in column 5.

Step 11—Transfer the weights of cement, pozzolan, and amount of air-entraining agent from column 2 to column 5. Determine the

total weight of mixing water by adding the weights of water in
column 2 and in column 3B and record in column 5.

Step 12—Compute the weight per cubic yard of each ingredient by
dividing each corrected batch weight (in column 5) by the volume
of the batch in cubic yards. The volume of the batch is the combined batch weights (total of either column 2 or column 5), divided
by the product of the unit weight of fresh concrete and the factor
27. Space is provided in the lower right-hand corner for this
computation.

Step 13—Compute the solid volume of each mix ingredient (weight
per cubic yard divided by the product of the specific gravity of the
ingredient and 62.3, the weight of a cubic foot of water) and record
in column 7. Total the solid volumes and record at the bottom of
column 7. The volume of air is equal to the difference between 27
cubic feet and the total solid volume of the ingredients. The percentage of air is equal to the volume of air in cubic feet divided by
27. Space is provided for this computation at the bottom of form.

Step 14—Compute the percentage of each size of aggregate based
on clean separation from the weights in either column 5 or column
6, and record in column 8.

Step 15—Compute the water-cement ratio from the weight of water
and cement (plus pozzolan if used) in either column 5 or column 6.
Space is provided for this computation in the lower right-hand
corner of the form.

Step 16.—After checking computations, the information marked with
an asterisk (*) should be recorded on the form shown in figure 68.
The use of an electronic computer program for performing these
calculations is described in section 80.

62. Evaluation of Test Data.—Production of concrete meeting Bureau
requirements for uniformity involves testing ingredients and concrete. The
strength of standard test cylinders not only indicates the strength of a
structure but also reflects other properties of the mix such as materials
used, durability, watertightness, and ability to resist abrasion. Uniform
strength of test cylinders signifies uniform control. Large variations in test
results necessarily require increased average strengths to satisfy design
criteria; furthermore, excessive variations permitted in quality always suggest the danger of inferior concrete.

The quality of control may determine to a large extent the useful life of
a structure. The 28-day strength of standard 6- by 12-inch cylinders varies
above and below the average, depending on how well the job is controlled,
and falls in some pattern of a normal probability curve as illustrated in
figure 70. When there is good control, strength values vary little from average, and the curve is steep. With poor control the values are spread
laterally, and the curve is flattened. The radius of gyration of points about

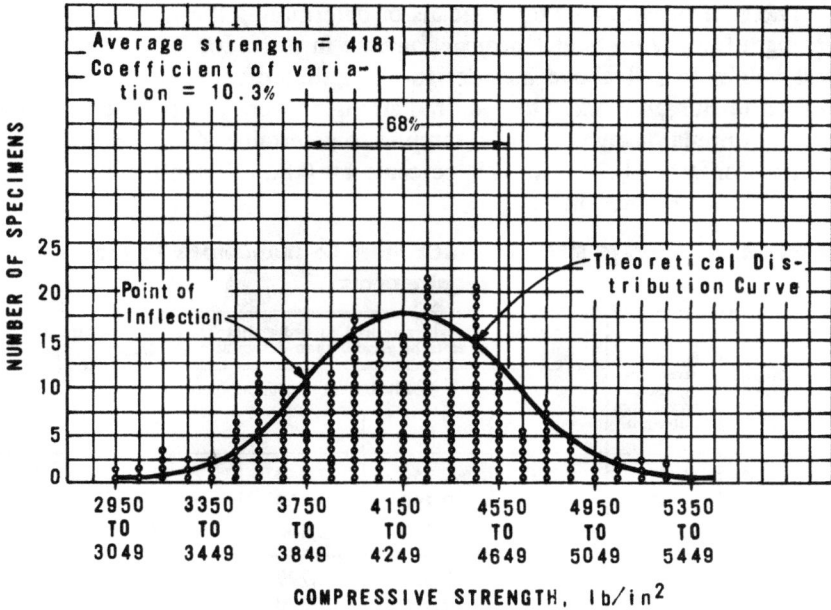

Figure 70.—Typical frequency distribution of the strength of a series of 6- by 12-inch control cylinders. 288–D–2639.

the center is called the standard deviation. Standard deviation divided by the average value is called the coefficient of variation. That coefficient is a statistical tool established to indicate the amount of variation. A high coefficient indicates poor control, and a low coefficient indicates good control.

Criteria generally accepted by Bureau designers require that the strength of 80 percent of the test specimens be greater than the design strength. Table 21 shows average strengths, based on 10 or more tests, that the project must maintain for several coefficients of variation and for the requirements that 75, 80, and 85 percent of the tests be greater than design strengths ranging from 2,000 to 6,000 pounds per square inch. The values were computed by the following formula:

$$f_{cr} = \frac{f'_c}{1 - tV}$$

where:

f_{cr} = average strength required, pounds per square inch,

f'_c = design strength, pounds per square inch,

t = a constant—depending upon the proportion of tests that may fall below f'_c and the number of samples used to establish V, and

V = coefficient of variation expressed as a decimal.

Table 21 provides a simple means of determining the required average strength for any particular project concrete in order for 80 percent of the test values to fall above any design strength ordinarily specified or required in Bureau work. The required strengths were determined from the above formula, using at least 10 samples to determine V and the following values for t: $t = 0.703$ for 75 percent; $t = 0.883$ for 80 percent; and $t = 1.100$ for 85 percent.

Table 21.—Average strength which must be maintained to meet design requirements

Design strength, lb/in ^{2} (f'_c)	Percent of strengths greater than design strength	Average strength required in order that 75, 80, or 85 percent of tests be greater than design strength (f_{cr}) Coefficient of variation, percent				
		5	10	15	20	25
2,000	75	2,070	2,150	2,240	2,330	2,430
	80	2,090	2,190	2,300	2,430	2,570
	85	2,120	2,250	2,400	2,560	2,760
2,500	75	2,590	2,690	2,790	2,910	3,030
	80	2,620	2,740	2,880	3,040	3,210
	85	2,650	2,810	3,000	3,200	3,450
3,000	75	3,110	3,230	3,350	3,490	3,640
	80	3,140	3,290	3,460	3,640	3,850
	85	3,180	3,370	3,590	3,850	4,140
3,500	75	3,630	3,760	3,910	4,070	4,250
	80	3,660	3,840	4,040	4,250	4,490
	85	3,700	3,930	4,190	4,490	4,830
4,000	75	4,150	4,302	4,470	4,650	4,850
	80	4,190	4,390	4,610	4,860	5,140
	85	4,230	4,490	4,790	5,130	5,520
4,500	75	4,660	4,840	5,030	5,240	5,460
	80	4,710	4,940	5,190	5,470	5,780
	85	4,760	5,050	5,390	5,770	6,200
5,000	75	5,180	5,380	5,590	5,820	6,070
	80	5,230	5,480	5,760	6,070	6,420
	85	5,290	5,620	5,990	6,410	6,900
5,500	75	5,700	5,920	6,150	6,400	6,670
	80	5,750	6,030	6,340	6,680	7,060
	85	5,820	6,180	6,590	7,050	7,590
6,000	75	6,220	6,450	6,710	6,980	7,280
	80	6,280	6,580	6,920	7,290	7,700
	85	6,350	6,740	7,190	7,690	8,280

Intermediate values may be determined by interpolation or can be calculated.

For example, it is assumed that concrete for a particular project will have a coefficient of variation of 15 percent and that the designated design strength is 3,000 pounds per square inch; then from the table an average strength of 3,460 pounds per square inch must be maintained in order that 80 percent of the tests will fall above 3,000 pounds per square inch.

When aggregate larger than 1½ inches is used, the 6- by 12-inch cylinders must necessarily be fabricated with concrete from which the plus 1½-inch aggregate has been removed. The screening equipment shown in figure 46, although primarily designed for screening concrete aggregate or soils, can be used to remove the aggregate larger than 1½ inches from a mass concrete sample. A simpler apparatus for screening fresh mass concrete is shown in figures 71 and 72. (If fabrication details are desired, request drawing No. 288–D–3273 from the Chief, Division of Research, Engineering and Research Center, Denver Federal Center, Denver, Colorado, 80225.)

The 1½-inch-square opening screen is removable from the metal frame for easy cleaning. The frame folds to provide a clear 4- by 5-foot work area. Specimens used to measure the strength of mass concrete are either 12- by 24-inch or 18- by 36-inch cylinders seal-cured through an expected temperature cycle and contain the full mix. The comparative 6- by 12-inch cylinders are fog-cured at 73.4° F. Therefore, it is necessary for control purposes to correlate the strength of the 6- by 12-inch cylinders with the mass concrete cylinders. This correlation is usually accomplished during the concrete mix investigations conducted in the Denver laboratories for each project. Past experiences with mix investigations have shown that the age-strength relationships between 6- by 12-inch and mass concrete specimens can be predicted with a fair degree of accuracy. Figure 73 shows these relationships. The age of the wet-screened 6- by 12-inch cylinders is given on the x-axis, and the ratio of the mass concrete cylinder strengths to that of the wet-screened 6- by 12-inch cylinders is shown on the y-axis. For example, the ratio of 180-day mass concrete strength to 28-day strength of wet-screened 6- by 12-inch cylinders is 1.07; therefore, if the 28-day strength of 6- by 12-inch cylinders is 3,000 lb/in², the 180-day strength of the mass concrete cylinder will be about $1.07 \times 3,000 = 3,210$ lb/in². Another way this figure can be used is, for example, if the desired strength of mass concrete at 180 days' age is 4,000 lb/in², then the 7-day strength of the 6- by 12-inch cylinders should be about $\dfrac{4000}{1.52} = 2630$ lb/in².

Some of the many factors that influence the strength of standard 6- by 12-inch cylinders may be controlled; other influences are caused by fac-

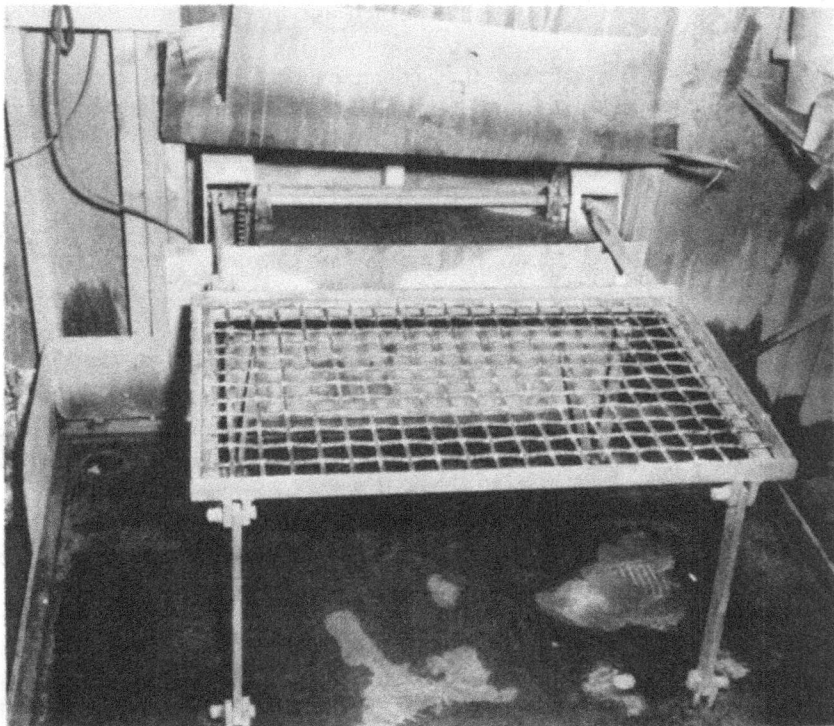

Figure 71.—Drive mechanism and wet screen used to remove plus 1½-inch aggregate from fresh concrete. Batch plant control laboratory at Pueblo Dam, Fryingpan-Arkansas project, Colorado. P382–706–10407.

tors beyond the control of project personnel. Unavoidable variables include changes in climate and variations in raw materials, such as cement fineness and aggregate absorption.

Effects of avoidable variables caused by differences in the maximum sizes of aggregate used in different structures and changes in water-cement ratio for different exposure conditions and types of concrete can be eliminated by proper classification of the mixes involved as outlined in section 61. Data for each type of concrete should be reported separately.

Included among the controllable factors that contribute to variation of concrete are changes in water-cement ratio for a given type of concrete, differences in air content, poor control of moisture content in the aggregate, poor control of mixing proportions, sampling and testing procedures that do not conform to Bureau standards, and haphazard handling of test cylinders. Concrete cylinders improperly handled on the job are not repre-

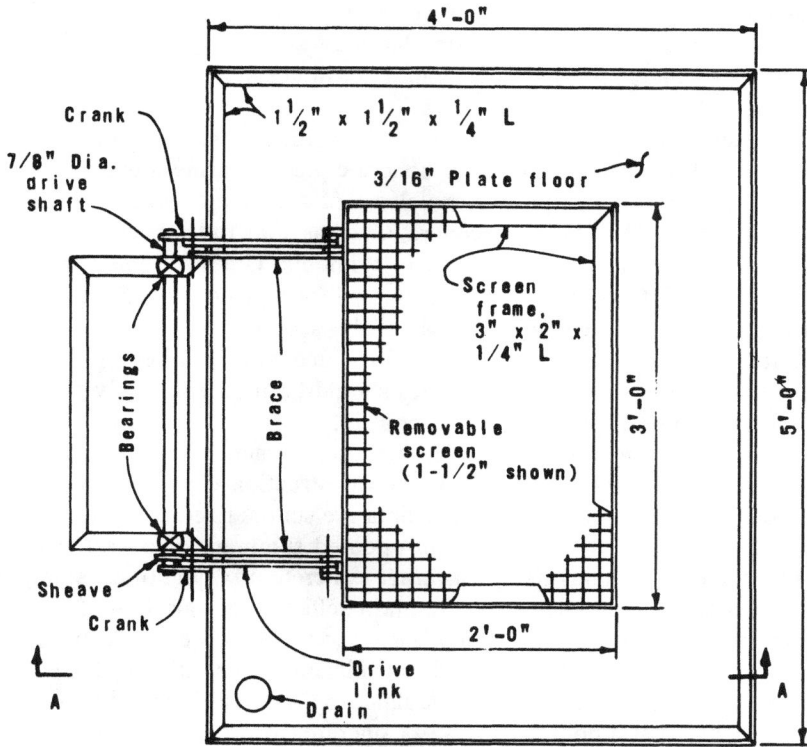

PLAN

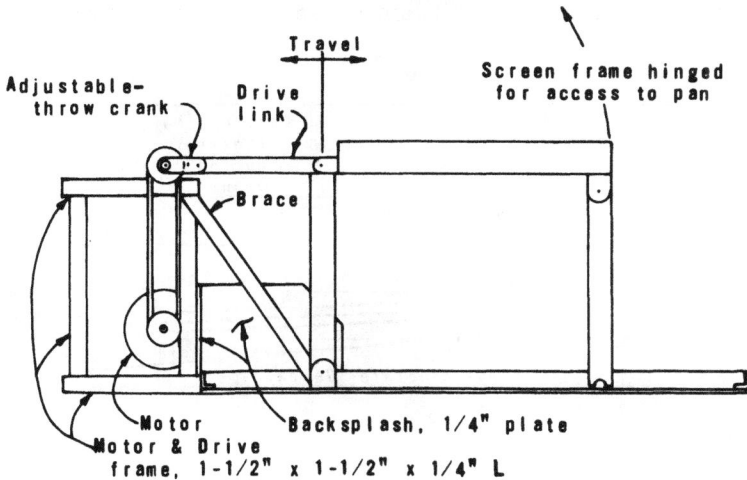

SECTION A-A

Figure 72.—Fresh concrete screening device for removing aggregate larger than desired. 288–D–3252.

sentative of concrete in the structure. Therefore, special care should be exercised to maintain uniform standards for fabricating and handling 6- by 12-inch cylinders so that these cylinders will not indicate more variation in strength than exists in the concrete.

Concrete cylinders, when made in accordance with the provisions of section 90, are representative of concrete placed during a given period of time rather than for a particular volume of concrete on small jobs involving only a small yardage of concrete. It has sometimes been the practice to make only a few concrete control cylinders. Unless at least ten 28-day 6- by 12-inch test cylinders are fabricated for a given project regardless of the yardage involved, a statistical analysis of strength data is not warranted. When a project or activity is provided with facilities and personnel for concrete testing, these resources should be used effectively to provide sufficient data for proper concrete control.

For large construction jobs, it is present practice to process the data contained in Monthly L–29 Concrete Construction Control Reports by an electronic computer program. The data are summarized on a 1-page print-out for each month along with an updated summary of overall averages. The printout contains data of average concrete mix quantities, slump, entrained air, compressive strength, and coefficient of variation of the compressive strength tests for each class of concrete and each maximum size aggregate. On smaller projects, the coefficients of variation of the concrete control cylinder strengths are calculated by the field laboratory personnel and included as a note in the L–29 Concrete Construction Control Report.

Formerly, for a number of years, an annual summary report was prepared which included the compilation and analysis of all concrete-mix data

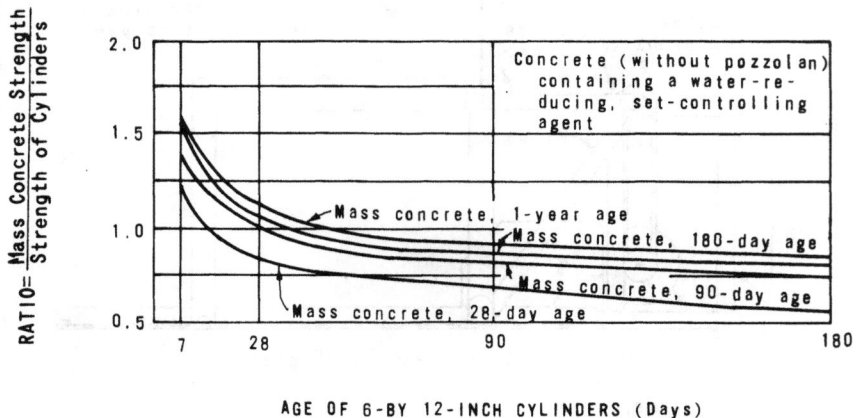

Figure 73.—Ratios of mass concrete compressive strengths in seal-cured cylinders to compressive strengths of 6- by 12-inch fog-cured cylinders fabricated from minus 1½-inch MSA wet-screened concrete. 288–D–3253.

EVALUATION OF CONTROL STANDARDS

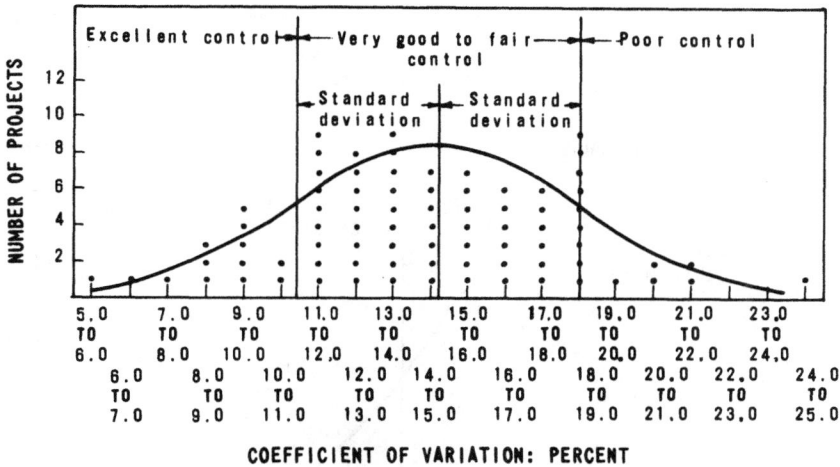

Figure 74.—Frequency distribution of coefficients of variation of concrete strengths for Bureau projects—1952. 288–D–1538.

for Bureau of Reclamation projects during any calendar year. This summary provided an overall comparison of quality of concrete control for all Bureau projects. Figure 74 reveals the relative standard of control for concrete strengths on each project during the construction season of 1952. Each point represents one Bureau contract under construction during the period. As might be expected, the points fall into the normal pattern of the probability curve—the majority of the points are located near the average, with a few of them in the range of excellent control and a few on the other side in the range of poor control. Projects with excessive concrete strength variation must sometimes compensate by requiring 400 to 500 pounds per square inch more strength than the average project to meet the established criterion that 80 percent of the strengths fall above the design strength. For this reason, many of the projects that are in the category of poor control are also listed in figure 75 with the group of projects that do not meet design strength requirements.

Figure 75 shows the percentage of the required concrete design strength that each project produced in 1952. A large majority of projects obtained strengths in excess of those required, and the average strength of Bureau concrete placed during that year was about 110 percent of those required. A few projects produced strengths that were unnecessarily high; and unless the cement content was required for durability, or other factors, it could normally have been reduced at a saving. An excess of strength is much preferred to strengths that are too low. Low strength is the most serious problem encountered in concrete control because failure to obtain the design strength reduces the safety factor. Unless concrete of adequate strength can be produced uniformly, design stresses must be lowered to increase the safety factor. This procedure would cause a general increase in construction costs of Reclamation projects.

EVALUATION OF HOW BUREAU PROJECTS CONFORM WITH STRENGTH REQUIREMENTS

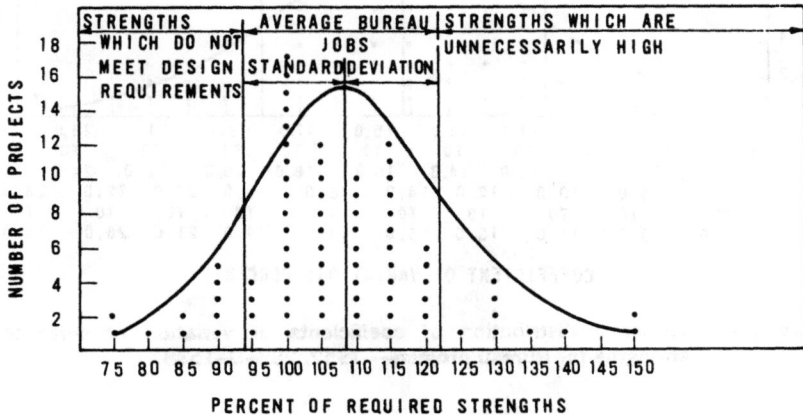

Figure 75.—Frequency distribution of the variation of concrete strengths for Bureau projects from those required by design—1952. 288–D–1539.

Chapter V

CONCRETE MANUFACTURING

A. Materials

63. Aggregate Production and Control.—The control of production and handling of aggregates is largely a field problem. The lack of uniformity in sources of supply and the difficulty in maintaining uniformity in the finished product require positive actions by the contractor and constant vigilance by the inspector. Finish screening at the batching and mixing plant corrects gradings that exceed specifications limits because of mishandling and breakage and removes, just before batching, uncontrolled amounts of minus 3/16-inch undersize which contribute to problems in controlling uniformity of fresh concrete. As aggregates must meet specifications requirements when batched, the aggregates should be well processed initially. Deleterious materials must be removed, either by washing or by special processing; unsatisfactory grading must be corrected by wasting some parts or by supplying deficiencies; segregation and breakage must be minimized; and moisture content of the aggregate must be kept as uniform as practicable. The required degree of control depends to some extent on the size and importance of the work, and the necessary frequency of adjustments of equipment depends on the uniformity of raw materials with respect to quality and, gradation. (See sec. 18.)

The primary function of an aggregate processing plant is to produce clean and properly sized materials. Prior to installation of plant and aggregate handling equipment, contractors on Bureau projects involving large amounts of mass concrete are required to submit for review by the Chief, Division of Research, drawings showing the general arrangement of the plant and a detailed description of equipment proposed for use. Early submittal of this information, together with comments and recommendations from the project office, will expedite approval by the Chief, Division of Research. It may also prevent the contractor's purchase or installation of unsuitable equipment or forestall plans for use of undesirable operating methods.

64. Sand Production.—The gradation of sand as it comes from the pit does not usually conform to the specifications, and some form of processing is required. Defects in grading may be corrected by adding suitable blending sand, by crushing a portion of the excess of larger sizes, by removing portions of sizes present in excessive amounts, or by a combination of methods. Bureau specifications generally require that only natural sand be used, except as allowed by specifications or when permission is obtained from the contracting officer to supplement sand gradings with crushed material from the excess of other aggregate sizes. Figure 76 shows typical results of three of these methods of sand processing.

Use of sand manufactured by crushing or grinding rock or gravel may result in a harsh mix and should be accepted only when it is impracticable to obtain suitable natural sand at reasonable cost. Since the angular shape of crushed sand is its only inherent disadvantage, it is important that crushing equipment be used that will produce the best practicable shape of particles. Sand produced by crushing in rolls is generally unsatisfactory because of the high percentage of thin and elongated particles. The product of a rod mill is much better in this respect. Oscillating cone-crushing equipment has been developed which, through an adjusted grinding action, will economically produce relatively cubical shaped sand particles from almost any material. Impact-type equipment, commonly known as the hammer mill, excels in producing sand particles approaching the cubical shape from softer rock such as limestone.

Usually, screening on the No. 8 or No. 10 screen is necessary to remove excess amounts of the larger fractions in crushed sand. Washing is commonly required to remove excess material that would pass the No. 100 or No. 200 screen. The No. 50 to No. 100 size will frequently be short, and correction must be made by appropriate grinding or by blending with fine natural sand.

A blending sand should be batched separately, fed into the processing plant through an adjustable feeder, or blended with the natural plant-run product on the belt leading to the sand stockpiles. It may also be spread over the gravel source area and processed with the natural feed. Unless adequate classifying equipment is incorporated in the plant when blending in this manner, most of the blend sand is lost during processing. Also, blending at the storage piles by conveyor belt, clamshell buckets, or bulldozer is usually not sufficiently uniform or dependable.

A deficiency of fines in the sand at Marshall Ford Dam was remedied by the reduction of a part of the sand in a rod mill which was operated in closed circuit with one of two rake classifiers. At Shasta Dam, soft particles in the sand were disintegrated in a ball mill, and the resulting fines were removed by classifiers. To make up the deficiency caused by the removal of these fines, small gravel was ground in a rod mill.

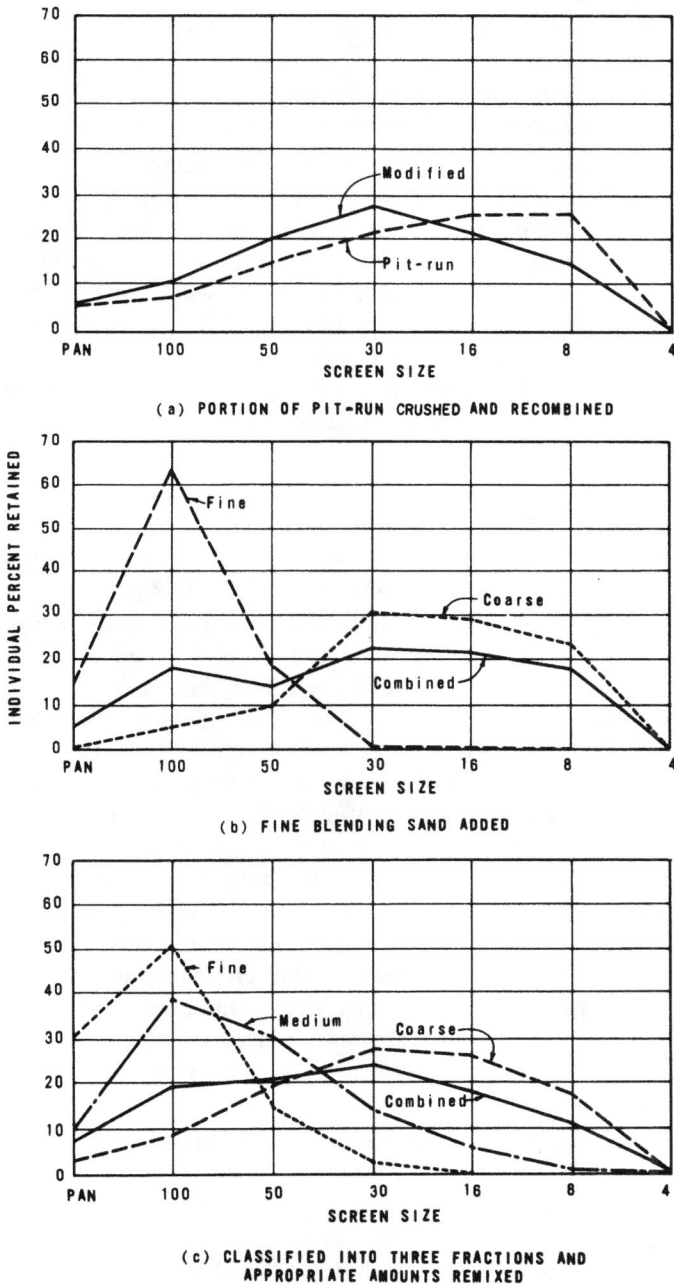

(a) PORTION OF PIT-RUN CRUSHED AND RECOMBINED

(b) FINE BLENDING SAND ADDED

(c) CLASSIFIED INTO THREE FRACTIONS AND
APPROPRIATE AMOUNTS REMIXED

Figure 76.—Three methods of correcting defects in sand grading. 288–D–1540.

The fineness modulus of coarse sand on the Roza division, Yakima project, was reduced by wasting from 20 to 25 percent of the fraction between No. 4 and No. 8 screens. As wasting an excess in parts of the sand is the most common form of processing, the equipment and methods used for this purpose will be discussed in some detail.

On major projects, improvements in sand grading are accomplished by large classifying units arranged in series or series-parallel as may best suit the requirements. Such units may be spiral, rake, bowl, or different types of water scalping tanks or a combination of types. At Hoover Dam, two rake classifiers and one bowl classifier were used in making one finished-sand product. Each of the three bowl classifiers at Grand Coulee Dam separated a different size range, and the fine, medium, and coarse sands thus obtained were recombined to produce a uniform product from pit-run material of very erratic grading, as shown by the diagram at the bottom of figure 76.

There are different types of classifiers suitable for use on the smaller jobs. The equipment described in subsequent paragraphs includes small commercial machines of moderate cost for wet classification and washing and devices for wet screening and for dry screening.

Careful selection of sand processing methods and equipment will frequently avert excessive overexcavation and unnecessary wasting of otherwise usable aggregate. This selection becomes more critical in sources where the availability of sand is marginal with respect to need. The spiral and rake classifiers are most effective as washing and dewatering equipment. They have been adequately used as classifiers in producing an acceptable sand grading, usually when it was necessary to remove the finer sizes. It is very difficult to eliminate excess coarser sizes with such equipment. If the sand is processed only with this type of equipment, the coarse sand sizes may be partially removed by selected screens or carried over with the coarse aggregate into the secondary screening plant and wasted there. The objection to this procedure is that some of the other usable sand sizes are lost as well. Equipment is now available to separate sand feed into several sizes, portions of which can be wasted or saved, and thus it is possible to readily and economically produce sand within close gradation limits.

When processed sand is divided into different size fractions, it is important that it be mixed thoroughly before being batched. Usually, mixing is accomplished in stockpiling and handling operations. When the finished product is stockpiled, the side slopes should be kept flatter than the angle of repose of the sand (a slope of 7 inches per foot is satisfactory) to prevent accumulation of the coarser material at the bottom of the slope. A spray of just enough water to moisten dry-processed sand will materially aid in preventing objectionable segregation during handling operations.

65. Wet Processing of Sand.—As sand is commonly damp in the deposit and often requires washing, wet processing is a much more effective means than dry processing for improving quality and grading.

It is important that the quantity and velocity of water through washing and classifying equipment be maintained in proper balance with the capacity of the units to avoid excessive turbulence and loss of valuable fines.

(a) *Spiral Classifiers.*—This commercial type of classifier, illustrated in figure 77, consists of an inclined trough or tank containing a revolving

Figure 77.—Spiral classifiers for washing and dewatering sand. PX–D–32058.

helix, generally referred to as a spiral. The machine is made in both single spiral and double spiral models and in a range of sizes to meet the requirements of small and large jobs. The location of classifiers in the aggregate processing tower used at Friant Dam is shown in figure 78. Sand from the hydroseparator was fed into two of the four spiral classifiers, from which the overflow carried off a fine sand and the spirals discharged a washed and dewatered coarse sand. The fine overflow product was separated by the other pair of classifiers into a silty overflow, which was wasted, and a clean fine sand which was blended with the coarse sand from the primary classifiers to produce a fineness modulus of approximately 2.75.

**Figure 78.—A well-designed plant used for screening and washing aggregates.
PX–D–32059.**

The operation of the spiral classifier is as follows: When sand and water are fed into the tank, the coarser particles settle and are removed by the spiral at its upper end together with a small amount of water. At the lower end is an adjustable weir over which the finer and lighter materials are discharged with the greater part of the water. The feed enters through the side or sides of the tank at or near the water level and at some distance from the weir. The spiral revolves at a speed just sufficient to remove the settled sand at the desired rate and moves the sand along the bottom and one side of the trough, leaving the other side above water level comparatively free for drainage of moisture and slime. The correct operation for any desired separation is effected by varying (1) the rate of feed of the sand, (2) the depth of the pool, (3) the inclination of the tank, (4) the speed of rotation of the spiral, (5) the number, size, and arrangement of water jets and the jet pressure, and (6) the width and depth of the overflow. Water added on the slope above the pool aids greatly in washing and in removing fines. By using two or more machines in series, and, if necessary, wasting undesirable portions of the product from one or more of them, a finished washed sand of any desired gradation can be produced.

(b) *Reciprocating Rake Classifiers.*—The reciprocating rake classifier, also used for classifying and washing sand, and the spiral machine are similar in fundamental principle but are quite different mechanically. A quadruplex rake classifier is pictured in figure 79. The rake classifier, which is manufactured in a range of sizes, consists of an inclined tank containing transverse reciprocating blades, with a settling pool in the

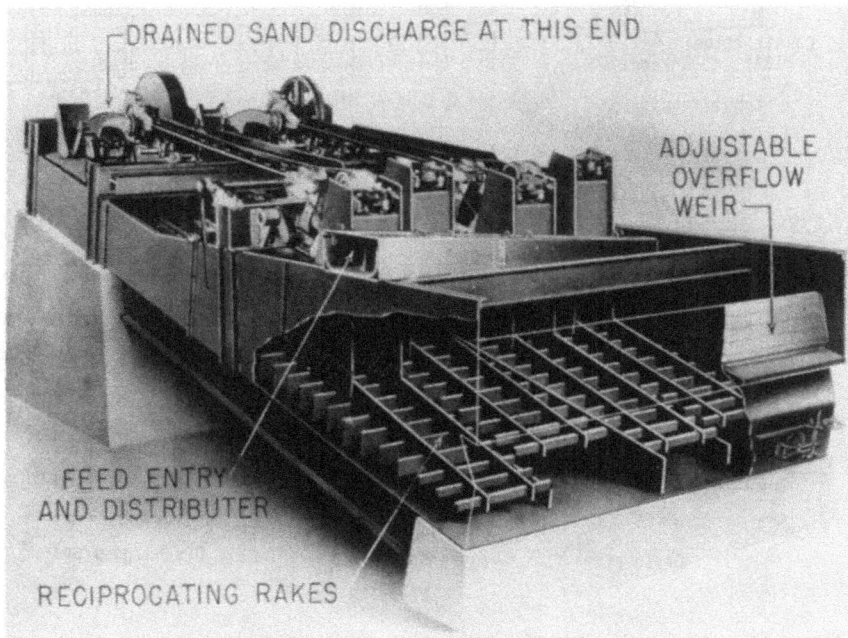

Figure 79.—Quadruplex rake classifier. PX–D–32060.

lower end into which sand and water are fed. Upward strokes of the blades drag the settled sand out of the pool and up the inclined tank bottom to the discharge lip and, in so doing, wash the sand and permit the excess water to drain from it. During the return strokes, the reciprocating cycle is completed with the blades in the elevated position so as to clear the sand bed. Fine material in suspension flows over the crest of an adjustable weir at the end of the pool. Operating control of the rake classifier is similar to that of the spiral type.

(c) *Hydraulic Classifiers.*—For many years, this type of classifier has been used in ore dressing to separate relatively small, closely sized material into fractions having widely different specific gravities. The hydraulic classifier is also capable of separating, into size groups, material consisting of grains that are fairly uniform in specific gravity. The latter function is receiving more general recognition as a practicable and relatively economical means for improving the gradation of concrete sand.

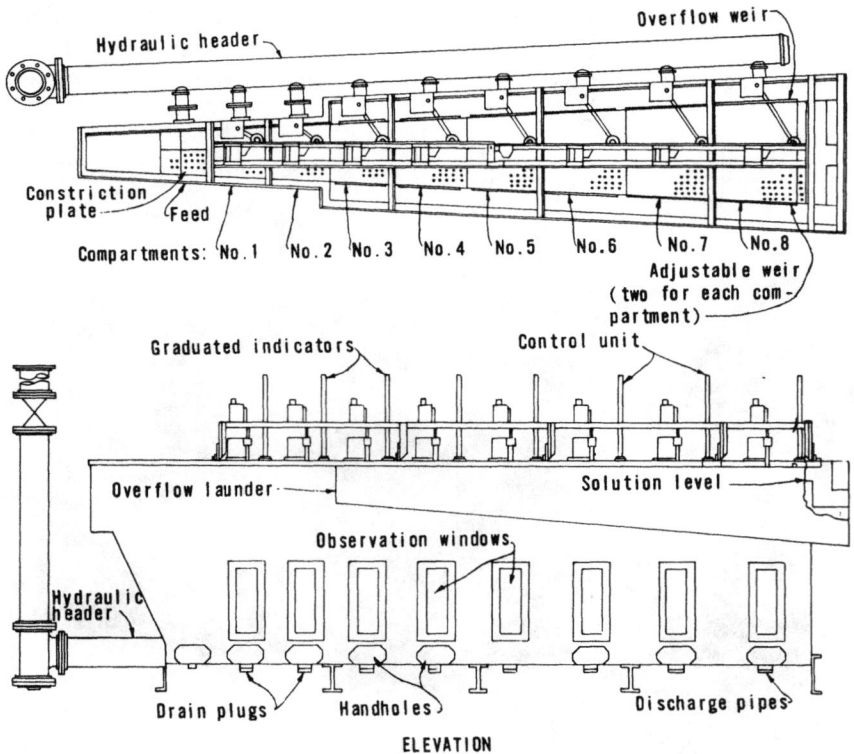

Figure 80.—Hydraulic sizer of a type used with considerable success at Hungry Horse Dam in Montana. 288–D–1541.

Hydraulic classification, as applied to size separation of particles having reasonably uniform specific gravity, is essentially a process in which the material to be treated is fed into a vertically rising current of water having such velocity that the smaller grains are lifted and discharged through an overflow at the top of the classifier; the larger grains sink and are drawn off through a spout at the bottom. The principle involved in this operation is known as hindered settling.

Efficient hydraulic classifiers, equipped with automatic controls and having single or multiple compartments, are sold by several manufacturers. This type of equipment is moderate in cost and has a large capacity in relation to the floor space required.

(d) *Hydraulic Sizers.*—A relatively new type of equipment for wet processing of sand which also operates on the hindered settling principle is the hydraulic sizer (see fig. 80). The sizer is essentially a trapezoidal tank divided into several compartments or pockets. Water enters from a pressure pipe through perforated construction plates at the bottom of each compartment. Minus No. 4 material is introduced into the narrow end of the sizer, and, as it is carried by water over baffles from compartment to compartment, coarser fractions of the successively finer and finer remaining sand settle to the bottom and are withdrawn through discharge pipes. The coarsest fraction settles in the first compartment, the finest in the last compartment. The size of material settling in each compartment is controlled by the velocity of water entering through constriction plates. Silt and deleterious materials are removed by an overflow launder.

Two hydraulic sand sizers, each having eight compartments, were used at Hungry Horse Dam. Approximately 25 percent of all sand was passed through these sizers. Portions of resulting fractions were blended with pit-run sand as necessary to meet specifications requirements. Typical gradings of sand obtained from each of the eight sizer pockets at Hungry Horse Dam are shown in the following tabulation:

Screen size	Sizer pocket No.							
	1	2	3	4	5	6	7	8
No. 4	2	0	0	0	0	0	0	0
No. 8	40	7	0	0	0	0	0	0
No. 16	25	32	1	0	0	0	0	0
No. 30	21	31	12	0	0	0	0	0
No. 50	11	29	84	56	4	1	0	0
No. 100	1	1	3	44	87	68	50	12
No. 200	0	0	0	0	9	30	49	83
Pan	0	0	0	0	0	1	1	5
F.M.	3.98	3.15	2.11	1.56	0.95	0.70	0.50	0.12

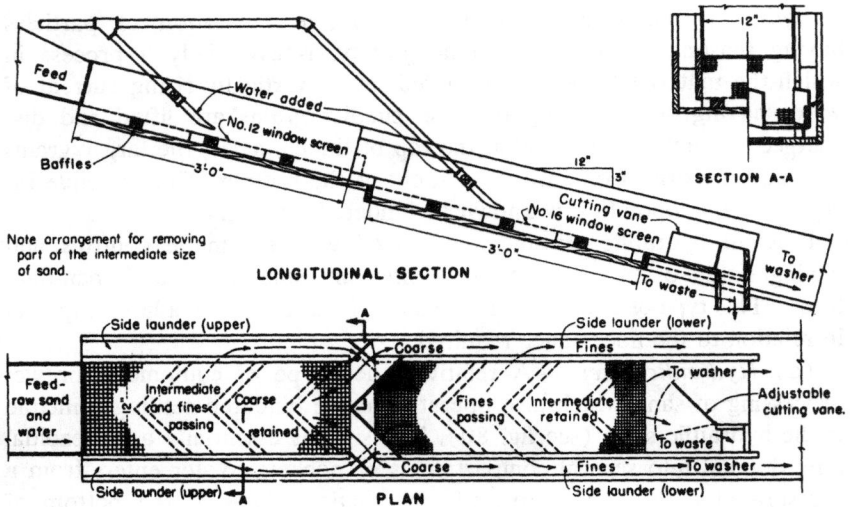

Figure 81.—A simple wet-screening device for removing surplus portions of a sand. 288–D–140.

(e) *Wet Process Screens.*—The arrangement of screens shown in figure 81 was developed to remove an excess of intermediate-size (No. 50 to No. 16) material. Subsequent installations of similar apparatus gave excellent results on several jobs. In operation of the apparatus illustrated, sand and water are fed onto a No. 12 window screen, which roughly separates the coarse particles of sand. The coarse material passes through chutes into the lower side launders and becomes a part of the finished product. Baffles beneath the screen divert the undersize to the upper side launders and also check the flow of water so as to keep the screen surface flooded. (The flooding facilitates movement of the material on the screen and brings about a more effective separation.) The material passing the No. 12 screen is discharged by the upper side launders onto a No. 16 window screen, effecting a separation between material of intermediate size and fine material. The undersize from this screen passes into the lower launders where it joins the coarse material retained by the first screen. The intermediate-size material retained on the No. 16 screen is divided by an adjustable cutting vane, diverting the undesired portion to waste and permitting the remainder to join the balance of the finished product. Water is added on the screens in amounts necessary to keep the sand in motion at the proper rate. The screens are fastened on individual frames so that they may be removed and replaced by spare screens without serious interruption in operation. Window screen is readily obtainable in the two mesh sizes shown and because of its flatness gives better results than the heavier and rougher double-crimped screen ordinarily used. Screens with dimensions and slopes as shown in figure 81 handle 7 to 10 cubic yards per hour.

Other problems in sand classification can be solved by variations of this method. To remove only coarse particles, the first screen and the cutting vane and waste chute shown in figure 81 could be used without the second screen. To waste finer material in the intermediate-size range, the No. 12 screen could be replaced with a shorter No. 16 screen, and the openings in the second screen, No. 16, could be reduced by distorting the mesh into a diamond shape. To waste more intermediate-size material, the first screen could be lengthened and more water used on the second screen. Worn or split screens should be promptly replaced.

Availability of improved wet sand classification equipment has led to further simplified fine material processing. Hydraulic sizers, which automatically direct portions of the sand sizers to waste while retaining and combining only the amounts necessary to control the overall grading, are being more frequently used for efficient sand processing. The incorporation of supplemental equipment, such as cyclones, centrifuges, and jigs to separate and retain fine sand sizes which were formerly inadvertently wasted in the washwater, makes it possible to save more of all sizes of sand and thereby increase overall plant processing efficiency.

66. Dry Processing of Sand.—In some localities where sand is naturally dry and water is not economically available for processing, the impracticability of using screens much finer than No. 8 for volume production in dry sand processing has led to development of other methods. During construction of the Colorado River Aqueduct in the southwestern desert, excess fines passing 100 mesh were successfully removed (when the moisture content was well below 1 percent) by specially designed equipment that drew air through thin sheets of sand cascading over a stepped series of steel angles.

Fairly effective control of the coarser part of a sand may be obtained by use of suitably arranged, adjustable baffles placed under the sand screen to deflect undesired portions to waste. Such separation is based on the fact that the coarser material progresses farther along the screen before falling through it.

The tendency of dry sand to segregate in handling, particularly as it is dropped onto stockpiles or into bins, is very much greater than that of moist or wet sand. For this reason special precautions are necessary. At Parker Dam, sand was moistened by spray on the conveyor belt to the stockpile. Although this virtually eliminated segregation, some difficulty was encountered in obtaining uniformly the 5 to 8 percent moisture needed to prevent excessive segregation. On the Colorado River Aqueduct work, segregation of dry sand was held to a practicable minimum by reducing the maximum size of particles to one-eighth of an inch.

Beneficiation of sand by reducing the angularity of the particles through agitation or attrition has been achieved in the laboratory, resulting in reductions of from 6 to 18 percent in water and cement requirements of concrete containing these sands with no appreciable reduction in compressive strength. Investigations indicate that, generally, any scrubbing action on a sand results in a somewhat smoother, less angular sand having a lower water requirement when it is used in mortar or concrete mixes.

67. Production and Handling of Coarse Aggregate.—The greater portion of the coarse aggregate used on Bureau work is natural gravel. On some projects, however, where gravel is not economically available, crushed aggregate is used. Although the shape of individual particles is important, it is not so critical for coarse aggregate as for sand. Use of corrugated roll crushers to produce the smaller sizes of coarse aggregate and of gyratory crushers or cone crushers with curved breaking plates to produce the larger sizes results in production of the least amount of flat and elongated pieces. Where crushed aggregate is unusually harsh and sharp edged, an improvement in particle shape, and consequently in workability of the concrete, may be obtained at low cost by passing the crushed material through a revolving cylinder equipped with lifters. The tumbling action reduces the sharp edges and also avoids much of the potential undersize that would otherwise result from subsequent handling of the aggregate. The undersize is screened out and washed.

In addition to necessary washing and crushing, coarse aggregate for Bureau work is separated into several sizes so that the concrete can be properly proportioned and uniform grading of successive batches assured.

Because of the segregation and breakage which results during normal handling and stockpiling operation, much aggregate that was well screened at the aggregate plant is far from being cleanly separated when it reaches the mixer. The more often the material is handled and the more breakable the stone, the greater the departure from uniformity. Improper methods of handling and stockpiling aggravate segregation of undersize. An accurately graded gravel may segregate during a single improper stockpiling operation to such an extent that the workability and quality of the concrete in which it is used will vary to an unacceptable degree. For instance, tests made at Hoover Dam showed that when $\frac{3}{16}$- to $\frac{3}{4}$-inch gravel was dropped from a belt on a conical pile and recovered through a single gate beneath the center of the pile, the fineness modulus of material withdrawn increased uniformly, during a period of one shift, from 6.58 to 6.94, as indicated in the following tabulation:

Screen	Sample No. 1		Sample No. 7	
	Total retained	Each size	Total retained	Each size
	Percent	*Percent*	*Percent*	*Percent*
¾ inch	2	2	10	10
⅜ inch	57	55	84	74
³⁄₁₆ inch	99	42	100	16
F. M.	6.58	(99)	6.94	(100)

Such segregation can be largely remedied by drawing simultaneously from two or more gates beneath each pile, thus effecting a remixing action.

For batching and mixing plants when a reclaiming tunnel is not used, some thought should be given to preparation of the stockpile area. A hard, clean base should be provided to prevent contamination from underlying material. It is good practice to place sand, gravel, or rock bedding materials over the area before stockpiling. This will reduce the effect of truck travel and wind which blows loose topsoil into the stockpiled aggregate. Overlap of different sizes should be prevented by providing ample separation or by suitable walls to separate the piles. If cranes are used, swinging of buckets of one size of aggregate over stockpiles of other sizes should be avoided.

Stockpiling and other handling operations result in a certain amount of chipping and breakage of the aggregate, particularly in the larger sizes. Breakage is reduced by decreasing the number of handlings and the heights of drop. Use of rubber belting as a drag to reduce the velocity of the moving material and as a resilient lining for areas of impact will reduce breakage extensively. Rock ladders for retarding the fall are also effective in minimizing breakage (fig. 82). Design of the steps should incorporate flat and sloping sections as shown in figure 83. The rise between flat sections of the steps should not exceed 2½ feet and preferably should be less.

The presence of sharp, angular rock chips in the gravel, usually as undersize, reduces concrete workability. Furthermore, it is difficult to adjust a mix to compensate for angular undersize because such material does not remain uniformly distributed but collects in pockets and in the lower parts of the bins and stockpiles, and enters the mix in intermittent heavy concentrations. For these reasons, coarse aggregates should be finish screened at the batching plant, as discussed later.

All handling and stockpiling operations cause some segregation and breakage of coarse aggregate, but such difficulties can be minimized by

Figure 82.—Rock ladders used at Grand Coulee Third Powerplant in Washington. These effectively reduce breakage of coarse aggregate. P1222–142–10039.

separating the aggregate into a number of sizes and by use of well-designed and properly operated handling systems. Some methods of handling helpful in limiting segregation are illustrated in figures 83, 84, and 85. On the All-American Canal, where finished aggregates were handled many times, difficulties caused by segregation were largely avoided by separating the 1½-inch gravel on ½- and 1-inch screens.

INCORRECT METHODS OF STOCKPILING AGGREGATES
CAUSE SEGREGATION AND BREAKAGE

PREFERABLE

Crane or other means of placing
material in pile in units not
larger than a truck load which
remain where placed and do not
run down slopes.

OBJECTIONABLE

Methods which permit the aggregate
to roll down the slope as it is
added to the pile, or permit
hauling equipment to operate
over the same level repeatedly.

PERMISSIBLE BUT NOT PREFERABLE

Pile built radially in horizontal
layers by bulldozer working from
materials as dropped from convey-
or belt. A rock ladder may be
needed in this setup.

Bulldozer stacking progressive
layers on slope not flatter
than 3:1

STOCKPILING OF COARSE AGGREGATE WHEN PERMITTED

(STOCKPILED AGGREGATE SHOULD BE FINISH SCREENED AT
BATCH PLANT. WHEN THIS IS DONE NO RESTRICTIONS ON
STOCKPILING ARE REQUIRED)

Uniform about
center

CORRECT

Chimney surrounding material falling
from end of conveyor belt to prevent
wind from separating fine and coarse
materials. Openings provided as required
to discharge materials at various
elevations on the pile.

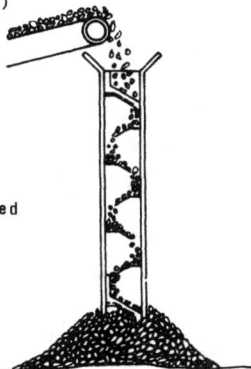

Wind

Separation

INCORRECT

Free fall of material from high
end of stacker permitting
wind to separate fine from
coarse material.

UNFINISHED OR FINE AGGREGATE STORAGE
(DRY MATERIALS)

When stockpiling large-sized
aggregates from elevated
conveyors, breakage is minimized by use of a rock ladder

FINISHED AGGREGATE STORAGE

Figure 83.—Correct and incorrect methods of stockpiling aggregates.
288–D–2655.

Figure 84.—Aggregate plant and stockpiles, Flaming Gorge unit, Utah.
P591–421–2559.

Aggregates for Colorado River Aqueduct were separated by screens having net square openings of ⅛, ½, 1 or 1¼ or 1½, and 2 or 2½ or 3 inches, the choice of sizes depending on the maximum size required.

Despite all reasonable precautions on the average job, the 2-percent maximum limit for significant undersize, a practicable limit for coarse aggregate *as screened,* is often exceeded in the materials *as batched,* especially when aggregates are taken from stockpiles. To eliminate excess undersize, finish screening of coarse aggregate over vibrating screens at the batching plant is generally required by specifications. The screened material passes directly to the batching plant storage bins. With properly operated vibrating finish screens, it is practicable to keep significant undersize within the specified 2-percent limit in the batched aggregate. A $\frac{3}{16}$-inch slotted screen is often used instead of a $\frac{3}{16}$-inch square opening screen when screening well-drained, but moist, $\frac{3}{16}$- to ¾-inch aggregate. However, care must be exercised in depositing the aggregate in storage bins; otherwise, undersize may accumulate. To minimize the formation of excess undersize through high drops, it is advisable in some instances to provide rock ladders for larger size aggregates.

Vibrating screens are most effective when mounted over the batching plant storage bins; however, they should be so mounted that vibration of the screens will not affect operation of batching scales. Where an existing plant is not structurally suited to this arrangement, the finish screening equipment may be installed adjacent to the batching plant either at

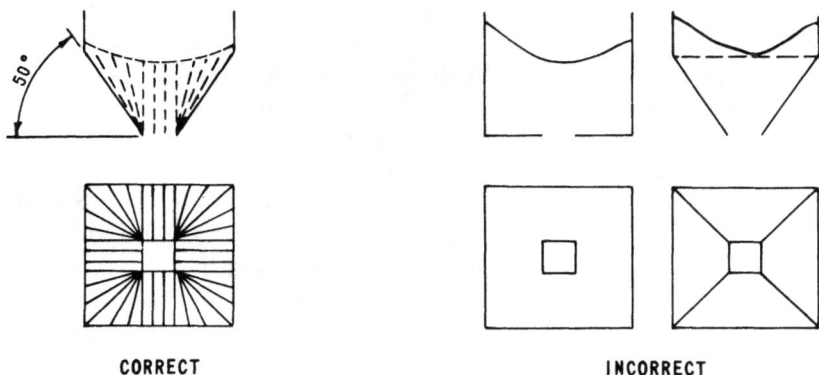

CORRECT

Full bottom sloping 50° from horizontal in all directions to outlet with corners of bin properly rounded.

INCORRECT

Flat-bottom bins or those with any arrangement of slopes having corners or areas such that all material in bins will not flow readily through outlet without shoveling.

SLOPE OF AGGREGATE BIN BOTTOMS

CORRECT

Material drops vertically into bin directly over the discharge opening, permitting discharge of more generally uniform material.

INCORRECT

Chuting material into bin on an angle. Material that does not fall directly over outlet is not always uniform as discharged.

FILLING OF AGGREGATE BINS

Figure 85.—Correct and incorrect methods of handling aggregates. Use of improper methods causes segregation which results in lack of uniformity in the concrete. 288–D–2648.

ground level or above the bins. In any case, the screened material should be conveyed directly to the proper bin.

Wet or dirty material may require use of wash-water sprays on finish screens. Where this is necessary, difficulty may arise from variations in water content of material entering the batchers if the operation is intermittent. If it is necessary to wash at the finish screens, a dewatering screen should be used before finish screening. However, on canal lining jobs in the Columbia Basin, where batching was quite regular, wash-water sprays were used on finish screens as needed with no appreciable difficulty reported in batching operations. Of their own volition, contractors adopted finish screening in construction of Seminoe, Island Park, and Scofield Dams and the Colorado River Aqueduct; they thereby avoided the usual difficulties and variations in mix arising from segregation and undersize accumulations. The volume of concrete required for Scofield Dam was less than 4,500 cubic yards. Figure 86 illustrates a finish screen mounted above a small batching plant.

Experience indicates that the cost of finish screening at the batching plant is often more than offset by (1) elimination of the need for careful screening and handling prior to finish screening, (2) avoidance of wasting rejected aggregate, (3) assurance that material will not be placed in the wrong bin over the batchers, (4) easier handling and placing because the concrete will be more uniform in consistency and workability, and (5) a savings of 2 to 10 percent or more in cement made possible by greater uniformity of the concrete. Ready-mix suppliers often find that installation of finish screens on their plants results in cement savings—sometimes as much as one bag per cubic yard.

To keep the different sizes balanced in the batching plant bins, the feed to the finish screens should be a regulated mixture of all the coarse aggregate sizes; otherwise, the rough-screened sizes must be finish screened alternately for relatively short periods. Only the undersize passing the 3/16-inch screen needs to be wasted during finish screening.

It is questionable whether bins should be drawn down to empty once each day. With some aggregates, accumulation in bin bottoms of undersize from breakage and permissible undersize that escapes removal in screening may occasionally be reduced by daily drawdown. However, additional undersize created during filling from the low point may negate the advantage of the drawdown. Judgment as to the amount of undersize that is collecting and its effect on concrete uniformity should establish the timing of the drawdown or its necessity.

For smaller jobs, coarse aggregate is sometimes screened over a stationary, sloping screen having 3/16-inch slotted openings and mounted above the batching bins. This arrangement does not have all the advantages of a vibrating screen, but it does much to improve uniformity of concrete by removing most of the fine undersize when the material is not too wet. Where this simple screening is permitted, a 3 percent maximum significant undersize limit is specified for the material as batched.

Figure 86.—Top view of a finish screen deck erected on an independent tower over the batching plant, Cachuma project, California. 368–SB–1810–R2.

Smaller size screens may occasionally have adequately removed the minus $\frac{3}{16}$-inch undersize material, but generally the stationary, slotted, sloping screens should be 4 feet wide and 6 feet long. The slots should be $\frac{3}{16}$ inch by 4 inches long in the direction of the slope. The slope should range between 30° and 50° from the horizontal and should be readily adjustable within these limits to provide maximum effectiveness.

Figure 87.—Stationary screen with 3/16-inch by 4-inch slotted openings for finish screening coarse aggregate used on the Columbia Basin project, Washington. Note pile of undersize material visible through screen. CB–3702–4.

The slope is dependent on rate of feed and on the material. The screens should be mounted above the batcher bins with some means of carrying away the rejected undersize material (fig. 87).

Stationary, slotted, sloping screens are usually sufficiently effective when rounded, natural dry aggregate is screened. However, this is not so with crushed aggregates (especially limestone) used in some areas or when moist aggregate is being screened. Also, overloading or too rapid feeding is a common problem in the use of these screens. Constant attention is usually required to keep the slots clean. For these reasons, a single-deck, vibrating, $\frac{3}{16}$-inch slotted screen is preferred.

68. Screen Analyses.—Periodic grading analyses of aggregate materials should be made at the aggregate plant to determine relative percentages of various size fractions. These also serve as a part of the production record enabling comparison with test pit data included in specifications for large mass concrete dams. The Bureau requires grading analyses at the batching plant because specified grading limits are for aggregates *as batched* and because grading *as batched* provide the best information on which to make the adjustments in batch weight necessary to maintain uniform graduation in the mix. These analyses also constitute a valuable record of concrete mix and cylinder test data.

It is sometimes helpful to obtain a complete history of aggregate plant production, as discussed in section 56, Daily Inspection Reports. The frequency of sand tests at the aggregate plant will depend on the uniformity of the pit-run material and the rate of plant production as well as on the condition and effectiveness of the processing equipment. On large jobs, it may be necessary to determine the grading as often as once an hour so that the equipment may be adjusted to keep the sand as uniform as practicable and within specified limits. Tests of sand at the batching plant, except for moisture content, are usually required less frequently.

The contractor's operations in rough-processing coarse aggregate at the aggregate plant need not be subject to rigid inspection. However, a sufficient number of tests should be obtained to establish rate and gradation of raw plant feed, total amount of material processed, amount of finished sand and coarse aggregate produced, quantity of crushed aggregate, and amounts of individual size fractions wasted. Such records may be important when closing out a contract. These tests are also helpful to the contractor during pit operation. At the batching plant, tests of coarse aggregate are made for grading, moisture content, and specific gravity. The frequencies of these tests should be sufficient to assure that aggregates conform to specifications requirements. The grading analyses include determination of undersize and oversize (sec. 18(g)). This information makes possible better control of concrete proportioning and better understanding of the nature of undersize problems. Tests for the nominal fraction and significant undersize and oversize should be made for compliance with limits as outlined in specifications and submitted as part of the L–29 Concrete Construction Data, form 7–1317, which summarizes grading test data recorded on form 7–1316. A graphic representation and description of the nominal fraction, oversize, and undersize limits are shown in figure 214, designation 6, in the appendix. Methods of making screen analyses are given in the appendix, designations 4, 5, and 6.

69. Deleterious Substances in Aggregate.—The quantity of deleterious substances in sand and gravel may often be reduced by avoiding contaminated areas in the deposit and by wasting strata of clay, shale, or silt. If contaminations are present in objectionable amounts and are not satisfactorily removed by washing or other means, the finished sand and gravel should be wasted if retreatment is not practicable. Extremely fine material occurring in aggregates requires relatively large amounts of mixing water. Also, the fines and inert material tend to work to the surface of the concrete, producing cracking and checking. Tests for determination of the amounts of clay lumps, organic matter, material passing No. 200 screen, and shale are given in the appendix, designations 13, 14, 16, 17, and 18. Chunks of clay or shale are readily visible, but quantitative tests are necessary. Both the presence and amount of organic matter and silt must be determined by test.

Other local deleterious substances, such as alkali, mica, coated grains, soft flaky particles, and loam, must be identified and the amounts determined by appropriate chemical tests or petrographic analyses. Requirements for quality of wash water are usually the same as those for mixing water. (See sec. 19.) Loose coatings of clay or calcareous material may often be softened by keeping the materials moist for some time and later removed during the washing operation. Special scrubbing and processing equipment for removing coatings and excessive quantities of soft particles from aggregate is commercially available. The processing should be so arranged that coatings and softer material are removed and not merely degraded or changed into smaller nominal sizes that may be retained.

Wood fragments, such as tree roots and driftwood, are a contaminating substance commonly found in aggregate deposits. These fragments are usually waterlogged and, therefore, cannot be removed by normal processing. However, there is available commercial mechanical equipment of a type used in mined-coal processing which, by a reversed waterflow, economically and effectively removes wood fragments, as well as other lightweight material, from the finished coarse aggregate. Because of the wide variety of sizes and shapes, it is impracticable to specify maximum limits for such material. Frequently, the amount of wood fragments may be materially reduced by deeper stripping or selective digging. Handpicking has been resorted to in a few instances but is expensive. In the construction of Bull Run Dam in Oregon, excessive quantities of roots that could not be removed by handpicking from conveyors were successfully removed by use of airblasts directed through the gravel as it fell from chutes into bins. In determining whether wood fragments are objectionable, the following may be helpful:

(1) For walls that are designed to withstand hydraulic pressures, sticks of any size having lengths two-thirds or more of the thickness of the wall are objectionable.

(2) For formed surfaces where a good appearance is important and for unformed surfaces where a high degree of durability of the surface is necessary, it is desirable that all wood fragments be removed.

(3) For other types of structures, wood fragments may not be particularly objectionable; but as a matter of good construction practice, the amount of such contaminating substances should be reduced to a practicable minimum.

70. Beneficiation of Aggregates.—Aggregates should be hard, to resist grinding action; tough, to withstand impact; strong, to stand up under heavy loads; and sound, to remain whole during freezing and thawing and other changes in weather conditions. Laboratory tests usually will eliminate for consideration and use most aggregates containing excessive quantities of deleterious, lightweight, and porous materials. However, freeze-thaw tests have shown that some aggregates which otherwise meet specifications requirements may, because of excessive amounts of these undesirable constituents, produce concrete that is susceptible to surface scaling or other deterioration during freezing and thawing. Where these are the only aggregates readily available, their quality frequently can be improved by reducing the quantity of porous, lightweight, deleterious particles present. These deleterious particles can be economically removed from a given source in many instances by one or more of the following methods: Heavy media separation, hydraulic jigging, or elastic fractionation. Another means of beneficiating aggregates, but one not presently developed to the point of being economically feasible, is through particle shape improvement accomplished by hydraulic attrition.

(a) *Heavy Media Separation.*—Heavy media separation was developed originally for beneficiation of low-grade mineral ores. However, the application of this process, or HMS as it is commonly known, has been expanded to include use by producers of aggregate and other nonmetallic minerals to separate, for one reason or another, various constituents occurring in the raw materials. HMS is a sink-and-float process whereby a medium consisting of water, finely ground ferrosilicon, and magnetite mixed to a selected specific gravity causes the undesirable lightweight aggregate constituents to float and be wasted and permits the desirable heavy portion to sink and be recovered for use.

Several types of HMS equipment are available, differing only in minor details. All are similar in their main principles and operate essentially as follows:

The feed is normally processed by washing and preliminary classification prior to heavy media separation. The specific size or sizes to be treated are fed into a separatory vessel which may be a cone (fig. 88), a revolving drum, or a spiral separator (fig. 89). A slurry having sufficient density to remove the objectionable material circulates through the machine and the complete sink-and-float separation takes place within the vessel. The sink-and-float products are removed continuously, the float being removed over a weir or lip on the separatory vessel, and the sink material reclaimed by augers, lifters, or an airlift. Submersion of feed into the heavy medium is gained by gravity fall. Mild agitation is

Figure 88.—Laboratory model of cone-type heavy media separation plant used to beneficiate concrete aggregates: (1) feed hopper, (2) adjustable vibrating feeder, (3) gear-head motor, (4) separation heavy media cone, (5) airlift for sink material, (6) wash-water spray nozzle, (7) vibrating screen, (8) media drain sump, (9) wash sump, (10) sink product spout, (11) float product spout, and (12) partition to separate sink and float products. PX–D–14811.

Figure 89.—Laboratory model of spiral-type heavy media separator used for beneficiating aggregates. Lightweight particles are floated off over the weirs shown in foreground and are wasted. Heavy particles sink to the bottom of the heavy media pool, are reclaimed by the spiral, and rewashed to remove media. PX–D–14646.

applied to maintain proper density of slurry during operation of the machine.

The sink-and-float portions are discharged from the HMS plant onto divided sections of a vibrating screen. During initial travel on the screen, the heavy medium is drained by gravity and returned to the machine. As the processed material travels farther along the screen, sprays of water remove those portions adhering to sink-and-float products. These products are then chuted from the screens to their respective storage areas. The diluted medium removed during washing is fed to magnetic separators for recovery of the solid portion of the medium and then conveyed to a densifier that dewaters and stores the cleaned medium for return to the circuit as needed. The test method for control of the HMS process is outlined in designation 41.

(b) *Hydraulic Jigging.*—This process also depends upon the difference in specific gravity of the retained material and that to be wasted. Deleterious particles conducive to removal from low-grade sand and gravel are mainly coal, chert, wood, sandstone, clay, chalk, and shale.

These materials range in specific gravity from 1.6 to 2.4; the specific gravity of good quality sand and gravel usually varies from 2.6 to 2.8.

A hydraulic jig employs vertical pulsations of water through a layer of aggregate in a nearly horizontal flat bed. The difference in specific gravity of deleterious and sound particles of aggregate is sufficient to bring about stratification of the material during upward pulsations of water through the aggregate as the material moves across the bed. The lightweight material is scalped over a weir and both the light and heavy material chuted to respective storage areas. Because of the relatively small capacity of this type of equipment, multiple installation is usually necessary.

(c) *Elastic Fractionation.*—The principle of operation of this method of aggregate beneficiation is based on variations in elastic modulus rather than specific gravity, wherein hard, dense aggregates are more elastic and consequently bounce farther than soft, porous aggregates.

Aggregates to be processed by this method are conveyed to an overhead hopper. This hopper discharges to a vibratory feeder that controls the feed and drops the aggregates onto an inclined hardened steel plate. Depending on height of drop and angle of inclination of steel plate, the dense hard, elastic particles will rebound a considerable distance, whereas the porous, soft, friable particles will rebound short distances. Adjustable bins can be located at suitable distances from the plate to receive the product of the quality desired. By positioning the adjustable bins, the beneficiated product can be made to conform to almost any specifications requirement for quality. Obviously, particle shape will affect the bounce of individual particles, and recirculation of intermediate material is generally provided to increase recovery of desirable material.

(d) *Sand Attrition.*—Equipment is available for processing angular sand into more acceptable subrounded shapes by abrasion. It is also sometimes possible to remove soft, friable particles through vigorous scrubbing and agitation in a circulating sand-water slurry. Unless the soft particles readily break down into fines that can be eliminated by washing or classification, sand quality improvement is often negligible if the larger soft particles only degrade to a smaller size. Hydraulic jigging might then become a more effective method for improving an unacceptable quality of sand. With the usual concrete mixing time, a surprisingly small amount of sand particle breakdown occurs because of a cushioning effect, and thus a friable sand is usually not as detrimental as is frequently believed. However, every effort should be made to find and use the best quality of sand available.

71. Control of Surface Moisture in Aggregate.—There are two reasons for determining surface moisture in aggregate: First, to permit compensation in batch weights; and second, to permit the computation of water-cement ratio. Each is sufficiently important to warrant the making of frequent moisture tests. Designation 11 in the appendix gives a procedure for determining surface moisture in aggregate. In plants that rapidly mix large quantities of concrete, it is always required that an electrical resistance device be installed in the sand batcher to indicate rapid changes in moisture content. These moisture meters are available commercially and are effective in indicating abrupt free moisture changes.

Specifications require that the sand delivered to the batching plant shall have a reasonably uniform and stable moisture content. (One percent change in surface moisture in the sand will change the slump of the concrete approximately 1½ inches.) This requirement may be met by providing suitable drainage facilities and allowing adequate time for drainage—at least 24 hours. It may be necessary to arrange for two or three piles of sand near the processing plant. Where three piles are used, one receives the wet sand, one is in process of draining, and the other is drained and ready for use. The amount of moisture that can be held by sand with reasonable stability depends on grading, particle shape, surface texture, and storage practices and usually ranges from 3 to 6 percent.

72. Specific Gravity.—Methods for determination of specific gravity are described in the appendix, designations 9 and 10. Tests should be made as required or found desirable during the period of concrete operations, particularly during the early stages and when there is reason to suspect a change in the aggregate. Accurate knowledge of the specific gravity is important for use in initial air-entrained mix design computations and in the determination of air content and yield quantities of each material in the batch.

73. Miscellaneous Tests of Aggregate.—Tests for organic matter, material passing the No. 200 screen, clay, and shale are made when there is a question concerning compliance with specifications. The particular tests to be made in the field as the work progresses are dependent on job conditions and should be determined by the control engineer. The tests enumerated, together with grading, specific gravity, Los Angeles abrasion, petrographic properties, absorption, and sodium sulfate soundness tests of aggregates, freezing and thawing tests of the concrete, determination of potential benefits of heavy media separation, and other tests as necessary, are initially made in the Denver laboratories during the investigational period, and copies of the reports are sent to the project.

74. Aggregate Purchased.—Specifications for the quality of aggregates obtained from commercial plants and inspection of such materials are the same as for aggregate processed on the job.

75. Cement.—The discussion in this section applies principally to portland cement but could also apply to blended hydraulic cements, such as portland blast-furnace slag cement, portland-pozzolan cement, and slag cement. As requested by the Portland Cement Association, cement quantities are now expressed in bags or tons. Finely divided pozzolanic material is sometimes used as a separate ingredient. It will be recognized that some but not all of the following statements are applicable to pozzolan.

The quality of cement and pozzolan for Bureau construction is monitored by the Corps of Engineers. Either the cement is accepted from commercial bins of prequalified manufacturers, or it is accepted from sealed silos, previously tested by the Corps of Engineers and reserved for Government use. Nevertheless, instances have occurred where excessive bleeding, abnormal stiffening, delayed setting, or low strength of concrete have been attributed to the cement. Any abnormal performance of cement should be reported immediately to the Chief, Division of Research. Also, a bag of the suspect cement should be shipped to the Denver laboratories, and a sample identified for future-reference should be preserved on the project.

When the cement is supplied from a prequalified manufacturer, Bureau personnel sample the cement at the jobsite and send it to the Corps of Engineers laboratory in Vicksburg, Mississippi, for testing to monitor the quality of cement supplied by the manufacturer. If the cement fails to meet specifications standards, the manufacturer may be removed from "prequalified" status. If cement is supplied from a bin reserved for Government use, cement samples for testing by the Corps of Engineers are normally taken at the cement plant as the bins are being filled. Bins containing the tested cement are reserved for exclusive Government use. Tests for false set are normally made on samples taken at the latest time prior to shipment. On jobs where only small quantities of cement are involved, generally in the range of 10 to 300 tons, the cement is usually accepted on manufacturer's certification that the cement conforms to specification requirements. In such cases, Bureau specifications require that the certification be accompanied by a mill-test report. This test report should contain sufficient data to show that the cement meets the specified physical and chemical requirements. On very small jobs, generally involving less than 10 tons of cement, testing or certification is normally not required.

Cement is inspected for contamination and for lumps caused by moisture after arrival on the job. There have been been instances in which tramp iron in bulk cement has seriously damaged the contractor's unloading equipment. Apparently most of this material comes from the cement manufacturing plant; however, it may, at times, come from railroad cars.

Placing effective dust collectors on the vents of bins and silos in which bulk cement is stored reduces loss of cement during handling and discomfort to workmen. Loss of cement during transfer from cement truck to cement silo can be minimized by an arrangement similar to that shown in figure 90.

Figure 90.—Connection used in dumping cement from transport trucks to receiving hoppers of cement storage silos at Glen Canyon Dam, in Arizona. This arrangement minimized cement loss. P557–420–04599.

When cement is delivered in bags, sufficient check weighings should be made to give reasonable assurance that the bag weights conform with the Federal specification requirements. The specification requirements provide that packages that vary more than 3 percent from the stated weight may be rejected and that if the average weight of 50 packages taken at random from any shipment is less than that stated, the entire shipment may be rejected. The bags should also be inspected for rips, tears, and other defects. Records of departures from specified weight requirements will serve as a basis for adjustment of payment for the cement.

To reduce the possibility of damage to bagged cement from unfavorable storage conditions, specifications require the contractor to use such cement in the chronological order in which it is received on the job. They also require that bulk storage bins be emptied and cleaned when directed, but such operations are not usually required at intervals of less than 4 months. In ordering cement, account should be taken of expected periods when cement will not be needed, such as seasonal shutdown, so as to avoid unnecessary carryover and possible deterioration in storage. Damage is indicated by lumpiness or falling off in concrete strengths (if strength variations are not known to be attributed to other causes). Denver laboratory tests for deterioration include determinations of specific surface, percentage passing a 325-mesh screen, strength, and any other tests that appear necessary. In the event that cement shows evidence of deterioration in storage, a 1-bag sample should be sent to the Denver laboratories for testing.

Temperature of the cement is important on large mass concrete jobs because high temperatures tend to increase the rate of reaction between cement and water and also increase the possibility of false set. (See discussion of abnormal set in section 16.) False set and its effects, slump loss and increased shrinkage cracking, have been associated with some cements that arrived on the job at very high temperatures. Cement temperature is dependent on conditions of storage, season of the year, and time that has elapsed since manufacture. Cement temperatures should be recorded when they are limited by specifications or may be a factor in cement performance.

Although the test for false set in cement is normally performed by the cement manufacturer immediately prior to shipment, it is sometimes advisable that the cement be tested on the project. Designation 8 of the appendix describes the procedure.

Cement for contraction-joint or foundation grouting should be sufficiently fine to penetrate very thin joints and sufficiently slow setting to prevent plugging of the grouting system before the joints are completely filled. For these reasons and to prevent stoppage of pumping operations caused by plugging of equipment, the cement should be completely free of lumps of hydrated cement, raw clinker, tramp metal, or any other foreign material. Generally, an air-separated cement of a fineness and composition comparable to type II, with 100 percent passing a No. 30 sieve, will meet the requirements for use in either contraction-joint or foundation grouting. However, as an additional control, it is required that 99.7 percent will pass a No. 100 mesh sieve. Use of freshly ground cement is desirable to forestall any possibility of partial hydration which can cause formation of unsuitable lumps in the cement. If this occurs, screening through a No. 30 sieve may be necessary at the jobsite. Grouting cement should be packaged in waterproof bags to prevent hydration from exposure to moist atmospheric conditions.

76. Water.—Requirements with respect to quality of mixing and curing water for concrete, and reasons for such requirements, are discussed in sections 19 and 42. Procedures for sampling water may be found in the appendix, designation 3. After construction is underway and the sources of water have been established, regular testing and inspection are not necessary unless the water becomes contaminated with excessive amounts of suspended matter because of abnormal streamflow or with objectionable concentrations of soluble salts during dry seasons. If the water is clear and does not have a brackish or saline taste, testing is unnecessary.

77. Admixtures.—The use of admixtures in concrete is discussed in detail in section 20. Procedures for sampling and testing are outlined in section 41 and the applicable test designation in the appendix. On the job, care should be taken in the handling and storage of any admixture.

It is important that calcium chloride in storage be protected from moisture and moist air. The amount of this admixture used should not exceed 2 percent of the cement, by weight. Generally, 1 percent is sufficient. Before being added to the mix, the calcium chloride should be dissolved in a part of the mixing water because undissolved flakes or lumps may severely damage the concrete.

Use of an air-entraining agent requires at least daily tests during concreting operations to determine the air content of the concrete. The dosage of the agent must be adjusted as necessary to maintain air content within desired limits. In Bureau work, entrained air content varies from a minimum of 3 percent for mass concrete to a maximum of 8 percent for concrete in which the maximum nominal size of aggregate is one-half inch. (See table 14 in chap. III.)

The amount of air entrained is governed by the amount of solids contained in the air-entrained solution. Solutions are available in single, double, and triple strengths. It is often advisable to dilute the extra-strength solutions to single strength to facilitate accurate measurement. Variations in amounts of solids can be detected by testing with a hydrometer. Should dosage amounts fluctuate excessively, such checking may indicate the source of variation.

Water-reducing, set-controlling agents are normally required when the quantity on a contract exceeds 2,000 cubic yards, except in certain types of construction. These agents have been discussed in detail in section 20(c). Like air-entraining agents, their effect on a concrete mixture is proportional to the amount of solids introduced. However, there is an amount above which no additional water reduction will be achieved, and the minimum amount of WRA required to obtain the desired effects should be used. Low ambient temperatures extend the retardation period which can result in a delay in concrete finishing. During cold weather, it may be advisable to reduce the dosage somewhat. With normal behavior of the WRA with the cement, the admixture will extend the setting time, which is advantageous in hot weather concreting, but will not extend the time for corresponding slump loss. A sample of the WRA the contractor selects to use should be tested with materials and mixes used on the job.

B. Batching and Control Facilities for Large Concrete Jobs

78. Weight vs. Volume Batching.—Expressing mix proportions by volume and batching of materials on a volume basis have been discarded in favor of the weight system. The amount of solid granular material in a cubic yard is variable; a volume of moist sand in a loose condition weighs much less than the same volume of dry compacted sand. A ton of aggregate, on the other hand, is a definite quantity, which, for precision, usually requires qualification as to moisture content only. Also, the weight of any concrete ingredient is directly related, through specific gravity, to the solid space that the material occupies in the concrete. Use of the weight system in batching makes for accuracy, flexibility, and simplicity.

79. Batching Equipment.—Requirements for submittal, by the contractor, of drawings and a description of the batching equipment proposed for installation and for preliminary approval of the equipment are similar to those in section 63 applying to aggregate processing and handling equipment.

Improved equipment for accurate and quick weighing of concrete ingredients has been developed to keep pace with facilities for rapid placing of concrete in large dams. A wide variety of electronic controls is available to fit individual requirements for maximum ease and efficiency of operation, simple mix selection, and proportional-batch control. Figure 91 shows a part of the batching equipment used at one of the mixing plants at Grand Coulee Dam. The two scales are for weighing two sizes of aggregates. Note that there are five cutoff beams with sliding poises in each scale box. This scale arrangement for each of the ingredients makes it practicable to set the scales so that the various materials for any one of five different mixes can be weighed simultaneously at the push of a button. In the upper left-hand corner of each scale box can be seen the two mercoid switches—one for stopping the initial rapid feed and the other for stopping the final dribble feed. These switches are actuated by the movement of the long balancing beam seen just below them. Such equipment is furnished and operated by the contractor, and its satisfactory operation in conformance with the specifications is the contractor's responsibility.

Specifications for large Bureau concrete dams require the cement, pozzolan, sand, and each size of coarse aggregate entering each batch of concrete to be weighed separately. Water and admixtures may be separately weighed or measured volumetrically. The contractor is required to provide convenient facilities for readily obtaining samples of concrete ingredients from the discharge streams between bins and batch hoppers or between batch hoppers and the mixers. Figure 92 shows an arrangement

Figure 91.—Batching equipment used at the mixing plants for Grand Coulee Dam, in the State of Washington. Batching equipment is usually installed on one floor of the mixing plant. PX–D–38758.

for obtaining samples of aggregate materials entering the weighing hoppers.

Each weighing unit should be equipped with a visible springless dial which will register the scale load of any stage of weighing operation from zero to full capacity. Full capacity dials not only show whether the hoppers are properly charged but also whether the hoppers are completely discharged, which is equally important. In addition, these dials reveal the nature and extent of any irregularities in flow of the materials caused by arching, jamming, leaky gates or valves, and formations of incrustations in the hopper.

Figure 93 is a view of the concrete batching and mixing plant at Morrow Point Dam. Suspended screens for finish screening were placed above the batching bins. Aggregates were cooled by refrigeration units on the outside of the bins. In this plant, automatically controlled hydraulic batching gates supplied six 4-cubic-yard mixers. Weight-indicating dials, control panel, and recording instruments were located conveniently in the control room.

Common features of modern batching and mixing plants include automatic batching with interlock controls which prevent double batching, batch selectors as well as punched cutoff adjustments, automatic recorders

Figure 92.—Aggregate sampling bucket in batcher at mixing plant. Bucket is on rollers for easy removal of sample. PX–D–25253.

of batch weights, and moisture meters. Scale dials and a recorder located conveniently in the control cabinet are shown in figure 94.

The use of ice as a part of the mixing water, to lower the temperature of concrete as placed, is common practice. Ice absorbs 144 Btu per pound

Figure 93.—Concrete batching and mixing plant at Morrow Point Dam in Colorado. P622B–427–2945.

in melting; and cold water is about 4½ times as effective, per unit weight, as cold aggregate or cement in reducing the temperature of the mix. However, coarse aggregate and sand comprise three-fourths of the total weight

Figure 94.—Control panel (lower view) and recording chart (upper view) of the mixing plant at Hungry Horse Dam, in Montana. P447–D–20638.

of a cubic yard of concrete, or 10 to 15 times the weight of the usual amount of mix water. Obviously then, the temperature of the aggregate is the most influential factor in controlling the temperature of the fresh concrete.

It was necessary to use ice as well as precool the aggregates to maintain the required 50° F concrete placing temperature during hot weather at Glen Canyon Dam. Because of the difference in weight per unit volume and its tendency to clog the discharge lines, ice should preferably be batched by weight, separately from the water. At Glen Canyon Dam, ice for the mixing water was supplied by 12 refrigerator units. The ice was passed through an ice flaker and the crushed ice transported to the batching and mixing plant on a belt conveyor. In short-time mixing, when aggregates are precooled substantially, the ice may not completely melt in the mixer. If this is the case, additional mixing time may be required or the amount of ice limited to approximately 30 percent of the mixing water.

Methods and equipment for batching air-entraining agents and WRA's are discussed in section 84.

80. Checking Scales.—In accordance with contract specifications, concrete ingredients to be weighed must be weighed with approved equipment that operates within certain specified limits of accuracy. To ensure proper compliance, the weighing apparatus is inspected by the Government prior to its use and periodically throughout the construction period. Periodic checks of the accuracy of the equipment are required to be made by the contractor under the supervision of a Government representative.

Standardized test weights, furnished by the contractor, are used in checking scales. The scale should be loaded in increments, making successive use of test weights available, until a total weight has been reached that is equal to the maximum range of the scale dial.

If there is evidence that scales are not operating properly because the mechanism is not clean or because of interference by other objects, the objectionable operating conditions should be corrected.

Testing of the batching system in a large plant is a more involved procedure than checking of platform scales because the equipment usually includes not only full reading dials but also automatic feeding and cutoff facilities and graphic recorders. The usual procedure is to check the weighing accuracy of the scales and at the same time determine the accuracy of the batch-weight recording. Figure 95 shows a typical batching hopper with the test weights suspended. Figure 96 illustrates a method which can be used to tabulate the data during the scale calibration. The operator's control dial readings are recorded because the batch determination selector is located in the operator's control room and actuates both the weighing hopper scales and the indicating dials on the control panel. The weight in the weighing hopper actuates the graphic recorder.

In the batching operation, it is necessary to weigh material in the weighing hopper as the hopper is being filled from the bins, and for this reason, automatic timing devices for closing the bin gates are provided. The limits of accuracy for these batch-weighing cutoffs are established in

Figure 95.—A typical arrangement for checking batching scales. Scales are checked by suspending test weights from hopper. P557—420—05353.

the specifications, and thus it is necessary that regular and systematic checks of the accuracy of this equipment be made during plant operation.

Specifications require that weighing equipment shall conform to the

BATCHING PLANT SCALE CHECK

DATE _11-1-60_ SHIFT _Day_ INSPECTOR _MacDonald_

SPEC. No. _DC-4825_ PLANT _Perm._ MATERIAL _Sand_ CAPACITY, LBS _5000_

ACTUAL TEST WEIGHT, POUNDS	OPERATOR'S CONTROL DIAL		BATCH RECORDER		WEIGHING HOPPER SCALE DIAL		ALLOWABLE ERROR IN WEIGHING HOPPER, 0.4 PERCENT OF ACTUAL WEIGHT
	READING	ERROR	READING	ERROR	READING	ERROR	
0	0	0	0	0	0	0	0
100	100	0	100	0	100	0	0
200	200	0	225	+25	200	0	1
300	295	−5	320	+20	300	0	1
400	390	−10	410	+10	400	0	2
500	485	−15	510	+10	500	0	2
600	585	−15	615	+15	600	0	2
700	685	−15	720	+20	700	0	3
800	780	−20	825	+25	800	0	3
900	880	−20	920	+20	900	0	4
1000	985	−15	1010	+10	1000	0	4
1100	1085	−15	1110	+10	1100	0	4
1200	1190	−10	1200	0	1200	0	5
1300	1290	−10	1300	0	1300	0	5
1400	1390	−10	1400	0	1400	0	6
1500	1495	−5	1510	+10	1500	0	6
1600	1600	0	1610	+10	1600	0	6
1700	1700	0	1720	+20	1700	0	7
1800	1805	+5	1820	+20	1800	0	7
1900	1905	+5	1920	+20	1900	0	8
2000	2005	+5	2015	+15	2000	0	8
2100	2110	+10	2115	+15	2100	0	8
2200	2210	+10	2215	+15	2200	0	9
2300	2310	+10	2315	+15	2300	0	9
2400	2410	+10	2410	+10	2400	0	10
2500	2505	+5	2520	+20	2500	0	10
2600	2605	+5	2625	+25	2600	0	10
2700	2700	0	2725	+25	2700	0	11
2800	2795	−5	2820	+20	2795	−5	11
2900	2895	−5	2920	+20	2895	−5	12
3000	2990	−10	3020	+20	2995	−5	12
3100	3085	−15	3110	+10	3095	−5	12
3200	3180	−20	3210	+10	3195	−5	13
3300	3280	−20	3310	+10	3295	−5	13
3400	3380	−20	3410	+10	3395	−5	14
3500	3480	−20	3510	+10	3495	−5	14
3600	3585	−15	3610	+10	3595	−5	14
3700	3685	−15	3710	+10	3690	−10	15
3800	3785	−15	3810	+10	3790	−10	15
3900	3885	−15	3910	+10	3890	−10	16
4000	3985	−15	4010	+10	3990	−10	16
4100	4090	−10	4110	+10	4090	−10	16
4200	4190	−10	4210	+10	4190	−10	17

Figure 96.—A typical form for recording data and computations for the checking of a batcher scale. 288–D–3279.

applicable requirements of Federal Specification AAA–S–121D for such equipment, except that accuracy to within 0.4 percent for any increment of test weight up to the capacity of the scale is considered satisfactory. Specifications also stipulate that the combined feeding and weighing (scale) errors shall not exceed specified limits, which differ with the material to be weighed. More accurate batch weights can be obtained if scale errors, even though within permissible limits, are taken into account in making subsequent scale settings. For convenience, the corrections required at 500-pound intervals for aggregate and at 250-pound intervals for cement may be recorded on cards when the scales are checked.

The method as shown in figure 97 can be used to tabulate data for setting the desired weights on the batch selector. Figures 98 and 99 are used to calculate the batch weights, taking into consideration the original mix design and the existing variation in aggregate grading as being batched. It is more convenient to set the batch selector for the desired weights on the operator's control panel dial and check the actual weights during operation on the weighing hopper dial. The test data are then tabulated as shown in figure 100, and from these the combined weighing and feeding errors during plant operation are computed to determine compliance with the specifications. This form can also be used to tabulate readings to determine the accuracy of the operator's control dial and the batch recorder during plant operation.

An automatic data processing program has been developed and applied that will perform all calculations used in concrete mix design, mix adjustments, scale settings, clean separation gradings, and yield. The last two items are used in preparation of the monthly L–29 Concrete Construction

CONCRETE BATCH DETERMINATION

DATE _11-17-60_   SHIFT _Swing_   MIX No. _6-1-16_   SELECTOR No. _3_

CONCRETE DESIGNED FOR _Interior mass_   CU. YD./BATCH _4.0_   COMPUTED BY _L.D.R._

MAX. AGGR. _6"_   MAX. W/(C+P) _0.59_   MAX. SLUMP _2.0"_   ENT. AIR, PERCENT _3.5_

CEMENT, LBS./CU.YD. _198_   POZZ.,LBS./CU.YD. _94_   WATER,LBS./CU.YD. _172_   RETARDER BY WT. OF CEM.+POZZ.,PERCENT _0.37_

MATERIAL	CEMENT	POZZ.	SAND	$\frac{3}{4}$"	$1\frac{1}{2}$"	3"	6"	WATER	ICE	W.R.A.	CaCl₂
Aggregate, percent			21	23	24	22	10				
Weight required (Sat.Sur.Dry)	792	376	2960	3240	3392	3112	1420	688			
Grading correction			−102	+229	−13	−87	−27				
Corrected weight			2858	3469	3379	3025	1393				
Percent moisture, net		0.6	4.0	2.0	0.2	0	−0.2				
Moisture, pounds		2	114	69	7	0	−3	189			
Batch weight, pounds	792	378	2972	3538	3386	3025	1390	268	192	11	28
Control dial correction	+6	+1	−10	0	+5	+30	−4	+2	+3	0	0
Control dial setting	798	379	2962	3538	3391	3055	1386	270	195	11	28
Weighing hopper scale correction	−2	0	−5	−15	−5	+5	−2	0	0	0	0
Desired reading on scale weighing hopper	790	378	2967	3523	3381	3030	1388	268	192	11	28

Figure 97.—A typical form for tabulating data for control dial and batch selector setting. 288–D–3280.

CONCRETE MIX DESIGN DATA   DATE _10-24-60_

MIX No. _6 1 16_   FOR USE IN _Interior mass_   COMPUTED BY _O.W.P._

W/(C+P) _0.59_   BASED ON _2"-inch slump, 3.5% Air_

CEMENT, LBS./CU.YD. _198_   POZZ., LBS./CU.YD. _94_

CoCl₂ BY WEIGHT OF CEMENT, PERCENT _--_   WRA BY WEIGHT OF CEMENT + POZZ., PERCENT _--_

MATERIALS	CEMENT	POZZ.	WATER	SAND	No. 4 to $\frac{3}{4}$"	$\frac{3}{4}$" to $1\frac{1}{2}$"	$1\frac{1}{2}$" to 3"	3" to 6"	AIR
Batch weights	198	94	172	740	810	848	778	355	--
Proportion, percent	--	--	--	21	23	24	22	10	3.5
Specific gravity	3.16	2.37	1.00	2.61	2.61	2.62	2.62	2.63	--
Unit weight	196.87	147.65	62.4	162.60	162.60	163.23	163.23	163.85	--
Volume, cu. ft.	1.006	0.637	2.756	4.548	4.981	5.197	4.764	2.166	0.945

UNIT WEIGHT _147.96_

Figure 98.—A typical form for recording concrete mix design data. 288–D–3281.

Data Report. The program is written in Fortran. Access to the central computer at the Engineering and Research Center is gained by telephone using either a portable terminal with an acoustical coupler or a permanent terminal. The adoption of this equipment will considerably reduce time required to perform such calculations as well as reduce the incidence of errors.

81. Graphic Recorders.—Bureau specifications require the use of combined autographic recorders for concrete batching and mixing operations for major concrete dams. The recorder is required for making a continuous, visible record on a single roll of paper, of the amount of each material (including mixing water) measured for every batch of concrete and of the consistency of the concrete. The recording mechanisms should be so designed that the recording pens indicate the weight in the batching bins until they are discharged. With this feature, any additions to the batch or incomplete discharge will be indicated on the recorder roll. Provision is also made for registering time of day, usually at 15-minute intervals.

Some of the recording charts are as wide as 6 feet and record the movements of 15 pens. In addition to the 15 moving pens, there are adjustable limit-line pens. A valuable feature of the recording apparatus is a printing device that stamps gradation lines and numerals, time-interval lines, and designations for the graphs on a roll of blank, unperforated paper during the recording operation, thus providing the benefits of factory-ruled-and-printed paper without its shortcomings.

Operation and adjustment of the recorder are responsibilities of the contractor; but if the apparatus does not function accurately or is not given

COMPUTATIONS FOR CONCRETE BATCH DETERMINATIONS USING AVERAGE OF 10 PREVIOUS GRADINGS

DATE 11-17-60   COMPUTED BY O.W.P.

MIX NO. 6 I 16   CHECKED BY L.D.R.

MATERIAL	BATCH WEIGHT, POUNDS	SCREEN SIZE	PER-CENT	WEIGHT, POUNDS	COR-RECTED WEIGHT, POUNDS	COR-RECTED BATCH WEIGHT, POUNDS	COR-REC-TION	SAT. SUR. DRY DESIGN WEIGHT	BATCH WEIGHT, POUNDS
(1)	(2)	(3)	(4)	(5)	(6)	(7)	(8)	(9)	(10)
SAND	(+) 2860	PASSING NO. 4	100.0	2860	2860	(−)			
		RETAINED ON NO. 4	0.0	0	102	2962	(−) 102	2960	2858
NO. 4 TO 3/4"	3512	PASSING NO. 4	2.9	102	0				
		NO. 4 TO 3/4"	91.1	3199	3199				
		RETAINED ON 3/4"	6.0	211	83	3282	(+) 230	3240	3470
3/4" TO 1 1/2"	3329	PASSING 3/4"	2.5	83	211				
		3/4" TO 1 1/2"	90.1	3000	3000				
		RETAINED ON 1 1/2"	7.4	246	132	3343	(−) 14	3392	3378
1 1/2" TO 3"	3011	PASSING 1 1/2"	4.4	132	246				
		1 1/2" TO 3"	90.7	2731	2731				
		RETAINED ON 3"	4.9	148	121	3098	(−) 87	3112	3025
3" TO 6"	1409	PASSING 3"	8.6	121	148				
		3" TO 6"	91.4	1288	1288				
		RETAINED ON 6"	0.0	0	0	1436	(−) 27	1420	1393

Col. 2: Corrected Sat. Sur. Dry Batch Weight from previous day operation. Average of 10 tests.

Col. 3 thru Col. 7: Same procedure as Cylinder Work Sheet (U.S.B.R. No. 7-1557).

Col. 8: Algebraic sum, Col. 2 minus Col. 7. Enter this correction on line 3 of Batch Determination Card (LF 420-95).

    (A) If weight in Col. 2 is greater than weight in Col. 7, add correction (Col. 8) to Sat. Sur. Dry Design Weight (Col. 9). Enter in Col. 10

    (B) If weight in Col. 2 is less than weight in Col. 7, subtract the correction from Sat. Sur. Dry Design Weight and enter in Col. 10.

Col. 9: (Sat. Sur. Dry Design Weight) and Col. 10 (New Batch Weight) recorded on line 2 and line 4 respectively of Batch Determination Card (LF 420-95). Line 4 and Col. 10 represent adjusted batch weight for next day operation.

Figure 99.—A typical form for computing sand and coarse aggregate batch weights to adjust for clean separation. 288-D-3282.

proper care, the inspector should report the faults immediately. The equipment should include suitable means to assure proper alinement, rate of travel, and tension of the paper and to prevent wrinkling. The recorder-pen lines should be sharp and clear. The inspector should see that zero-line and zero-load pens are in proper position at all times and that the contractor's forces keep them inked. He should make suitable notations on the recorder chart, over his name or initials, to mark the date, time,

SCALE WEIGHING AND FEEDING ERROR

DATE 11-17-60 SHIFT Swing INSPECTOR L.M.M. CONCRETE BATCHING PLANT Noble SPEC. No. DC-4825

MIX No. (1)	BATCH No. (2)	MATERIAL (3)	DESIRED BATCH WEIGHT, POUNDS (4)	OPERATOR'S CONTROL DIAL, POUNDS			WEIGHING HOPPER SCALE DIAL, POUNDS		ACTUAL SCALE READING, POUNDS (10)	BATCH RECORDER WEIGHT, POUNDS (11)	FEEDING ERROR, DIFFERENCE OF COLS. 9 AND 10 (12)	COMBINED ERROR, COLS. 8 AND 12 (13)	SPECIFICATIONS LIMIT, COL. 4 TIMES (*) (14)
				ERROR (5)	SETTING (6)	READING (7)	ERROR (8)	SETTING (9)					
6-I-16	59	Ice	192	+2	194	195	0	192	193	200	1	1	2
	60								194	200	2	1	2
	61								193	200	1	1	2
	62								193	200	1	1	2
6-I-16	63	Pozzolan	378	+1	379	377	0	378	373	372	5	5	6
	64					378			370	373	5	5	6
	65					382			378	380	2	2	6
	66					380			375	380	3	3	6
6-I-16	67	Cement	792	+6	798	798	-2	790	788	800	2	4	12
	68					796			790	800	0	2	12
	69					794			784	800	6	8	12
	70					796			788	800	2	4	12
6-I-16	71	Water	307	+3	310	310	0	307	306	306	1	1	3
						313			308	306	1	1	3
						307			303	307	4	4	3
						313			309	306	2	2	3
6-I-16	82	AEA	28.0 OZ.		28.0 OZ.		0	28.0	28.0	28.8	0.8	0	1
	133		28.0		28.0			28.0	27.0	21.0	1.0	1.0	1
	245		28.0		28.0			28.0	28.5	14.4	0.3	0.3	1
	358		30.5		30.5			30.5	31.0	22.7	0.5	0.5	1

*SPECIFICATIONS LIMITS:

Water and Ice, 1% Sand, No.4 to 3/4" and 3/4" to 1½", 2%

Cement and Pozzolan, 1¼% 1½" to 3" and 3" to 6", 3%

Figure 100.—A typical form for tabulating data in determining the combined error of feeding and weighing. 288-D-3283.

name of plant operator, and serial numbers of the batches at the start and end of each shift; also, if feasible, he should note other pertinent information of value in interpreting the record, such as reasons for plant irregularities and delays.

Examination of recorder rolls exposes errors in counting batches, errors in batching not noted by either the inspector or operator, errors in scale settings, shortages in mixing time, and waste cement chargeable to the contractor. The simple apparatus shown in figure 101 is an effective aid in examining recorder rolls.

Figure 102 is a reduced reproduction of sections of a graphic record made in the mixing plant at Friant Dam. The concrete sampler was wired to the clock mechanism to indicate on the chart (see ⊙, fig. 102) which batches were sampled. Information on measured slump and notes on batches used for making test cylinders were added when the recorder rolls were later examined. Notes and symbols have been inserted to facilitate explanation. Batches that differ from the regular cobble mix No. AXPH are indicated among the symbols. Proceeding across the chart from left to right, the indicated weights, to scale, of the five sizes of aggregate are shown, then those of cement, pumicite, and water, and finally the concrete consistency record for each of the four 4-cubic-yard mixers, designated north, east, south, and west. A note has been inserted in the batching record to indicate the mixer in which each batch was prepared.

Batching sequence indicated by this record is as follows. As a weigh batcher is charged, its recorder pen moves from the zero line on the left to the right until flow is cut off; then with the pen stationary, and the strip

Figure 101.—An illuminated frame used to expedite examination of mixing plant recorder rolls. PX–D–32061.

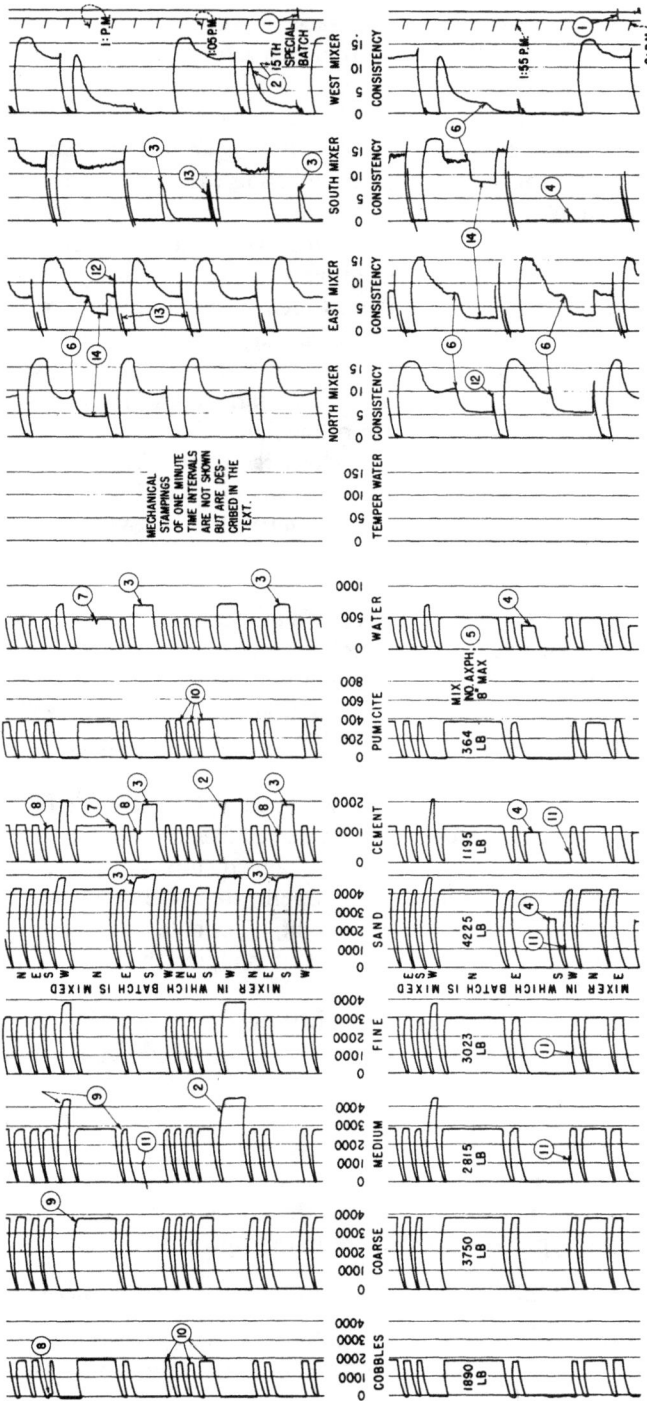

Figure 102.—Sections of recording chart, Friant Dam mixing plant. 288-D-2649

chart traveling upward, the pen traces a vertical line downward. When the batcher is discharging, the recorder pen returns to the zero line on the left. As a mixer is charged, its consistency-recorder pen moves from the zero line on the left to the right until the tendency for the mixer to leave its normal inclination, caused by the building up of unmixed concrete in the rear of the drum, is balanced by reacting torque (torsional resistance) in the support of the dumping ram, as described in section 88. As mixing proceeds, the concrete levels out and reduces the concentration of eccentric load at the rear of the drum, and the pen gradually finds a fairly uniform intermediate position from which it moves quickly to the left as the batch is dumped. For the sake of clarity, the mechanical stamping of 1-minute time intervals as described elsewhere in this section, has been omitted on this reproduction of the chart. These intervals are indicated, however, at the right-hand side of the figure.

Portions of the Friant Dam chart in figure 102 illustrate an unusual number of batching irregularities. Points of special interest are denoted by encircled numbers and are briefly described as follows:

(1) Concrete sampler wired to clock mechanism to give a record of which batch was sampled.

(2) Special batch, inspector's marking.

(3) Two-cubic-yard batch of mortar.

(4) One-cubic-yard batch of mortar.

(5) Inspector's note of mix number and batch weights of 8-inch-maximum cobble concrete with 25 percent pumicite.

(6) Mixers stopped.

(7) Outside interference with pens, perhaps while filling them.

(8) Batcher momentarily stopped filling.

(9) The beveled corner or changed slope at the top of the fill lines is caused by the change from open gate to dribble feed.

(10) Variation in the weight of cobbles caused by difficulty of close cutoff and in the weight of pumicite because of irregular flow by moisture.

(11) Outside interference with recorder.

(12) Inertia and static friction of the mixer cause this initial force to be expended as dumping begins.

(13) Same as (12) but prior to raising the mixer from the dumping position.

(14) The dry concrete tends to build up in the back of the mixer and does not level off when the mixer is stopped. As a result, the recorder pen does not return to or approach the zero line.

C. Batching Methods and Facilities for the Average Job

82. Central Batching by Weight.—Use of bagged cement and measurement of water by suitable volumetric means are not inconsistent with the weight system of batched assembly inasmuch as cement is bagged by weight at the mill and unit volume is as definite a measure of water as is unit weight. Bagged cement is subject, however, to weight variations of sufficient extent at times to upset seriously concrete control; consequently, it is important that periodic field checks be made to detect weight variations greater than permitted under specifications for the purchase of cement. (See sec. 75.) Partial bags used in batches should be weighed.

In order that full advantage of accurate weigh batching may be realized, the weighed materials must be properly and carefully handled to the end that batches reaching the mixer will be as uniform and complete as when released by the measuring equipment. This is particularly important where dry batches are transported by truck or other means to portable mixers near the work. Following are some examples of objectionable conditions, with suggestions for correction or avoidance:

(1) Intermingling of batches transported in multiple-batch trucks may be avoided by use of higher partitions and greater care in loading and discharging.

(2) Loss of materials in transferring batches from trucks to mixer skips will not occur if the skips are sufficiently large and the trucks are properly maneuvered into place and dumped. Skips that are too small can be enlarged by adding steel side pieces.

(3) Incomplete discharge of a batch from the batcher or a truck occasions a deficiency in that batch and an overcharge in a subsequent batch. The remedy is more careful handling or sometimes revamping of equipment.

(4) Loss of cement caused by dusting or scattering when it is allowed to fall freely from the batcher should be prevented by use of canvas tremies or curtains. (See fig. 103.)

(5) Loss of cement during truck haul will not occur if a separate section or compartment is provided for it in each batch compartment or if it is loaded simultaneously with the aggregate so as to be completely covered. If the cement is simultaneously loaded and mixed with damp aggregates and delays of 2 to 6 hours occur between filling and emptying the compartments, additional cement must be added to the batch, the amount depending on the delay. Separate cement compartments are preferable and can readily be attached to the batch release gates (fig. 103). Generally, the cost of separate compartments for the cement will soon be compensated by elimination of the need for additional cement to offset the effect of prehydration and of the need for wasting batches that stand more than 6 hours.

(6) Hang-up of cement in charging hoppers at stationary mixers is apt to occur when there is free moisture in the aggregate. This condition may be eliminated by feeding the cement into the lower portion of the hopper through a vertical or steeply sloping steel pipe or closed

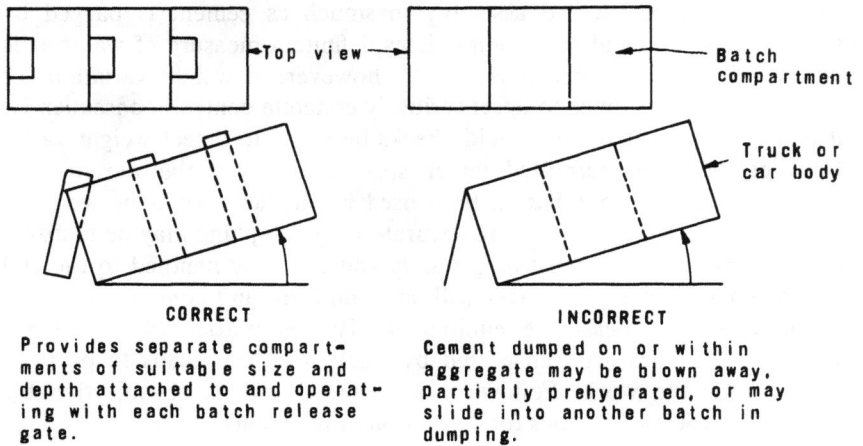

Top view → ← Batch compartment

Truck or car body

CORRECT

Provides separate compartments of suitable size and
depth attached to and operating with each batch release
gate.

INCORRECT

Cement dumped on or within
aggregate may be blown away,
partially prehydrated, or may
slide into another batch in
dumping.

**PROVISION FOR CEMENT IN DRY-BATCH
COMPARTMENTS**

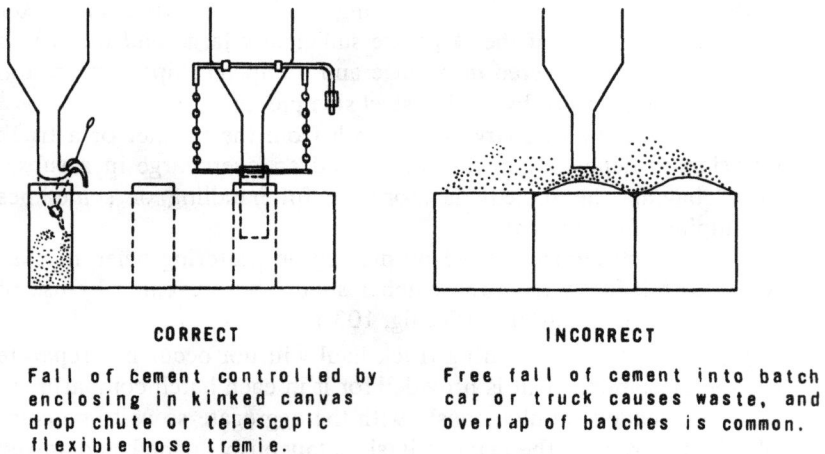

CORRECT

Fall of cement controlled by
enclosing in kinked canvas
drop chute or telescopic
flexible hose tremie.

INCORRECT

Free fall of cement into batch
car or truck causes waste, and
overlap of batches is common.

LOADING CEMENT FROM BATCHER INTO BATCH TRUCKS

**Figure 103.—Correct and incorrect methods of handling batched bulk cement.
Use of proper methods prevents waste and dust and results in more uniform
cement. 288–D–839.**

chute. The cement is introduced after the lower end of the pipe or
chute is buried in aggregate. This procedure also minimizes dust loss
and allows ribbon feeding of the cement.

The comparative advantages of three arrangements of storage bins and
weigh batchers, one involving individual weighing and the others cumulative weighing, are illustrated in figure 104. Central batching, using transit
mixers at the forms, is increasing in favor among engineers and contractors

for all types of scattered work. The haul is the same, and central batching provides a flexibility of operation, a freedom from waste and contamination of material, and an accuracy of proportioning that cannot be equalled by batching at the forms from piles of material dumped on the ground.

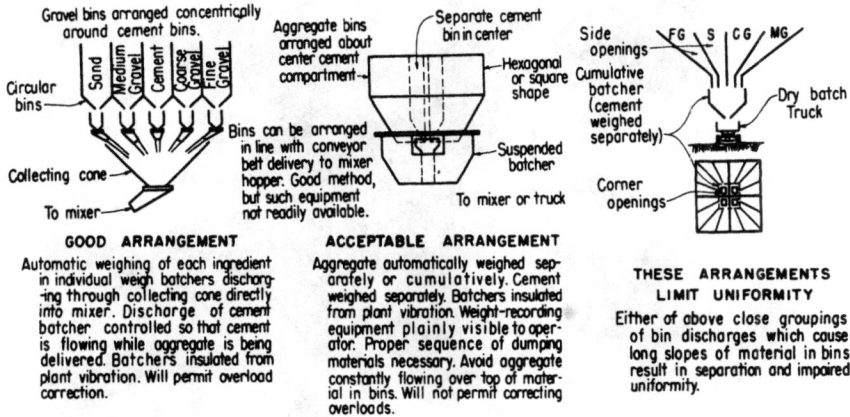

GOOD ARRANGEMENT

Automatic weighing of each ingredient in individual weigh batchers discharging through collecting cone directly into mixer. Discharge of cement batcher controlled so that cement is flowing while aggregate is being delivered. Batchers insulated from plant vibration. Will permit overload correction.

ACCEPTABLE ARRANGEMENT

Aggregate automatically weighed separately or cumulatively. Cement weighed separately. Batchers insulated from plant vibration. Weight-recording equipment plainly visible to operator. Proper sequence of dumping materials necessary. Avoid aggregate constantly flowing over top of material in bins. Will not permit correcting overloads.

THESE ARRANGEMENTS LIMIT UNIFORMITY

Either of above close groupings of bin discharges which cause long slopes of material in bins result in separation and impaired uniformity.

Figure 104.—Arrangements of batcher-supply bins and weigh batchers. The arrangement affects uniformity of the concrete. 288–D–2651.

Small portable weigh batchers are often used which discharge directly into the mixer skip by means of a trolley arrangement under the bins. From the standpoint of concrete control, this procedure is no better than the use of platform scales and wheelbarrows, unless the aggregates are finish screened.

83. Weighing Equipment.—Field weighing equipment may conveniently be divided into two general types: (1) weigh batchers, which usually are fed from storage bins; and (2) scales of the platform type for weighing materials in wheelbarrows or carts. Two weigh batchers are illustrated in figures 105 and 106. The manually operated weigh batcher in figure 105 is for cumulative weighing of three sizes of aggregates. The close-mounted arrangement of weigh batchers for aggregate and bulk cement shown in figure 106 results in a desirable mixing of cement and aggregate as the batch goes to the mixer.

On large jobs, where graphic recordings are made of the weights of concrete ingredients, each weighing unit should preferably include a visible springless dial which will register the scale load at any stage of the weighing operation from zero to full capacity. Over-and-under indicators may be used, however, if they show the scales in balance with no load or when loaded at any desired beam setting and if the scales are interlocked so that a new batch cannot be started until the weighing hoppers have been emptied. A full-capacity dial permits the checking of completeness

Figure 105.—Typical manually operated, cumulative weigh batcher with dial scale and gates for three aggregates. PX–D–32758.

of batcher discharge and also the detection of any irregularities in operation of weighing equipment or flow of materials. A full-capacity dial is of special value in weighing cement and water because of the importance of and difficulty frequently encountered in accurate control of these materials.

Weighing equipment of the platform type is shown in figures 107 and 108. Similar equipment can be employed for weighing aggregates from stockpiles for concrete batches containing one or more bags of cement where the mixers are equipped with charging hoppers or power-operated

Figure 106.—An automatic cement batcher mounted at the center of a manually operated, cumulative aggregate weigh batcher. This combination provides an advantageous mixture of cement and aggregate as the batch goes to the mixer. PX–D–32759.

skips. The so-called wheelbarrow scale is shown in figure 107. The particular scale illustrated has only two weighing beams (one for sand and one for gravel), an insufficient number for most Bureau work; however, scales of this type are available with beams in any desired number and with platforms of various sizes. The platform sizes commonly used are 30 inches and 42 inches square, for wheelbarrows and carts, respectively. The upper beam shown is a tare bar for compensating the weight of the empty wheelbarrow or cart, and the tell-tale dial or balance indicator on top of the beam box is a convenient means for indicating when the correct amount of material has been added. In operation, the weigh-beam poises are set to correspond with the desired gross weights (including surface moisture) of the aggregates; and when any size of aggregate is to be weighed, the proper beam is engaged by releasing the trigger on the outside of the box. It is not essential that the scale be at the stockpiles,

Figure 107.—Batching concrete by means of a portable wheelbarrow scale. Use of such equipment permits securing uniform concrete on the small job. 466–602–604.

Figure 108.—Practical arrangement used for weight proportioning of concrete for a small job. PX–D–32979.

for by keeping a small supply of each aggregate at the scale, the wheel-barrow or cart load can be trimmed or supplemented as required. With a little experience, workmen become proficient in approximating the correct load.

More modern mobile, truck-mounted materials transporters, combined with proportioning and mixing equipment, are now available. Such modern equipment will eventually replace the laborious hand method of mixing and placing concrete on small isolated jobs. However, the batching and mixing methods shown in figures 107 and 108 include the fundamental requirement for quality concrete of batching by weight for small amounts of concrete.

The arrangement of equipment used on a job where small amounts of concrete were placed is shown in figure 108. A platform scale was mounted in the rear of a truck body in which separate spaces were provided for sand, gravel, and cement; a skip was arranged on the scale so that it could be tipped directly into a small mixer drawn by the truck.

84. Batching of Liquids.—Accuracy, dependability, visibility, ready access for repair and adjustment, and ease of adjustment are important requisites of equipment for measuring the mixing water and solutions of admixtures such as accelerators, air-entraining agents, and retarders. It is also desirable that equipment be fitted with valves and connections necessary to conveniently divert the water measured for a batch, so that accuracy of measurement or adjustment can be quickly verified. Larger jobs justify water measurement for mixes being made with automatic meters or automatic batchers.

Compliance with requirements for water measurement on most Bureau work is afforded by use of a vertical tank equipped with a gage glass, a graduated scale for reading in terms of weight, and a suitable overflow for regulating the filling of the tank. Such a device provides the flexibility needed by the mixer operator in controlling concrete consistency and permits direct reading of the amount of mixing water. Use of horizontal water tanks on portable mixers is not permitted because such tanks are incapable of retaining their calibration under even slight changes of inclination.

Meters of the type shown in figure 109 are so constructed that the flow may be cut off automatically at a predetermined discharge or stopped manually at any time. Before each delivery is started, the two dial hands (one fast and one slow) are set back to zero by turning the projecting knob at the side of the dial housing. The total water passing through the meter is registered on a totalizer on the dial face. The totalizer is particularly useful in determining the average water requirement and water-cement ratio for a group of concrete batches.

A watermeter, whether manually operated or automatic, should be selected only after a careful study of all the conditions under which it op-

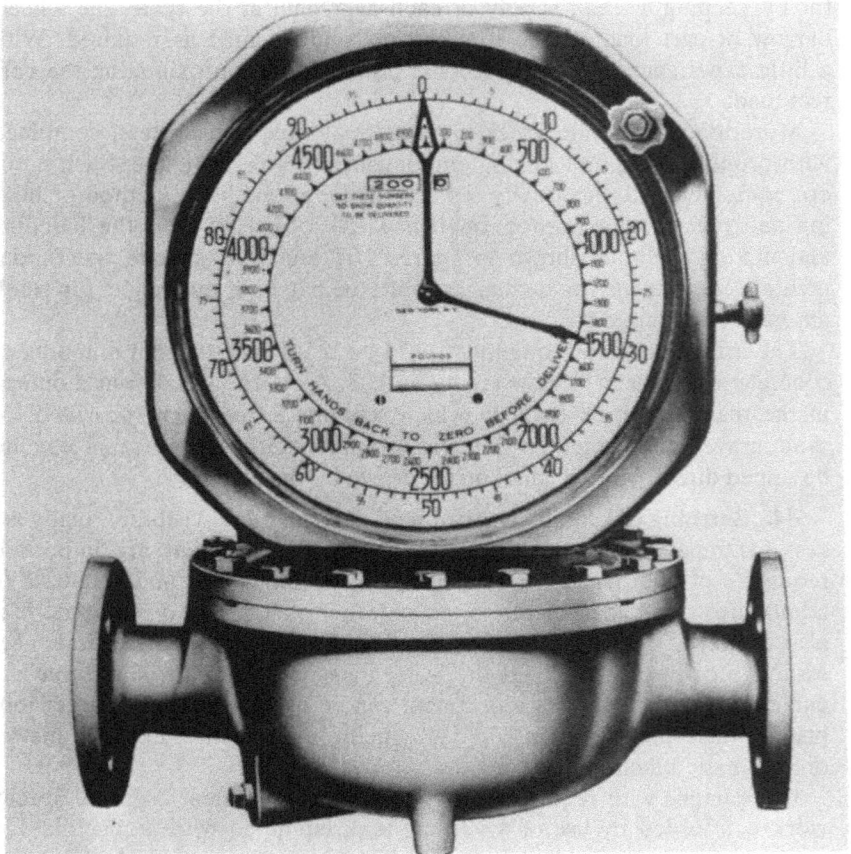

Figure 109.—A watermeter for batching the mixing water. Such a meter of suitable construction is a reliable means of batching water. PX–D–32760.

erates. The effects of variations in operating pressures on the rate of delivery and accuracy of the meter should be determined. Careful investigation should be made of the effects of scale and sediment often found in construction water supply systems and of higher than ordinary water temperatures which may be necessary in winter or may occur in summer. Difficulties from these sources are generally greater and more common under conditions that exist when portable mixers are used. In these circumstances, automatic metering equipment, in particular, is less suitable than reliable vertical tanks with center-siphon discharge.

On the other hand, satisfactory results have been obtained from watermeter installations for stationary mixers and central batching plants. In these installations it is sometimes necessary or desirable to add an auxiliary tank into which the metered water for the batch may be discharged and held ready for the mixer. Flow from such a tank, rather than directly from the meter, allows closer regulation of the time, amount, and rate of flow into the mixer. (See fig. 110.)

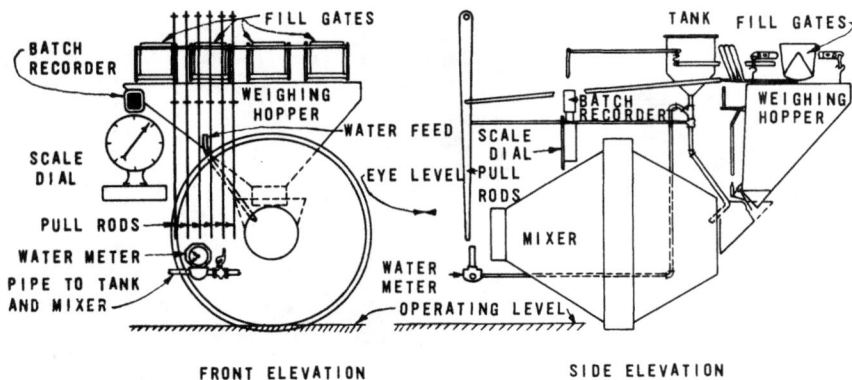

FRONT ELEVATION   SIDE ELEVATION

Figure 110.—Schematic diagram of batching and mixing facilities that proved very satisfactory in the Government batching plant at Hoover Dam. 288–D–157.

Measurement of the correct amount of water for concrete is so intimately related to the amount of water in the sand, and particularly to variations in sand moisture, that means should be provided for quickly adjusting the weight of water to compensate for moisture variations. It is occasionally necessary to add a small amount of water toward the end of the mixing period to avoid too low a slump. If the water measuring equipment is not suitable, supplementary facilities should be provided which will permit quick and convenient addition of a determined amount of water. This water should be discharged well into the mixer when needed so as to blend quickly with the batch. See section 79 for use of ice as part of the mixing water to reduce temperature of the concrete ingredients during batching.

Calcium chloride should always be batched in the form of a solution and this solution should be introduced into the mixing water. Any method that meets this requirement and provides accurate measurement is satisfactory. Where small portable mixers are used, the correct quantity of solution may be poured directly into the mixer after the skip or hopper is discharged. Precautions to be observed in handling solutions of calcium chloride are (1) avoid use of containers, valves, etc., in which the solution would be in contact with dissimilar metals, and, being an electrolyte, would cause corrosion as the result of galvanic action; and (2) avoid mixing solutions of calcium chloride and air-entraining agents in batching or storage tanks, as these materials, when combined in solution, form a precipitate of gummy, adherent calcium resinate which plugs valves, interferes

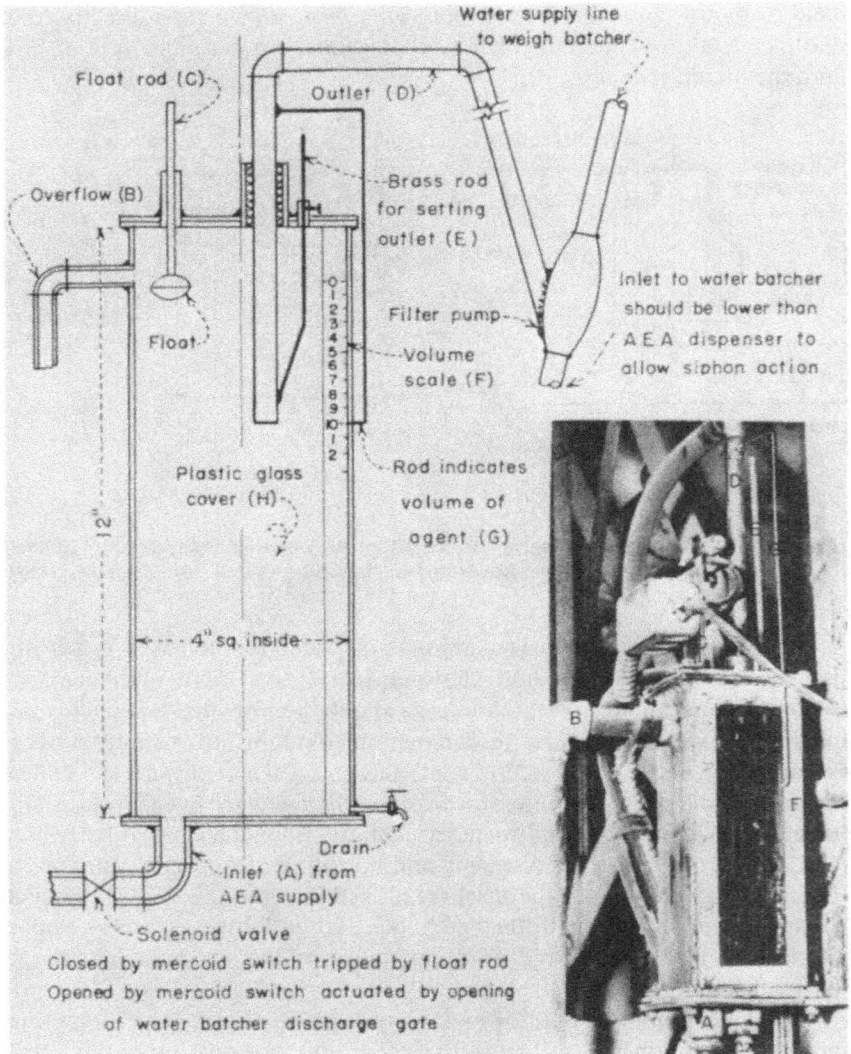

Figure 111.—Dispenser for air-entraining agent. PX–D–32761.

with operation of meters, and reduces the volumetric capacities of calibrated containers.

A commercial visual type of air-entraining agent and WRA dispenser used for manufacture of concrete under the initial construction contract for Pueblo Dam, Colorado, is illustrated in figure 111. The dispenser may be constructed to operate either as an automatic siphon or by gravity. When constructed for automatic operation, the dispenser fills while the water batcher empties, and empties while the water batcher fills. Thus the dispenser remains empty as long as the water batcher is full. This provision allows for adding tempering water to the mixer from the water batcher without causing any unwanted siphon action from the dispenser. A switch or stop-and-go button in the circuit to the inlet solenoid valves will enable the operator to use the water batcher without using the dispenser, as in washing the mixer. A gravity-type dispenser is made by providing an adjustable outlet pipe through the bottom of the batcher, instead of through the top, and without the filter pump.

Satisfactory commercial dispensers can be obained from suppliers of air-entraining agents and water-reducing set-controlling agents. To ensure that these admixtures are evenly distributed throughout the batch, they should be in solution, and the solution should be added to the mixing water at the time of batching (see fig. 112).

Figure 112.—Visual-mechanical batchers for dispensing an air-entraining agent and a water-reducing, set-controlling agent. The left photograph shows visual-mechanical batchers for measuring the dosage of air-entraining agent and water-reducing agent. The dispenser on the left contains water-reducing agent and the one on the right contains air-entraining agent. Controls for activating the visual-mechanical batchers for dispensing air-entraining agent and water-reducing agent are shown in the right photograph. P382–706–10550. P382–706–10551.

D. Mixing

85. General.—The conditions governing preliminary approval of the contractor's mixing equipment for large construction projects are the same as those stated in section 63 for aggregate processing and handling equipment.

Both stationary and portable, nontilting and tilting mixers are in common use on small jobs. The nontilting type usually has a cylindrical drum and is equipped with either a cable-operated loading skip or a charging hopper and a manually operated swinging discharge chute. Tilting mixers usually have conical or bowl-shape drums. These machines are obtainable in a large range of sizes. Tilting mixers of 6-cubic-foot capacity and larger are available with loading skips. A portable mixer with a skip is more conveniently charged than one equipped with a feed hopper unless an elevated charging platform is provided.

Nontilting-type horizontal mixers (often referred to as turbine mixers) of up to 3-cubic-yard capacity have been used on several Bureau projects. This type of mixer consists of a flat-bottom, cylindrical drum which may be either stationary or rotating clockwise. Mixer blades may be mounted on a single shaft, rotating counter-clockwise, or may have several sets of blades within the cylindrical mixing drum driven by gears from the central shaft and each rotating counterclockwise. Water may be sprayed onto the concrete mix through a circumferential water ring at the top or otherwise introduced during charging. Some types are fully enclosed, except for openings for introducing the concrete materials, to reduce dust emission. The mixed concrete is discharged through a gate opening in the bottom of the drum. Rapid and efficient mixing can be accomplished with this type of mixer. Three or four cubic yards of concrete can be mixed in about 60 seconds; a cycle of charging and discharging can be completed in about 1½ minutes.

In general, tilting mixers are currently considered to be more efficient than other types largely because they can be discharged quickly with a minimum of segregation regardless of slump or size of aggregate. Tilting mixers also have the advantage of being cleaned more easily. The effectiveness of the mixing action of any mixer (except the turbine type) depends primarily on the shape of the drum, shape and arrangement of the blades, method of charging, and sequence by which the materials are introduced into the drum. Drum mixers should have a combination of blade arrangement and drum shape such as to ensure an end-to-end exchange of materials parallel to the axis of rotation, as well as a rolling, folding, or spreading movement of the mix over on itself as the batch is mixed.

Considerable amounts of hardened concrete around the blades or on the inner surface of the drum detrimentally affect the mixing action and may cause inaccuracies in the records from consistency-measuring devices. Such hardened concrete should be removed as rapidly as it accumulates.

Loss of materials during charging or mixing operations should be remedied immediately by suitable repairs. If the consistency of concrete produced in the specific mixing time is not reasonably uniform throughout the batch, the blades are probably worn or are poorly designed with respect to shape and arrangement or that the sequence in which the concrete ingredients are charged needs to be altered. Where a concrete plant has several mixers located concentrically around a collecting cone with the weighing hoppers above, considerable experimentation may be necessary to set the proper discharge sequence to enable the mixers to meet mixer performance requirements. Mixers should not be loaded in excess of 10 percent more than the manufacturer's rated capacity. For mixers handling large batches and equipped with adequate charging control, the following practices are desirable:

(1) The ingredients (cement, pozzolan, if any, and fine and coarse aggregate) should be fed into the mixer simultaneously and in such a manner that the period of flow of each is about the same.

(2) Except when the mixing water is heated (see sec. 93), a portion of the water (between 5 and 10 percent) should precede, and a like quantity should follow, introduction of the other materials. The remainder of the water should be added uniformly with other materials.

(3) The minimum mixing time generally specified on Bureau work is as follows—the timing starts after all ingredients, except the last of the water, are in the mixer:

Capacity of mixer, cubic yards	Time of mixing, minutes*
2 or less	1½
3	2
4	2½
5	2¾
6	3

* Mixing time may be adjusted as need is indicated by the mixer performance test, designation 26, appendix.

Overmixing is objectionable because the grinding action increases fines, thereby requiring more water to maintain consistency of concrete. Also, overmixing may drive out entrained air. It is therefore recommended that mixing time not exceed three times the number of minutes given in the tabulation. Mixing equipment should be so designed that mixing can be discontinued and resumed with a full load in the mixer.

(4) The ability of a mixer to mix concrete properly is determined by mixer efficiency tests. (See designation 26.) Mixers should be capable of mixing concrete so that the unit weight of air-free mortar of samples taken from the first and last portions of the batch as

discharged from the mixer will not vary more than 0.8 percent from the average of the two mortar weights. The average variability for six batches should not exceed 0.5 percent. Also, the weight of coarse aggregate (aggregate retained on a No. 4 screen) in a cubic foot of concrete from the first and last portions of the batch should not vary more than 5 percent from the average of the two weights of coarse aggregate. On jobs involving large quantities of concrete, Bureau specifications provide maximum limitations on average variability, as follows:

Number of tests	Average variability, percent, based on average mortar weight of all tests
3	0.6
6	.5
20	.4
90	.3

Excessive variation in unit weight of air-free mortar indicates that mixing time should be increased. Excessive variation in weights of coarse aggregate in a cubic foot of concrete indicates that the mixer is improperly designed or that the blades are excessively worn. Mixer efficiency tests should be made at the start of a job and at such intervals as may be necessary to ensure compliance with specification requirements for effective mixing. Minimum mixing times listed in subsection (3) may be reduced if mixer efficiency tests confirm that the decreased time still permits satisfactory mixing.

Regardless of the size of a job, at least one mixer in each mixing plant should be so constructed and arranged that the operator or his assistant may look into the mixer drum to inspect consistency of the concrete. This mixer is conveniently used for concrete containing 1½-inch, or smaller, maximum size aggregate. The importance of this requirement for mixers equipped with consistency meters depends on effectiveness of the meters.

Precautions should be taken to prevent the concrete discharged from the mixer from becoming segregated because of the uncontrolled chuting effect as it drops into buckets, hoppers, carts, etc. This effect is particularly noticeable with nontilting mixers employing discharge chutes through which the concrete passes in relatively small streams. With tilting mixers, the batch usually slides out in a bulkier mass which has less opportunity to segregate. Where required to prevent objectionable segregation, a baffle, or preferably a section of down-pipe, should be provided at the end of a discharge chute so that the concrete will fall vertically or nearly so into the center of the receiving container. (See fig. 113.)

86. Truck Mixers and Agitators.—Truck mixers consist of equipment mounted on trucks and capable of mixing concrete en route between the batching plant and the forms. Agitators are portable machines designed to prevent segregation of the constituent parts of mixed concrete by imparting

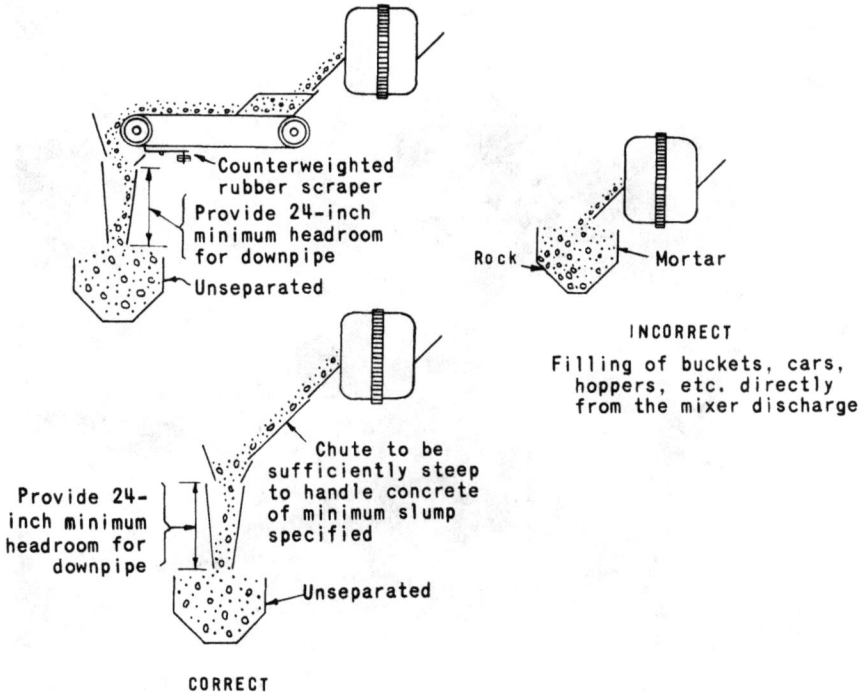

Counterweighted
rubber scraper
Provide 24-inch
minimum headroom
for downpipe
Unseparated

Rock   Mortar

INCORRECT

Filling of buckets, cars,
hoppers, etc. directly
from the mixer discharge

Chute to be
sufficiently steep
to handle concrete
of minimum slump
specified

Provide 24-
inch minimum
headroom for
downpipe

Unseparated

CORRECT

Either of the arrangements at the left prevents separation
regardless of length of chute or conveyor, whether discharging
concrete into buckets, cars, trucks, or hoppers.

Figure 113.—Correct and incorrect methods of discharging concrete from a mixer. Unless discharge of concrete from mixers is correctly controlled, the uniformity resulting from effective mixing will be destroyed by separation. 288–D–846.

to the concrete an occasional mild mixing action en route to the work. Truck mixers can be used for this purpose as well as open-top revolving blade or paddle-type agitators (fig. 114). When used as agitators, truck mixers are rotated at a slower speed than when used for initial mixing and when so used can handle batches a third larger.

Control of the quality of truck-mixed concrete presents some problems not common to other types of concrete mixing. Unless adequate precautions are taken, troublesome conditions such as segregation and variations in consistency may occur to such an extent that control of the water-cement ratio may be lost. Concrete having the proper consistency is adequately plastic and workable, is readily placed, and is not subject to objectionable segregation due to normal transportation and handling.

Improvement in the design of transit mixers has mitigated some difficulties formerly encountered in transit mixer operations. Larger tilting

Figure 114.—Dumpcrete bed being used to discharge concrete into feed hopper of a belt conveyor. Note agitator in bed which largely eliminates segregation of material. 214—TD—5010—CV.

drum mixers with larger drum openings that permit faster charging and discharging and easier inspection of the mixed concrete are worthy improvements. Also, these mixers are capable of faster mixing speeds as well as being otherwise mechanically improved.

The amount of water required to provide proper consistency is affected by factors which also influence temperature increase in the concrete. These factors are characteristics of ingredients, length of haul, amount of mixing, time required for unloading, climatic conditions, and others. Under unfavorable conditions such as irregular delivery, long haul distances, small and slow placements, and warm weather, the problems of maintaining some degree of uniformity and quality are considerably increased. Additions of water in excess of 3 percent of the initially designed amount required for the established water-cement ratio are not permitted. Precautions should be taken in advance to minimize loss of slump by speeding up delivery and placement and eliminating delays from other causes. The use of a retarder is sometimes advantageous. In warm weather, the concrete temperature should be maintained in the 70° to 80° F range so far as practicable. This may be accomplished by:

(1) Using cold mixing water or ice.

(2) Keeping materials as cool as practicable by shading or spraying aggregate piles to provide evaporative cooling.

(3) Avoiding the use of hot cement.

(4) Painting mixer drums white and keeping them white.

There is often considerable slump loss in truck-mixed concrete, especially in warm weather. Such loss can be kept to a minimum by stopping the initial mixing at 30 revolutions and by avoiding overmixing. In some situations where the loss of slump cannot be offset by these measures, the difficulties can be minimized by adding all of the water or doing all of the mixing at the jobsite or by doing all mixing in a suitable mixer at the forms using centrally dry-batched materials. To prevent slump loss in extremely hot weather, concrete may be mixed and placed at night or early in the morning.

Where truck mixers are used on Bureau projects, there are certain procedures which should be followed as well as certain precautions taken to assure that the concrete will be uniform from batch to batch. These are:

(1) Each mixer should be equipped with an accurate watermeter between supply tank and mixer, the meter to have indicating dials and totalizer.

(2) Each mixer should be equipped with a reliable revolution counter, with reset features, for indicating the amount (not speed) of mixing.

(3) Mixers should be charged with a ribbon-fed mixture of aggregates, cement, and water while the drum is rotating.

(4) A partial amount of the initial mixing water can be withheld so as to preclude any possibility of exceeding the proper slump. The remainder should be added at the placement site and thoroughly intermixed.

(5) It should be determined prior to use that the mixer will meet the requirements of the mixer efficiency test (designation 26) when mixing for not less than 50 or more than 100 revolutions at mixing speed.

(6) An optimum number of mixing revolutions in relation to batch size should be established by mixer-efficiency tests.

(7) The batch should be mixed only 75 percent of the required number of revolutions at mixing speed prior to inspection of consistency at the point of delivery, then mixed the additional revolutions required, with additional water not exceeding the amount required

for the established water-cement ratio being added to obtain the proper slump.

(8) Condition of the concrete uniformity and slump can be observed through the discharge opening in the drum. This should be carefully done while the drum is rotating.

(9) Every effort should be made to ensure that the same proportions of mortar and rock are maintained throughout the discharging operation.

(10) Wear of mixer blades should not exceed 1 inch.

If the truck mixer does not meet the variability limits outlined in the mixer efficiency test, designation 26, a reduction of the quantity being mixed will often correct the deficiency. In some cases, it may require a reduction to two-thirds or one-half of the mixer capacity.

Approval of truck mixers should be based on ability to mix and discharge uniform concrete throughout the entire batch. Specifications limit the degree of variability cited in section 85. Mixer performance tests should be made on each truck before they are permitted to be used on a job; later, they should be made at timed intervals during their use, depending on mechanical condition and observed performance. Large truck mixers should not be used for very small scattered structures that require less than a truckload of concrete. The last-used portion of the batch is subject to objectionable slump loss.

As with stationary types of nontilting mixers, separation of mortar from coarse aggregate will sometimes occur in truck mixers. This condition may be alleviated by arrangement of the discharge chute as shown in figure 113 so that the concrete will fall vertically into the receiving container. However, it is also important that the mixer blade arrangement and discharge mechanism be such that throughout the discharging the aggregate is well distributed from coarse to fine. Should the last portion of the batch contain an excessive amount of coarse aggregate, the truck mixer is not meeting the mixer performance requirements. As was discussed earlier, the batch size should be reduced to determine whether this action will correct the condition. Otherwise, the mixer should be repaired or taken out of service. With some truck mixers, this type of separation may be reduced somewhat by reversing the direction of rotation of the mixer drum for 10 to 12 revolutions prior to discharge.

Some truck mixers and agitators for transporting centrally mixed concrete do not readily and uniformly discharge concrete having low slump and containing large aggregate (for example, 2-inch slump and 3-inch-maximum aggregate), although such concrete is usually specified for relatively massive Bureau structures. Therefore, because of the tendency to use greater slumps to expedite discharge, use of truck mixers and agitators for massive concrete work should be discouraged. If used, they should not be a factor in choice of slump or aggregate size as these should be determined only by the requirements for placement and consolidation using modern concrete vibrators.

Horizontally rotating agitators mounted on rail cars have been effectively used for transporting concrete in large tunnels. These consist of a barrel-type rotating container with an attached helical blade extending full length. Openings at the ends permit interconnecting, and several agitators can be used in tandem. The agitators are loaded through openings in the barrel. Their capacity is 6 cubic yards of concrete. When discharging several connected in tandem, all agitators should be rotated in unison. While en route the agitator does not turn, but the helical blade imparts some mixing when discharging, which keeps the aggregate uniformly distributed (fig. 115).

E. Quality Control of Concrete

87. General.—The production of uniform and economical concrete is largely dependent on inspection at the batching and mixing plants. Mix adjustments are made using results of gradation and moisture tests of aggregates. Quantities and sequence of each ingredient entering the mixer and mixing time are frequently checked to ensure minimum variations. Mixed concrete is tested for consistency, temperature, air content, and unit weight; concrete cylinders are made for compression tests. Where practicable, samples for aggregate and concrete tests should be obtained from the same batch.

Figure 115.—Horizontal agitators used for transporting concrete for Clear Creek Tunnel, Central Valley project, California. 214–TD–4115–CV.

The frequency of sampling, number of cylinders made, and ages at which the cylinders are tested will vary with the type and size of the job. In general, each type of concrete is sampled once each day that concrete is placed. When large quantities are involved, however, each type of concrete is sampled once each shift. Samples of concrete aggregate and concrete should be representative of the materials used and the concrete placed during the day or shift that samples are taken. Because extra handling of air-entrained concrete reduces the air content, concrete samples should be sufficiently large to make all tests without reuse of concrete in other tests or in cylinders.

Bureau specifications require the contractor to provide adequate mechanical facilities for procuring and handling representative test samples of concrete at the mixing plant. Figure 116 illustrates a sampling device installed under the discharge hopper of the mixing plant at Friant Dam. Similar samplers have been used on many Bureau mass concrete projects. The sample is obtained from the full cross section of the discharge with little or no delay to concreting operations. Mechanical equipment for handling, transporting, and screening the sample should also be devised and installed to reduce the manual lifting of the concrete sample to a minimum.

The location of sampling devices for both aggregate and concrete should be such that sufficiently large representative samples can be obtained without involving unnecessary labor. It is also advisable that the sampling of concrete be performed as close to the mixer discharge as can be conveniently arranged. This is required to eliminate the effects of segregation that might occur when sampling from the discharge of a large receiving hopper. Mixer efficiency tests (designation 26), which are useful in establishing the sequence in which the materials should enter the mixer, as well as mixing time and adequacy of mixing equipment, should be performed on samples of concrete taken as close to the mixer mouth as possible.

88. Consistency.—The term concrete consistency is used in this manual to denote the fluidity of concrete as measured by the slump test. The aim in controlling the slump is to control directly the consistency and workability necessary for proper placement, and indirectly the water-cement ratio, the principle being that repeated batches of the same mix brought to the same consistency will have the same water content and consequently the same water-cement ratio, provided factors such as batch weights, aggregate grading, and temperatures of materials are uniform. Variations in water content have a much more pronounced effect on slump than do normal variations in the factors mentioned; hence, on jobs where grading and batching are reasonably well controlled, slump variations will reflect variations in water content and water-cement ratio. On the other hand, if the aggregate moisture content is uniform and fixed quantities of water are added at the mixer, the slump test should indicate grading varia-

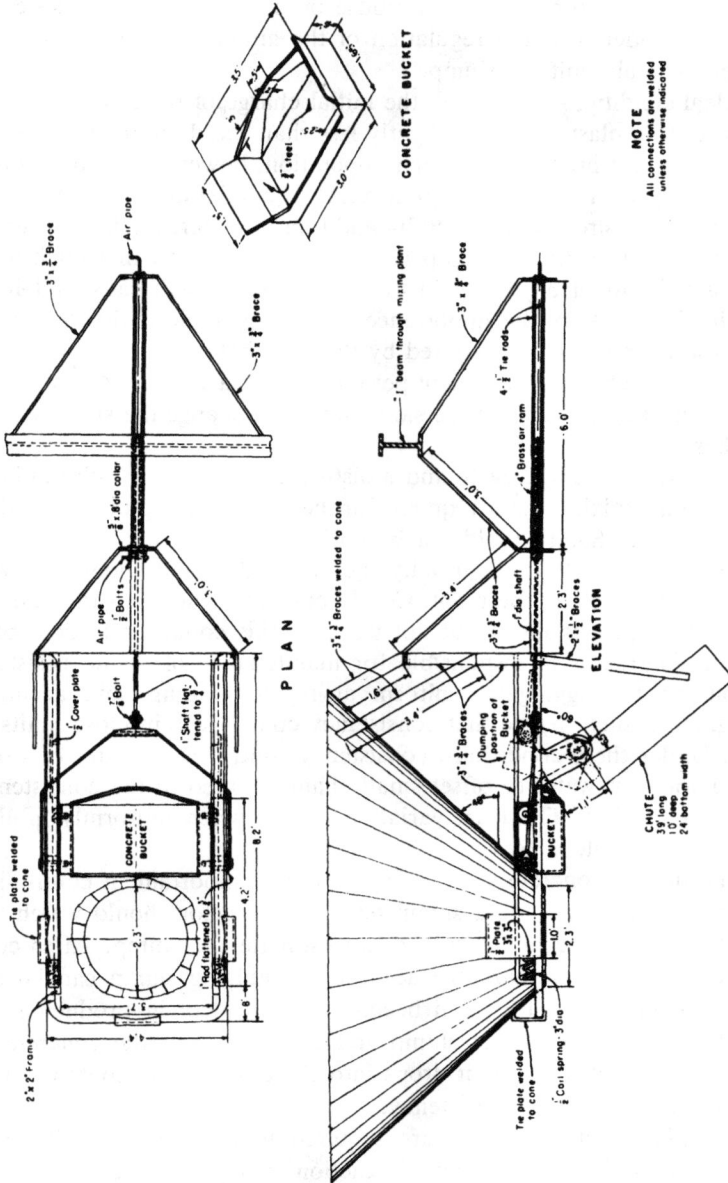

Figure 116.—Sampling device for fresh concrete. PX–D–25249.

tions or batching errors. For the purpose of this discussion, consistency control is considered as the regulation of the amount of water added to the mixer to obtain uniform slump.

For ideal conditions of control, the initial charge of water would bring the mixture to a plastic state at slightly less than the desired slump. Such a procedure is not practicable in plant operation; however, this procedure can effectively be used for transit mixers. The mixture would then be brought to the desired consistency by addition of water. Adding water to a batch to obtain a certain slump is a delicate operation inasmuch as a change of only about 3 percent in the water content will cause a 1-inch change in slump. Also, the importance of having stable moisture conditions in the aggregates is illustrated by the fact that for a mix having 38 percent sand and 5.3 cwt of cement per cubic yard, a 1-percent change (by weight) in surface moisture in the sand alone will change the slump about 1½ inches.

On jobs where the aggregate and moisture conditions are substantially uniform, all the mixing water required for the desired slump may be added in one operation. Sand should not be moistened at the mixing plant; in fact, if wet processed—as is usually the case—the sand should be well drained before reaching the plant. Conditions are ideal when the surface of coarse aggregate is moist but not wet, but this condition seldom prevails. The contractor is responsible for maintenance of stable moisture conditions in the aggregate, uniform aggregate grading, and accurate batching, all in such degree that consistency control within close limits is feasible. Under these conditions, consistency control is a matter of varying the amount of water to offset small changes noted in the consistency of previous batches. The least variation occurs when uniformly graded surface-dry aggregate is used.

Observation of concrete while being mixed is important in controlling consistency. Before making a slump test, the inspector should watch the action of the concrete in the mixer and estimate the slump. After considerable experience, remarkable accuracy in judging slump can be acquired, and mixer operators can become highly adept in producing the slump desired. Correlation of slump and appearance of the concrete as it moves down a chute or as it drops into the transporting container is a further aid in controlling consistency.

Mixing plants on large jobs are required to be equipped with consistency meters. Stationary pivoted reaction cones or vanes inside the drum have been used in nontilting mixers at some commercial plants. To avoid the mechanical problems that would be involved by use of this system in a tilting mixer, contractors have used electrical torque-measuring meters that indicate variations in power consumption of the mixer motor. Such meters can be designed so that transmission line variations do not cause objectionable errors in consistency indications.

The torque type of consistency meter lacks sensitivity in registering slump changes in stiff, lean concrete. During the original construction of Grand Coulee Dam, this difficulty stimulated development of a type of meter that indicates changes in the overbalancing effect of the concrete in a tilting mixer of the bowl type. In its normal operating position, the axis of rotation of the mixer is inclined, and any buildup of depth of the concrete in the rear of the bowl tends to upset the balance of the mixer on its trunnions and increases the inclination of the axis. The drier mixes build up higher than those of more fluid consistency, tending to overturn and thus affect the meter reading. The meter consists of an arrangement of links and levels attached to the tilting frame and connected through a cushioning air cylinder to a spring coil, which deflects under the overturning pressure from the drum and actuates a solenoid. As the solenoid core moves, there is a change in the electric current in the fixed coil, and the change is transmitted to a companion solenoid that operates a recording pen. Accurate batching and reasonable uniformity in aggregate grading are important in the use of this meter, as it is a means of measuring the fluidity of the batch, and the changes indicated may be caused by other than changes in water content.

The consistency meters installed in the mixing plants for Hungry Horse, Shasta, and Friant Dams (fig. 117) were a modification of the Grand Coulee Dam meter.

In later years, attempts to duplicate the results obtained with the tilting type were not successful on Yellowtail Dam. The shape of the mixer and reportedly a higher pressure in the hydraulic system may have contributed to the problem. There have been many advancements in electronic circuitry since the days of the original Grand Coulee Dam construction, and it now appears that effective torque-measuring electronic equipment is available or can be readily fabricated with the use of the proper electronic devices.

Although consistency meters indicate changes in consistency of the mix and provide information for maintaining uniformity in the concrete, their use must be supplemented by frequent slump tests, the results of which should be noted on the recorder roll opposite the corresponding consistency graphs.

89. Slump.—Bureau specifications require that the slump be the least at which the concrete can be consolidated satisfactorily by vibration and state that use of equipment incapable of handling and placing such concrete shall be prohibited. It is of primary importance that the inspector have a thorough knowledge of conditions in the forms affecting concrete placement and of what can be expected of a properly functioning modern vibrator; he can thus determine the practicable minimum slump of the concrete for each portion of the work. The consistency of the driest con-

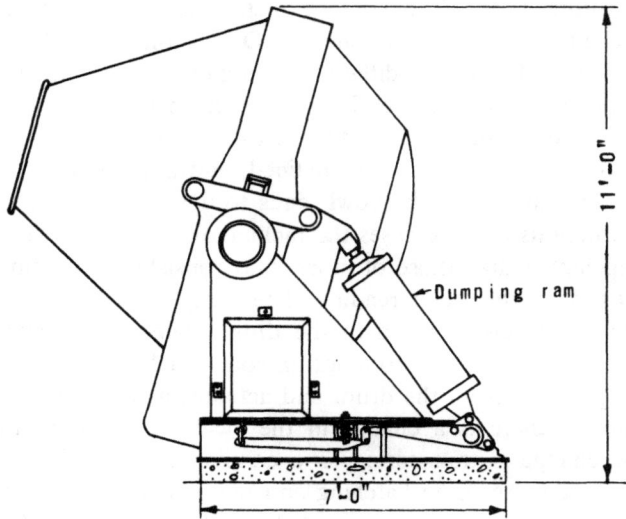

11'-0"

Dumping ram

7'-0"

ELEVATION

Recorder chart

Adjustment weights

Pen

110 Volts A.C.

Transmitting solenoid

Recording solenoid

Shock spring

Dash pot

Torsion tube connecting rocker arm to both dumping rams

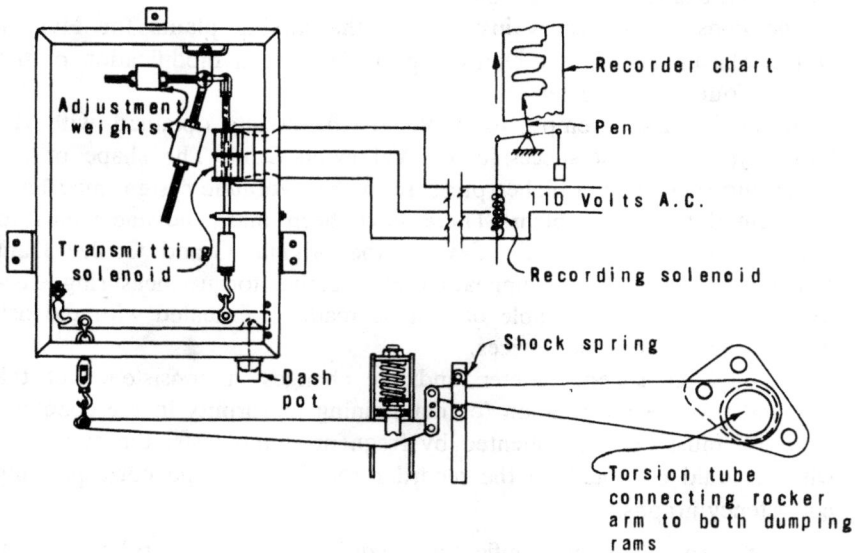

DETAIL OF CONSISTENCY METER ASSEMBLY

Figure 117.—Consistency meter installation. Consistency meters have aided materially in producing concrete of uniform slump at several Bureau projects. The installation shown is for a 4-cubic-yard mixer. 288–D–847.

crete practicable for placement with full vibration should be determined and concrete of this consistency used.

Specifications as to slump require that slump tests be made at the point of placement, but proper control of batching operations requires frequent checking of consistency at the mixing plant. Slump tests are also made at the mixer on samples from which cylinders are cast. When the mixer is at a considerable distance from the forms, slumps should be taken occasionally on the same batch at the mixer and at the point of placement to determine the slump loss in handling.

During handling between mixer and forms, the loss of slump may sometimes exceed the maximum of 1 inch allowed by Bureau specifications. Occasionally this excessive loss is caused, at least in part, by abnormal set of the cement; but regardless of the cause, there are several controls that may be applied to reduce the loss. Of these, the most important are short haul and prompt handling. In general, it is good practice to mix concrete near the forms, especially if the weather is hot, and to handle it in compact units such as cars and buckets rather than in thin streams in long chutes or on conveyor belts. Wherever long chutes or belts are necessary, they should be well shaded and protected from wind. Good results have been obtained by shading concrete-pump lines or by wrapping them in burlap kept damp from a parallel, perforated water pipe. Any measure that will reduce the temperature of the concrete, as discussed in section 127, will reduce the loss of slump. Compensating for excessive slump loss by allowing increased slump at the mixer, with consequent higher water and cement contents and increased segregation in transit, should not be permitted.

Sometimes in warm weather the water requirements of established mixes appear to increase for reasons other than slump loss. When it is certain that such increases are caused only by higher temperatures, the cement content should be increased only part of the amount needed to sustain the water-cement ratio since evaporation at the higher temperature will tend to compensate for the difference. Inexpensive water-reducing and set-controlling admixtures are commonly specified for all concrete except under special conditions. They can be beneficial in compensating for increased water requirement and lower strengths that usually occur in plain concrete mixes during warm weather.

When the slump test is made in accordance with designation 22 of the appendix, it is more than a measure of consistency. Plasticity is indicated by bulging of the slumped specimen. If the specimen does not break or crumble when tapped with the tamping rod, the concrete is satisfactorily cohesive (see fig. 2).

90. Compressive Strength.—Properly interpreted, the compressive strength of concrete test cylinders is a valuable criterion of the potential quality. Strength tests at early ages are desirable during the first stages of concreting operations to establish the proper mix. Cylinders broken at 7 days will usually serve this purpose, although tests at even earlier ages often serve equally well. Strength tests at later ages are more indicative of the actual strength of concrete in the structure and are, therefore, more valuable for reference purposes. After characteristics of the concrete have been established, strength tests at early ages may be discontinued except when new mixes are used or if unusual conditions are encountered.

When construction is progressing normally, two cylinders for testing at 28 days should be made from each sample of concrete taken (see sec. 87). On projects involving large quantities of concrete or where pozzolanic materials are used, cylinders for testing at 90, 180, and 365 days should be made from approximately each tenth sample taken; on other projects one cylinder per week should be made for testing at 90 days. Also, specially sized cylinders of 12 by 24 or 18 by 36 inches should be made of the full mass concrete mix at the construction site for comparison (compressive strength) with similar laboratory specimens. Strengths of these cylinders also can be compared with strengths of diamond drilled cores extracted at other ages. Such studies aid in correlating concrete mix design investigations with construction concrete mixes and establish other physical properties of the mass concrete. When there is considerable variation in the mix, additional cylinders may be required. If curing-room space is not available, long-time specimens may be removed anytime after 28 days and stored in damp sand where they will not freeze.

When curing facilities and testing equipment are available on the project, the cylinders are commonly made in cast-iron molds (appendix, designation 29). Cylinders should not be moved more than necessary before final set is attained and preferably should not be handled during the first 24 hours. They should be protected from temperatures considerably above or below 70° F and from loss of moisture.

On projects where construction activities are widely dispersed and involve isolated structures, it is sometimes advisable to use tin-can molds for fabrication of compressive strength specimens. Experiments have shown that cylinders cast in tin cans (appendix, designation 30) and kept at or near 73.4° F during the curing period give strengths close to those attained by specimens cast in cast-iron molds and cured under moist conditions at approximately 73.4° F. Two types are available. One type is made with a friction lid; another type has a full-opening top with circumferential and vertical metal stripping tabs. Plastic cylinder molds for one-time use are also available. The plastic type can be stripped by applying a cold chisel along the seam and tapping lightly to cause separation. Both the open-top tin and plastic types are more economical than the cast iron. The open-top types can be sealed by use of polyethylene dish covers. Extra care should be taken in placing, finishing, and curing test specimens.

When using the open-top type of can mold, it is advisable to provide flat bases that will fit inside of the crimp edges and a circular jig of proper diameter which fits over the top to maintain the shape of the can during casting. Cylinders from disposable molds, with proper base area calibration, will provide an accurate compressive strength at much less cost than cast-iron molds because of the labor involved in stripping, cleaning, and assembling the latter.

Cans having friction lids are advantageous when cylinders are to be shipped or when climatic conditions make it impracticable to maintain the specimens in a reasonably damp condition. Projects not equipped with compression testing machines should make a practice of using can molds and should ship the cylinders to the nearest machine-equipped Bureau laboratory for testing.

Cardboard or paper molds should not be used for molding test cylinders for concrete control. Laboratory tests and field experiences indicate that cylinders made in these molds give much lower compressive strengths than those obtained from cast-iron molds. It has also been reported that cylinders cast in such molds and immediately subjected to steam curing cracked horizontally.

Approved methods of curing specimens are described in the appendix, designations 29, 30, and 31. Loss of moisture from specimens should be prevented, and cylinders should be kept as close to 73.4° F as possible. Compressive strength specimens should not be job cured in lieu of standard curing. Specimens cured at uncontrolled temperatures near the structure are useful for indicating the time at which concrete is sufficiently strong to remove forms or support traffic. However, specimens thus cured are not suitable for measuring strengths on which designs are based and concrete specifications written. Strengths of job-cured specimens are misleading because they are much lower than strength in the structure, mainly because of the greater ratio of exposed surface to volume. Although job-cured specimens give lower strengths than standard-cured specimens, the data in table 3 indicate that cores from structures are usually higher in strength than comparable standard-cured control cylinders.

Approved methods for capping and breaking test cylinders are described in the appendix, designations 32 and 33.

A simple and rapid check for quality of concrete may be made by use of the concrete impact test hammer, sometimes referred to as the Swiss hammer. It is a hand-operated instrument for nondestructive testing of concrete, based on an empirical relationship between the strength of concrete and the rebound of a steel plunger which is impinged against the face of the concrete under controlled conditions.

The impact test hammer must be calibrated for each particular mix to ensure reasonable accuracy. This calibration is obtained by correlating hammer readings with compressive strengths of cylinders made from the same mix used in the structure. Should there be changes in the properties of that mix, such as aggregate type or content, water-cement ratio, cement content, or slump, it will be necessary to recalibrate the instrument with a new set of specimens. When these calibrations are made, the cylinder should be restrained in a testing machine to approximately 15 percent of the ultimate cylinder strength to prevent physical displacement of the specimens.

To obtain uniform results, it is necessary first to provide a smooth concrete surface. A carborundum stone is furnished with the instrument for this purpose. Reliable readings cannot be made on honeycombed sections or on a single piece of aggregate that protrudes from the structure.

Laboratory tests have shown that the impact hammer has the following limitations in its use:

(1) Hammer readings on specimens selected at random are not reliable indications of strength because of wide variations in other properties which also affect hammer rebound.

(2) Restrained specimens result in higher rebound readings.

(3) Hammer readings vary with different aggregates and concrete surface textures and between flat and rounded surfaces.

(4) The hammer is not usable for determining the earliest time that forms can be stripped.

(5) There is no significant difference in rebound readings between air-entrained and plain concrete.

(6) There is no significant difference between wet and dry surface concrete at 1 day's age but as much as 10 points difference when the concrete is 3 years old.

(7) Readings taken on unrestrained specimens, thin slabs, wall sections, or long beams are lower because some of the hammer impact physically displaces the subject area.

The uniformity of concrete from a particular mix can be checked with this instrument.

91. Air Content and Unit Weight.—The air content of the concrete should be determined for each sample from which cylinders are made (see sec. 87). Several reliable airmeters that operate by pressure methods are commercially available. Instructions for determining air content by the pressure methods are contained in the appendix, designation 24.

The determination of unit weight of concrete is an important control test and should also be made on each sample from which cylinders are cast. It can be made either in conjunction with the determination of air content (designation 24) or as a separate test (designation 23). Together with the specific gravity of the ingredients, the unit weight furnishes a convenient basis for computing unit cement and water content and volume of concrete per batch, and it also provides a check on indicated air content as determined by the airmeter. Generally, there is a slight difference in the amount of entrained air in a given concrete mix as determined by pressure or gravimetric methods, especially when using a ¼-cubic-foot container or smaller. This difference is usually small and may be caused by inaccuracies in calibration of equipment, in striking off, or in weighing. A large discrepancy between the two methods may be an indication of an error in mix design, very possibly specific gravity determination of materials, or batch weights.

F. Hot and Cold Weather Precautions in Concrete Production

92. Hot Weather Precautions.—Most Bureau specifications require that concrete, as deposited, shall have a temperature no higher than a stipulated value—usually 80° F for concrete to be placed in hot arid climates and 90° F for most other concretes. For mass concrete dams, temperature studies have shown the need for considerably lower maximum placing temperatures. Such placing temperatures, as low as 50° F at Glen Canyon Dam, are established to control cracking in the structure. On some jobs in desert regions of the Southwest, concrete placing has been prohibited during the extremely hot weather between June 1 and October 1.

Limitations on maximum temperature and on the placing of concrete during hot weather have been imposed because of the impairment of quality and durability resulting when concrete is mixed, placed, and cured at high temperatures. This impairment affects several different properties of the concrete: First, the ultimate strength of concrete mixed and cured at high temperatures is never as great as that of concrete mixed and cured at temperatures below 70° F; second, cracking tendencies are increased because of the greater range between the high temperature at the time of hardening and the low temperature to which the concrete will later drop; third, the unit water requirement for mixing is increased by higher concrete temperatures, contributing to greater shrinkage on drying (see fig. 118); and fourth, concrete that is mixed, placed, and cured at high temperatures has been found to fail sooner, as a result of repeated cycles of moisture and temperature changes above the freezing point, than concrete that is mixed, placed, and cured at lower temperatures.

Various means employed to lower the temperature of concrete as

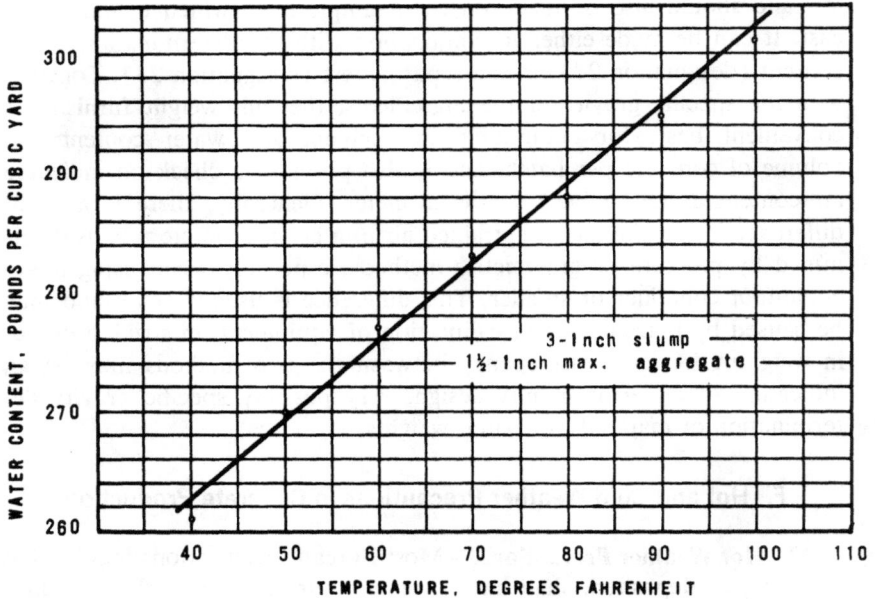

Figure 118.—Water requirement for a typical concrete mix as affected by temperature. The increase of water content accounts in part for greater shrinkage of concrete that is mixed and cured at high temperatures. 288–D–2653.

mixed include (1) using cold mixing water, even to the point of adding large quantities of ice; (2) avoiding, so far as practicable, the use of hot cement; (3) insulating water-supply lines and tanks, or at least painting exposed portions white to reflect heat; (4) cooling coarse aggregate with refrigerated water by sprinkling or inundating or with cold air blasts; (5) insulating mixer drums or cooling them with sprays or wet burlap coverings; (6) shading materials and facilities not otherwise protected from the heat; and (7) working only at night. To obtain lower placing temperatures for mass concrete, different combinations of these measures may be required. Use of ice for mixing water should be carefully controlled to ensure complete melting before mixing is completed. For information on concrete placing, protecting, and curing during hot weather, see section 127.

93. Cold Weather Precautions.—When there is danger of freezing, certain minimum temperatures of concrete, as placed, are specified because much of the heat generated during hydration of the cement is not immediately available. Specifications require that the temperature of the concrete be not less than 40° F in moderate weather or 50° F when the mean daily temperature drops below 40° F (see table 22). As an addi-

Table 22.—Effect of temperature of materials on temperature of various freshly mixed concretes

[Temperatures in degrees F]

	Thin sections		Mass concrete	
Approximate maximum size of rock	¾ inch	1½ inches	3 inches	6 inches
Approximate percent of sand	40 percent	35 percent	30 percent	25 percent
Weight of sand for batch	1,200 pounds	1,100 pounds	1,000 pounds	900 pounds
Weight of rock for batch	1,800 pounds	2,100 pounds	2,400 pounds	2,700 pounds
Weight of water for batch	300 pounds	250 pounds	200 pounds	150 pounds
Weight of cement for batch	600 pounds	500 pounds	400 pounds	300 pounds
Minimum temperature of fresh concrete AFTER PLACING and for first 72 hours	55	50	45	40
Minimum temperature of fresh concrete AS MIXED, for weather:[2] Above 30° F	60	55	50	45
0 to 30° F	65	60	55	50
Below 0° F	70	65	60	55

See footnotes at end of table.

Table 22.—Effect of temperature of materials on temperature of various freshly mixed concretes—Continued

[Temperatures in degrees F]

	Thin sections								Mass concrete							
Minimum temperature of materials to produce indicated temperature of freshly mixed concrete. Cement [3]	35	10	10	−10	35	10	10	−10	35	10	10	−10	35	10	10	−10
Added water	140	140	140	140	140	140	140	140	140	140	140	140	140	140	140	140
Aggregate water [4]	38	95	50	61	35	100	46	55	33	105	43	52	33	113	40	48
Sand	38	95	50	61	35	100	46	55	33	105	43	52	33	113	40	48
Rock	38	[1]10	50	61	35	[1]10	46	55	33	[1]10	43	52	33	[1]10	40	48
Temperature of freshly mixed concrete	60	65	65	70	55	60	60	65	50	55	55	60	46	50	50	55
Maximum allowable GRADUAL drop in temperature in 24 hours at end of protection, degrees F	50				40				30				20			

[1] Rock at temperatures below freezing is assumed to be surface dry and free of ice.
[2] For colder weather a greater margin is provided between temperature of concrete as mixed and the required minimum temperature of fresh concrete in place.
[3] Cement temperature has been considered the same as that of average air and of unheated materials.
[4] The amount of free water in the aggregate has been assumed equal to one-fourth of the mix water.

tional precaution, when the mean daily air temperature is lower than 40° F, 1 percent calcium chloride by weight of cement may be used to bring the concrete to a stage of greater maturity at the end of the specified period of protection. However, use of calcium chloride is never justification for reducing the amount of protective cover, heat, or other winter protection which would normally be used. Calcium chloride should not be used in concrete that will be subjected to sulfate attack or which contains embedded galvanized metal parts; where sulfate conditions exist, the water-cement ratio should be reduced to a maximum of 0.47 in addition to using type II or V cement.

The temperature of concrete leaving the mixer should be no higher than necessary to assure that the concrete after exposure during transportation and placing will have a temperature not more than a few degrees higher than the specified minimum.

Overheating of concrete is objectionable because it may accelerate chemical action, cause excessive loss of slump, and increase the water requirement for a given slump (see fig. 118). Moreover, the warmer the concrete as placed, the greater the drop to ultimate low temperatures, with corresponding decrease in volume.

To obtain the required temperatures for freshly mixed concrete in cold weather, it is often necessary to heat mixing water or aggregates, or both, depending on severity of the weather. Heating the mixing water is the most practicable and efficient procedure. Water is not only easy to heat, but each pound of water heated to a given temperature has roughly five times as many available heat units in it as are in a pound of aggregate or cement at the same temperature. The temperature at which water and other constituents should enter the mixer, to produce a given temperature of concrete, is indicated in table 22. Temperatures are computed from the formula in the appendix, designation 35.

Mixing water should be heated so that appreciable fluctuations in temperature from batch to batch are avoided. Very hot water should not be allowed to touch the cement because of the danger of causing quick or "flash" set. If hot water and the coldest portion of the aggregate can be brought together in the mixer first so that the temperature of the mixture does not exceed about 100° F, the possibility of flash set will be minimized and advantage can be taken of high water temperatures. The maximum temperature that can be used with given materials and charging methods without causing noticeable increase in water requirement or flash setting may be determined by experiment.

When heating of aggregates is used as an alternative method, the aggregates should be heated uniformly and carefully, eliminating all frozen lumps, ice, and snow, and avoiding overheating or excessive drying. Unless temperatures are uniform, noticeable variations in water requirement and slump will occur. Average temperatures should not exceed 150° F, and maximum temperatures should not exceed 212° F as higher temperatures may crack the aggregate. As indicated in table 22, these temperatures for freshly mixed concrete are considerably higher than those required if hot mixing water is used.

Heating aggregates is preferably accomplished by steam or hot water in pipes. Heating with steam jets is objectionable because of the resulting variable moisture in the aggregate. Experience has demonstrated that variable moisture can be exceedingly troublesome. On small jobs aggregates can be thawed by heating carefully over metal culvert pipe in which fires are maintained. Exposed surfaces of aggregates in stockpiles, bins, etc., should be covered with tarpaulins during heating to obtain a uniform distribution of heat and avoid frozen crusts.

Concrete placing, protecting, and curing during freezing weather are described in section 128. Additional information on cold weather concreting appears in many articles in the technical press.

Chapter VI

HANDLING, PLACING, FINISHING, AND CURING

A. Preparations for Placing

94. Foundation.—The procedures necessary for satisfactory preparation of foundation surfaces upon or against which concrete is to be placed are governed by design requirements and by the type and condition of the foundation material.

(a) *Rock.*—Where it is necessary to provide tight bond with rock foundations, the rock surface should be prepared by roughening, where necessary, and thorough cleaning. Loose and drummy rock, dried grout, flaky and scaly coatings, organic deposits, and other foreign material must be removed. Open fissures should be cleaned to a suitable depth and to firm rock on the sides. Cleaning may be done by use of stiff brooms, picks, jets of water and air applied at high velocity, wet sandblasting, or any other effective means, followed by thorough washing. Accumulations of wash water in depressions must be removed prior to placing the concrete. The surface to be bonded should be completely surface dried by air jets. The presence of any free surface water, which may be indicated by shininess, will prevent proper bonding to rock surfaces.

Trimmed surfaces of rock subject to air slacking or raveling present special problems. Various expedients, such as a coat of shotcrete or a covering of wet burlap, have been used successfully to keep surfaces intact until concrete is placed against them.

A sprayed coating of stabilizing chemicals or of a bituminous or other sealing compound may be used where tight bond of concrete to the rock is not required.

(b) *Earth.*—Many Bureau structures are constructed on earth foundations. Before concrete is placed for these structures, the foundation surface and subgrade should be inspected for adequacy of design loading. The Bureau's Earth Manual should be consulted for additional information about earth foundations. Earth subgrades should be damp, but not wet, when concrete is placed. Free-draining subgrades in hot, arid regions should be wetted to a depth of several inches to provide a reservoir of moisture in contact with the concrete. This practice should compensate for lower efficiency of curing procedures under such climatic conditions.

(c) *Porous Underdrains.*—These are constructed of sand and gravel or crushed rock or gravel, suitably sized and graded to provide rapid drainage. There is no reason for concern in regard to loss of mortar from the concrete or plugging of the drain by penetration of mortar because mortar will not drain from concrete that satisfies the maximum slump limitation of present specifications. For this reason, covering of drains with tar paper or burlap is unnecessary.

95. Construction Joints.—Bureau specifications define construction joints as the contact between newly placed concrete and existing concrete surfaces that have become so rigid that the new concrete cannot be incorporated integrally by vibration with that previously placed. Unformed construction joints are horizontal or approximately so and are sometimes spoken of as fill planes. Although this section has special application to joints in massive structures, the principles and methods described are largely applicable to joints in other types of structures.

A high quality of bond and watertightness in a horizontal construction joint is best assured when the concrete, and especially that in the upper portion of the lift, has the least slump that will permit proper working and consolidation. Wet mixes particularly should be avoided. Their tendency to segregate and bleed badly results in weak concrete and a heavy layer of laitance at the surface which makes cleanup difficult. Inferior joints, such as that shown in figure 119, are the result of wet concrete; they may also be produced in low-slump concrete by excessive working of the surface or subjecting it to traffic before it hardens. Plank walkways or other suitable means should be provided to take care of job traffic as soon as the concrete is in place. Workmen embedding coarse aggregate or setting form anchors and other embedded parts should be furnished snowshoes similar to those shown in figure 120.

The quality of a joint depends on quality of the concrete and on cleanup of the joint surface. A rough surface, in itself, does not assure a good construction joint. Footprints, protruding pieces of large aggregate, or depressed keys interfere with accomplishment of good cleanup. Such features also make the necessary complete removal of free water difficult which, if not achieved, will prevent good bond even though the surface is otherwise properly cleaned. Proper use of a vibrator usually will leave the surface suitably even. Rock that does not settle sufficiently when vibrated can be walked down flush with the joint surface, as illustrated in figure 120. When embedding mortar has lost slump, protruding rock should be pressed down only during supplemental vibration; otherwise, some loosening may result.

Figure 119.—Segregation in overwet concrete at fill planes, resulting in weak, porous joints subject to early weathering. PX–D–11747.

Any unconsolidated hardened concrete should be removed before cleanup is started.

Experience and investigations demonstrate that where bond and watertightness at construction joints are desired, the surfaces of existing concrete should be wet sandblasted and washed thoroughly and completely dried immediately prior to placement of fresh concrete (see figs. 121 and 122). This method is characterized by: (a) simplicity, since there is no repetition of treatment; (b) dependability, in that uniformly good results are readily obtainable; and (c) economy, as a result of improvement in sandblasting equipment and use of properly graded and dried sand.

Construction joint cleanup by wet sandblasting, followed by thorough washing, is standard Bureau practice for placement of mass concrete and structural concrete where watertight joints are required. Commercial sandblasting equipment is readily available; however, details of an easily fabri-

Figure 120.—Workman wearing "snowshoes" on a fresh joint surface. Snowshoes minimize undesirable working of the concrete and facilitate cleaning of the surface. PX–D–33053.

Figure 121.—Final sandblasting of construction joints at Glen Canyon Dam, in Arizona. This is an expeditious and effective means of construction joint cleanup. P557–420–05101.

Figure 122.—Final washing with water jets just prior to placing next lift of concrete. P1222–142–11967.

cated type are shown in figure 123. Figure 124 illustrates a multiple-purpose air gun that can be assembled readily on most jobs. When the gun is used as an air-water jet, the suction line is connected to the water-line. For wet sandblasting, small streams of water can be admitted to the sand through a suitable attachment on the nozzle. This device has a limited capacity for sandblasting but gives results superior to those obtainable with an air water jet alone. The multiple-purpose gun has been used to advantage for minor cleanup work. Wet-sandblasting equipment should be operated at an air pressure of approximately 100 pounds per square inch.

Blasting sand should be dense, hard, not easily broken, and sufficiently dry to permit free passage through the equipment. Careful selection to

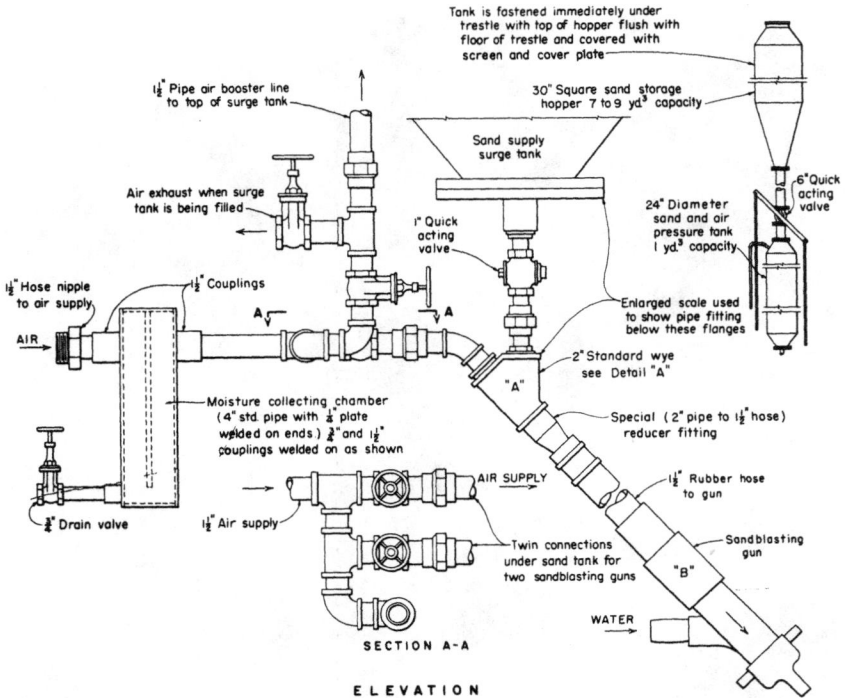

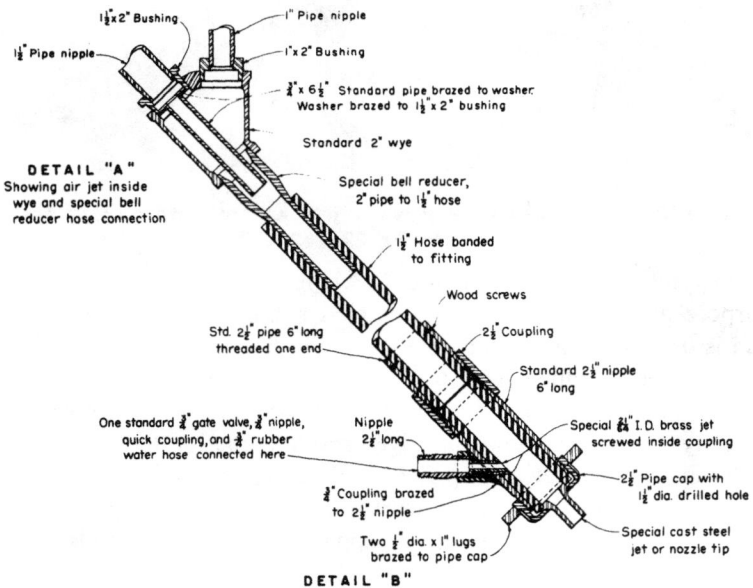

Figure 123.—Sandblasting equipment used at Grand Coulee Dam, Columbia Basin project, Washington. 288–D–1542.

SUCTION HOSE AND NOZZLE SIZES FOR VARIOUS USES OF SUCTION GUN

USE TO BE MADE OF GUN	1½ INCH 45° Y		NOTES
	HOSE	NOZZLE	All pipe, tubing and fittings are standard.
Air water jet	¾"	¾" x 24"	Standard hose sizes for suction fit nipples of these pipe sizes fairly tightly.
Dry sand blast	1" to 2"	¾" x 24"	Suction hose should be of convenient length (6 to 10 ft.). For vacuum cleaning and dry material, 2 in. vacuum hose is preferable.
Wet sand blast	¾" or 1"	¾" x 24"	Nozzle lengths shown have worked well. Other lengths may be better suited for other conditions.
Plastic mortar gun (Mix 1-2 plastic) *	¾"	¾" x 10"	The nozzle may be fitted with elbow for buttering perimeter of concrete replacements with plastic mortar.
Dry mortar gun (Mix 1-4 damp) *	1" to 2"	¾" x 10"	Gun with modified water-ring nozzle for repairs by dry mix shotcreting of areas larger than 1 ft.[2]
Small-scale shotcrete application (Mix 1-3 dry) *	1" to 2"	¾" water-ring	A good nozzle for dry mortar is the rubber lined tip of a standard small shotcrete nozzle with an orifice of about ⅝ in. diameter.
Vacuum cleaner or low lift pump	1½" to 2"	1¼" or 1½" Hose to waste	*Mix proportions are cement to sand by dry volume or weight. Air pressure should be 60 to 80 lbs./in.[2] measured about 8 ins. back of "Y" with valve open.

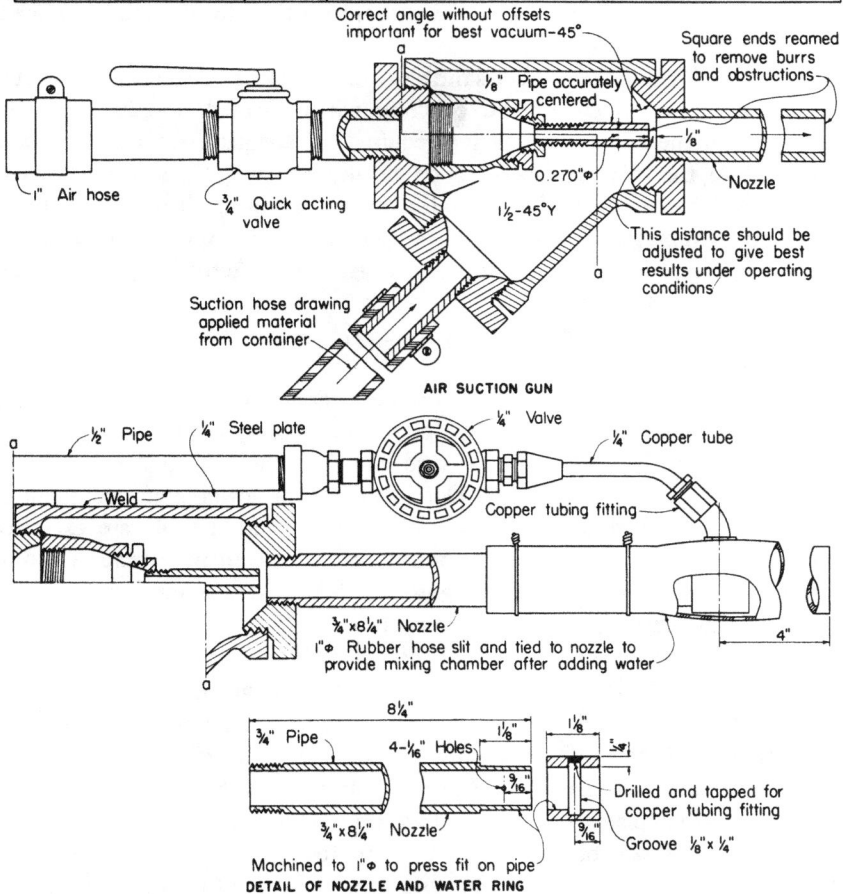

Figure 124.—Air-suction gun, with details of nozzle and water ring, for dry sandblasting, washing, mortar application, and vacuum cleaning. 288–D–1543.

obtain an exceptionally good quality sand can be economically justified
because such sand may sometimes be reclaimed, reprocessed, and reused.
Ordinary concrete sand may be used on smaller jobs, but on large jobs
equipment for removing fine material and for drying the sand should be
installed or special sand should be purchased from a commercial producer.
At Grand Coulee Dam, concrete sand was passed through a cylindrical,
oil-fired drier and was screened on a No. 16 screen to eliminate the finer
material. At Friant Dam, sand was wet screened at the gravel plant, and
the No. 16 to No. 4 material was drained and dried for use as blasting
sand. At Hungry Horse Dam, blasting sand consisted of concrete sand
dried and screened so that approximately 26 percent was retained on the
No. 8, 23 percent on the No. 16, 23 percent on the No. 30, and 21 percent
on the No. 50 screen.

The performance of wet-sandblast equipment varies considerably
depending on hardness of the concrete when cleaned, roughness of the
surface, and the amount of cutting required for adequate surface treatment.
In cleaning the joint surface of good quality concrete, it is only necessary
to remove the laitance film from the mortar covering. Tests have shown
that no advantage is gained by cutting the joint surface to a depth
necessary for exposing underlying coarse aggregate. Actually, the joint may
be weakened when the surface is cut so deep as to imperil anchorage
of the larger aggregate pieces. Figure 125 offers visual comparison of
construction joint treatments.

Cleaning capacity of the wet-sandblast equipment adopted at Friant
Dam was about 500 square feet of surface per gun per hour, using a
little less than 1 cubic yard of sand per hour and 21 gallons of water
per minute. At Hungry Horse Dam, because of lower concrete strength
development at early ages with use of pozzolan and special care exercised
in preparation of joint surface during concrete placement, surface cleaning
of up to 1,500 square feet per gun per hour was achieved using about
three-fourths cubic yard of sand per hour.

The so-called green cutting of horizontal construction joints consists
of cutting the surface of fresh concrete with an air-water jet to remove
the surface layer and expose a clean surface of sound concrete. The
operation must be performed after the concrete has stiffened but before
it has become too hard for effective cutting; usually this is from 4 to
12 hours after placement, depending on temperature and other factors
that affect the rate of hardening. Cutting too early will loosen aggregate,
remove too much good material, and leave milky water which will form
a film. This method usually is not approved for Bureau work because
the treated surface, when not specially protected, almost invariably becomes
so contaminated before time to resume concreting that sandblasting is
necessary to ensure a satisfactory joint. After green cutting, the

Figure 125.——Comparison of construction joint treatments. Top: construction joint surface before sandblasting. Middle: after correct sandblasting. Bottom: after oversandblasting, causing aggregate to be undercut. PX–D–33033.

condition of the surface can be preserved by keeping it covered with a 2-inch layer of wet sand. In rare cases, it has been kept free of contamination up to 48 hours by flushing with an air-water jet at 3- to 4-hour intervals. With these expedients, however, green cutting usually costs more than one final cleanup operation.

In recent years, extremely high-pressure water-jet blasting equipment has been developed which will exert 10,000-lb/in² pressure. With this equipment, satisfactory cleanup of concrete surfaces can be obtained at 7 days' age with about 7,000-lb/in² pressure. The objection to initiating cleanup several days before the next lift is placed, as in dam construction, concerns the deposit of carbonates during curing on the surface which should be removed before the next lift is placed. With high-pressure equipment, cleanup can be accomplished just prior to placing the next lift, provided there is no excessive delay between the two placements.

Because of more difficult placing conditions, structural concrete usually contains more water, has greater slump, and requires more working and vibration than does mass concrete. Consequently, bleeding and coatings of laitance and other inferior material at joint surfaces are more prevalent. Steel erection, form construction, and installation of reinforcement, piping, conduit, etc., interfere with cleanup procedures and increase the time between lifts and the chances of contamination. On this class of work, however, it is seldom practicable to sandblast after forms for both sides are erected. As a result, there is a greater interval between sandblasting and placing of new concrete than in the case of mass concrete, and the sandblasted surface must be washed thoroughly with water, air, or an air-water jet just before concrete is placed. This procedure, with the added precaution of covering the sandblasted surface with damp sand, has been used with good results. The sand, with sawdust and other debris, is washed off just prior to placing new concrete. Loose nails, which tend to lodge along the edges of the joints against the forms and would cause unsightly blemishes and rust stains, can be removed by a magnetic device.

Sometimes the contractor may wish to eliminate or change some of the horizontal joints shown on the drawings. He should be urged to study the plans with this in mind as early as possible so there will be ample time for a review of any requests submitted. Ordinarily, as many joints as practicable should be eliminated. With modern placing equipment there has been a trend to place high walls, as much as 40 feet, in one monolithic placement.

Formed structural concrete construction joints are treated differently than horizontal joints in dams. These joints should be avoided wherever practicable as they result in planes of weakness susceptible to formation of cracks and to passage of water unless effective waterstops are provided.

Surfaces of construction joints to which concrete is to be bonded should be sandblasted, washed completely, and surface dried. No concrete should be placed during rain, and particular care should be used to ensure intimate contact. A richer concrete of ¾-inch-maximum size aggregate can be used on the horizontal construction joint to obtain a better bond and improved appearance. This concrete need not have excessive slump, 2 to 4 inches being adequate. The concrete should be well vibrated at the joint surface.

96. Forms.—Although Bureau specifications do not require contractors to submit drawings of form and form support details for approval before the forms are constructed, they do require several important items of performance for forms and form supports to obtain specified finish surfaces. It is desirable to review these performance requirements with the contractor at a very early stage while his plans are still on paper. Changes in form systems after initiation of the work, because of unsatisfactory performance, are costly and cannot correct poor results already obtained because of inadequacy.

It is not particularly economical to use poor form lumber. Too often any savings from injudicious reuse of form lumber are negated by hand labor in final dressing of the structure to an acceptable appearance. Such efforts are not always permanent, especially in areas of severe climatic exposure.

After forms for concrete structures have been set to line and grade, they should be inspected as to their adequacy. If the forms are not tight, there will be a loss of mortar which may result in honeycombing or a loss of water which may cause sand streaking. If the forms are not strong enough to hold the concrete or are not braced sufficiently to stay in alinement, the inspector should report the defects to the engineer in charge. Tolerance limits specified are for finished concrete—not for the forms. Use of internal vibrators for consolidation requires that formwork be tight and and strong. While concrete is being internally vibrated, the pressure against the forms in the immediate area of the vibrator approaches the full pressure of a liquid weighing about 150 pounds per cubic foot. Lateral pressure of concrete is influenced by many variables. The major ones are usually rate of placement, temperature of concrete, and effect of consolidation by vibration. Form pressure tests in the laboratory showed that the use of a WRA did not increase form pressure over that of concrete without the agent. Figure 126 presents data on the effect of placement rate on form pressure at different depths.

Bureau specifications stipulate the types of finish required for various formed surfaces, and the placing inspector should assure that surfaces of the forms are satisfactory. If more than one type of form will produce the required result, the choice rests with the contractor, who ordinarily furnishes all form material. For Flaming Gorge Dam, the contractor elected to use steel forms in lieu of No. 2 common or better pine shiplap.

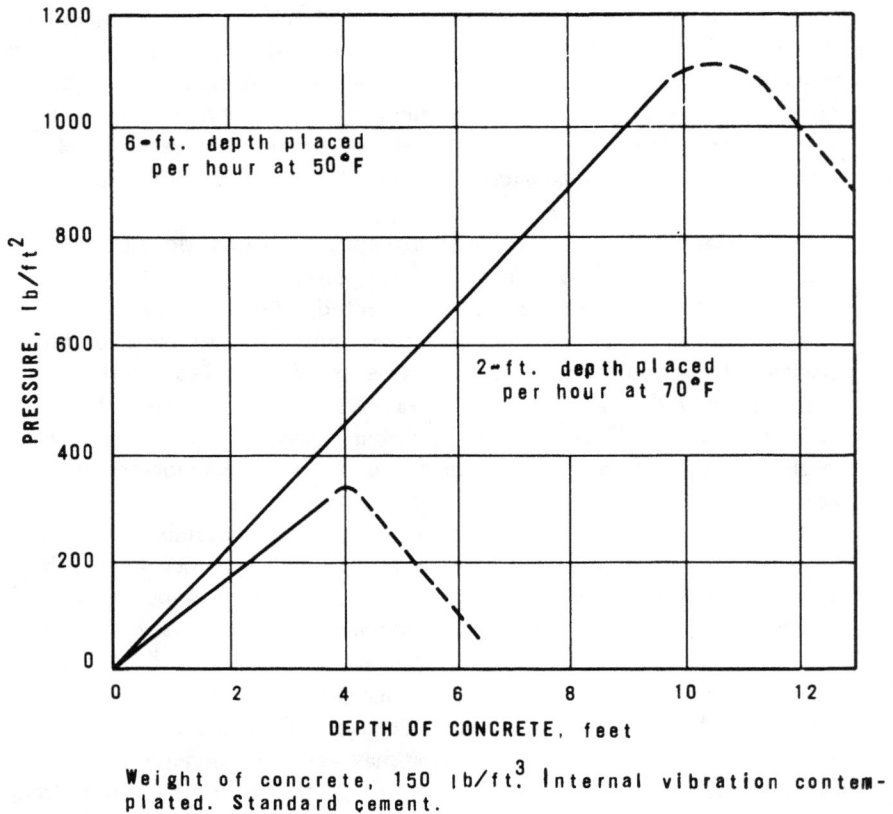

Weight of concrete, 150 lb/ft.^{3} Internal vibration contemplated. Standard cement.

Figure 126.—Pressure on forms for various depths of concrete. 288-D-2658.

The use of steel forms considerably enhanced the appearance of placed concrete.

Stability is a very important consideration in construction of forms. Some common deficiencies resulting in form failure are: (1) inadequate cross-bracing of shores; (2) inadequate horizontal bracing and poor splicing of double-tier multiple-story shores; (3) failure to regulate rate of placement without regard to drop in temperature; (4) poor regulation of the horizontal balance of the form filling; (5) unstable soil under mud sills; (6) abnormal form displacements during and after placing; (7) lack of complete and adequate inspection by capable personnel; (8) no provision for lateral pressures; (9) shoring not plumb; (10) locking devices on metal shoring improperly locked, inoperative, or missing; (11) inadequate bracing for wind pressure; (12) vibration from adjacent moving loads; and (13) nearby embankment slippage.

Immediately before concrete is placed, the forms should be inspected for cleanness. Also, it is important that they be properly treated with a suitable form oil or other coating material to prevent sticking of the concrete. Oil or other coating should be applied by brush or spray to cover the forms evenly without excess or drip and should not be allowed to get on construction joint surfaces or reinforcement bars. The oil or other form coating should not cause softening or permanent staining of the concrete surface; moreover, it should not impede the wetting of surfaces to be water cured or proper functioning of sealing compounds used for curing. Form oil used on surfaces of wood forms should be a straight, refined, pale, paraffin-base mineral oil.

For wooden forms, almost any commercial form oils will satisfactorily meet these qualifications. Products other than oil are sometimes used for treating wooden forms to preserve the form material itself, as well as serve the purpose of form oil. Shellac applied to plywood is more effective than oil in preventing moisture from raising the grain and thus detracting from the smoothness of the surface. Several commercial lacquers and similar products are available for this purpose. Occasionally, lumber delivered to the project will contain sufficient tannin or other organic substance to cause softening of the surface concrete. When this condition is recognized, it can be remedied by treating the form surfaces with white-wash or limewater prior to applying form oil or coating.

On steel forms, good results can seldom be obtained with oils that are satisfactory for wooden forms, especially if there is sliding or movement of the concrete against the forms as in tunnel lining. When an oil does not produce good results, use should be discontinued as suitable oils usually are readily obtainable. Specially compounded petroleum oils are satisfactory and are marketed under trade names by most major oil companies. Synthetic castor oil and some marine-engine oils are typical examples of compounded oils good to use on steel forms for tunnel lining. Sticking of concrete to metal forms may be the result of (1) too vigorous use of wire brushes or use of a sandblast for cleaning, (2) abrasion of form surfaces opposite ports or other areas where streams of concrete are directed against the forms, or (3) use of an unsuitable form oil. Abrasive cleaning methods should be avoided and a suitable oil used. Rough surfaces where sticking occurs may sometimes be conditioned by rubbing in one or more treatments of a liquid solution of paraffin in kerosene. Allowing the cleaned and oiled forms to stand in the sun a day or two during a weekend or other layoff period often helps to prevent sticking. Abrasion of forms caused by contact with the entering stream of concrete can be prevented by use of protective sheets of metal, plywood, or rubber belting.

Occasionally, spalling will occur from the face of the concrete when forms are removed. This is often caused by rough spots on the forms

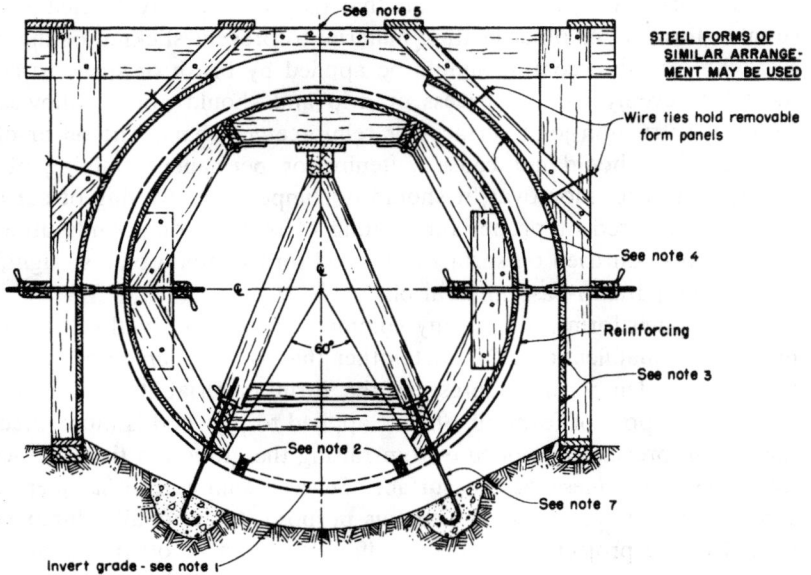

See note 5

STEEL FORMS OF
SIMILAR ARRANGE-
MENT MAY BE USED

Wire ties hold removable
form panels

See note 4

Reinforcing

See note 3

60°

See note 2

See note 7

Invert grade - see note 1

Screw jack

Steel plate

I beam

Welded pipe frame

Horizontal pipes
connecting bents

See note 6

Invert
grade

See note 8

Subgrade

**CROSS SECTION THROUGH GANTRY
FOR SETTING AND REMOVING OF INSIDE FORMS**

NOTES

1. Invert concrete is placed direct from inside
   barrel after placing first lift in both side forms.
   This avoids detrimental results of over vibration
   necessary to fill invert from the sides.

2. 2 by 4 pieces on edge, and wired to reinforcing bars
   a few inches from edge of inside form, prevent
   excessive inflow of concrete from under forms and
   prevent concrete from pulling away from under
   forms while screeding invert. They are removed
   when no longer needed as the concrete approaches
   initial set, and the grooves are filled.
   Where slope of barrel is steeper than 5 inches per foot,
   use of temporary, readily removable short panels may
   be necessary to hold invert shape from bulging. At
   initial set, panels should be promptly removed and the
   surface finished.

3. Openings are left near the bottom of outside forms of
   larger siphons to provide access for placing and
   vibrating concrete.

4. Removable form panels, 18" to 24" wide, for the outside form
   above springline are laid on top of barrel until needed,
   then, one at a time are slipped down, tied in place, and filled.

5. Outside form is taken apart along ₵ at top and moved to a
   new position with dragline or crane.

6. Inside forms are moved as a unit with gantry after pulling
   them downward and inward with turnbuckles.

7. Anchors should be firmly embedded in concrete pads or sills
   and fastened to inside forms to prevent floating.

8. Gantry runs on track with ties cut to fit invert during
   removal of inside forms.

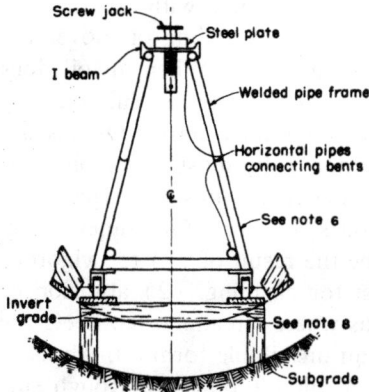

Figure 127.—Forms for siphons 8 feet in diameter and larger. 288–D–1556.

where mortar adheres strongly enough to overcome the tensile strength of the green concrete. Such areas on the forms should be cleaned, polished, and then covered with a suitable form oil. Wire brushing or sandblasting in most cases will only aggravate the difficulty.

In form construction, it is important that ready access be provided for proper placement, working, and vibration and for inspection of these operations. Figures 127 and 135 contain suggestions relative to positions and types of openings in forms. Figure 127 shows details of forms for siphons 8 feet or more in diameter. The surface of the invert is unformed for 30° on each side of the vertical centerline in order that the concrete may be more readily placed and hand finished. Other openings in inside forms for siphons are unnecessary.

Forms for tunnel lining should be provided with openings in each sidewall about midheight of the tunnel and with openings in the arch on alternate sides of the centerline so as to clear the central area of the arch for the discharge pipe of placing equipment. Openings should be at least 24 inches in the least dimension and should be spaced at a distance along the forms equal to or less than $d/2$, where d is the diameter of the tunnel in feet, but in no case more than 8 feet center to center.

For large-size tunnels, 12 feet in diameter or larger, more wall openings may be necessary for adequate placing, vibration, and inspection. In the construction of the 18.5-foot-diameter Spring Creek Tunnel No. 1 on the Central Valley project, openings along two longitudinal lines in each sidewall were necessary. The openings were staggered and spaced on 8-foot centers in each longitudinal line.

It is best to place large-diameter tunnels in two steps. In this method, the invert, with embedded bolts to provide anchorage for the arch form, is placed first. Such anchorage provides form stability and makes alinement of forms easier. The unformed portion of the invert should not exceed an arc of 60°. The maximum arc is limited by the slope of the concrete at the outer edges. When this slope becomes too steep, the concrete slumps and it is difficult to screed. The use of a full circle form for large tunnel lining often results in a poor invert finish because of entrapment of air beneath the relatively horizontal portions of the form. The full circle form may also more easily become misalined with the usual type of support. In the method whereby the invert is placed first, it can be readily finished to meet the specifications requirement.

Small-size tunnels, 12 feet in diameter or less, are frequently placed with full circle forms. With this type of placement, numerous small indentations, or bug holes, develop on the formed concrete surfaces below spring line. These are not usually objectionable in free-flowing tunnels or upstream from the gate in outlets; however, in reaches of high-velocity flow, a smooth finish is necessary. Where a surface free from bug holes is required, about 60° of the invert can be placed, first with a wood float finish, followed by the arch filling. When the full circle form is used, the form should be so designed that the bottom 90° arc section of the form can be removed immediately after final set of the concrete but not later than 8 hours after placement. Then any indentations or bug holes can be filled and the surface given a wood-float finish. Finishing criteria may sometimes be relaxed through the use of properly designed and located circumferential aeration slots. The special finishing requirements for concrete surfaces subject to high-velocity flow are described in section 122.

A very common blemish on formed concrete surfaces is the offset often found at horizontal construction joints where the forms have given way a fraction of an inch at the bottom of the new lift. For surfaces where appearance and alinement are of considerable importance, these offsets can be prevented by setting the forms so they fit snugly against the top of the concrete in the previous lift and then securing them so they will remain in tight contact during placing operations. This anchoring can be done by using an ample number of form ties or bolts, above and within a few inches of the construction joint (see fig. 128). Ties in the top of

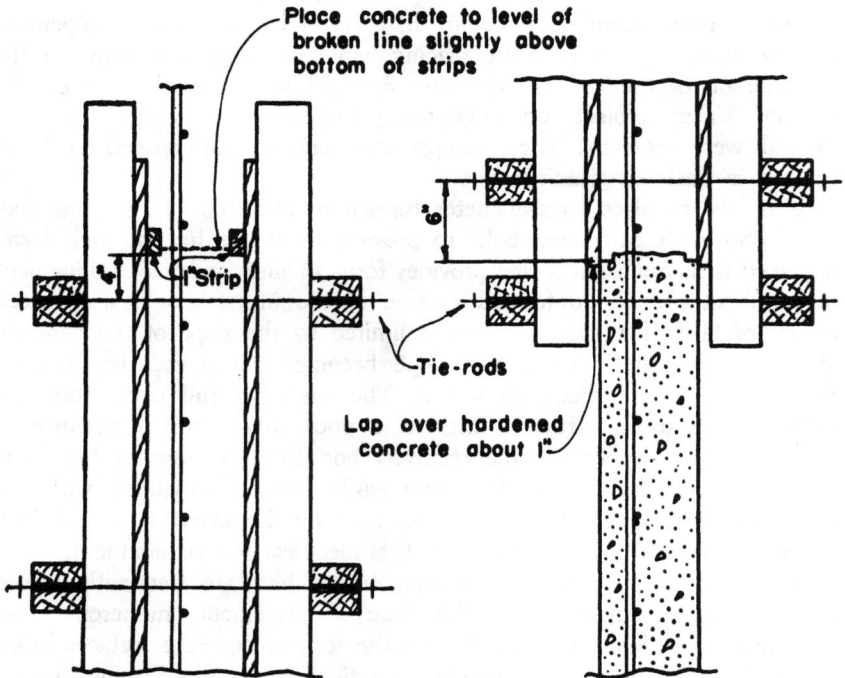

Figure 128.—Construction joint treatment at formed concrete surfaces. Bulges and offsets are avoided when tierods are close to the joint. 288–D–3275.

the previous lift cannot be relied on to prevent a slight spreading of the forms at the joint. The amount of sheathing lap on the previously placed concrete should be kept to a minimum. One-inch lap is considered sufficient.

Snap-form ties sometimes are placed so close to the corner or edge of a structural member that the concrete adjacent to the tie spalls off when the tie is released and snapped. The correct method is to place the tie away from the corner or edge.

The unsightly appearance of offsets can be minimized by use of formed grooves at the construction joints. Typical groove dimensions are shown in figure 129. Grooves permit more flexibility in form construction and reduce leakage of mortar onto surfaces at lower elevations. They also provide ideal locations for anchor bolts. Grooves should be straight and continuous across the face of the structure and should be spaced so as to create a uniform pattern. Special attention should be given to placing and consolidating concrete around groove strips.

Forms for vacuum processing are discussed in chapter VIII.

97. Marking Dates on Concrete Work.—To facilitate future studies of performance and service, concrete should be marked showing the month and year in which it was placed. For canal linings, parapet walls, flumes, channels, paving, or other linear form of construction, the markings should be placed at the start and finish of each month's work. Each canal structure should be clearly marked. In large structures, such as dams, buildings, and powerhouses, the monthly markings should be sparingly used but so placed as to show with certainty in which of the four seasons the concrete in any principal portion of the structure was placed.

A convenient method for marking concrete is to set brass benchmark plates, or similar plates, while the concrete is still plastic. Before installation the center lettering on the plates should be removed by grinding to expose an area for the markings. After installation, numerals and letters

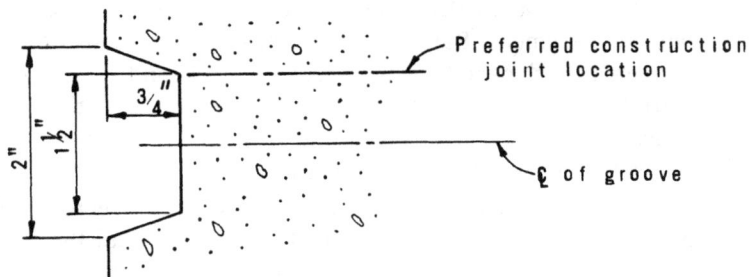

Figure 129.—Typical groove dimensions for construction joints. Horizontal grooves at construction joints obscure the joints and improve the architectural appearance. 288–D–1544.

can be easily stamped on the surface of the plate with appropriate dies. This method of marking is preferred where the markings are readily accessible for close reading. For areas where the markings must be read from a distance, markings about 3 inches high and at least one-quarter inch deep should be cast in the concrete surface by figures of metal or wood attached to the forms. Markings on unformed surfaces may be made by pressing dies into the plastic concrete at the proper stage of hardening. Markings should be placed so that letters will drain.

Patterns of letters and figures cast in place have been made of wood and also have been cast of lead in steel or well-oiled plaster of paris molds made on the job. Some plastic house numerals are satisfactory. Figures attached to a form must have a flat surface and must be held tightly against the form so that the concrete will not spall when the forms are removed.

98. Reinforcement Steel and Embedded Parts.—The concrete placing inspector on Bureau work receives from his superior any special instructions regarding inspection, handling, and installation of reinforcement steel and other embedded parts, including special instruments such as strain meters and resistance thermometers.

Each shipment of reinforcement steel should be checked for conformance to specifications. If the steel does not meet specifications requirements, the inspector should request instructions. It may be necessary to send samples to the Denver laboratories for examination and test.

According to Building Code Requirements for Reinforced Concrete, ACI Standard 318, all bars should be bent cold, unless otherwise permitted by the contracting officer. No bars partially embedded in concrete shall be field bent, except as shown on the design drawings or permitted by the contracting officer.

Some years ago, the Bureau required that reinforcing steel be cleaned to a metal-bright condition. Since this was, at times, an expensive process, laboratory tests were made to determine the effect of rust on bond in reinforced concrete specimens. Bars with four different conditions of rust were used: untreated, burlap-rubbed, wire-brushed, and sandblasted. Results of these studies corroborated the findings of previous investigations by other persons and organizations that some rust is not detrimental to bond. The following conclusions were reached:

(1) Some rust is not harmful to the bond between concrete and steel, and no benefit appears to be gained by removing all the rust. However, any rust and mill scale which is not firmly attached should be removed to assure the development of good bond.

(2) Bond is determined by the size and number of deformations.

(3) Rust increases the normal roughness of the steel surfaces and

consequently tends to augment the holding capacity of the bar, but it may reduce the effective cross-sectional area of the bar.

(4) Usually normal handling is sufficient for removal of loose rust and scale prior to embedment of reinforcement steel. However, in some instances it may be necessary to rub with a coarsely woven sack or to use a wire brush.

Bars that appear to have rusted beyond usefulness may be checked by cleaning and weighing them for conformance with Federal Specification QQ–S–632.

Some steels become brittle in cold weather and therefore require careful handling, especially when bending, to avoid breakage when temperatures are below 40° F.

Most of the reinforcement used on Bureau work is intermediate grade, deformed, billet steel, although hard grade billet or rail steel bars are permitted. Specifications require that the surface deformations of reinforcement steel be of the high-bond type. Reinforcement for prestressed concrete must be of high-stress steel. Wire for tying reinforcement is usually soft, annealed steel. Concrete blocks, metal supports, and spacers are used for holding reinforcement in place, as discussed in the following paragraphs.

Bar splices not indicated on drawings or by the specifications should not be made without approval of the engineer in charge. Tests on beams, columns, and pull-out specimens have indicated that deformed bars develop full bond resistance at tied splices in good-quality concrete that has been well vibrated. Splices may also be made by welding, provided the connection develops strength equal to that of the bar itself.

Sufficient concrete coverage should be provided to protect reinforcement from corrosion and from injury by fire where such a possibility exists. Special protection is required where concrete is submerged in or exposed to alkali or salt water. The position of reinforcement steel and minimum depth of embedment are ordinarily indicated on drawings.

Before reinforcement bars are fixed in position, the inspector should see that they are of the specified sizes and are cut and bent in accordance with drawings and specifications. All reinforcement should be supported rigidly in accurate position by means of concrete blocks, metallic supports, or other suitable devices. After the bars are in place they should be checked for positions, spacings, and splice lengths.

Bureau experience gives no indication that use of calcium chloride in concrete, under the conditions set forth in section 20(a), results in any corrosion of reinforcement steel. Published data have indicated that prestressed steel strands embedded in concrete containing calcium chloride were badly corroded after 3 years, and calcium chloride should not be used in prestressed concrete. Evidence of corrosion has also been found on structural steel embedded in concrete exposed to varying humidity conditions when the concrete is porous or only moderately dense. In all instances where this corrosion was observed, the extent of the action was not progressive, and the use of calcium chloride would be permitted. However, where embedded metal is galvanized, the use of calcium chloride is not permitted.

Some nonferrous metals do corrode if they are not effectively protected. Zinc, aluminum, and cadmium-plated parts are particularly susceptible and should be protected by an unbroken film of asphalt, varnish, pitch, or other inert material. Lead is subject to corrosion and needs a protective coating if it is not sufficiently thick to allow for initial attack. As corrosion continues, a coating is formed which protects the metal from further attack. Copper may be embedded in concrete without danger of corrosion if chlorides are not present.

Dissimilar metals should not be embedded in direct contact or close proximity with each other unless there is assurance that the conditions are such that there will be no serious galvanic action.

Where necessary to prevent damage to the concrete or unsightly rust stains on its exposed surfaces, specifications require that metal bar supports and spacers be made of noncorrodible metal. Many grades of stainless steel satisfy this requirement.

99. Final Inspection.—The final inspection is made immediately before concrete is placed and is a thorough examination of all preparations for placing concrete. The first requirement is that the forms be checked for position, alinement, and grade; their stability and adequacy of support in accordance with the placing rate should be established. The examination will include a detailed inspection of the foundation cleanup, construction joint cleanup, draintile and water stops, grout and cooling-water pipe and fittings, and reinforcement and other metalwork and equipment to be embedded. All these features should be carefully examined to make sure they are in accordance with drawings and specifications and with any special instructions that have been issued. A checklist of items of inspection is provided in the appendix following test designations.

100. Contractor's Preparations.—Before notifying the contractor that concrete placing may begin, the placing inspector should make certain that the contractor is prepared to conduct the work in a satisfactory manner. This involves an inspection of the transporting and placing equipment to see that it is clean and in proper repair and that it is adequate and properly arranged so that placing may proceed without undue delays. When concrete is placed at night, the lighting system should be sufficient to illuminate the inside of the forms. The contractor's operating force should be sufficient to assure proper placing and finishing of the concrete, and equipment should be arranged to deliver the concrete to its final position without objectionable segregation.

The number and condition of concrete vibrators for use and standby during placement should be ample for the requirements. Concrete placing should not be started when there is a probability of freezing temperatures occurring unless adequate facilities for cold-weather protection have been provided. Also, if there is a possibility of rain, the contractor should have protective coverings available for fresh concrete surfaces. Facilities for prompt commencement of water curing or for application of curing compound should be in readiness for use at the proper time. It is also advantageous, where practicable, for the contractor to provide radio or telephone communication between sites of major placements and the batching and mixing plant. Such rapid communication can result in better control of delivery schedule and prevent wasting of mixed concrete as well as provide better control of workability.

B. Transporting

101. Plant Layout and Methods.—Although concrete is carefully proportioned and properly mixed, its quality may be seriously impaired by use of improper or careless methods in transporting and placing. The contractor has the option, within the limitations of the construction specifications, to select the methods and facilities to be used; nevertheless, he is responsible for adequacy and suitability of such facilities and methods. It should not be necessary to change the design of a concrete mix to meet the requirement imposed by use of certain equipment. Specifications for dams and other work involving placement of considerable quantities of concrete require that the contractor submit drawings or detailed descriptions of handling and placing layouts and equipment he proposes to use. These data, together with project comments and recommendations, are reviewed by the Director of Design and Construction in determining adequacy and suitability of the equipment.

During the contractor's preparation of plans for plant layout and methods, project engineers often find opportunities to cooperate with the contractor to the end that undesirable features will be avoided. Some of these features are discussed in the following sections of this chapter. There should be clear understanding, by all concerned, of the specifications provision that approval of the plant layout and procedures does not release the contractor from responsibility for fulfillment of the specifications requirements. Requirements for minimizing segregation and slump loss should be strictly observed in all handling and placing operations, and any equipment incapable of producing acceptable results should be promptly modified or replaced.

102. Transporting.—Some methods of transporting concrete for major construction work are discussed in the following paragraphs with particular reference to points of interest to the concrete inspector.

(a) *Buckets.*—When designed for the job conditions and properly operated, buckets are a satisfactory means for handling and placing concrete. They should not be used, however, where they have to be hauled so far by truck or railroad that there will be noticeable separation, settlement, or a loss of slump greater than 1 inch. (See (b), following, and sec. 89.) The capacities of steel buckets for handling concrete range from 1 and 2 cubic yards for structural work up to 12 cubic yards for larger mass concrete placements (fig. 130). Use of buckets should comply with the following requirements:

(1) Less segregation will occur if the bucket capacity conforms to the size of the concrete batch, or a multiple thereof, so that there will be no splitting of batches in loading buckets.

(2) Buckets should be capable of prompt discharge of low-slump, lean-mix concrete. The dumping mechanism should permit discharge of as little as a ½-cubic-yard of concrete in one place. Also, the discharge should be controllable so that it will cause no damage to or misalinement of forms. Buckets should be equipped for attachment and use of drop chutes or collapsible elephant trunks.

(3) Buckets should be filled and discharged without noticeable separation of coarse aggregate. (See figs. 114 and 131.)

(b) *Cars and Trucks.*—Rail cars are often used to transport concrete for tunnel linings and for short hauls from plant to bucket dock for highline pickup. Trucks are convenient for the distribution of concrete from a central mixer to small, scattered structures. Ordinarily, the haul should not exceed 2 or 3 miles, unless remixing of some type is done at the placing point or agitation is provided in transit. On one section of the San Luis Canal, Central Valley project, California, the contractor was able to haul concrete up to 17 miles in bottom-dump trucks along a rough haul road without difficulty. The concrete was remixed by distributing screws in the paver before being placed as lining. The concrete contained entrained air and a WRA, but most importantly the concrete slump was controlled within narrow limits. Segregation during the filling and discharging of these units may be avoided by following the correct methods shown in figures 114, 131, and 132.

Free water should not be on the surface of the concrete as delivered, nor should there be an objectionable amount of settlement of coarse aggregate or caking at the bottom of the load. Such stratification or settlement can be considerably reduced by (1) use of drier mixes, (2) use of air entrainment, or (3) an appropriate remixing as the concrete passes through the discharge gates of trucks, hoppers, and cars. Of the above measures, the most effective are the use of drier mixes and air entrainment.

Figure 130.—A 12-cubic-yard bucket which will readily discharge concrete of low slump, permit slow or partial discharge, and dump the concrete in relatively low twin piles requiring a minimum of lateral movement during consolidation. P557—420—05480.

CORRECT

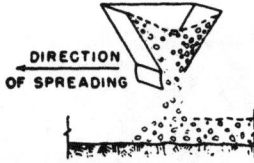

TURN BUCKET SO THAT SEPARATED ROCK FALLS ON CONCRETE WHERE IT MAY BE READILY WORKED INTO MASS

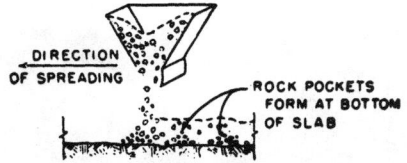

INCORRECT

DUMPING SO THAT FREE ROCK ROLLS OUT ON FORMS OR SUBGRADE

DISCHARGING CONCRETE

CORRECT

DROPPING CONCRETE DIRECTLY OVER GATE OPENING

INCORRECT

DROPPING CONCRETE ON SLOPING SIDES OF HOPPER

FILLING CONCRETE HOPPERS OR BUCKETS

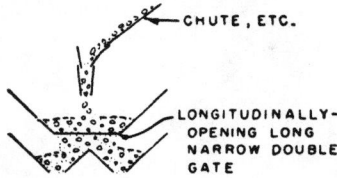

CORRECT

THE ABOVE ARRANGEMENT SHOWS A FEASIBLE METHOD IF A DIVIDED HOPPER MUST BE USED. (SINGLE DISCHARGE HOPPERS SHOULD BE USED WHENEVER POSSIBLE.)

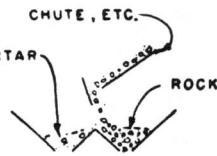

INCORRECT

FILLING DIVIDED HOPPER AS ABOVE INVARIABLY RESULTS IN SEPARATION AND LACK OF UNI- FORMITY IN CONCRETE DELIVERED FROM EITHER GATE.

DIVIDED CONCRETE HOPPERS

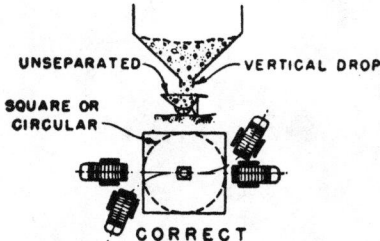

CORRECT

DISCHARGE FROM CENTER OPEN- ING PERMITTING VERTICAL DROP INTO CENTER OF BUGGY ALTERNATE APP- ROACH FROM OPPOSITE SIDES PERMITS AS RAPID LOADING AS MAY BE OBTAINED WITH OBJECTIONABLE DIVIDED HOPPERS HAVING TWO DISCHARGE GATES.

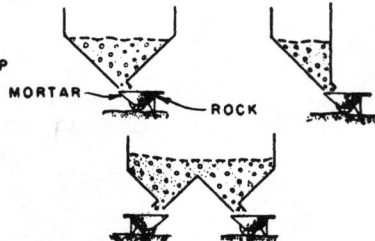

INCORRECT

SLOPING HOPPER GATES WHICH ARE IN EFFECT CHUTES WITH- OUT END CONTROL CAUSING OBJECTIONABLE SEPARATION IN FILLING THE BUGGIES.

DISCHARGE OF HOPPERS FOR LOADING CONCRETE BUGGIES

Figure 131.——Correct and incorrect methods for loading and discharging concrete buckets, hoppers, and buggies. Use of proper procedures avoids separation of the coarse aggregate from the mortar. 288–D–3276.

Provide 24-
inch min.
headroom
for down-
pipe

Counterweighted
rubber scraper

No separation

Mortar

Baffle   No baffle

Rock   Mortar

CORRECT

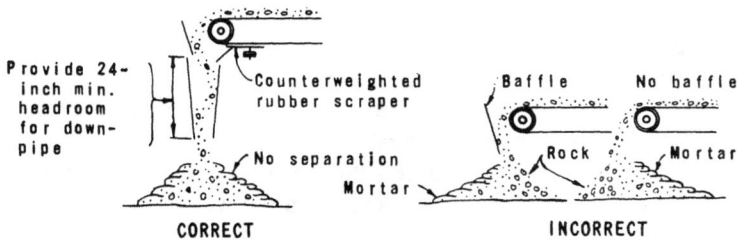

The above arrangement prevents
separation of concrete whether
it is being discharged into
hoppers, buckets, cars, trucks,
or forms.

INCORRECT

Improper or complete lack of control
at end of belt.
Usually a baffle or shallow hopper
merely changes the direction of
separation.

CONTROL OF SEPARATION OF CONCRETE AT THE END OF CONVEYOR BELT

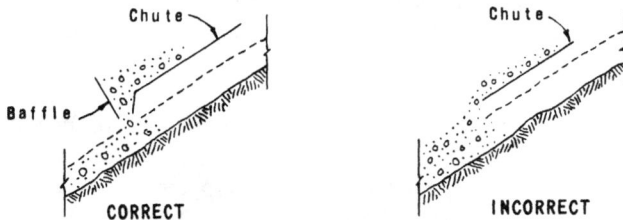

Chute

Baffle

Chute

CORRECT

Place baffle and drop at end of
chute so that separation is
avoided and concrete remains
on slope.

INCORRECT

Concrete discharged from a free
end chute on a slope to be paved.
Rock is separated and goes to bottom
of slope. Velocity tends to carry concrete down slope.

PLACING CONCRETE ON A SLOPING SURFACE

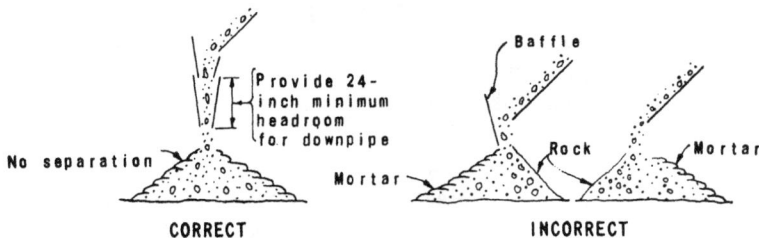

Provide 24-
inch minimum
headroom
for downpipe

No separation

Mortar

Baffle

Rock   Mortar

CORRECT

The above arrangement prevents
separation, no matter how short
the chute, whether concrete is
being discharged into hoppers,
buckets, cars, trucks, or forms.

INCORRECT

Improper or lack of control at end
of any concrete chute, no matter
how short.
Usually a baffle merely changes
direction of separation.

CONTROL OF SEPARATION AT THE END OF CONCRETE CHUTES

This applies to sloping discharges from mixers, truck mixers, etc. as well
as to longer chutes, but not when concrete is discharged into another chute
or onto a conveyor belt.

**Figure 132.—Correct and incorrect methods of concrete placement using
conveyor belts and chutes. Proper procedures must be used if separation at
the ends of conveyors and chutes is to be controlled. 288–D–854.**

(c) *Chutes.*—This equipment, as ordinarily used, is one of the most unsatisfactory devices for transporting concrete. Use of chutes is not prohibited on Bureau work, but the operations must be so controlled that segregation and objectionable loss of slump will be avoided. To meet these conditions, the following requirements must be fulfilled:

(1) The chute must be on a slope sufficiently steep to handle concrete of the least slump that can be worked and vibrated and must be supported so the slope will be constant for varying loads.

(2) If more than a few feet long, the chute must be protected from wind and sun to prevent slump loss.

(3) Effective end control that will produce a vertical drop and prevent separation of the ingredients must be provided, preferably in the form of two sections of metal drop chutes. (See figs. 114 and 132.) This end control is of primary importance as segregation results not from the length or slope of the chute but from the lack of such control.

(d) *Belt Conveyors.*—There is no objection to the use of belt conveyors if segregation and objectionable slump losses are prevented and there is no loss of mortar on the return belt. Slump loss is largely preventable by protecting the belt from the sun and wind. Segregation, which may occur chiefly at transfer points and at the end of the conveyor, can be avoided by use of a suitable hopper and drop chutes. (See figs. 114 and 132.)

Several types of high-speed commercial belt conveyors are available. One type, manufactured in several lengths, has the maximum handling capability of about 150 cubic yards of concrete per hour. Although actual placing rates will vary from 50 to 130 cubic yards per hour depending on the type of placement, specially designed equipment at the transfer points eliminates segregation. The end section is usually equipped with a telescopic cantilevered swing section which facilitates easy positioning of the discharge.

(e) *Pneumatic Methods.*—Pneumatic methods are commonly used for placing concrete in tunnel linings, often with good results. However, pneumatic methods have certain disadvantages. In most instances, the end of the discharge line (slick line) should be buried in the fresh concrete in the tunnel arch; otherwise, separation of coarse aggregate results from the impact of the violently discharged concrete. There have been instances where thin concrete linings were placed in machine-bored tunnels in which complete filling of the arch could not be obtained with the slick line buried. In order to obtain complete filling, the slick line was placed about 2 to 5 feet from the face of the fresh concrete.

Separation is most pronounced at the beginning of concrete placement in bulkhead sections, and it is especially objectionable if the concrete is to be molded around waterstops at the bulkheaded joints. Specifications, therefore, require that equipment used in placing tunnel concrete, and the method of operation, will permit introduction of the concrete into forms without high-velocity discharge and resultant separation. Some air guns have been operated in compliance with this requirement, but constant and firm inspection is required to ensure compliance until the end of the discharge pipe is covered with concrete in the arch. Where an air gun is used without velocity control, difficulty with separation at the start of placement can be reduced by using richer concrete with smaller aggregate and allowing ample time for proper vibration of the concrete in the forms, particularly until a slope is established that may then be advanced as in the continuous method.

A further objection to the pneumatic method is the loss of slump which occurs in the shooting process. Slump losses as great as 3½ inches between mixer and forms have been observed, and a loss of 2 to 3 inches is not uncommon. This means that if the mixed concrete requires addition of water to offset loss in slump between the mixer and the forms, the cement content needs increasing as much as one-tenth of a barrel per cubic yard to obtain the required water-cement ratio. In addition to the cost of extra cement, there is the disadvantage of increased water content causing greater drying shrinkage and rendering the concrete less resistant to freezing and thawing. (See figs. 1 and 17.)

Bureau specifications permit placement of concrete in the sidewalls and arches of tunnels by pneumatic equipment. Its use in placing tunnel inverts is also permitted with the limitation that the discharge from the placer line must be controlled so that segregation will not occur with high velocities. With slight modifications, whatever equipment is used to charge an air gun can place the concrete directly in the invert. In making such direct placement, separation should be avoided by dropping the concrete through one or two sections of vertical downpipe at the end of the chute or belt.

(f) *Pumping.*—One of the most satisfactory methods of transporting concrete where space is limited, as in tunnels, on bridge decks, and in some powerhouses and buildings, is by pumping through steel pipelines. (See fig. 133.) The normal effective pumping range will vary from 300 to 1,000 feet horizontally or 100 to 300 feet vertically. There have been isolated cases in which concrete has been pumped horizontally in excess of 2,000 feet and vertically 500 feet. Curves, lifts, and harsh concrete materially reduce the maximum pumping distance. A 90° bend, for example, is equivalent to about 40 feet of straight horizontal line, and each 8 feet of line is equivalent to 1 foot of head.

There have been reported difficulties with the use of extensive lengths of aluminum pipe for pumplines. The difficulty is caused by abrasion of metallic aluminum from the pipe and its subsequent reaction with lime being released during the hydration of cement, producing hydrogen gas.

Figure 133.—Pumpcrete equipment used in placing concrete for gate chamber in Sugar Loaf Dam, Fryingpan-Arkansas project, Colorado. P382–706–5404.

The gas causes expansion of the concrete, and the resulting voids reduce the concrete strength excessively. The Bureau does not now permit the use of aluminum pipes for pumplines.

Although some manufacturers rate their largest pumping equipment as capable of handling concrete containing aggregate as large as 3 inches, Bureau experience indicates that operating difficulties will be materially lessened if the maximum size of aggregate pumped through such equipment is limited to 2½ inches. The maximum size of well-rounded aggregate should not exceed 40 percent of the pipe diameter. Frequently, however, such ideally shaped aggregates are not available but contain considerable amounts of flat and elongated shapes. Thus for a 2½-inch-maximum size aggregate it may be advisable to use an 8-inch-diameter pumpline. In some instances, depending on the thickness of the lining, quantities involved, strength desired, and spacing of reinforcement steel, use of aggregate larger than 1½-inch-maximum size is not advisable. Normal rated pump capacities range from 10 to 90 cubic yards per hour. Specifications limit the slump of concrete for tunnel arch lining to a maximum of 4 inches (table 13, chap. III), and a pump will normally make good progress with concrete having a slump of 3 to 4 inches.

It should be noted that acceptable performance will be obtained only with pumps in good mechanical condition and with valves in proper adjustment. When operating difficulties arise with good workable mixes, mechanical features of the pump are usually at fault and should be corrected before changes in either mix or slump are considered. However, particularly for successful delivery through long lines, it should be noted that air-entrained concrete may require somewhat more sand and perhaps an inch greater slump than would likely be necessary without air entrainment. These are required to offset reduction in workability as a result of compression of the entrained air while the concrete is moving in the pipeline. For short lines, as used in most tunnel lining operations, extra sand and water are not as likely to be required for successful operation and should not be used unless necessary.

Concrete as discharged from pumplines is generally free from segregation; in fact, concrete that segregates badly cannot be handled satisfactorily by pumping. Best performance requires use of a standard accessory agitator in the pump feed hopper. (See fig. 134.) It should also be kept in mind that inflexibility of the system and amount of concrete in a long

Figure 134.—Equipment and method used for pumping concrete into spillway tunnel at Trinity Dam, Central Valley project, California. TD–5008–CV.

line seriously hamper quick adjustments necessary for good control. Because of difficulty resetting the heavy delivery lines, there is a tendency toward pumping too much concrete into the forms at one delivery point and making it flow in the forms to such an extent that noticeable segregation occurs. The result is a serious lack of coarse aggregate in concrete that has flowed a considerable distance. This condition may be avoided by pumping concrete into a centrally located hopper, from which it is transported in buggies to points where it can be deposited and vibrated without excessive flow.

A constant supply of fresh, plastic, unseparated concrete of medium consistency is essential for satisfactory operation of the pump. A plugged pumpline not only involves delay in pumping operations but will result in additional loss of time, extra expense, and detriment to the work if placing operations are suspended long enough to cause cold joints. Plugged lines are usually occasioned by segregated, harsh, or lean concrete; by too dry or too wet a consistency; or by interruptions caused by improper or inadequate servicing. In starting a pumping operation, the concrete should be preceded by mortar consisting of the regular concrete with the coarse aggregate omitted. About 1 cubic yard of mortar is sufficient to lubricate 1,000 linear feet of pipe irrespective of size. Thereafter, lubrication will be maintained as long as pumping continues. To minimize friction and plugging of the pipe, pumplines should be as straight as the work will permit. Flexible rubber pipe sections directly connected to the pump, especially if curved, are often the cause of pumping difficulties. Standard 5-foot radius bends of $11\frac{1}{4}°$, $22\frac{1}{2}°$, 45° and 90° are available. When disconnecting an inclined pipeline for cleanout, a suitable pin-type valve at the lower end will provide control of the outflow of the concrete.

Compressed-air boosters attached near the end of the pumpline in tunnel lining operations have been used to permit injection of intermittent shots of air when filling the arch. The air clears the end of the pipe, makes pumping easier, and aids materially in thrusting the concrete laterally into the shoulders of large tunnel arches. Compressed air should be used only when the discharge end of the line is well buried in concrete, and the amount of air used should be held to a minimum to keep down slump loss.

Care must be taken to prevent leakage of water into the concrete from the pump water jackets through which a stream of water is passed to keep the cylinder and piston clear of mortar. Leakage occurs when the rubber seal on the piston becomes worn (after pumping about 1,000 cubic yards of concrete) and may be detected by an increase in amount and discoloration of the stream as it leaves the bleeder pipe under the cylinder. An increase in the slump of the concrete between entrance to the pump and discharge from the pipeline, or free water found in the end of the operating cylinder when the pipe is detached for cleaning, is also evidence of leakage around the rubber seal. When the cylinder liner and piston are so badly worn as to need replacement or repair, renewal of the seal will seldom stop the leakage. Replacement of these parts is usually necessary after pumping 4,000 to 7,000 cubic yards of concrete. Leakage will rarely be serious if the equipment is properly maintained and if the water fed through the water jackets is not under pressure. This water should flow by gravity from a tank mounted above the pump, and not more than a few gallons per hour should be required.

If water is used behind a go-devil to clean concrete from the pipe at the end of a run, progress should be followed by tapping the pipe and stopping just short of spilling into the forms. When pumping is resumed, the pump may be relieved of considerable shock and strain if a foot of the pipe ahead of the go-devil is filled with wadded burlap or excelsior. When the line is several hundred feet long, two go-devils should be used in tandem, each preceded by the shock-absorbing material.

Lightweight concrete is being increasingly used for roof decks and other structural members. Concrete made with lightweight aggregates can be readily pumped with proper precautions. Water loss from the mortar paste through absorption by the aggregate significantly impairs the fluidity during pumping. This may be corrected to some extent by presoaking the fine and coarse aggregate for 2 or 3 days; however, the soaking time can be determined by trial. Following this, additional time should be allowed for draining. Under atmospheric conditions some lightweight aggregates will absorb only a small percentage of their total absorption capacities. Under pressure during pumping, more water is forced into the aggregate as the remaining air in the voids is compressed.

Lightweight concrete has been pumped successfully for many years in large lines 6 to 8 inches in diameter. It becomes difficult to pump such concrete with small- to medium-size pumps that use lines less than 5 inches in diameter.

Vacuum saturation of coarse aggregate has reduced the problem of using small pumplines to a minimum. Saturation by this method can be accomplished in 30 to 45 minutes. Early freezing and thawing tests of concrete made with vacuum-saturated aggregate indicate a low resistance; however, if the lightweight concrete is allowed to dry properly for several weeks, resistance to freezing and thawing is greatly increased.

Another method of presoaking lightweight aggregate is by thermal saturation. In this method, the heated aggregates are immersed in water. To avoid damage to aggregates from thermal shock, the optimum combination of aggregates and water temperatures should be determined by saturation. Moisture absorption by this method is equivalent to that obtained by dry-vacuum processing.

The optimum slump range for pumping lightweight concrete is from 2 to 5 inches. In concrete mixes of high slump, the aggregate will separate from the mortar and may plug the pumpline. Also, overly wet mixes have excessive bleeding and volume change characteristics. It is much more important to obtain a truly plastic mix by proper proportioning of materials than to try overcoming deficiencies by adding more water. Mix design and physical properties of lightweight concrete are discussed in sections 140, 141, 142, and 143.

C. Placing

103. The Mortar Layer.—When concrete is to be placed on and bonded to rock foundations, a mortar or grout layer should be scrubbed into the joint surface only when the rock surface is horizontal or nearly so and the surfaces are porous or absorptive. Mortar layers are not practicable or used on rock surfaces in tunnels, as any spaces behind the rock are subsequently grouted. The rock surfaces should be surface dry and free of standing water when the mortar coat is applied. The mortar should be of the same proportions as those in the concrete with coarse aggregate omitted, and should include the same quantity of air-entraining agent. The mortar should be soft enough to be readily spread on the joint surface at a thickness of approximately three-eighths of an inch. This soft consistency is usually possible well within the specified limitation for water-cement ratio. Overwet mortar tends to segregate and to flow into lower elevations. The mortar coat should be broomed thoroughly into the joint surfaces wherever possible. On inaccessible joints in structural concrete, care must be taken to assure distribution of mortar onto the entire joint. Air jets have been used to spread mortar in such places, and the air-suction gun shown in figure 124 has been found effective for this purpose.

Treatment of construction joints prior to placing new concrete is described in detail in section 95. Watertightness between the new and the old concrete can be obtained only by complete contact and bond throughout the joint surface. Under ideal conditions, with a well-cleaned joint surface and with thorough vibration of new concrete throughout the area of contact, a mortar coat would not be needed. However, to ensure watertightness of such construction joints in thin structural walls, mortar coating is specified. On the other hand, in mass concrete structures it has been demonstrated that effective bond can be obtained on horizontal construction joints with good cleanup and a dry surface without a layer of mortar.

104. General Discussion of Concrete Placement.—Properly placed concrete is free of segregation, and the mortar is intimately in contact with coarse aggregate, reinforcement, and other embedded parts. If any one detail of the placing inspector's many duties deserves special emphasis, it is that of guarding against objectionable segregation during concrete placement.

The basic reason for segregation is the nonhomogeneous character of

CONCRETE WILL SEPARATE SERIOUSLY UNLESS
INTRODUCED INTO FORMS PROPERLY

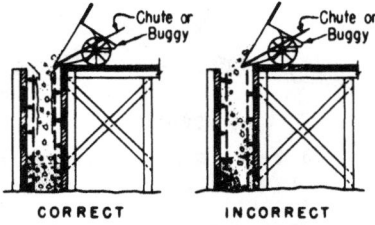

CORRECT
Discharge concrete into light hopper feeding into light flexible drop chute. Separation is avoided. Forms and steel are clean until concrete covers them.

INCORRECT
To permit concrete from chute or buggy to strike against form and ricochet on bars and form faces causing separation and honeycomb at the bottom.

PLACING CONCRETE IN TOP OF NARROW FORM

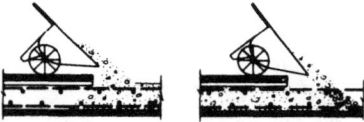

CORRECT
Necessarily wetter concrete at bottom of deep narrow form made drier as more accessible lifts near top are reached. Water gain tends to equalize quality of concrete. Settlement shrinkage is minimum.

INCORRECT
To use same slump at top as required at bottom of lift. High slump at top results in excessive water gain with resultant discoloration, loss of quality and durability in the upper layer.

CONSISTENCY OF CONCRETE IN DEEP NARROW FORMS

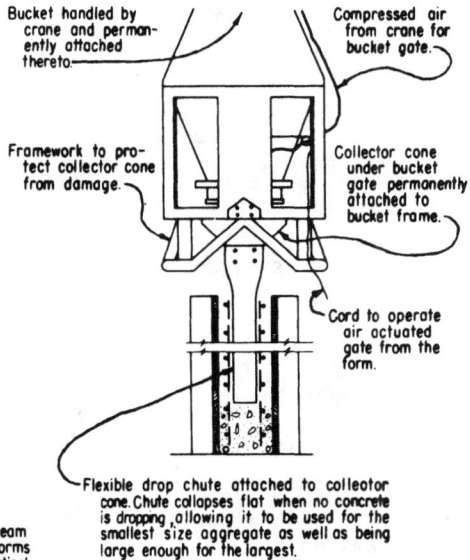

CORRECT
To dump concrete into face of concrete in place.

INCORRECT
To dump concrete away from concrete in place.

PLACING SLAB CONCRETE FROM BUGGIES

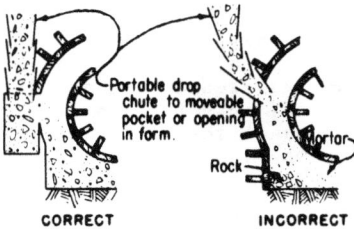

CORRECT

INCORRECT

PLACING CONCRETE IN DEEP NARROW FORMS

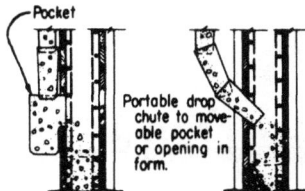

CORRECT
Drop concrete vertically into outside pocket under each form opening so as to let concrete stop and flow easily over into form without separation.

INCORRECT
To permit high velocity stream of concrete to enter forms on an angle from the vertical. This invariably results in separation.

PLACING IN DEEP OR CURVED WALL THRU PORT IN FORM

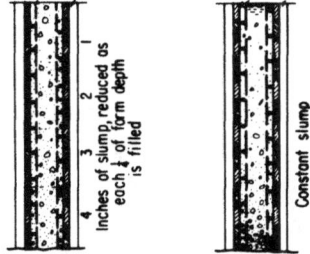

Figure 135.—Correct and incorrect methods of placing concrete in deep, narrow forms and slabs. 288–D–2657.

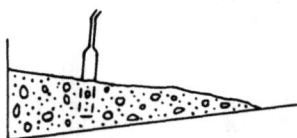

CORRECT

Start placing at bottom of
slope so that compaction
is increased by weight of
newly added concrete as
vibration consolidates.

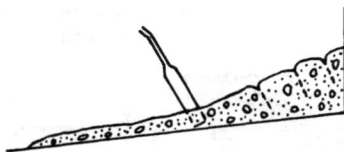

INCORRECT

To begin placing at top of
slope. Upper concrete tends
to pull apart, especially
when vibrated below, as vibration starts flow and removes
support from concrete above.

WHEN CONCRETE MUST BE PLACED IN
A SLOPING LIFT

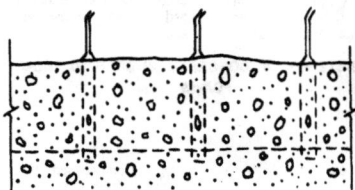

CORRECT

Vertical penetration of
vibrator a few inches into
previous lift (which should
not yet be rigid) at systematic regular intervals found
to give adequate consolidation.

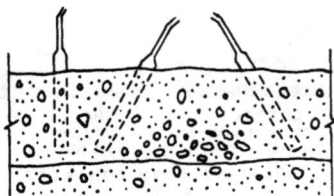

INCORRECT

Haphazard random penetra-
tion of the vibrator at all
angles and spacings without
sufficient depth to assure
monolithic combination of
the two layers.

SYSTEMATIC VIBRATION OF EACH NEW LIFT

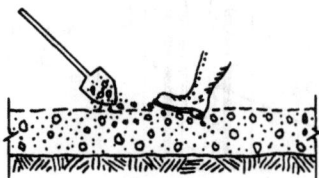

CORRECT

Shovel rocks from rock
pocket onto a softer,
amply sanded area and
tramp or vibrate.

INCORRECT

Attempting to correct rock
pocket by shoveling mortar
and soft concrete on it.

TREATMENT OF ROCK POCKET WHEN PLACING
CONCRETE

Figure 136.—Correct and incorrect methods of vibrating and working concrete.
Use of proper methods ensures thorough consolidation. 288–D–856.

the product, an aggregation of materials widely different in particle size and specific gravity. As a consequence, from the time concrete leaves the mixer there are internal and external forces acting to separate the dissimilar constituents. If there is lateral movement, as in dumping at an angle or depositing continuously at one point and allowing the concrete to flow, the coarse aggregate and mortar tend to separate; or if the concrete is confined laterally, as in a container or in restricting forms, there is a tendency for the coarser and heavier particles to settle and for the finer and lighter materials, especially water, to rise. Separation of coarse aggregate from the mortar may be minimized by avoiding or controlling the lateral movement of concrete during handling and placing operations, as illustrated schematically in figures 114, 132, 135, and 136.

Bureau specifications require that concrete be deposited as nearly as practicable in its final position to preclude use of placing methods that would allow or cause the concrete to flow in the forms. These methods result in concentration of less durable mortar in ends of walls and corners, where durability is most important, and encourage use of a mix that is wetter than necessary for adequately vibrated concrete.

Bureau specifications also provide that concrete be placed in horizontal layers (except in linings of small-diameter tunnels where such procedure is impracticable). Each layer should be soft when a new layer is placed upon it. This requirement may determine the depths of the layers. Practicable depths of layers for mass concrete are 15 to 24 inches and for structural concrete 12 to 20 inches. Precautions should be taken to avoid entrapment of air within partially enclosed spaces to be filled.

Hoppers and drop chutes may or may not be required for placement in walls. When the placement can be completed before mortar dries on forms and reinforcement, drop chutes will not be needed for preventing encrustations. If the concrete can be placed directly from a crane bucket moving along the top of the forms as it drops the concrete vertically between curtains of reinforcement, separation will be less than if the concrete is put through a hopper and drop chute, and placement will be faster. Any method is acceptable if it places the concrete close to its final position without objectionable separation of coarse aggregate. Scattered individual pieces of aggregate should give no concern because they will be readily embedded as more concrete is placed and vibrated. As long as groups and clusters of separated aggregate are not produced, unconfined fall should be permitted and the height of fall should not be limited. Methods producing objectionable and excessive separation as herein described should be prohibited at any point from the mixer to final placement.

Hoppers for drop chutes should have throat openings of sufficient area to readily pass concrete of the lowest slump that it is practicable to work and vibrate. If drop chutes are discharged directly through form ports, considerable separation results, and rock pockets and honeycomb will probably be formed. Provision of an outside pocket below each port (fig. 135) will check the fall of the concrete and permit it to flow into the form with a minimum of separation.

In placing concrete at surfaces of construction joints where adequate vibration is impracticable, the concrete should be worked by hand floating or other effective means to ensure thorough compaction of concrete at the surfaces of joints. Such hand working may be necessary at joints in thin, heavily reinforced walls, particularly where the joint is sloping, and in stub walls where the base of the wall is not totally confined.

When slabs and beams and the supporting walls and columns are cast monolithically, the concrete in the top 2 or 3 feet of walls and columns should be of the lowest slump that can be vibrated adequately and should be fully consolidated at the surface. Before placing concrete in the top fillets, slabs, and beams, the concrete in walls and columns should be allowed to settle as long as possible without allowing it to become so hard that a running vibrator will not penetrate it of its own weight. This will be 1 to 3 hours or more, depending on temperature and other conditions. During this time, care should be taken to keep the concrete surfaces clean and free of loose or foreign material. Such material is often blown or swept off the slab forms into the walls. Except where the amount is excessive or

Figure 137.—A chute lining that failed because of almost complete lack of consolidation except at the surface. Most of the reinforcement steel was found to be in a useless location at the bottom of the slab. PX–D–20776.

is contaminated with dirt and other debris, it is generally permissible to ignore green concrete droppings as they will be absorbed into the fresh concrete placed on them. After placing the fillet, beam, and slab concrete, the vibrator should penetrate and revibrate the concrete in the tops of walls and columns.

When placing an unformed slab on a slope by hand methods, there is a tendency to place the concrete at the full slab thickness, using a stiff mix that will not slough. Drill cores have shown that the placement of such low-slump concrete without thorough vibration usually results in considerable honeycomb on the underside, especially when the slab is reinforced. Figure 137 shows chute lining that was badly honeycombed because the concrete was too dry when placed and was poorly consolidated. To avoid such results, the consistency for this purpose should not be stiffer than a 2½-inch slump. Concrete with this consistency will barely stay on the slope, but it should not be drier. After spreading, concrete should be thoroughly and systematically vibrated, preferably just ahead of a weighted, unvibrated, steel-faced slip-form screed working up the slope, as shown in figure 146. During vibration, surface concrete will move downward on the slope, and the excess should be returned to where it is needed. Slabs placed in this way with concrete of medium slump will be free of voids and rock pockets. Placing, vibrating, screeding, and strokes of finishing operations should proceed in an upward direction on the slope.

Effective control of concrete placing operations makes possible the use of economical concrete mixes. The most economical concrete is that which contains the least amount of water and cement compatible with sufficient workability and adequate quality. Highest quality is associated with lowest water-cement ratio, but this should be attained as a result of low water content and not as a result of high cement content. With the cement content held at a normal amount, the lowest water-cement ratio can be attained only by designing the mix to include: (1) least slump practicable, (2) largest maximum size of aggregate usable, (3) lowest percentage of sand consistent with good workability, (4) entrainment of air in the proper amount, and (5) use of water-reducing, set-controlling agents. Good judgment must be used in selection of the criteria expressed in (1), (2), and (3). Workability is a very important consideration and should not be sacrificed in striving for economy and quality in mix design. Unless the concrete can be readily handled with proper equipment, any savings in materials costs are quickly dissipated.

A reasonable and practical judgment of how low the slump should be will not be influenced by such criteria as the ability of the concrete to flow out of a certain bucket, down a certain chute, or through the gate of a certain hopper. If slump is sufficient to enable good consolidation in the forms by ample vibration, it is the responsibility of the contractor to provide handling and placing equipment that will put such concrete into the forms ready for vibrator action. In fact, specifications prohibit the contractor from using buckets, chutes, hoppers, or other equipment that will not readily handle and place concrete of lower slump.

The contractor is also required to handle and place concrete containing aggregate of maximum practicable size, within the specified size range. Since all oversanded mixes are workable, it will not be certain that the mix is not oversanded until successive reductions in sand content have been made and the percentage of sand determined that represents the borderline at which there is a definite hazard to workability and workmanship. With the restoration of 1 or 2 percent of sand for a margin of workability, the proper mix will have been established. The advisability of changing the maximum size of aggregate or amount of sand for parts of a concrete placement having different dimensions, reinforcement, or accessibility must be determined on the job. It is generally feasible to reduce slump for the more accessible portions of the work. This is especially true of wall and column placements, where the slump should be reduced as the level of concrete rises (see fig. 135), to offset water gain that would weaken the upper portion of the concrete and make it less durable.

An inspection of existing structures discloses the fact that weathering and disintegration of concrete are most severe at the tops of walls, piers, and parapets, and in curbs, sills, ledges, copings, exposed corners, and those portions in intermittent contact with water or spray during freezing weather. All available means of improving the durability of such concrete should be employed during construction, including those in the following list. Where these measures should be followed in part of a placement and it is not practicable to make the indicated changes during the placing operation, items 1, 3, and 4 should be observed from start to finish, and the slump (item 2) should be made as low as practicable at all stages of the work.

(1) Use the maximum allowable amount of entrained air in critically exposed portions of the work (see sec. 20 (b)); the resulting small loss of strength is relatively unimportant.

(2) Lower the slump to the minimum that can be vibrated well in critically exposed portions, usually readily accessible. At this low slump, which rarely will need to be more than 1 inch, the usual amount of air-entraining agent must be increased to obtain the required percentage air entrainment.

(3) Reduce the water-cement ratio to 0.45 ± 0.02 by weight. Additional cement required should be substituted for sand in the regular mix. There should be no increase in the unit water content. Essentially, these changes will result in a new mix having close to 5 or 6 percent air, a 1-inch slump, and a water-cement ratio of 0.45, for the critically exposed parts of the work.

(4) In placing concrete in walls, curbs, and slabs, work from cor-

ners and ends of the forms toward the center, rather than toward the corners and ends, thus avoiding accumulation of mortar and wetter concrete in those parts of structures where exposure is most severe.

(5) Exposed unformed surfaces should be sloped to provide quick, positive drainage and to avoid puddles in low spots.

(6) Outdoor unformed surfaces should be finished with only a minimum of wood floating.

(7) The concrete should receive thorough curing beginning with water as soon as it (formed or unformed) has hardened, and the surface should be kept continuously wet for the specified 14 or 21 days (see sec. 124) or until it can be thoroughly sealed with approved pigmented sealing compound (see sec. 125). The concrete should be kept saturated with water during the time the forms are in place. Forms are not a substitute for curing, particularly in the summertime.

(8) Where it is probable, after the structure is in service, that seepage will come in contact with or flow over the concrete, permanent drainage should be established, preferably by back-filling with gravel over a suitable installation of drains.

Note that items 1, 2, 3, 4, 7, and 8 are also important factors (along with the use of type V sulfate-resisting cement) in improving the resistance of concrete to corrosion and disintegration by aggressive sulfate waters.

The engineer in charge of concrete placement is at times confronted with the problem of deciding whether concreting may be continued during precipitation in the form of rain, hail, sleet, or snow and if continued, what precautions are necessary. As a quick decision must be made, trouble may be forestalled by thoughtful planning based on job conditions and current weather reports. A reasonable supply of protective covering should be available for immediate use. If periodic or sustained precipitation is imminent, the scale and arrangement of operations should be so devised as to facilitate protection of the work. The greatest difficulty in placing mass concrete while precipitation is occurring is the problem of maintaining the surface of the previous lift in a surface-dry condition. Covering with polyethylene sheets and drying with compressed air as the first layer is completed across the block may be a solution. However, this process would need to be continued until the entire block is covered with a first layer of plastic concrete. Because of the variety of job conditions, a comprehensive, definite procedure for concreting in inclement weather cannot be included in this manual. It may be said, however, that work should be discontinued when the precipitation is so severe that it is infeasible to prevent water from collecting in pools or washing the surface of the fresh concrete.

Dust storms in semiarid regions occur without warning and may seriously impair the quality of finished surfaces of unformed concrete. Usually, concrete may be placed in forms of a limited exposed area during a dust storm without serious damage, but fine dust settling on more exposed work rapidly adulterates the surface concrete and makes a proper finish virtually impossible. Under such conditions, work should be promptly discontinued.

Concrete placing should be continued without avoidable interruptions until the placement is completed or until satisfactory construction joints can be made. Concrete should not be deposited faster than the vibrator crew can properly consolidate it; however, the faster it can be placed without damage to the forms and with ample vibration, the better are the results generally obtained. In placing thin members and columns the rise of the concrete should not be so rapid as to result in movement or failure of the forms. Where concrete production facilities are adequate and it is otherwise practicable, it is desirable to place concrete to full height in one lift and thus avoid construction joints and cleanup problems. As long as the rise in the forms does not exceed about 5 feet an hour in warm weather and 3 feet an hour in cold weather, concrete hardens at a sufficient rate to permit placement to any height in forms without creating excessive fluid pressures.

105. Mass Concrete.—The small amount of mortar in mass concrete is the source of many of its special advantages for massive work. These are mainly lower cost and lower generation of heat and temperature rise gained from lower cement content. The corresponding lower water content reduces shrinkage on drying and improves durability. However, to obtain the full measure of these benefits, it is necessary to keep sand content at a minimum. The percentage of sand required is surprisingly low, especially with entrained air. Tests should be made to determine whether the practicable minimum percent of sand is being used by observing when further decrease produces concrete that cannot be worked satisfactorily.

The practicality of maintaining low sand content depends largely on the uniformity with which the concrete can be produced, handled, and placed. Factors that contribute toward this end are (1) finish screening at the batching plant, (2) drainage of sand for 72 hours to provide uniform moisture content, (3) effective and well-designed rock ladders, (4) self-cleaning bins, (5) accurate batching, (6) good mixing, (7) handling without appreciable separation or slump loss, and (8) strong, effective vibration. With these requirements satisfied, only narrow margins in sand and mortar content are needed to offset variations in workability.

A construction procedure used to increase durability of large concrete dams in localities where freezing occurs is the placement of a somewhat richer concrete (higher in cement content) in the upstream and downstream faces of the dam than in the interior. Where this procedure has been used on several Bureau dams, the extra cement content, as compared to that of the concrete in the interior of the dam, ranged from

0.8 to 1.5 cwt per cubic yard, and the thickness of the concrete facing ranged from 4 to 12 feet. The thickness obtained is usually a matter of construction expediency influenced by the placing and dispatching facilities, the size of buckets used, and whether the buckets can dump a part of a batch. Although thicker facings are easier to place, any savings in placement cost tend to be offset by the cost of extra cement required. In general, it is more satisfactory to keep the thickness of face concrete to a practicable minimum because the lower cement content will foster better temperature control. For instance, a 2-foot facing has been regarded as ample to provide protection against freezing. However, because of batter of the upstream and downstream forms, it is often impossible to hold the exterior concrete to a 10-foot thickness.

Current Bureau practice for mass concrete dams requires the contractor to place the concrete in 7½-foot lifts. The layers generally do not exceed 18 inches; thus, five would be necessary for each lift. For concrete placements other than mass concrete, layers of 20 inches are suitable. These criteria permit satisfactory penetration of vibrators. Good construction practice requires that the layers be placed without cold joints. The deeper layers require extremely careful and thorough vibration. It is generally required that the exposed area of mass concrete be maintained at a practicable minimum by first building up the concrete in successive, approximately horizontal layers to the full width of the block and to the full height of the lift over a restricted area at the downstream end of the block and then continuing upstream in similar progressive stages to the full area of the block. (See fig. 138.) The minimum elapsed time between placing operations on successive lifts in any one block is usually specified as 72 hours.

The manner of discharge of concrete-placing buckets is extremely important. The concrete should fall vertically and should be discharged fast enough so that a cohesive, bulging, and growing mass is formed without significant separation as the concrete is discharged. If the discharge is not vertical or is too slow and the bucket too low, low-slump mass concrete will stack slowly in a cone and large coarse aggregate and cobbles will separate and cluster at the toe of its slopes. When this occurs, rock pockets are almost a certainty, particularly where reinforcement has been placed.

As mass concrete is placed with a relatively dry consistency of 1- to 2-inch slump, it is important that it be adequately and thoroughly vibrated into place. Figures 139 and 140 show an 8-cubic-yard batch of concrete immediately after depositing and after proper consolidation, respectively.

Mass concrete should be so vibrated that there will be no doubt as to its thorough consolidation, batch by batch, as it is placed. Care should be taken to vibrate and consolidate concrete in the area where a newly placed layer of concrete joins previously placed concrete. Usually, it is customary to revibrate the concrete against the forms around the perimeter of each block, and as a result, few rock pockets are found when the forms are removed. However, the absence of rock pockets on the face is not proof that full consolidation is obtained within the block. This can be determined only by watchful inspection as the concrete is placed and vibrated and by establishing procedures for vibration that will eliminate the possibility of incomplete consolidation.

In construction of some European dams, placing mass concrete differs from methods used by the Bureau in that gang vibrators mounted on a bulldozer blade are used for vibration. The blade levels and spreads the concrete as dumped in the pile. In this method, no-slump concrete is used, and the concrete is placed in horizontal layers covering the entire block instead of being placed in steps. Areas adjoining embedded materials are not amenable to this type of vibration and will require special consolidation with manual vibrators.

Figure 138.—Step method of placement as used at Monticello Dam, Solano project, California. SO–1038–R2.

Figure 139.—Eight cubic yards of concrete immediately after being deposited. PX–D–25254.

Portions that sometimes may be suspected of being poorly consolidated are the outward edges of the batches as deposited. Occasionally these edges are left unvibrated until concrete is placed against them during the next advance in the placing operations. Often by that time the unvibrated concrete at the edges becomes too hard to consolidate properly or there is lack of systematic vibration that is required to assure thorough consolidation at the junction between the batches. It is best to vibrate fully all parts of each bucketful when it is dumped, sloping the forward edge about 4 to 1, or flatter as necessary, to avoid flow on the slopes and overrunning of lower slopes.

Delays in placement may occur which result in cold joints within a lift. When placement is resumed while the concrete is so green (and therefore capable of ready bonding) that it can readily be dug out with a hand pick, the usual construction joint treatment will not be required if (1) the surfaces are kept moist and (2) the concrete placed against the surfaces is thoroughly and systematically vibrated over the entire area adjacent to the older concrete. When the delay is short enough to permit penetration of the vibrator into the lower layer during routine vibration of successive layers, the vibration will assure thorough consolidation.

Before the top surface of each lift sets, it should be gone over by a man wearing wooden snowshoes (fig. 120) which tend to prevent foot-

Figure 140.—Concrete after proper consolidation. Note that workman stands on concrete despite its appearance of wetness. P557–420–08212.

print depressions by providing him with an area of support two or three times that of ordinary boots or shoes. Using a small immersion-type vibrator as he steps on protruding large pieces of rock, he can embed them to the level of the surface of the lift. Cleanup is more effective and economical if the surface of a lift is reasonably free of protruding large aggregate, footprints, and other irregularities.

106. Tunnel Lining.—(a) *Preparation for Lining.*—The requirements for preparing a tunnel for lining depend on the purpose of the lining, whether it is for (1) support of the opening, (2) creating smooth conditions for flow of water, (3) sealing off inflow of water, or (4) containing flow under pressure. Usually, the lining serves more than one purpose and it is desirable that all timber not essential for support be removed so that the completed lining will be, to the greatest practicable extent, a mass of solid, continuous concrete in close contact with surrounding rock. Backfill radial grouting fills the voids and achieves the complete contact. Such construction gives best assurance of adequate strength, firm contact with surfaces of the material in which the tunnel is excavated, minimum leakage, minimum requirement for grouting, and maximum service. The practice of using shotcrete in lieu of steel sets for tunnel support is now accepted. In some tunnels, this support will also serve as a lining. This subject is discussed in detail in sections 173 through 180 of this manual. When every precaution must be taken to avoid flow or seepage of water along the junction between the lining and the rock, the bond to the rock should be enhanced by washing the latter with water jets so as to remove rock dust or mud.

The inverts of tunnels in canal systems (except pressure tunnels) are usually prepared for lining by removing all loose material between sidewall bases including rock protruding inside the "B" line. (The "B" line, indicated in specifications drawings, is the outside neat limit to which payment for excavation and placing of concrete will be made.) Any steel support ribs and sets which have been displaced inside the "B" line must also be reset to line and grade, and all loose material must be removed to clean undisturbed surfaces under the sidewalls. Any broken material remaining on the invert must be compacted. These tunnels are commonly of horseshoe section.

Pressure tunnels and spillway and outlet tunnels are required to be cleaned thoroughly of all oil, objectionable coating, loose, semidetached or unsound fragments, mud, debris, and standing water before concrete lining is placed. It is also necessary to remove rocks protruding inside the "B" line (tights) as well as steel support sets that were not installed properly or have shifted because of rock pressure. It may be necessary to use explosives in some instances and special care must be taken not to overshoot and to assure that adequate safety preparations are made to prevent injury to nearby workmen.

(b) *Control of Seeping or Dripping Water.*—Seepage water must be well handled or the concrete lining will be severely damaged before it sets. Water can be kept from the concrete in the sidewalls and arch and guided down to the invert by corrugated sheet metal appropriately fastened close to the arch and sidewalls where water is entering the tunnel. In the invert, water can be controlled by a longitudinal drainage trench filled with coarse gravel. The trench should be in the lowest part of the invert and should have branches to any springs and to points where water comes down from the sides. If the flow is heavy, uncemented tile pipe may be placed under the gravel. If this is insufficient to keep the water below the level of the subgrade, the water should be led to a sump beyond the section where concrete is placed and pumped to where it will flow harmlessly away.

In tunnels where considerable waterflow is encountered, it may be necessary to install temporary dams and pipe the water through the working area in addition to operating a suction system designed to remove water from the invert area ahead of the concrete placing. Any remaining drains other than those included in the structure design are usually grouted after the concrete has hardened. Records should be maintained of all water-control features, including accurate locations of all piping and connections and a description of drains, so that these features may be effectively grouted.

Sometimes large inflows of water can be materially reduced or diverted from the work by chemical grouting in the area ahead of the rock face.

(c) *Concrete for Tunnel Lining*.—Concrete for the arch portion of tunnel linings must be somewhat more workable than most formed concrete because of the lack of opportunity to vibrate the material in place. Concrete of 1½-inch-maximum size aggregate should have a slump of about 4 inches.

The sand content should be increased 2 to 4 percent or more in order that the concrete may readily mold and work itself around any supporting ribs and sets and into the irregularities of the tunnel roof. Because entrained air reduces the tendency of concrete to segregate and increases workability, it is an important factor in obtaining good placement. The maximum size aggregate will depend on thickness of lining and amount of reinforcement. The size of concrete pump or air gun used should not be a consideration in selecting the maximum size aggregate to be used. There are few tunnels where the maximum cannot be at least 1½ inches; the largest pumps and air guns have successfully placed 2½-inch-maximum size aggregate.

In tunnels larger than 12 feet in diameter, the required practice or procedure is to place the invert and arch separately, the invert usually being placed first. To preclude slumping of invert concrete at the outboard edges, the number of degrees of arc included in the invert section is limited by the slope of the concrete at the edges. With slumps normally used in tunnel lining concrete, the angle subtended by the invert should not exceed 60°. Concrete for the invert need not be different from that which is suitable for unformed concrete placed on nearly horizontal subgrade. No additional sand is needed and the slump should be about 1½ inches. Concrete having this slump is very responsive to good consolidation by vibration, which can be readily applied in invert concrete. Moreover, this low slump will aid materially in holding the shape of the invert to the upward slope from the centerline usually required in tunnel designs. Also, at the lower slump, bleeding will be less and thus interfere less with finishing.

(d) *Placing Concrete in Tunnel Lining*.—The selection of the method and equipment to be used in a tunnel lining operation will be governed by physical dimensions of the tunnel, construction schedule and program, extent of reinforcement, and specifications requirements for spacing of construction joints and waterstops. Placing of concrete in the arch is usually accomplished with a 6- to 8-inch-diameter pipeline (attached to the crown of the tunnel) from a concrete pump or an air gun.

After concrete has been placed in sufficient quantity to submerge the discharge end of the pipeline, the concrete flows alternately down the advancing slopes in the sidewalls. As the sidewalls fill, support is provided, somewhat back of the top of the slope, for the concrete being placed in the crown. Unless the end of the delivery pipe is embedded at least 5 to 10 feet in the concrete in the crown (the depth of embedment depends on the thickness of the lining at this point), the arch will probably not be filled.

When the end of the discharge line is well buried in concrete in the arch in conventionally drilled tunnels with the normal overbreak, the performance of most concrete pumps and most air guns is much the same as far as finished results are concerned. However, because of the invariably high discharge velocity of the air gun, there is tremendous impact and considerable separation at the start of each bulkheaded length of tunnel lining, and this continues until enough concrete is placed to fill the sidewalls at the end of the form and to cover the end of the pipe. Since concrete in the sidewalls must have a forward slope between 3 to 1 and 5 to 1, depending on thickness of the lining and spacing and strength of supporting sets, a large proportion of the concrete between bulkheads is subject to this separation before the arch begins to fill. For this reason, the air gun normally has been looked upon with less favor for tunnel lining than the pump, which discharges concrete more slowly and without separation. Bureau specifications, therefore, require that tunnel lining equipment and its operation shall not involve high-velocity discharge and resultant separation.

Forms for the arch and sidewall lining should be provided with ample openings through which concrete may be worked and inspected as it moves into place (see sec. 96). It is difficult to achieve optimum consolidation in tunnel arches. Such areas require special effort to achieve a quality structure. If sufficient headroom is available, internal vibrators can be used. Working through form openings with vibrators, portions of the arch can be consolidated. Form vibrators are a useful tool if they are properly attached and not permitted to run for extended periods at the same location. Vibration of the slope should proceed in the upslope direction. Flexible-shaft, internal-type vibrators have also been used to good advantage through the side doors. Vibration above the side doors should be systematically obtained with high-speed vibrators rigidly attached to the forms back of the advancing slope of the concrete (see fig. 141). Care should be taken to see that the vibration is not lost at hinge points and also that the rib to which the vibrator is attached is connected to the skinplate securely enough to effectively transmit the vibrations to the concrete. Only by maintaining constant observation through openings in the crown can the operation be managed and the discharge pipe moved so as to obtain the most complete filling of the arch.

Figure 141.—A commercial air-operated vibrator clamp used to attach a high-speed vibrator to a tunnel form. The air-operated feature facilitates attachment of the clamp. SB–1758–R2.

The exact positions of high overbreaks and points of doubtful filling should be recorded accurately so that, if the lining must be grouted, holes for grouting can be drilled at the most advantageous points. If quantities of sand mortar or neat cement are to be used in grouting, grout pumps are preferable for this low-pressure grouting. Air vents should be installed in high pockets above the lining. Grout should be made to travel until it appears at holes as far in advance of the point of introduction as possible without exceeding the allowable pressure at the insert.

The same factors previously discussed for arch placement govern the selection of equipment and methods for placing the invert. Placement of concrete in the invert follows the usual procedure for unformed concrete, but it is made more difficult by the necessity of transporting the concrete from its delivery point to the section where it is to be deposited—sometimes over a length of invert in which track is being removed and the subgrade prepared. In some tunnels the track is elevated so that cars can pass over the section being prepared and dump directly in the invert, but this method leaves cramped space for placing, screeding, and finishing operations. The concrete pump or conveyor belt is probably most satisfactory for distribution without segregation.

The transverse section of the invert is usually curved to a prescribed shape. The best way to hold such a shape and at the same time obtain good vibratory consolidation of higher areas along the sides is to use a heavy, weighted slip form equal in length to the width of the tunnel. (See fig. 142.) The slip form is moved forward by winches as the concrete is placed and vibrated ahead of it. In small tunnels the surface may be formed by a transverse screed or template operated longitudinally. For wide, curved inverts, better results are obtained by setting transverse screed guides, curved to the proper shape, at 10- to 15-foot intervals and screeding with a straightedge parallel to the centerline. Finishing should conform to the procedure described in section 121, the principal precaution being to assure that any free water accumulating on the bottom of a curved invert is removed or reabsorbed by the concrete before finishing tools are used.

If the tunnel invert is placed first, treatment of the longitudinal construction joints depends on whether the lining is to be watertight against hydrostatic pressure, whether structural requirements call for a joint securely bonded, or whether a close-fitting but unbonded joint will suffice. If the joint is close fitting, it will be sufficient to wash it with a jet of air and water just before the arch concrete is placed to ensure removal of loose dirt and debris. For the other conditions, the joint should be wet-sandblasted prior to washing.

107. Monolithic Siphons.—In monolithic siphons 8 feet and larger in diameter (see fig. 127) the inside form is omitted in the 60° arc at the

Figure 142.—Placing concrete with slip form in invert section of Spring Creek Tunnel No. 1, Central Valley project, California. P416–229–9608.

bottom to permit hand finishing and thus avoid the usual surface imperfections found under such forms. In locations where the profile requires the siphon to lie on a steep slope, temporary panels are required on those portions of the invert sides that are steeper than about 1½ to 1. These temporary panels are removed in an hour or two and the invert surfaces are finished by hand. When the siphon is 20 feet or more in diameter, it is usually more economical to place invert concrete from cars run into the siphon on tracks or by introducing it through openings in the top of the inside forms. For diameters under 20 feet, the concrete may be passed through openings near the bottom of the outside forms; when vibrated, it will flow downward to its position in the invert. With an air-entrained mix, there should be no segregation. In no event should a special mix (smaller aggregate, more sand, more water) be used to expedite placement. When the first layer of concrete above the invert is placed in the side forms and vibrated, there will be a tendency for it to boil up in the invert below the edge of the inside form. This action may be restrained by temporarily tying a 2- by 4-inch board on edge to the reinforcement just below the inside form. Flow of the concrete into the invert has also been restrained by wedging a plank flat on the surface of the invert concrete adjacent to the edge of the inside form. When placing concrete in siphon barrels and similar structures, it is usually important to keep freshly placed concrete in opposite sides of the form at fairly even elevations so as to prevent distortion and misalinement of forms. Two concrete distribution methods are illustrated in figures 143 and 144.

Siphons are provided with metal or rubber waterstop strips at transverse joints, and close attention is required to see that these are properly embedded. Particular care should be exercised to obtain thorough filling under the strip at the top and bottom of its vertical diameter. In molding concrete under the waterstop at these critical points, the concrete should

Figure 143.—Placing concrete in circular siphon with conveyor belt and drop chutes on Delta-Mendota Canal, Central Valley project, California. PX–D–33034.

Figure 144.—Monorail concrete distribution for box siphon 65 feet wide at Big Dry Creek, Friant-Kern Canal, Central Valley project, California. PX–D–34583.

be well vibrated so that it will flow in a direction nearly parallel to the plane of the joint. Also, the waterstop should be vibrated if it is sufficiently heavy to withstand the vibration without distortion or displacement. It is important that the ends of the forms be braced rigidly so that they cannot distort out of round when they become filled with concrete. Such distortion causes offsets and difficulty in fitting the forms for the adjacent length of barrel; it also tends to loosen the waterstop strips and necessitate expensive repair at the bottom of the circle.

Forms for cast-in-place siphons may be built of wood or steel. The latter is most common for the inside forms. There should be no openings in the inside forms, except the 60° gap in the invert and openings in the top to admit concrete for direct placement in the invert. Outside forms are required only at the sides up to about 60° above the horizontal centerline. Openings must be provided in the outside forms for introducing the concrete to be placed in the lower portion of the barrel and for access for vibration and inspection. Boxlike receptacles, as shown in figure 135, can be used at the lower openings to receive the concrete from the drop chutes, break its fall, and let it move easily into the forms at relatively low velocity and with considerably less separation. The lowest row of openings should be as low as practicable to facilitate introduction of concrete into the invert and to assure minimum velocity and separation.

108. Canal Lining.—An ideal canal lining would be watertight, moderate in cost, prevent growth of weeds, resist attack of burrowing animals, be strong and durable, provide maximum hydraulic efficiency, and have a reasonable amount of flexibility. No canal lining material will possess all of these characteristics; however, concrete composed of selected aggregates with proper control of placing, finishing, and curing, on adequate subgrade, will require minimum maintenance, satisfactorily carry water at relatively high velocities, and have a long service life.

(a) *The Concrete Mix.*—Concrete for canal lining must be plastic enough to consolidate well and be stiff enough to stay in place on the slopes. Special care is necessary to obtain good consolidation under reinforcement steel when specified for the lining. For hand placing and for placing with the lighter machines where the concrete is screeded from the bottom to the top of the slope, the consistency should be such that the concrete will barely stay on the slope. A slump of 2 to 2½ inches is usually satisfactory. Use of drier concrete with these methods of placing is apt to result in honeycomb on the underside of the slab, as shown in figure 137. For the heavier, longitudinally operating slip-form machines, best results are obtained with a slump of about 2 inches. A close control of consistency and workability is important, as a variation of an inch in slump can upset the established operating adjustments and interfere with progress and quality of the work.

In placing canal lining with a subgrade-guided slip form similar to the Fuller form, the slump has a critical effect on slip-form operation. If the concrete is not sufficiently plastic, it is difficult to control thickness of the lining. A 2-inch-thick lining may require as much as 3¼- to 3¾-inch slump; a 2½-inch-thick lining from 3- to 3½-inch slump; and a 3-inch-thick lining, 2½- to 3-inch slump.

Concrete for canal lining should include enough well-graded sand to ensure a reasonably good finish with the minimum treatment specified. Use of more sand than necessary for this purpose should be avoided. Entrainment of from 3.5 to 5.5 percent air also helps materially in securing a satisfactory finish for concrete containing 1½-inch-maximum size aggregate. Another factor that will considerably improve the finishability of the concrete is the reduction of the pea gravel (No. 4, or $\frac{3}{16}$- to $\frac{3}{8}$-inch) content of the mix to about 5 percent. This reduction is possible only when the pea gravel is batched separately. Bureau specifications for canal lining usually stipulate this separation where sufficient quantities are required.

The maximum size of aggregate should ordinarily not be greater than one-half the thickness of the lining. However, Bureau specifications require the use of ¾-inch aggregate in a 2½-inch-thick, or less, lining. Since consolidation of the concrete below reinforcement does not involve appreciable lateral flow under the steel, but rather a closing together from both sides, a decrease in the maximum size aggregate because of reinforcement is not justified. Proper consolidation is chiefly dependent on the mix, consistency, and placing procedure.

(b) *Reinforcement.*—Reinforcement in canal linings is normally not needed, and since about 1946 concrete linings constructed by the Bureau have been unreinforced except in specific canals where structural considerations are involved in the design.

When the lining is reinforced, consolidation of the concrete is both difficult and uncertain unless the steel is firmly held in its proper position in the middle of the slab and not permitted to sag during the placing operation. This is not easily accomplished, and many cores and other examinations reveal that steel is often much lower than it should be (see fig. 137). Where it sags during concrete placing, there is poor consolidation under the steel. As a consequence, steel is often exposed and corrodes.

To ensure proper positioning and prevent displacement, reinforcement must be adequately tied and supported, as illustrated in figure 145. Rocks or precast concrete blocks are commonly used as supports. These are

Figure 145.—Job-built slip form being used on sloping apron of left abutment of Nimbus Dam, Central Valley project, California. Reinforcement is systematically tied and supported on concrete blocks. Notice pattern formed by using V-grooves at construction joint of abutment in background. AR-1523-CV.

satisfactory if adequate in size and spaced at proper intervals. For firm ground, 3- by 4-inch blocks at 3-foot centers are sufficient, but for average soil conditions the blocks should be 5 inches square. For less favorable conditions, the blocks should be thicker to permit embedment in the subgrade. Bearing blocks should be made of concrete of a quality at least equal to that in the lining; they are usually provided with grooves or wires so that they may be secured in position under the bars.

When a general downward displacement of the reinforcement cannot be entirely avoided, an allowance should be made in setting the steel to compensate for the displacement. In some canals, mesh reinforcement is held in position by a special pipe cradle, suspended at the sides, which is moved ahead under the mesh as the lining machine progresses.

(c) *Placing the Lining.*—Subgrade preparation should be performed far enough in advance to avoid delay of the lining operations. At the time concrete is placed, the subgrade should be thoroughly moist (but not muddy) for a depth of about 6 inches (see sec. 94)). Some leeway is given the contractor in the degree of refinement to be used in trimming the subgrade. A comparison of canal linings and lining methods used on various projects is given in table 23.

Placing methods range from the hand method commonly used on small canals or laterals to the longitudinally operating slip-form machine. The simplest hand operation is placing unreinforced lining in small laterals and farm ditches where the concrete is dumped and spread on the sides and bottom. Screed guides are laid on the subgrade, and the concrete is screeded up the slope to proper thickness. Ten-foot screed panels are practicable for two-man operation. These thin slabs are consolidated mainly in the screeding operation. One or two passes with a long-handled steel trowel complete the finishing. Transverse grooves are cut at 6-foot intervals, and the lining is cured by use of sealing compound. Mixes for this method should be well sanded to simplify the labor of placing and finishing.

When constructed by hand, the larger linings are usually placed in alternate panels to facilitate placing, finishing, and curing operations. There may also be some reduction in overall shrinkage cracking if enough time elapses before placing the intervening panels. In this method, it is best to place the bottom slab first to provide support at the toes of the side panels. The panels are screeded up the slope, the concrete being vibrated ahead of the screed as described in section 104.

Most efficient placement of concrete on slopes is accomplished by use of a weighted, unvibrated steel-faced slip-form screed about 27 inches wide in the direction of movement. The screed may be pulled up the slope by equipment on the berm as in figure 146 or by airhoists mounted on the slip form as in figure 147. Concrete should be vibrated internally just

Table 23.—Comparison of canal linings and lining methods

Project	Yakima, Kittitas Division	Boise, Payette Division	Yakima, Roza Division	Central Valley, Delta Mendota	Columbia Basin, Pasco Pump Ltls.	Columbia Basin, West Canal	All-American Canal, Coachella	Central Valley, Friant-Kern	Gila, Wellton-Mohawk	Central Valley, San Luis Reach 3	Central Valley, Coalinga Reach 1	Central Valley, Folsom South Reach 1
Location	Washington	Idaho	Washington	California	Washington	Washington	California	California	Arizona	California	California	California
Year Constructed	1927–28	1937	1938–39	1947	1947	1947–48	1947–48	1950–51	1952–53	1963–1967	1970	1971–1972
Specifications No.	—	—	—	1183	1230	1286	1512	3056	DC–3618	DC–6148	DC–6612	DC–6833
Bottom, feet	12	14	12	48	8	12	8	24	5	75	12	34
Perimeter, feet	37	37	45.6	112	37	82	41	85	16	223	72	106
Side, Slope	1¼:1	1¼:1	1¼:1	1½:1	1½:1	1½:1	1½:1	1¼:1	1¼:1	2:1	1½:1	1½:1
Thickness Lining, inches	3	4	4	4	1½	4½	3½	3½	2½	4	3	4
Max. Size Aggr., inches	1¼	1½	1½	1½	½	1½	1	1½	1	1½	1½	1½
Concrete Linear ft., cu. yd.	0.4	0.7	0.6	1.37	0.17	1.14	0.18	0.92	0.12	0.004	0.67	1.33
Longitudinal Steel	⅜ rd @ 12'' ctrs.	½ rd @ 12'' ctrs.	½ rd @ 12'' ctrs.	—	—	⅜ rd @ 12'' ctrs.	6 GA 8x8 Mesh @ 31.9 lb.–100 sq. ft.	—	—	—	—	—
Transverse Steel	⅜ rd @ 12'' ctrs.	½ rd @ 24'' ctrs.	½ rd @ 24'' ctrs.	—	—	⅜ rd @ 12'' ctrs.	6 GA 8x8 Mesh @ 31.9 lb.–100 sq. ft.	—	—	—	—	—
Method of Placement	Steelplate Slope First	Metal Box	Long-operated Slip-Form	Long-Operated Slip-Form	Pneumatically App'd in 2 Layers	Channel Iron Screed Slopes First	Long-Operated Slip-Form	Long-Operated Slip-Form	Long-Operated Slip-Form	Long-Operated Slip-Form	Long-Operated Slip-Form	Long-Operated ½ Sec. Slip-Form
Length of Section	14 ft.	12 ft.	Continuous	Continuous	Panels-20 ft.	Panels-14 ft. 6 in.	Continuous	Continuous	Continuous	Continuous Slope, Inv-Sep.	Continuous	Continuous
Contraction joint, longit.	—	—	—	5 on Bottom 2 on Each Slope	1 on Bottom	1 on Bottom	1 on Bottom	1 on Bottom 2 on Each Side	—	15' 0''	12' 0''	13' 0'' Bottom 15' 0'' Slope
Contraction joint, trans.	—	—	—	15'6'' Max.	9' 0''	—	15' 0''	12' 0''	10'0''	15' 0''	10' 0'	15' 0'' Max.
Rate of Travel, ft. per min.	0.3 Avg.	0.4 Avg.	1.2 Avg.	1.6	0.92 Max.	—	1.7	—	—	5 Aug.	2	3.3

Type of Compaction	Internal Vibrator Tubes	Internal Vibrator	Internal Vibrator	Vibrator	Vibrator	Vibrator	Vibrator		Vibrator	Vibrator Tubes	Immersion Vibrator	Hand
Mix. by Weight, Average	1:2.6:4.2	1:2.6:3.9	1:2.6:4.1	1:1.90:3.37	1:1.96:3.64	1:2.54:3.86	1:2.19:4.08	1.4	1:2.56:3.53	1:2.7:4.1	1:2.5:3.7	1:2.8:4.2
Percent Sand	39	37	39	40	35	38	35	—	42	—	—	—
Percent No. 4 to ⅜"	6	7	5	—	8	—	25	—	—	—	—	—
Percent Air	5.0	4.2	5.6	4.4	4.1	—	4	—	4	—	—	—
Slump, inches	2	3	3	3½	1¾	2½	2½	—	2	2¼	2 to 2½	2
W/C by Weight	0.50	0.48	0.50	0.53	0.54	0.56	0.51	0.63	0.57	0.57	0.57	0.55
Cement, cwt/yd³	4.88	5.05	4.68	5.64	5.41	5.04	5.23	—	5.08	5.00	5.11	4.74
Comp. Strength, lb/in², 28 day	4340	4000	4420	3335	3250	3255	3300	—	3546	4900	4400	4700
Curing	White Pigmented	White Pigmented	White Pigmented	White Pigmented	White Pigmented	White Pigmented	White Pigmented	White Pigmented	White Pigmented	1 Coat Coal Tar 1 Coat White Wash	Two Coats Clear Compound	Water Sprinkling Asphalt Cutback
Length per 8 hr. Shift, ft.	1,600/ ½ Sec.	944	Invert—2776 Slope—2527	—	—	800	—	444 Max.	757	445 Avg.	160 Avg.	112 Avg.
Finish	Power Trowel, Long-Handled Trowel	Suppl. by Hand Finish	Suppl. by Hand Finish	Suppl. by Hand Finish	Long-Handled Trowel Suppl. by Hand Finish	Long-Handled Trowel Suppl. by Hand Finish	Long-Handled Trowel Suppl. by Hand Finish	None	Long-Handled Trowel Suppl. by Hand Finish	Hand Floats and Trowels	Hand Finish	Hand Finish
Drainage	Finger, Toe, and Flap Weeps	Finger, Some Areas	Finger, Some Areas	—	Sewer Pipe in Gravel Trench	—	Sewer Pipe in Gravel Trench as Directed	—	Sewer Pipes in/or Gravel Trenches as Directed	6" Open Joint Pipe in Gravel Trench	—	4" to 12" Open Joint Pipe in Gravel Trench

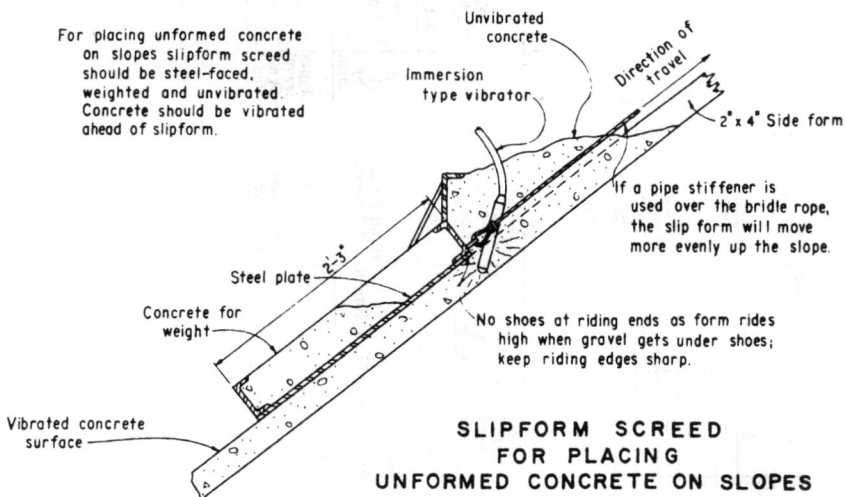

Figure 146.—Placing concrete on a canal slope. Slip-form placing concrete in panel on right slope of canal for Delta-Mendota Canal relocation. Concrete should be well vibrated ahead of the screed as it is pulled up the slope. P805-236-7608 and 288-D-3284.

Figure 147.—Modified slip-form screed ready to "pull itself" up the slope on the wooden guides. PX–D–33036.

ahead of the slip form. Under proper conditions of operation the surface made by the slip form will require no further screeding and very little finishing. The slip form itself should not be vibrated, as this procedure causes a swell in the concrete emerging from the lower edge. This excess concrete is not only laborious to remove but it also emphasizes sags that tend to form at longitudinal bars.

Many improvements have been made in the longitudinally operating slip-form machines for lining canals of all sizes since the first machine of this type was used on the Umatilla project of the Bureau in 1915. Great progress has taken place in recent years in extensive canal lining operations on Bureau projects. Large lining machines have been developed which are hydraulically operated and controlled as well as some electrically controlled to line and grade. Preformed longitudinal plastic joints can be extruded in the lining. Similar transverse joints are placed during finishing operations. This is discussed further in section 108(d).

Elimination of reinforcement has already been discussed (sec. 108(b)). Another change in the direction of economy which has accompanied elimination of reinforcement is the relaxation of tolerances in alinement, grade, thickness, and finish of concrete linings. Current Bureau specifications permit departure from the established line of 4 inches on curves and 2 inches on tangents and departure of 1 inch on grade and allow a 10-

percent reduction in thickness, provided each day's placement averages full thickness.

A simplified type of construction has been adopted for linings in relatively small canals and in laterals and farm ditches to reduce costs and at the same time maintain the durability and serviceability of the work. This type of construction makes use of a subgrade-guided slip form (see fig. 149). With reasonable care in operation, no difficulty is experienced in placing linings with such forms to the tolerances currently specified. The quality of materials, proportioning and mixing, uniformity in placing, and curing of these linings should be equal to similar requirements for linings used on the larger canals. A reduction in these requirements would seriously impair the durability and result in very little decrease in cost.

Pit-run aggregates should not be used for these simplified linings. The cost of processing is usually compensated by better progress made with uniformly graded materials and reduced cracking resulting from lower water requirement of the concrete. Finishing of simplified concrete lining should be kept at a minimum. Except for gaps or rock pockets, surfaces are generally acceptable as they emerge from the slip form. Substantial savings can be realized only by high hourly and daily production resulting from a minimum of refinements in workmanship and from cooperative and practical inspection.

Construction of a canal having a depth of 5½ feet and a bottom width of 5 feet is illustrated in figures 148, 149, 150, and 151. The equipment is not carried on rails on the canal berm as is equipment for larger canals. The slip-form lining machine is held to grade and line by a steel pan shaped to fit the canal section. Back of the pan and immediately preceding the slip form is the transverse, compartmented trough for uniformly distributing the concrete mix. Truck-mounted mixers supply concrete to the slip form as the entire machine is pulled forward. These machines have been used with and without internal vibration. It has been difficult to control the lining thickness when internal vibration was used, apparently because of the lack of weight on the machine to hold it down to the bottom of the canal. Because of the tendency to place an increasingly thicker lining, the design of the mix and the slump are of critical concern as discussed in section 108(a). Because the machine rides on the subgrade, it can be used for placing unreinforced lining only. Good workable mixes have been so well placed as to require practically no hand finishing.

Slip-form canal-lining machines have long been used on large canals. The older type of lining machine, illustrated in figures 152, 153, and 154, consists of a framework traveling on rails on the canal bank which supports the working platform, distributor plate or drop chutes, compartmental supply trough, vibrator tube in the bottom of the trough, and slip form. The slip form is a steel plate, curved at the leading edge, extending

Figure 148.—Excavating a small canal on the Gila project, Arizona. The canal is excavated in a single pass. PX–D–33037.

across the bottom and up the slopes of the canal and shaped to conform to the finished surface of the lining. A distributor plate, when used, is fastened to the leading edge of the slip form and extends upward on a steep incline to the working platform (see fig. 154). On some machines, a continuous row of hoppers in the working platform feeds into drop chutes, each supplying one compartment of the trough below (see figs. 152 and 153). This construction has the advantage of permitting easy communication between the men at the front and those in the rear. Concrete is dumped, usually from a shuttle car on the working platform, and is guided to the trough below by the distributor plate or the drop chutes.

As the concrete is distributed through the bottom of the trough and under the slip form, it is consolidated by a vibrating tube parallel to and a few inches ahead of the leading edge of the form. The concrete must be consolidated as it passes under the slip form. Proper consolidation cannot be obtained by vibrating the slip form of a lining machine, apparently because of the lack of a way to supply additional concrete needed to fill the voids. The trailing edge of the slip form is usually adjustable to positions somewhat lower than that of the leading edge. This provision improves consolidation and tends to mold the concrete more closely to the

Figure 149.—A subgrade-guided slip-form concrete-lining machine following the excavator. PX–D–32056.

Figure 150.—Rear view of slip form of figure 149. The pressure plate eliminates most of the hand finishing. PX–D–33039.

Figure 151.—Spraying sealing compound on a new canal lining to conserve the moisture necessary for curing. PX–D–33040.

subgrade. Too low a setting of the trailing edge causes tearing, rather than smoothing of the surface. On some machines, the slip form is followed within a few feet by an ironer plate 18 to 30 inches wide (see fig. 154), which under favorable conditions leaves a surface that requires little or no hand treatment. Such an ironer on one job served to make less conspicuous the slight rolls or bulges caused by sagging of the concrete against the reinforcement bars on the side slopes.

The more modern type of traveling canal lining machine is shown in figure 155. Such machines were used for placing lining in the San Luis Canal, Central Valley project. The maximum cross section consisted of a 110-foot invert with 2 to 1, 80-foot side slopes. The lining machines were mounted on caterpillar tracks hydraulically driven and traveling on the subgrade. Feeler gages running along set wires controlled grade and alinement. One machine covered the one side slope and a 10-foot section of the invert. The invert paver, which spanned 45 feet, required two passes to complete the invert. These machines could pave at least 1,500 feet per day and frequently exceeded this rate.

The concrete was discharged from tilting mixers into a hopper from which the concrete was distributed down the slopes by a bucket conveyor. The multipurpose machine also extruded longitudinally embedded, formed

CONCRETE MANUAL

Figure 152.—A mammoth slip-form lining machine, with drop chutes, following the excavator—Delta-Mendota Canal, Central Valley project, California. DM-762-CV.

waterstop contraction joints of polyvinyl chloride (PVC) and at set intervals inserted a transverse waterstop contraction joint of similar material. This basic machine was modified by contractors on different reaches of the canal to suit their plan of operation. One machine was supplied with concrete by bottom-dump trucks which drove onto a traveling platform discharging into a hopper from which the concrete was moved by screw to distributing buckets.

A smaller version of this lining machine is used in the construction of intermediate and small size canals and laterals. It is very similar to the rail-supported type in regard to distribution of concrete. It is operated hydraulically and guided from a control station at the front. With experienced operators, it is capable of placing an excellent lining. Care must be taken to provide a reasonably smooth berm along which the

Figure 153.—A lining machine, with drop chutes, progressing along the Putah South Feeder Canal, Solano project, California. SO–2652–R2.

machine tracks must crawl, and close inspection should be maintained for deviations from line and grade. (See fig. 156.)

Efficient canal construction always requires careful coordination of the successive operations. The trimming machine should be closely followed by the lining machine with coordinated concrete supply and by separate grooving and curing jumbos just behind the lining machine.

(d) *Contraction Joints.*—Transverse contraction joints are provided in canal linings by inserting polyvinyl chloride (PVC) plastic contraction joint-forming waterstops or by cutting or forming grooves in the upper surface of the slab (see fig. 157) while the concrete is still plastic. Shrinkage cracks will then be largely confined to the location of the PVC strips or to any grooves where the thickness of the lining has been reduced. Both transverse and longitudinal contraction joints are recommended in canals having a lined perimeter of 30 feet or more, particularly those that are unreinforced. Even smaller canals may require two-way crack control if subbase material warrants it.

Transverse grooves are cut either by hand along a straightedge or by a mechanical knife or cutter impressed and vibrated into the concrete. Longitudinal grooves are cut by stationary or revolving cutters attached

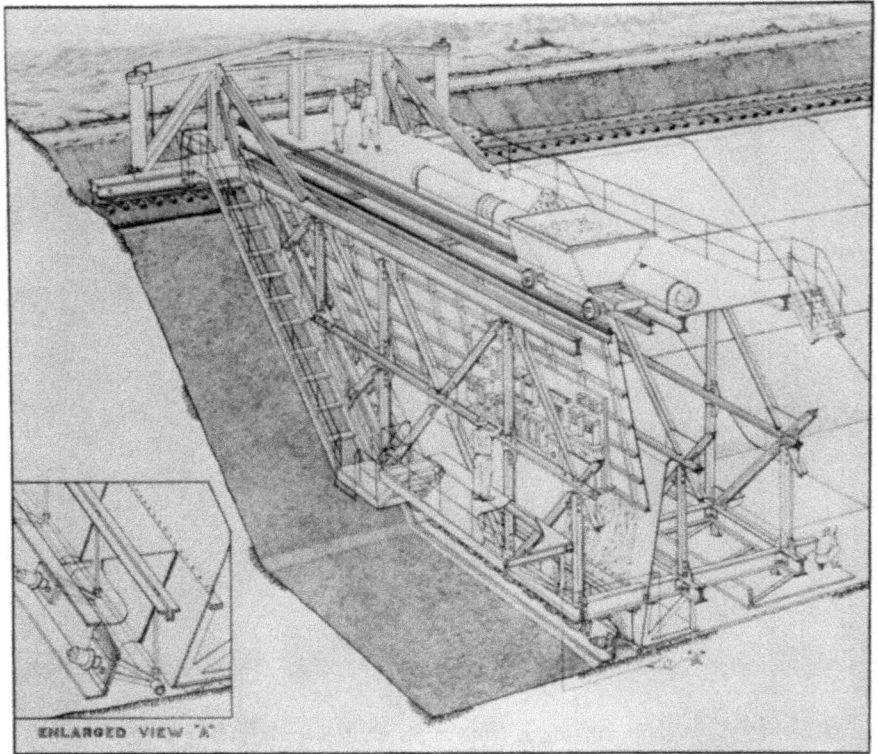

Figure 154.—A cutaway drawing of a canal lining machine with a distributor plate and ironer plate. PX–D–33038.

to the rear of the slip form. As plastic concrete will not always retain the groove shape after groove cutting equipment is withdrawn, grooves are reshaped to within specified dimensions after the concrete has stiffened sufficiently. Because of wear, the dimensions of grooving tools initially should be oversize. Inspectors should frequently check the depth and bottom width of the grooves. For some canal linings, particularly with more intricate groove shapes, it may be necessary to form rather than tool the groove to assure proper size and shape. Forms are usually fabricated from plastic or rubber, and with proper care they can be reused several times.

Previously, contraction grooves were often left open except where a high degree of watertightness was needed. This policy was dictated partly for reasons of economy but also because asphalt-based sealers were virtually the only materials obtainable, and these were subject to rapid deterioration from weathering. The asphalt sealers had often weathered so severely as to be almost unserviceable by the time water was introduced into the canal.

Figure 155.—Canal construction layout on San Luis Canal, Central Valley project, California. Lining operation progressing upstream on left slope, paving full height of slope and 10 feet of invert. Note the belly-dump truck and trailer on drive-over unloader supplying concrete to the lining machine. The truckload of sand is for finishing grout. Reels of PVC formed waterstop contraction joint material are shown in the left foreground. P805–236–9546.

A number of joint-sealing systems were studied in the laboratory and the field for replacing asphalt sealers. Of all those considered, two showed the greatest promise. One consisted of a plane-weakening vertical member added to a miniature waterstop (figure 157, alternatives 2 and 3). The second of the newer systems, a coal-tar extended polysulfide sealer (alternatives 1 and 4, figure 157), is a modification of a sealant developed earlier for airport runway and taxiway joint sealing.

The system consisting of a small waterstop with a vertical plane-weakening element is usually referred to as a contraction joint-forming waterstop or, more simply, as a PVC strip. Extruded from PVC plastic and inserted into the concrete during lining placement, this strip controls cracking effectively. Experience proves that the installation must be correctly made: the top of the strip must not be more than one-half inch below the concrete surface or the contraction crack might not develop at the desired location; and the strip must not be tilted sharply or the crack might lead to a sealing bulb, thereby destroying the waterstop effect.

Advantages of the PVC strip are:

(1) It forms a weakened plane in the lining, producing excellent crack control when properly installed.

(2) It seals by waterstop action so the seal is not dependent on bond to the concrete.

(3) It is buried so weathering is virtually eliminated.

Figure 156.—Caterpillar track-mounted lining machine in operation on the Putah South Canal, Solano project, California. SO–3694–CV.

(4) It is inert chemically and, therefore, not likely to be affected by extended water immersion.

(5) It is a manufactured item receptive to high quality control standards.

(6) It withstands high hydrostatic pressures (up to 26 lb/in^2).

(7) It enables lining and sealing a canal in a single operation.

(8) It tends to unitize a canal lining by tying the individual slabs together, but this effect is somewhat diminished by stress relaxation.

The second of the newer materials, the coal-tar extended polysulfide canal sealer (alternatives 1 and 4, figure 157), is used in much the same manner as the previously specified asphalt mastics but with markedly better results. It weathers well and will resist hydrostatic pressure up to 26 lb/in^2 over a ⅛-inch-wide crack. Moreover, it bonds well to concrete and remains bonded even after long periods of immersion in water. Being a two-component, quick-curing material, it requires large, expensive equipment for successful application. Similar, but slower setting, polysulfide sealants are also available for hand mixing and placement on small jobs. There is also now available a single-component polysulfide sealant which, although quite slow curing, appears to be effective when fully cured.

Alternative No. 4 in figure 157 shows a special case of the use of the coal-tar extended polysulfide canal sealant. Here the sealant is furnished as a preformed strip shaped in a triangular form as shown and fabricated

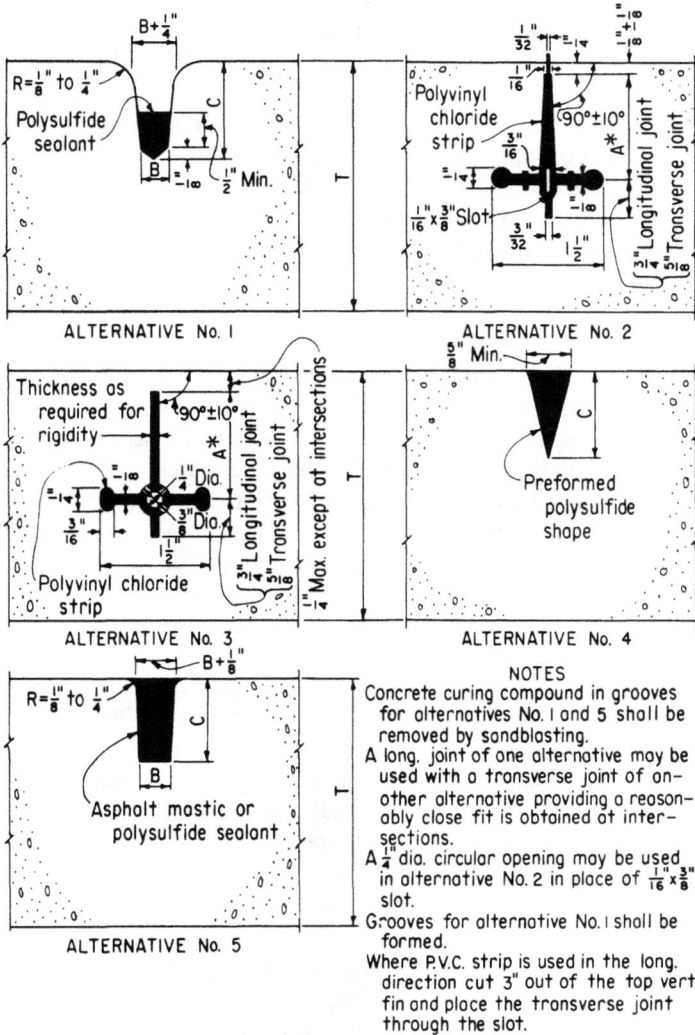

ALTERNATIVE No. I

ALTERNATIVE No. 2

ALTERNATIVE No. 3

ALTERNATIVE No. 4

ALTERNATIVE No. 5

NOTES

Concrete curing compound in grooves for alternatives No. I and 5 shall be removed by sandblasting.

A long. joint of one alternative may be used with a transverse joint of another alternative providing a reasonably close fit is obtained at intersections.

A $\frac{1}{4}$" dia. circular opening may be used in alternative No. 2 in place of $\frac{1}{16}$" x $\frac{3}{8}$" slot.

Grooves for alternative No. I shall be formed.

Where P.V.C. strip is used in the long. direction cut 3" out of the top vert. fin and place the transverse joint through the slot.

Groove and P.V.C. Strip Dimensions and Transverse Spacing

T Slab Thickness (inches)	A* P.V.C. Strip Height (inches)	B Groove Width (inches)	C Groove Depth (inches)	Approximate Groove Spacing (feet)
2	$1\frac{1}{4}$ to $1\frac{1}{2}$	$\frac{3}{8}$ to $\frac{1}{2}$	$\frac{5}{8}$ to $\frac{3}{4}$	10
$2\frac{1}{2}$	$1\frac{1}{4}$ to $1\frac{1}{2}$	$\frac{3}{8}$ to $\frac{1}{2}$	$\frac{3}{4}$ to $\frac{7}{8}$	10
3	$1\frac{1}{4}$ to $1\frac{1}{2}$	$\frac{3}{8}$ to $\frac{1}{2}$	1 to $1\frac{1}{8}$	12 to 15
$3\frac{1}{2}$	$1\frac{1}{4}$ to $1\frac{1}{2}$	$\frac{3}{8}$ to $\frac{1}{2}$	$1\frac{1}{8}$ to $1\frac{1}{4}$	12 to 15
4	$1\frac{1}{2}$ to $1\frac{3}{4}$	$\frac{3}{8}$ to $\frac{1}{2}$	$1\frac{1}{4}$ to $1\frac{3}{8}$	12 to 15
$4\frac{1}{2}$	$1\frac{1}{2}$ to $1\frac{3}{4}$	$\frac{1}{2}$ to $\frac{5}{8}$	$1\frac{1}{2}$ to $1\frac{5}{8}$	12 to 15

*Dimension of upper vert. member is decreased $\frac{1}{2}$" to $\frac{3}{4}$" for transverse joint.

Figure 157.—Details of transverse contraction joint-forming waterstops and contraction grooves for unreinforced concrete canal linings. 288–D–3254.

from the same material as regular canal sealant. This strip being pre-formed is suitable for installation only in plastic concrete. In this instance, a seal for the contraction joints results from the concrete bonding to the polysulfide.

Alternative No. 1, figure 157, is now the normally specified groove shape using coal-tar extended polysulfide sealer.

Alternative No. 5, figure 157, was the standard groove used for many years and is the easiest shape to make. Experience has shown that it is less effective in inducing contraction crack development at the bottom of the groove than are those shapes having a fairly sharp point such as alternative No. 1; generally, therefore, it is no longer included in Bureau specifications. It may prove satisfactory in small installations but only if it is completely filled with either asphalt mastic or polysulfide sealant as shown in the drawing.

Dimensions and spacings for various PVC strips and grooves for various thicknesses of unreinforced lining are shown in figure 157. Transverse joints in reinforced linings are similar to those for unreinforced linings except that minimum dimensions are usually specified and groove spacing may be increased to approximately 16 feet.

It is not feasible to establish fixed guidelines for spacing longitudinal joints either in reinforced or in unreinforced linings. This is particularly true when linings are constructed in stages by machines that place concrete on the side slopes first, then in the canal bottom. With this type of construction, the longitudinal grooves are located to form the construction joint between side slopes and invert lining. Concrete linings having perimeters of 30 to 50 feet placed without longitudinal construction joints are generally provided with longitudinal contraction joints near the bottom of each side slope. Linings with larger perimeters usually have additional contraction joints about one-third the distance up the side slopes; there is often a joint established about 3 feet below the design water level to provide stress relief at that location. The number of longitudinal joints will be decreased for reinforced lining, depending on the amount of reinforcement used. Dimensions and shapes of longitudinal joint strips and grooves are as shown in figure 157 except that alternatives No. 2 and 3 will usually have a ½- to ¾-inch longer upper vertical member than that shown to better accommodate the installation of the transverse strip at intersections.

Normally, it is left to the contractor to select which of the alternatives he will use to produce the specified contraction joints. One exception to this would be for linings thinner than 3 inches where alternatives No. 2 or 3 would be permitted in one direction only and alternative No. 1 (or possibly alternative No. 4) would be required for the other direction. (See fig. 158.)

Installation of PVC contraction joint-forming waterstops must be made in plastic concrete. Usually, the longitudinal strip is fed from reels

Figure 158.—Intersection of PVC plastic contraction joint-forming waterstop (longitudinal joint) with field-extruded, coal-tar extended, polysulfide canal sealant (transverse joint). Both plane weakeners developed contraction cracks; both joints are sealed. PX–D–67376.

mounted in front of the paver into the fresh concrete through guides and tension rollers so placed as to ensure proper depth and orientation of the strip (figure 155). Installation of the transverse strip is often made from a second jumbo into a rough groove. Sufficient vibration is required to produce thorough consolidation of the concrete around the strip and to provide continuous contact between the concrete and all surfaces of the strip. At intersections between longitudinal and transverse joints containing the PVC strip, the top vertical member of the first-installed strip is removed for 3 inches at the intersections and the second-installed strip is placed within the notch so formed. Some depression of the first-placed strip is to be expected in such an installation but should be permitted only to the extent necessary for placing the upper strip to the proper depth. The manner of making intersections should be such as to assure a reasonably close fit between transverse and longitudinal strips and to provide a nearly continuous weakened plane normal to the lining surface in both directions through the intersections (see fig. 159).

Equipment for mixing and placing rapid-set, coal-tar extended polysul-

Figure 159.—Intersection of transverse and longitudinal PVC contraction joint-forming waterstops. Upper vertical member on the longitudinal strip is one-half inch longer than that on transverse strip (left). Both strips developed contraction cracks; both joints are sealed; both contraction cracks are continuous across the intersection. PX–D–67427.

fide sealant is more elaborate than that used for asphalt mastics. The two components must be delivered at a temperature of 80 to 100° F in equal volumes to a mixing-head nozzle. Equal volumes is defined as a ratio of 1:1 plus or minus 10 percent. Accurate pressure gages of suitable ranges and inspection gages for flow measurement are necessary to permit convenient monitoring of pressures and proportioning. The proportioning should be checked at least every 2 hours. Flow rate through the mixing head should not exceed the equipment manufacturer's specified maximum. The sealant should be extruded at the bottom of the joint groove and should be tooled to work the sealant into intimate contact with the joint surfaces, eliminate air bubbles, and achieve the shape shown in figure 157, alternatives 1 and 5.

Any sealant that does not cure properly, fails to establish a satisfactory bond, or which is damaged by the contractor's operations must be removed and the joint recleaned and resealed.

Conditions for placing the polysulfide sealant are more critical than for asphalt mastic. The concrete must be cured at least 7 days, joints must be dry, temperature of air and concrete must be at least 50° F, and joint walls need to be sandblast cleaned to remove laitance, curing compound, or other bond inhibitors.

General and detailed requirements for the PVC contraction jointforming waterstop are given in Bureau construction specifications, and the test procedures are covered in U.S. Army Corps of Engineers Specifications for Polyvinylchloride Waterstops, CRD–C–572–74. The coal-tar extended polysulfide canal sealant is required to conform with Specifications for Polysulfide Canal Joint Sealer dated October 1, 1973. These latter specifications are based on some prepared by the California Department of Water Resources. They cover not only the rapid-set, machine-mixed-and-applied, nonsag sealant (Class R, Type II) which is most widely used in large-scale applications but also a rapid-set, self-leveling grade (Class R, Type I), a slow-set nonsag material (Class S), and a slow-set one-part sealant (Class O). Class R, Type I, is commonly used to fabricate the preformed polysulfide strips mentioned previously; Class S and Class O are used for smaller jobs, for repair work, or for sealing the intersections between PVC strips.

There is apparently still a need for a convenient, less expensive mastic sealant in some of the Bureau's operation and maintenance work. For this reason, specifications for previously used rubberized asphalt mastic were upgraded to assure procurement of a higher quality sealer and are now titled Specifications for Mastic Sealer, Rubberized, Cold Application, Ready-mixed for Joints in Concrete Canal Lining, dated March 1, 1971. Materials of this type were used in the past for sealing and resealing grooves. They are still used for resealing contraction grooves containing deteriorated asphalt mastics because there will be less surface preparation required and no compatibility problem which would result if a polysulfide were used.

Different types of equipment are available for placing mastic joint sealers in grooves and on random cracks. Hand-operated calking guns are suitable, but for larger operations there are hand-operated pumps designed for use on 5-gallon containers in a two-man operation, one working the pump handle, the other holding the nozzle. For larger jobs, a heavy-duty, 48:1 ratio, air-operated, double-acting pump, equipped with a drum cleaner is useful when 55-gallon drums of the material are used. The nozzle tip should be of such size that it can be easily inserted to the bottom of the groove to be sealed.

The nozzle tip should be kept near the bottom of the joint during the filling operation so that filling is done from the bottom to prevent trapping air bubbles. The groove should be uniformly and completely filled, and the sealer should hump up slightly above the surrounding concrete to compensate for subsequent shrinkage and subsidence. When joint sealer is applied at the time concrete is placed, the inspector should be sure that the joints conform to designed shape and cross section and that there is no free water in the grooves. When it is applied after the concrete has cured thoroughly, the grooves should be cleaned of all foreign substances, particularly concrete curing compound, before the sealer is placed. Sandblasting is usually the most rapid, effective, and economical method of proper cleaning of grooves.

Each shipment of polysulfide sealant should be sampled by the manufacturer prior to use; the samples should be taken during the canning operation. They should consist of a 2-gallon sample from each 500 gallons or less of each component of each batch or lot and should be accompanied by a certification stating that the sample is from the batch to be furnished.

Two 1-foot-long samples of the PVC contraction joint-forming waterstop should be taken from each batch of the strip. A certification for sampling and physical properties should be furnished with the samples.

Each shipment of mastic joint sealer should be sampled in accordance with designation 39 of the appendix. Samples and certifications of all types of joint sealers and joint-forming waterstops should be forwarded to the Denver laboratories for testing prior to use, as required by the particular specifications. Arrangements will usually be made for final inspection at the point of origin to prevent delay in operations after receipt of materials at the project.

Expansion joints in concrete and mortar canal linings are usually at junctions between linings and structures. When expansion joints are used, details of joints are included in the drawings.

109. Precast Concrete Pipe.—(a) *General.*—Extensive use is being made of precast concrete pipe for siphons in canal systems and underground distribution systems. The manufacture of such precast pipe should receive no less attention than that given to construction of monolithic concrete structures. For this reason Bureau specifications have been developed which permit plant control during manufacture of pipe. With proper attention given to mix design and placing and curing, there is less reason to be concerned about the pipe meeting specified physical tests, and need for expensive and undesirable repair of pipe will be minimized. Repairs, when required, should be made in accordance with the provisions given in section 138.

Where plant control is not feasible or is not justified because of small quantities involved, pipe may be accepted if it passes specified physical tests. Standard specifications for concrete pipe, such as those issued by the American Society for Testing and Materials (ASTM), are used to make and test concrete pipe. However, these specifications are often altered for specific reasons, such as changes in type of cement, reinforcement requirements, or curing.

Precast concrete pipe may be classified as cast, centrifugally spun, tamped, packerhead, prestressed, and roller-compacted asbestos cement.

(b) *Cast Pipe.*—In regard to the casting of concrete pipe (generally 48 inches or larger in diameter) in forms, there is little to add to what has been stated relative to final placement of concrete in ordinary forms. Most of the material included in this chapter under forms, placing, and vibration is applicable and should be reviewed in consideration of casting concrete pipe. Experience in the manufacture of such pipe for the San Diego Aqueduct and the Cachuma project demonstrated that high-quality pipe having a minimum of surface pits and air bubbles can be made by using an amplitude of form vibration at a frequency above 8,000 vibrations per minute (see fig. 160).

The tendency in casting concrete pipe is to use mixes that are oversanded, too wet, and have too small a maximum size of aggregate. Such mixes may have been necessary in the former method of working concrete with revolving square rods but are detrimental to production of the high-quality concrete now possible with use of adequate high-frequency vibration. Slump should not exceed 3 inches and should preferably be less than 3 inches, particularly at upper ends of pipe sections. Normal percentages of sand are adequate. In the 7-inch shell of 20-foot lengths of 69-inch pipe for the Salt Lake City Aqueduct, $1\frac{1}{2}$-inch-maximum gravel produced excellent results—better than when $\frac{3}{4}$-inch gravel was used. In the $1\frac{5}{16}$-inch inner shell and the $4\frac{3}{8}$-inch outer shell of the 16-foot lengths of 48-inch pipe for the first barrel of the San Diego Aqueduct, $1\frac{1}{8}$-inch-maximum gravel was used with complete success. Although the size of the gravel was 86 percent of the clear opening between the steel cylinder and the inside forms, no difficulty in casting was experienced, and the delay and inconvenience of using two mixes was avoided.

Concrete should not be placed rapidly in deep lifts but should be fed slowly into vibrating forms from the circumference of a plate or cone so as to distribute the concrete evenly around the pipe as it enters the forms. (See fig. 161.) Production can be maintained by casting several lengths of pipe at a time. Vibration should be continued while the concrete is entering the forms and as long as bubbles of entrapped air emerge from the concrete but no longer if overvibration is to be avoided. Particular attention should be given to vibration after filling the pipe form has been completed, as overvibration will cause coarse aggregate to settle and water and air to rise, resulting in a weak and fragile spigot. For this reason, entrained air in concrete for cast pipe should not exceed 2.5 percent. For long pipe sections, it may be advisable to discontinue use of lower vibra-

Figure 160.—Casting 20-foot lengths of 54-inch-diameter pipe. The concrete is consolidated by two form vibrators per pipe. PX–D–33052.

Figure 161.—Concrete placement by slow uniform flow over a conical surface. Manual operation of the butterfly valve at the outlet of the placing bucket and continuous vibration are used to control flow. The bucket at the top of this photograph is used for transporting concrete to the placing bucket. PX–D–33526.

tors when the level of concrete has risen well above them. A few surface pits and air bubbles at the upper end are less objectionable than inferior concrete caused by excessive vibration.

When forms are stripped, approximately horizontal settlement cracks are often visible in upper ends of cast pipe. These cracks usually occur where each hoop has been welded to longitudinal bars or where steel cylinders are welded to joint rings. Such cracks seldom appear in the lower portion of the pipe because of the revibration during placement of succeeding lifts. Cracks in the upper portion can be eliminated by revibrating the concrete just before it becomes so stiff that plasticity cannot be restored by vibration. When the concrete is revibrated too soon, some settlement again takes place and the results are not as good as those attained when revibration is delayed as long as possible.

With this emphasis on vibration, it is essential that forms be absolutely tight; otherwise, objectionable sand and gravel streaks will appear where mortar leaks occur at the gates closing the form cylinders and at the joint between the forms and the base ring. (See figs. 162 and 163.)

Information concerning early strength of concrete and steam curing will be found in sections 9 and 126, respectively.

Figure 162.—Forms for cast pipe. If tight gaskets are not used in form joints, gates in inside forms and joints with base rings should be sealed with 2-inch cloth tape that will adhere firmly throughout the placing operation. Paper tape is not fully satisfactory. PX–D–34610.

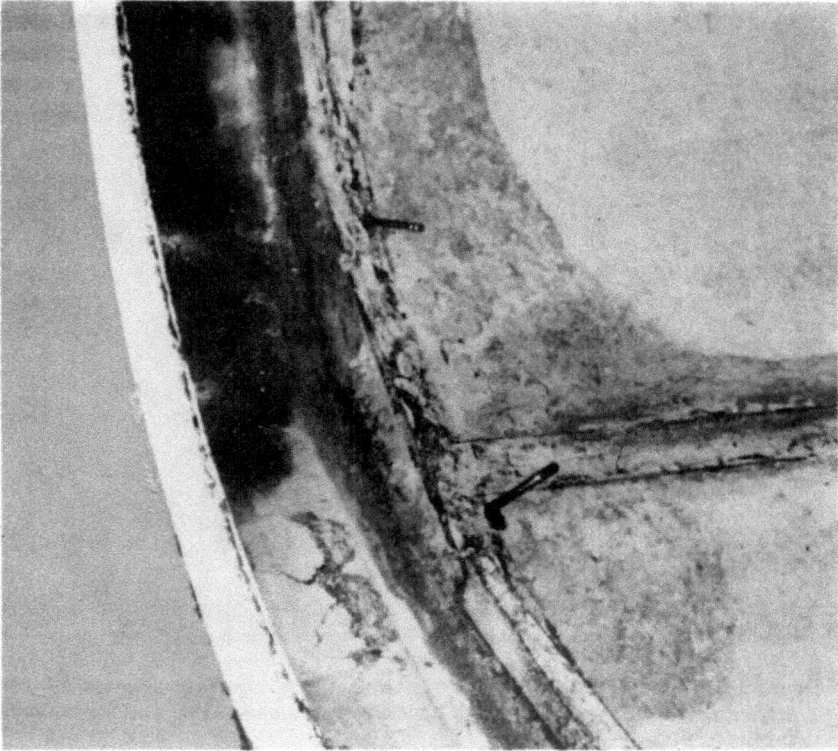

Figure 163.—Pencils inserted to a depth of 1 inch in holes caused by loss of mortar, illustrating the effect of leaky forms. PX–D–34611.

(c) *Centrifugally Spun Pipe.*—In one popular method of manufacturing reinforced concrete pipe (generally 42 inches or smaller in diameter), the outside form is rotated at a high rate of speed in a horizontal position so that concrete can be placed in the spinning form and compacted by centrifugal force. Common variations in compaction used in conjunction with centrifugal force are direct compaction of concrete at the interior surface with a steel roller and vibration during spinning (see figs. 164 and 165).

With the centrifugal method of manufacture, concrete is placed in the form during charging operations in such quantity and in such a manner as to ensure the specified wall thickness with minimum variation in inside diameter throughout the length of the pipe (see fig. 166). The duration and speed of spinning must be sufficient to prevent the concrete from sloughing when spinning is stopped; these are generally more than sufficient to ensure complete distribution of the concrete and produce an even interior surface. Laitance, clay balls, float rock, and interior fines, if found on interior surfaces, should be removed by light brushing, scraping, or

Figure 164.—A centrifugal method of manufacturing concrete pipe which also
employs both direct compaction of concrete with a steel roller and mechanical
vibration. P830–D–17221.

troweling. Hard brushing or troweling to the extent that deep marks appear in the cured pipe should not be permitted.

The maximum size of aggregate and mix proportions for centrifugally spun pipe are normally similar to those for cast pipe of comparable diameter. However, the consistency will be somewhat drier, usually ranging from zero to 1½-inch slump. Also, curing of spun pipe should be equivalent to that for cast pipe.

(d) *Tamped and Packerhead Pipe.*—Concrete pipe of these types is made by compacting very dry concrete into forms. The outside form is similar for both methods and consists of a split cylinder that can be easily removed without damage to the pipe as soon as the pipe is completed. For packerhead pipe, the base ring and outside form are stationary, the inside surface of the pipe being formed by a rapidly revolving shoe, or packerhead. As concrete is fed into the form, the packerhead packs the concrete into place and produces a smooth finish on the inside of the pipe. For tamped pipe, the base ring and outside form rotate, the interior surface being formed by a stationary cylinder. As concrete is fed into the forms, it is compacted by vertical tampers striking at the rate of 500 to 600 blows per minute. Concrete irrigation and drainage pipe is commonly

Figure 165.—A centrifugal method of manufacturing concrete pipe in which the concrete is also directly compacted by a steel roller. PX–D–34070.

made by the packerhead method, whereas large (up to about 36 inches) unreinforced pipe and reinforced culvert pipe is commonly made by the tamping method. (See figs. 167, 168, and 169.)

Unreinforced concrete pipe such as the tamped or packerhead type is commonly used for deep subsurface drainage pipe where high durability and long service life are required. Bureau specifications, Standard Specifications for Unreinforced Concrete Drainage Pipe, are followed for manufacturing and testing this pipe. The class of pipe to be installed under various field conditions is determined from the amount of soluble sulfate present or expected in the soil and/or the ground water. Where maximum sulfate resistance is required, fly ash is added to the concrete in an amount of not less than 25 percent by weight of cement. The amount of cement required is not less than 7.5 bags per cubic yard containing ⅜-inch or smaller maximum size coarse aggregate. If at least 50 percent of the total aggregate in the mix consists of material larger than ⅜-inch, the minimum cement content can be reduced to 6.5 bags per cubic yard. The following tabulation shows the conditions governing the class of drainpipe to be used and establishes the cementitious materials requirements for each class:

Class of pipe	Percent water soluble sulfate (as SO_4) in soil samples	P/m sulfate (as SO_4) in water samples	Cement required	Fly ash required
A	2 or more	10,000 or more	type V	Yes
B	0.2 or more but less than 2	1,500 or more but less than 10,000	type V	No
C	Less than 0.2	Less than 1,500	type II	No

Reinforced concrete pipe for culverts, storm drains, and sewers is manufactured and tested in accordance with American Society for Testing and Materials (ASTM) Designation C 76, Standard Specifications for Reinforced Concrete Culvert, Storm Drain, and Sewer Pipe. Items in which the pipe most often fails to meet the specified requirements are discussed below. Tables and sections hereinafter mentioned in this subsection are from those of ASTM Designation C 76.

Strength requirements are covered in tables I through V and sections 19 to 21, inclusive. When pipe that is dense and sound in appearance and otherwise satisfactory does not meet strength requirements, usually there is not enough cement in the mix or curing has been inadequate, and corrective measures become necessary. The specifications list minimum requirements for both cement content (sec. 8) and curing (sec. 16). The inspector should verify that these minimum requirements are being met at the start of the contract work and should see that they are later increased if necessary to meet strength requirements.

Reinforcement requirements are covered in tables I through V and sections 6, 9, 11, 12, 13, 15, and 23. Appropriate verification of compliance with these requirements should be made by the inspector. Circumferential reinforcement in bells and spigots of pipe is sometimes exposed. Displacement of the steel not only weakens the pipe but often interferes with proper consolidation of concrete in these portions. As such defects are not consistent with good workmanship or with requirements of specifications, pipe with exposed or displaced reinforcement should not be accepted. Thin-walled, mechanically tamped pipe with standard reinforcement sometimes develops spiral cracks caused by the release, when the forms are removed, of a twist induced in the cylindrical cage of reinforcement as the filling forms revolve against the tamping stick. This tendency

Figure 166.—Uniform placing of concrete in a rotating pipe form by use of a traveling belt conveyor visible at left center of form P830–D–17023.

should be guarded against and corrective measures are required if it occurs. Pipe containing fractures, large or deep cracks, or cracks passing through the shell should not be accepted regardless of cause (sec. 27 (a)).

Compaction of the mix throughout the full thickness of pipe shell is important and should be required even though less thorough compaction would produce pipe that passes strength tests and has interior surfaces that appear to be dense and watertight. This compaction should also be

Figure 167.—Equipment for the packerhead method of manufacture of unreinforced pipe up to about 15-inch diameter. PX–D–34071.

Figure 168.—Form removal from unreinforced packerhead pipe immediately after placement. PX–D–34072.

Figure 169.—Equipment for manufacture of reinforced or unreinforced tamped concrete pipe up to about 54-inch diameter. PX–D–34073.

applied to concrete in the bell and spigot. Specifications prohibit surface roughness, defects that indicate imperfect proportioning, mixing, and molding, and surface defects indicating honeycombed or open texture (secs. 24, 27(b), and 27(c), respectively). Porous and poorly compacted concrete, common in bells and on the outside surface of pipes, can be eliminated by proper proportioning of the mix, correct grading of the sand (with ample fines), sufficient cement, effective use of tamping devices, and some hand finishing as soon as forms are removed. Designating marks, numbers, or letters should not be scratched on interior surfaces of pipes.

Eccentricity of spigots in any appreciable degree should be avoided. Section 14 requires that joints "be so formed that when the pipe are laid together they will make a continuous and uniform line of pipe, compatible with the tolerances given in section 23(a)." This requirement permits very little eccentricity if proper space is left for a solid ring of mortar in the joint. Eccentricity of spigots is a defect indicating imperfect molding, and pipe having this interference to flow should not be accepted.

Acceptance inspection, whenever practicable, should be provided at the point of manufacture as pipe is produced and not after delivery of the product at some later date. At the very earliest date in his program of manufacture, the manufacturer should be notified if his pipe is unacceptable in any respect. In this way, the contract can be fulfilled with a minimum of rejections and a maximum of quality in the pipe accepted.

Reinforced concrete pressure pipe is manufactured and tested in accordance with Bureau specifications, Standard Specifications for Reinforced Concrete Pressure Pipe. These specifications also provide for the design of the pipe from 12 through 108 inches in diameter for hydrostatic heads of 25, 50, 75, 100, 125, and 150 feet measured to the centerline of the pipe and external loadings of 5, 10, 15, and 20 feet of earth cover over the top of the pipe.

(e) *Prestressed Pipe.*—Prestressed concrete pipe is commonly used for distribution of water under high pressures and is manufactured and tested in accordance with Bureau specifications, Standard Specifications for Embedded Cylinder Prestressed Concrete Pipe and Standard Specifications for Pretensioned Concrete Cylinder Pipe. The main difference between prestressed and other types of high-pressure pipes is that reinforcement steel in prestressed pipe is held in tension during manufacturing. Methods of manufacturing prestressed pipe usually employ centrifugal force to make the core by spinning the pipe form containing concrete and longitudinal reinforcement with or without a steel cylinder (see sec. 109 (c)). After curing, the core is wrapped with reinforcement steel under tension. Then the core and circumferential steel are usually coated with mortar to produce the outer shell of the pipe. In some methods of manufacture, the longitudinal steel is also prestressed. Additional adequate curing completes manufacture of the pipe. General information concerning handling and curing of pipe, contained in this section, also apply to this type of pipe.

(f) *Roller-Compacted Asbestos-Cement Pipe.*—Asbestos-cement pressure pipe, generally in sizes 4- through 42-inch diameter, may be used in lateral pipelines for heads up to about 700 feet. The manufacture and testing of this pipe are controlled by Bureau specifications, Standard Specifications for Asbestos-Cement Pressure Pipe. Acceptability of the pipe is determined from results of tests performed by the contractor for hydrostatic proof pressures, minimum crushing strength, joint leakage, and flexureproof tests for 4-, 6-, and 8-inch-diameter pipe, and through inspections during or after manufacture. Certified copies of test results should be furnished by the contractor. The Bureau reserves the right to perform acceptance tests for uncombined calcium hydroxide and sulfate resistance on samples of the pipe. If the Bureau elects not to perform these acceptance tests, the contractor is required to furnish certification that the pipe being furnished is made from materials identical to and in a manner similar to that of asbestos-cement pipe which has been tested and meets the requirements of Bureau specifications.

The most common method used to manufacture the pipe is to pass an intimately mixed slurry of water, portland or portland-pozzolan cement, silica, asbestos fiber free from organic substances, and curing agents, if used, between two oppositely rotating horizontal steel rollers under load to build the pipe gradually around the lower roller to the desired thickness. Pressure produced on the pipe wall during forming expels the excess water. It is then very important to cure the pipe properly by use of high-pressure steam to produce maximum durability. Sulfate resistance tests show that high-pressure steam curing reduces free lime content by formation of chemical compounds very resistant to sulfate attack. The use of silica in the mix aids formation of these compounds. Water curing and steam curing at atmospheric pressure result in an excess of free lime not yet chemically combined with other constituents in the pipe.

110.   Cast-in-Place Concrete Pipe.—(a) *General*.—Cast-in-place unreinforced concrete pipe has been used for conveyance of irrigation water under heads up to 15 feet, and Bureau specifications are used to control construction of these pipelines. Pipe sizes installed by this method range from 24 through 48 inches (see fig. 170).

Cast-in-place pipe is normally constructed in cohesive soils in which the excavation will stand vertically. However, in soils which tend to ravel or slough, overexcavation and refill with compacted cohesive soils is sometimes economically justifiable. Cast-in-place pipe may be constructed in one placement around the complete periphery or in two placements with longitudinal joints at approximately each end of the horizontal diameter. If a two-placement method is used, the second-stage concrete should be placed before first-stage concrete has initially hardened to provide ample consolidation and intermixing of first- with second-stage concrete by vibration or tamping.

(b)   *Concrete*.—Criteria for control of concrete and resultant high-quality cast-in-place pipe are similar with few exceptions to those used in fabrication of precast pipe by the cast method (sec. 109(b)).

(1)   Concrete is usually transported to the installation site by truck mixers, and appropriate requirements for this method of supplying concrete are included in Bureau specifications.

(2)   Minimum cement content is 5½ bags per cubic yard of concrete.

Figure 170.—Placement of cast-in-place concrete pipe. Concrete, transported by truck mixer, is being dumped from the left side into the placing machine. PX–D–34074.

(3) Percent of air entrained depends on the maximum size of aggregate used. (See sec. 20(b).)

(4) The concrete temperature should be controlled to about 50° F. and it is advisable to place the pipe during the coolest period of the day or at night if the pipe has no provisions for contraction joints. There have been instances of excessive cracking attributed to volume change caused by temperature differential.

(5) Immediately after concrete placement, the exposed surface should be coated with sealing compound. As soon as the concrete has attained sufficient strength to prevent damage from backfilling

operations, a 6-inch layer of damp earth backfill should be placed over the pipe and kept damp for a period of 7 days or until the trench is completely backfilled.

(6) After placement of each pipeline segment, the concrete should be protected from drying and freezing.

111. Vibrators.—The objective in consolidating concrete is elimination, so far as practicable, of voids within the concrete. Well-consolidated concrete is satisfactorily free of rock pockets and bubbles of entrapped air and is in close contact with forms, reinforcement, and other embedded parts. Accomplishment of this objective is easier if segregation and slump loss are avoided during transportation and deposition of the concrete.

Specifications require that concrete be consolidated with electric- or pneumatic-driven, immersion-type vibrators. For consolidation in structures and inverts of tunnel lining, the vibrators are required to be operated at an oscillation frequency of at least 7,000 vibrations per minute when immersed in the concrete. Concrete in the arch and sidewalls of tunnel lining is required to be consolidated by electric or pneumatic form vibrators rigidly attached to the forms, this type of vibration to be supplemented when practicable by immersion-type vibrators. (See fig. 136 and sec. 104.) Form vibrators are required to be rigidly attached to the forms and operate at speeds of at least 8,000 vibrations per minute.

Concrete in canal and lateral lining is normally required to be vibrated by internal-type vibrators operating at speeds of at least 4,000 vibrations per minute when immersed in the concrete. An exception exists for linings having a thickness of less than 3 inches; in these linings the concrete may be consolidated by external vibration if the contracting officer determines that the consolidation being obtained is equivalent to that produced by internal-type vibrators operating at speeds of at least 4,000 vibrations per minute.

For construction work involving large quantities of concrete containing 3- and 6-inch coarse aggregate and where large-diameter vibrators may be used, the concrete should be consolidated with vibrators having vibrating heads 4 inches or more in diameter operating at speeds of at least 6,000 vibrations per minute when immersed in the concrete. Each cubic yard of concrete should receive a minimum of 60 seconds of continuous vibration. Where adequate consolidation can be obtained with less vibration, this time may be slightly reduced. Vibrators having less than 4-inch-diameter heads should be operated at speeds of at least 7,000 vibrations per minute. For work inaccessible to immersion-type vibrators, such as precast concrete pipe and portions of tunnel lining, vibrators attached to the forms produce good consolidation if they are operated at a speed in excess of 8,000 vibrations per minute.

Some subgrade-guided slip-form concrete lining equipment has the tendency to float on the concrete when internal vibration is used. The use of this type of equipment without vibration can be permitted in placing linings of less than 3 inches in thickness if it is satisfactorily shown that suitable consolidation can be obtained.

Contractors proposing to use this type of equipment should be forewarned that use is contingent on whether good consolidation of the concrete and acceptable results, as determined by the contracting officer, are obtained. Internal vibration should be used in all linings having a thickness of 3 inches or more.

Immersion-type vibrators for consolidating mass concrete should be heavy-duty vibrators capable of readily consolidating large quantities of lean, low-slump concrete. One-man vibrators are now in general use. These vibrators, if operated in sufficient numbers and at proper speed, will produce results equivalent to those produced by two-man vibrators formerly used.

In placing and consolidating mass concrete, gang vibrators mounted on self-propelled equipment are permitted by Bureau specifications. Gang vibrators should be mounted so that they can be readily raised and lowered to eliminate dragging through fresh concrete. When vibration is performed by gang vibrators, hand vibration should be used near embedded equipment and at locations difficult to reach with the gang vibrators. The requirements for vibrating frequency and amplitude and penetration patterns should be met regardless of whether one-man, two-man, or gang vibrators are used.

Vibrator speeds should be regularly checked by inspectors. Pencil-size vibrating reeds to determine vibration frequency are available commercially for this purpose. When equipment will not run at the specified speed, it should be removed for servicing or replacement.

An immersion-type vibrator should be inserted vertically, at points 18 to 30 inches apart, and slowly withdrawn. However, in shallow or inaccessible concrete some consolidation can be obtained by using the vibrator in a sloping or horizontal position. Vibration periods of 5 to 15 seconds for each penetration are usually sufficient. The amount of vibration in one spot may be gaged by surface movement and texture of the concrete, by the appearance of cement paste where the concrete contacts nearby forms or embedded parts, by the approach of the sound of the vibrator to a constant tone, and by the feel of the vibrator in the operator's hands. Systematic spacing of the points of vibration should be established to ensure that no portions of the concrete are missed. Most of the common imperfections and rock pockets in concrete can be obviated by thorough vibration.

The entire depth of a new layer of concrete should be vibrated, and the vibrator should penetrate several inches into the layer below to ensure thorough union of the layers. (See fig. 136.) Under ordinary job conditions, there is little probability of damage from direct revibration of lower layers or from vibration transmitted by embedded steel, provided the disturbed concrete still is or again becomes plastic. Bonding of new concrete to concrete that has hardened and has been properly cleaned is essentially a matter of thoroughly vibrating the new concrete close to the joint surface.

There is little likelihood of overvibration when the slump of the concrete is as low as is practicable for placement using vibration. When overvibration occurs, the surface concrete not only appears very wet, but it actually consists of a layer of mortar containing little coarse aggregate. When overvibration is indicated, the slump, and not the amount of vibration, should be reduced. Efforts to avoid overvibration often result in inadequate vibration. Experience indicates that objectionable results are much more likely to be obtained from undervibration than from overvibration.

Considerably more vibration is sometimes required to satisfactorily reduce the amount of entrapped air and the number of surface bubbles than is necessary to eliminate rock pockets.

Revibration is beneficial rather than detrimental, provided the concrete is again brought to a plastic condition. Revibration may be accomplished by immersion-type vibrators, by form vibration, or by transmittal of vibration through the reinforcement system. Apprehension as to use of the last method appears unfounded as extensive observation has disclosed no instance in which damage to the concrete could be attributed to this cause. Revibration could well be more widely practiced to eliminate settlement cracks and the internal effects of bleeding and also as an aid in making tight concrete repairs in walls and other structures.

There should be no difficulty from cold joints if full advantage is taken of vibration and revibration. If the underlying concrete will still respond to revibration, the vibrator should be allowed to penetrate it deeply at each insertion in the new concrete. If this procedure is followed at close systematic spacing, the concrete at the joint will become monolithic. If the underlying concrete is too hard for revibrating and it is still very green concrete, thorough vibration close to the contact area will result in a good bond. Drill-core specimens have shown that the strength of such joints is equal to that of other portions of the specimens.

Experience has confirmed that the immersion-type vibrators that give the best results are amply powered of rugged construction, and of relatively high speed. (See discussion of speeds earlier in this section.) Air vibrators are adaptable over a wide range of service, but it is imperative that an adequate air supply be maintained. Freezing at the exhaust may be prevented by use of dry air from an adequate receiver or by trickling an antifreeze agent into the air line. However, antifreeze solutions with a glycol base are objectionable because of a tendency to gum the valves. Electric vibrators are highly effective, especially in medium and smaller sizes.

Small vibrators can handle from 5 to 10 cubic yards per hour, even in restricted spaces, and one large two-man, heavy-duty type can handle approximately 50 cubic yards per hour in spacious forms. One-man, heavy-duty vibrators handled a reported 80 cubic yards of concrete per hour working full time. This was accomplished with a well-designed mass concrete of 2-inch slump containing a WRA. Spare units and parts should always be on hand to take care of breakdowns and necessary repairs. The life of a vibrator may be prolonged considerably by systematic conditioning at short intervals. In determining the number of vibrators required in mass concrete placements, the effective vibrating time for each cubic yard of concrete is usually from 60 to 90 seconds.

112. Surface Imperfections.—Most imperfections in new concrete can be avoided by reasonable care in placing. Unfortunately, this is not altogether true in regard to air bubbles and surface pits, particularly those on surfaces where forms slope upward toward the concrete, as is the case for downstream faces of dams, battered walls and piers, and areas below the spring lines of tunnels, siphons, and conduits. Treatment of these surface imperfections is usually regarded as a matter of surface finish rather than repair; nevertheless, they can be reduced considerably by using proper precautions during concrete placement.

Most procedures for reduction of pits and air bubbles are based on the fact that, given the opportunity, a large bubble of *entrapped* air (not the minute bubbles of *entrained* air) will rise to the surface of the plastic concrete and escape. Such an opportunity is best afforded when sticky and oversanded mixes are avoided and when the newly placed concrete is deposited in relatively shallow layers and is amply vibrated and spaded along the forms. Avoidance of excessive coats of form oil and of viscous oils will diminish the tendency of bubbles to adhere to the forms. Occasionally, less surface pitting has been noted when forms were sprayed with lacquer. Paint applied to plywood forms, particularly when the grain is horizontal, produces improved results in this respect and preserves the forms in better condition for reuse.

The temporary fluidity of concrete resulting from vibration is probably the most important influence in the release of entrapped air. Fluidity must not be obtained by using high-slump concrete, because reduced quality, sand streaks, and other imperfections more objectionable than pitting would result. To be effective, vibration must be continued until the bubbles have had time to escape. Proper duration of vibration can be determined by noting when bubbles stop emerging from the concrete. To achieve maximum effect, the vibrator should be operated somewhat below the concrete surface, raising it as the concrete is placed above. Inserting a vibrator into concrete that is partially consolidated tends to compact the upper layers first, thus making the escape of air entrapped below more difficult.

Notable success has been achieved in reducing surface pitting on precast concrete pipe by continuous vibration of forms at speeds higher than 8,000 revolutions per minute during placing and by depositing the concrete in shallow layers. This was demonstrated during manufacture of precast concrete pipe for the Salt Lake, San Diego, and Second Mokelumne River Aqueducts. Experience has shown that extra internal vibration will achieve similar results on surfaces of architectural, structural, and other concrete cast in place.

Vibration does not drive bubbles to the form. Bubbles in a fluid medium do not move horizontally. They may move diagonally upward toward a more fluid portion of the mass and may move in this way toward the vibrator or toward the form if it is being vibrated. A certain amount of vibration will permit most of the entrapped air to escape without appreciable loss of valuable minute bubbles of entrained air.

Although some engineers are convinced that purposeful entrainment of air appreciably increases the number and size of air bubbles on formed surfaces of concrete, there is much evidence to the contrary.

From the preceding paragraphs it is evident that little can be done to eliminate pits and air bubbles from surfaces of concrete placed under a sloping form. Additional vibration which causes the bubbles to rise will only increase their numbers as they collect against the overhanging form; the minimum vibration necessary to prevent rock pockets will then result in the least surface pitting. Use of tightly jointed horizontal lagging, such as shiplap or tongue-and-groove boards, will often result in fewer surface bubbles than would occur if plywood or other sheet material were used for sheathing. Use of absorptive form linings or the vacuum process will tend to eliminate pits and air bubbles, but the cost of these procedures for this purpose alone is prohibitive, except where best appearance is very important or an exceptionally durable surface is a consideration.

Sand streaks are another common surface defect. Sand streaking is related to the concrete materials and their proportions, the tightness of forms, and the manner of handling the concrete. Lean, harsh, wet mixes with bleeding tendencies, poorly graded sand deficient in fines, coarsely ground cement, and leaky forms are all conducive to sand streaking.

113.   **Bond With Reinforcement and Embedded Parts.**—Surfaces of reinforcement and embedded parts should be free from contamination

such as mud, oil, paint, and loose, dried mortar. Removal of tight, adherent mortar is unnecessary. Loose rust or mill scale will generally be removed sufficiently in the normal handling of the bars (see sec. 98). During the earliest stages of hardening, after the concrete loses its plasticity, bond may be impaired if projecting reinforcement is subjected to impact or rough handling. Exposed portions of bars only partly embedded should not be struck or carelessly handled, and workmen should not be permitted to climb on bar extensions until the concrete is at least 7 days old. Forms to which embedded parts are fastened or through which they protrude should not be stripped until the concrete has hardened sufficiently to avoid damage to bond.

114. Waste Concrete.—Waste concrete is considered under two classes: fresh concrete which, because of inferior quality or some other undesirable condition, is rejected before it is placed; and defective concrete that must be removed after it has hardened.

One of the reasons for wasting a batch of concrete is arrival at the forms in such a stiff condition that proper placement cannot be assured. This condition may result from some unforeseen delay in transportation, from improper control of consistency at the mixing plant, or from premature stiffening. More often, an overwet batch arrives in such a segregated condition that it is unfit for use. Batching errors and overproduction are also responsible for waste concrete. Occasionally, when unsuitability is not detected earlier, concrete must be wasted after it has been deposited but before it has been consolidated.

Concrete that is not placed within a certain interval after mixing should not necessarily be wasted. It should be wasted only if it has stiffened to such an extent that it cannot be placed and consolidated properly by use of extra vibration.

After the forms are removed, the concrete is inspected. The contractor may be required, at any time before completion and acceptance of the work, to remove and replace any concrete that is found to be defective. This requirement includes concrete that is obviously unconsolidated or that has been damaged by accident or by freezing.

When any of the concrete materials are furnished by the Government, the inspector should keep a record of rejected batches and of concrete used in making replacements so that payments may be appropriately adjusted.

115. Shutting Down Concreting Operations.—At the beginning of a job, concreting operations may understandably be somewhat below standard, since the contractor may need time to break in new crews and correct equipment trouble. After initial difficulties have been eliminated, unsatisfactory concreting operations usually do not develop immediately but are the result of continually worsening conditions. The contractor should be required to correct such conditions before they reach the stage where operations become so unsatisfactory that they must be stopped.

116. Placing Concrete in Water.—In Bureau work whenever practicticable, the placing of concrete under water is avoided. Preferably, entry of seepage water to the working area should be stopped by diversion, well points, or other effective means. If there is shallow water over a solid subgrade, satisfactory concrete placement can be obtained by starting in a dry area and crowding the concrete toward the water, which will gradually be displaced with very little intermixing. This procedure should not be attempted, however, in deep or running water.

At times it is physically or fiscally impracticable to dry a foundation prior to concrete placement. In such cases, suitable underwater placing procedures such as pumping or use of tremies or special concrete buckets should be employed. Pumping is considered by the Bureau to be the best method of placing concrete underwater. A temporary plug should be placed in the end of the line before it is lowered into the water. To prevent the concrete from mixing with the water during pumping, the end of the delivery line is always kept submerged in the fresh concrete and is raised as the concrete rises. The surging action of the pumping will provide some consolidation as the concrete is being placed. No other consolidation should be undertaken.

A tremie is a pipe having a funnel-shaped upper end into which the concrete is fed. As with pumping, the discharge end should always remain buried in the fresh concrete. A tremie pipe should normally be eight times the maximum size of coarse aggregate. Pipes 10 to 12 inches in diameter and lengths of 10 feet are commonly used with the pipe sections bolted together using a gasket to prevent leakage. Pipe spacing varies depending on the thickness of placement and congestion from piles or reinforcement. Usual pipe spacing can be about 15 feet on centers or so spaced that one pipe will cover about 300 square feet in area. However, spacing has been increased to as much as 40 feet under ideal conditions using retarded concrete.

Concrete mix proportions for tremie concrete differ from the usual mix design for 1½-inch-maximum size aggregate since the concrete must flow into place by gravity and without any vibration. Thus, the cement content of the tremie concrete should range between 6.5 to 8.0 bags per cubic yard for 1½-inch-maximum size aggregate concrete with a slump from 6 to 9 inches. Rounded aggregate should be used to improve flow characteristics. Aggregate sizes less than 1½ inches maximum can be used in critical areas. Sand may comprise 40 to 50 percent of the total weight of the aggregate to obtain the desired flowability characteristics. The use of air-entraining agents, pozzolans, and water-reducing, set-controlling agents also help obtain flowability.

The tremie pipe should be equipped with a footvalve or other suitable device capable of sealing the pipe. Where such a device is not available, removal of the water from the pipe is accomplished by forcing a ball or plug or scraper through the pipe ahead of the concrete. In deep placements (70 feet or so) the buoyancy of the empty pipe may present difficulties in lowering it. In this case, use of the latter described sealing method is preferable as it permits the open pipe to fill with water as it is lowered into position.

Special underwater buckets are also sometimes used for placing concrete in water. These are bottom discharge buckets in which the discharge end of the bucket is lowered into the previously placed fresh concrete before the gate is opened for discharge.

When placing by pumping or by tremie, the placement must be started slowly to minimize scouring of the bottom. The bottom of the pump discharge line or tremie pipe should be placed as near as practicable to the surface against which the concrete is to be placed and not raised until a sufficient depth of seal has been established. The pipe should be lifted slowly to assure that the lower end of the pipe is not raised above the plastic concrete and that the seal is not broken. If this occurs, it is then necessary to reestablish the seal as if starting anew.

Initial placements should begin at the lowest points of elevation within the confines of the placement. The surfaces on which the concrete is placed should be free of mud, marine growth, or other materials that might prevent bond. The spacing of tremie pipes should be such that a minimum of hills and valleys will occur during placing. A slightly sloping surface is ideal, although difficult to achieve. The deeper the pipe is embedded in the plastic concrete, the flatter the slope. However, this also requires more slump.

Inspection of the concrete while being placed is difficult. The water is usually murky and the surface of the newly placed concrete will not support a diver. Thus, the control must be principally from above the water surface. This requires close inspection of equipment, constant checking and maintenance of the discharge seal, continued checking of concrete slope and height by sounding lines, and maintenance of a uniform rate of flow. If excessive hills and valleys occur, an airlift pump can be used to remove scum and laitance which collect in depressions before reestablishing the tremie seal to level the area.

D. Removal of Forms, and Finishing

117. Removal of Forms.—Determination of the time of form removal should be based on the effect of the removal on the concrete. When forms are stripped, there should be no measurable deflection or distortion and no evidence of damage to the concrete resulting from either removal of support or from the stripping operation. Supporting forms and shoring must not be removed from beams, floors, and walls until the walls are strong enough to carry their own weight and any superimposed load. Figure 171 depicts the probable early strength of standard-cured concrete

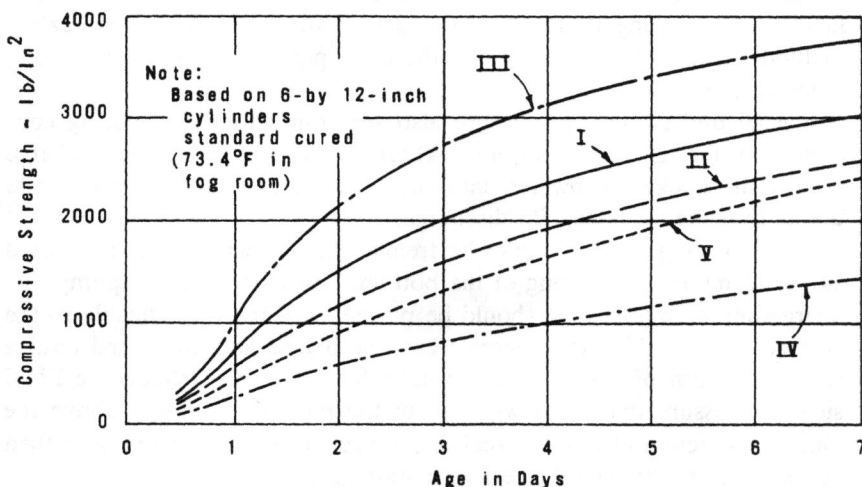

Figure 171.—Concrete strength gain at early ages for various types of cement. 288–D–1546.

cylinders made with various types of cement. However, the strength required and the time to attain it vary widely under different job conditions of temperature and materials, and the most reliable basis for the early removal of supporting forms is furnished by test specimens cured at job temperature. Figure 10 may be used to estimate, and correct for, the effect of temperatures on the strength of control specimens when they have not been stored at standard temperature. The figure may also be used as a basis for estimating 7- or 28-day strength from tests at intermediate ages (see sec. 9). In general, moderate temperatures are desirable for curing concrete. The importance of controlling the temperature of green concrete depends on the size of the section, necessity for early form removal, and likelihood of damage from overheating in hot weather or freezing in cold weather.

Warm weather, high concrete temperature, fast-hardening cement, a low water-cement ratio, light loads, and use of calcium chloride expedite early stripping. (See fig. 31 for effect of calcium chloride on the early strength of concrete.) Sufficient strength has been attained when test specimens indicate a safety factor of two for the stresses to be sustained.

On the other hand, experience has shown that even when the concrete in siphon barrels and tunnels is strong enough to show no distress or deflection from the load, it is still possible to damage the corners and edges during the stripping operation.

Forms should be removed at the earliest practicable time so that curing may proceed without delay. Forms are a poor medium for curing, as indicated by the dry concrete surfaces usually found when forms are removed. Another advantage of early form removal is that any necessary repairs or surface treatment can be done while the concrete is still quite green and conditions are most favorable for good bond. For these reasons, early stripping of inside forms of open transitions (1 to 3 hours after the concrete is placed) is advocated.

In cold weather, forms should not be removed while the concrete is still warm, as rapid cooling of the surface will cause checking and surface cracks. For the same reason, water used in sprinkling newly stripped surfaces should not be much colder than the concrete. Also, in cold weather the urgency for form removal to commence curing treatment is not great and, unless uninsulated steel forms are used, it may well be that the protection afforded by the forms warrants leaving them in place for the first few days.

118. Repair of Concrete.—Defects in new concrete that require repair may consist of rock pockets and other unconsolidated portions of various areas and depths, damage from stripping of forms, bolt holes, ridges from form joints, and bulges caused by movement of the forms. Accepted procedures for repairing concrete are described in chapter VII. These procedures are also applicable to the restoration and reconstruction of disintegrated portions of structures in service.

119. Types and Treatments of Formed Surfaces.—Except for occasional special finishes, formed concrete surfaces and finishes are designated in Bureau specifications as F1, F2, F3, F4, and F5. Surface irregularities permitted for these finishes are termed either abrupt or gradual. Offsets and fins caused by displaced or misplaced form sheathing, lining, or form sections, by loose knots in forms, or by otherwise defective form lumber are considered as abrupt irregularities. All others are classed as gradual irregularities. Gradual irregularities are measured with a template consisting of a straightedge for plane surfaces or its equivalent for curved surfaces. The length of the template for testing formed surfaces is 5 feet. Maximum allowable deviations are listed in table 24. Formed surfaces will generally require no sack rubbing or sandblasting. Except as required for special finishing of elbows of tunnel spillways and special treatment of offsets on surfaces of outlet works and spillways that will be in contact with flowing water having velocities of 40 feet per second or more, no grinding or stoning is generally required for formed surfaces other than that necessary to bring surface irregularities within specified limits. Repair of concrete surfaces is discussed in chapter VII.

(a) *Finish F1.*—This finish applies to surfaces where roughness is not objectionable, such as those upon or against which fill material or concrete will be placed, the upstream faces of concrete dams that will normally be under water, or surfaces that will otherwise be permanently concealed. The only surface treatment required is repair of defective concrete, correction of surface depressions deeper than 1 inch, and filling of tie-rod holes where the surface is to be coated with dampproofing or where the holes are deeper than 1 inch. Form sheathing may be any material that will not leak mortar when the concrete is vibrated. Forms may be built with a minimum of refinement.

(b) *Finish F2.*—This finish is required on all permanently exposed surfaces for which other finishes are not specified, such as surfaces of canal structures; inside surfaces of siphons, culverts, and tunnel linings; surfaces of outlet works other than high-velocity flow surfaces required to receive an F4 finish; open spillways; small power and pumping plants; bridges and retaining walls not prominently exposed to public inspection; galleries and tunnels in dams; and concrete dams except where F1 finishes are permitted on upstream faces. Form sheathing may be shiplap, plywood, or steel. Thin steel sheets (steel lining) supported by a backing of wood boards may be used on approval, but use of steel lining

Table 24.—Maximum allowances of irregularities in concrete surfaces

Type of irregularities	Finish (formed surfaces *)					Finish (unformed surfaces†)			
	F1	F2	F3	F4	F5	U1	U2	U3	U4
Depressions	1	.....	.....	.....	.....	.....	.....	.....	.....
Gradual	.....	½	¼	¼	¼	.....	.....	.....	.....
Abrupt	.....	¼	⅛	‡ ¼ § ⅛	¼	.....	.....	.....	.....
All surfaces	.....	.....	.....	.....	.....	⅜	¼	¼	.....
Canal surfaces, bottom slabs	.....	.....	.....	.....	.....	.....	.....	.....	¼
Canal surfaces, side slopes	.....	.....	.....	.....	.....	.....	.....	.....	½

* Allowance in inches—measured from 5-foot template.
† Allowance in inches—measured from 10-foot template.
‡ Allowance of irregularity or offset extending parallel to flow.
§ Allowance of irregularity or offset not parallel to flow.

is not encouraged. To obtain an F2 surface, forms must be built in a workmanlike manner to required dimensions and alinement, without conspicuous offsets or bulges.

(c) *Finish F3*.—This finish is designated for surfaces of structures prominently exposed to public view where appearance is of special importance. This category includes superstructures of large power and pumping plants; parapets, railings, and decorative features on dams and bridges; and permanent buildings. To meet requirements for the F3 finish it is necessary for forms to be built accurately to dimensions in a skillful, workmanlike manner. Occasionally tongue-and-groove boards or plywood sheets may be required for specific F3 surfaces. However, specifications usually permit either tongue-and-grove boards or plywood at the contractor's option. Steel sheathing or lining is not permitted. There should be no visible offsets, bulges, or misalinement of concrete. At construction joints forms should be tightly reset and securely anchored close to the joint, as described in section 96.

(d) *Finish F4*.—This finish is required for formed concrete surfaces where accurate alinement and evenness of surface are essential for prevention of destructive effects of water action. Such surfaces include portions of outlets, draft tubes, high-velocity flow surfaces of outlet works downstream from gates, and spillway tunnels of dams. The forms must be strong and held rigidly and accurately to the prescribed alinement. Any form material or sheathing that will produce the required surface (such as close-fitting shiplap, tongue-and-groove lumber, plywood, or steel) may be used. For warped surfaces, the forms should be built of laminated splines cut to make tight, smooth form surfaces after which the form surfaces are dressed and sanded to the required curvature.

(e) *Finish F5*.—This finish is required for formed concrete surfaces where plaster, stucco, or wainscoting is to be applied. As a coarse-textured surface is needed for bond, the concrete should be cast against rough-faced (S1S2E) form boards. Form oil should not be used. Steel lining or steel sheathing is not permitted.

(f) *Special Stoned Finishes*.—A special finish may be required on surfaces where offsets, bulges, and repair chimneys have been removed, on areas in tunnels and conduits where an especially smooth and even surface is necessary to prevent cavitation, and for stair risers in power and pumping plants.

The procedure for all surfaces except stair risers is as follows: The forms should be removed while the concrete is still green, but not sooner than twelve hours nor later than 24 hours after placing the concrete. Immediately after form removal, all patching and pointing, including filling of holes left by removal of fasteners from tie rods and openings left by removal of porous or fractured concrete, should be accomplished.

The surface that is to receive the special finish should be thoroughly cleaned with high-velocity water jets to remove loose particles and foreign material and then brought to a surface-dry condition, as indicated by the absence of glistening-free water, by clean air jet. A plastic mortar consisting of 1 part of cement and 1 to 1½ parts of sand, by weight, that will pass a No. 16 screen should be rubbed over the surface and handstoned with a No. 60 grit carborundum stone, using additional mortar until the surface is evenly filled. Stoning should be continued until the new material has become rather hard. After moist curing for 7 days, the surface should be made smooth and even by use of a No. 50 or No. 60 grit carborundum stone or grinding wheel. A flexible disc power sander may produce an acceptable surface. After final stoning, curing is continued for the remainder of the 14-day curing period. (See sec. 124.)

The procedure for stair risers is as follows: Forms are removed between 12 and 24 hours after the concrete is placed, and all required patching and pointing are performed. Surfaces of the risers are wet thoroughly with a brush and rubbed with a hardwood float dipped in water containing 2 pounds of portland cement per gallon. The rubbing is continued until all form marks and projections have been removed. The grindings from the rubbing operations are then uniformly spread over the riser surfaces with a brush to fill all pits and small voids. The brushed surface is allowed to harden and is then kept moist for at least 3 days, after which a final finish is obtained by rubbing with a silicone-carbide abrasive rubbing brick stone of approximately No. 50 grit until the entire surface has a smooth texture and is uniform in color. The time at which the wood-float finish is performed is critical. Wood-float rubbing should not be started so soon that the aggregate grains are easily dislodged nor so late that the surface is too hard to be readily dressed. Final rubbing with the abrasive rubbing brick should be just sufficient to produce the surface condition required without unnecessary cutting of the aggregate grains. After rubbing, curing is then continued for the remainder of the 14-day curing period. (See sec. 124.)

One type of special stoned finish which may be used where such an architectural finish is required is a stoned-sand finish. In texture and appearance it is somewhat similar to cement plaster, but it is much less costly. It may be either painted or unpainted. The procedure combines the prewetting and fine-sand mortar application of sack rubbing, a modified stoning from the special finish, and painstaking fog-spray curing and draft-free slow drying from the cement plaster procedure. The steps in the procedure are as follows:

(1) Prewet the concrete surface for several hours or overnight before treatment.

(2) Close the area of work to prevent drafts and reduce drying.

(3) Spread a thick, creamy sand mix, consisting of 1 part of cement to 2 parts of sand, thinly over the surface with a wood or rubber float or sacking. The sand should all pass a No. 12 mesh screen. The cement should be a light-colored brand or a mixture of standard cement with white cement.

(4) Stone-in at once with a carborundum float, working over the entire area and leaving only a minimum amount of material on the surface necessary to produce a sand texture, approximately 1/32 inch in thickness.

(5) Keep the surface continually damp with light fog spray for 7 days, then let dry slowly without air drafts.

(g) *Sack-Rubbed Finish.*—A sack-rubbed finish is sometimes necessary when the appearance of formed concrete, particularly of F3 surfaces, falls considerably below expectations. This treatment is performed after all required patching and correction of major imperfections have been completed.

The surfaces are thoroughly wetted and sack rubbing is commenced while they are still damp. The mortar used consists of 1 part cement; 2 parts, by volume, of sand passing a No. 16 screen; and enough water so that the consistency of the mortar will be that of thick cream. It may be necessary to blend the cement with white cement to obtain a color that will match that of the surrounding concrete surface. The mortar is rubbed thoroughly over the area with clean burlap or a sponge rubber float so as to fill all pits. While the mortar in the pits is still plastic, the surface should be rubbed over with a dry mix of the above proportions and material. This serves to remove the excess plastic material and place enough dry material in the pits to stiffen and solidify the mortar so that the fillings will be flush with the surface. No material should remain on the surface except that within the pits. Curing of the surface is then continued.

(h) *Sandblast Finish.*—Water stains, grout accumulations, and sealing compound can be effectively and economically removed from concrete surfaces by light sandblasting. Sand for this purpose should all pass a No. 30 screen. The sandblast equipment should be capable of controlling air pressures ranging from 15 to 45 pounds per square inch. Hose lengths should not exceed 200 feet. A ⅜-inch-diameter nozzle will normally be large enough, although other sizes may be used. Sandblasting should not be commenced sooner than 14 days after placement for concrete that has been water cured or 28 days after placement for concrete that has been membrane cured. Also, it should not be commenced until all concrete at higher elevations has been placed and cured. Walls should be sandblasted starting at the top and working downward using a horizontally oscillating motion. After a section of wall has been sandblasted, it should be washed with water to remove dust.

(i) *Vacuum-Processed Finish.*—The procedure for vacuum processing is described in chapter VIII.

120. Removing Stains from Formed Surfaces.—(a) *General.*—Formed surfaces sometimes become unsightly during construction operations because of accumulations of foreign materials, paint and oil drippings, rust stains, and drainage from concrete work at higher levels. These accumulations are required to be removed from F3 surfaces. Washing of surfaces below forms during and after concrete placing will reduce or eliminate much of the streaking that usually results from form leakage.

The first step in removal of stains, whether caused by construction activities or through exposure of the concrete during service, is to determine the source of the stain and then select the proper method for removal.

Common mechanical methods of removing some stains are sandblasting, grinding, steam cleaning, brushing, and light blowtorch application. Steel brushes, when used by themselves, wear at times in a manner that leaves iron deposits which can eventually rust and may later stain the concrete.

Chemical cleaning is more involved and requires application of specific chemicals. The action takes place by either dissolving the staining substance, which can then be blotted or driven deeper into the concrete surface, or by bleaching the discoloring agent chemically into a product having a color that blends with that of concrete.

With either mechanical or chemical methods, care should be taken to protect surrounding areas of materials other than concrete, such as glass and wood, from the effects of any treatment.

Chemicals may be brushed on or applied as a poultice. A poultice is a paste containing a solvent or reagent and a powdery inert absorbent material. Cotton batting or layers of white cloth may be soaked in chemicals and applied to the stain. The inert material may be diatomaceous earth, calcium chloride, lime, or talc. The selection of solvent depends on the stain to be removed. Enough of the solvent is added to inert material to make a smooth paste which is spread about one-half of an inch thick over the area with a trowel or spatula. The stain migrates with the evaporating solvent to the poultice surface. When the poultice is completely dry, the stain is contained in the powder and can be brushed or blown off. The poultice method also reduces the chances that the stain will spread, and it has the advantage of penetrating and extracting the stain from the concrete pores.

Many chemicals can be applied to concrete without appreciable damage to the surface, but strong acids or chemicals having a highly active reaction should be avoided; even weak acids such as oxalic acid may etch the surface if left for any length of time. It is advisable to saturate the surface with water before application of an acid so that the acid will not be absorbed too deeply into the concrete pores. A 10-percent solution of muriatic acid is often used to remove traces of staining. However, it may leave a yellow stain on white concrete. Any acid used should be completely flushed from the surface with water.

Most chemicals are toxic and hazardous and require safety precautions in their use. Skin contact or inhalation should be avoided. Rubber or plastic gloves as well as safety goggles and protective clothing should be worn. Adequate ventilation should be provided when the chemicals are used indoors, and manufacturer's directions for proprietary materials should be followed. Any unused portions of the acid or toxic materials should be properly disposed.

Some stains can be removed by more than one method. No attempt should be made until one is sure that the method or solvent selected will do the job. Experimentation with different bleaches or solvents is helpful. Such experimentation should be done in an inconspicuous small area since some bleaches or solvents can spread or drive stains deeper into the concrete. With careful experimentation the most effective method and materials can be selected.

(b) *Procedures for Stain Removal.*—Some of the more common staining substances and their treatment* are as follows.

Copper, Bronze, and Aluminum Stains.—Stains caused by copper and bronze are usually green; occasionally, the stains may be brown. To remove them, dry mix one part of ammonium chloride (sal ammoniac) or aluminum chloride with four parts of fine-powdered inert material. Add ammonium hydroxide (household ammonia) to make a smooth paste, apply over the stain, and allow to dry. Repeat if necessary, and finally scrub with clean water.

Aluminum stains appear as a white deposit that can be treated with dilute hydrochloric acid. Saturate the stained surface with water and scrub with a solution of 10-percent hydrochloric acid. Weaker solutions should be used on colored concrete. Rinse thoroughly with clear water to prevent etching of the surface and penetration of the dissolved aluminum salts into the concrete. Should this happen the salts may reappear as efflorescence.

Curing Compounds.—Generally, curing compounds will be worn off in a relatively short time by abrasion during normal use or by natural weathering. However, if an accelerated treatment is required

* Portland Cement Association Information Letter No. IS-142.03T.

or if the stained surface is not subject to abrasion, a removal treatment can be used.

Curing compounds have different chemical formulations. They may have a synthetic resin base, a wax base, a combination wax-resin base, a sodium-silicate base, or a chlorinated-rubber base. The base of the curing compound should be identified before an attempt is made to remove it.

Sodium-silicate-based curing compounds can be removed by vigorous brushing with clear water and a scouring powder. Wax resin or chlorinated-rubber curing compounds can be removed by applying a poultice impregnated with solvent of the chlorinated hydrocarbon type, such as trichloroethylene, or a solvent of the aromatic hydrocarbon type, such as toluene. A mixture of 10 parts methyl acetone, 25 parts benzene, 18 parts denatured alcohol, and 8 parts ethylene dichloride can also be used. Allow the poultice to stand for 30 to 50 minutes. Scrub the surface with clear water and a detergent at the end of the treatment. Old stains can be best removed by mechanical abrasive methods such as light grinding or sandblasting.

Fire, Smoke, and Wood Tar Stains.—Apply a trichloroethylene poultice. Scrape off when dry and repeat as necessary. Scrub thoroughly with clear water. As an alternative treatment for large areas, scour the surface with powdered pumice or grit scrubbing powder to remove surface deposit and wash with clear water. Follow this with application of a poultice impregnated with commercial sodium hypochlorite or potassium hypochlorite or any other effective bleach.

Grease Stains.—Scrape off all excess grease from the surface and scrub with scouring powder, trisodium phosphate, or detergent. If staining persists, methods involving solvents are required. Use benzene, refined naphtha solvent, or a chlorinated hydrocarbon solvent such as trichloroethylene to make a stiff poultice. Apply to stain; do not remove until paste is thoroughly dry. Repeat if necessary; this can be followed by scrubbing with strong soap and water, scouring powder, trisodium phosphate, or proprietary detergents specially formulated for use on concrete. Rinse with clear water.

Iron Rust Stains.—These are common and easily recognizable by their rust color or proximity to steel or iron in or out of concrete. Sometimes large areas are stained from use of curing water that contains iron. Rust stains resulting from water-curing operations can be minimized by using galvanized or aluminum pipe or soil soaker canvas hose as shown in figure 172. The appearance can be improved by mopping with a solution of 1 pound of oxalic acid powder per gallon of water. The action can be accelerated by adding one-half pound of ammonium bifluoride to the solution. (Caution: Ammonium bifluoride must be handled with great care as it is very toxic to the skin, eyes, and mucous membranes.) After 2 or 3 hours, rinse with clear water and scrub with stiff brooms or brushes. Remaining spots may require a second application.

For deep stains, saturate a bandage with a solution of one part sodium citrate in six parts lukewarm water and apply it over the stain for a half hour. The solution also can be brushed on the stain at 5- to 10-minute intervals. Following this treatment, where the stain is on a horizontal surface, sprinkle it with a thin layer of sodium hydrosulfite crystals, moisten with a few drops of water, and cover with a poultice made of powdered inert material and water. On a vertical surface, place the poultice on a trowel, sprinkle on a layer of sodium hydrosulfite crystals, moisten lightly, and apply to the stain in such manner that the crystals are in direct contact with the stained surface. Remove poultice after 1 hour. If the stain has not completely disappeared, repeat the operation with fresh materials. When the stain disappears, scrub the surface thoroughly with water and make another application of the sodium citrate solution as in the preliminary operation. The purpose of this last treatment is to prevent reappearance of the stain.

Occasionally, brown iron stains may change to black when they are treated with sodium hydrosulfite. This may also happen if the poultice is left on longer than 1 hour. If this occurs, the black stain should be treated with hydrogen peroxide until oxidized to the brown color. The sodium hydrosulfite treatment should then be resumed. (Caution: Unless adequate ventilation is provided, this method should not be used indoors as a considerable amount of toxic sulfur

Figure 172.—Water curing with soil-soaker hose. This prevents rust stains which may occur if iron pipe is used. PX–D–34609.

dioxide gas will be emitted when the sodium hydrosulfite comes in contact with moisture.)

Iron stains can be effectively removed by mechanical means such as stiff brushing or sandblasting if the stain is not too deep. Sandblasting is the more effective method. Water jet and soap powder brushing may be more applicable than chemical methods for removing light iron stains. The disadvantage of mechanical methods is that both result in the roughening of the surface.

In some instances, unsightly rust stains are potentially too extensive to be removed. These can be camouflaged by use of colored pigment in the concrete subject to staining. The treatment is used where a rapidly oxidizing and eventually stabilizing steel is used for maintenance reduction and achieving architectural effects in buildings and bridges. Initially, as oxidation occurs, copious amounts of iron rust form, which are washed from the surface and deposited as stain on the concrete. The oxidation rate, which depends largely on the presence of moisture, gradually decreases and finally stops altogether, leaving a pleasing, maintenance-free color on the steel and camouflaged stain on the concrete. In the relatively dry Western States of the United States the rate is slow, and considerable time, perhaps years, may be required to reach a stabilized condition.

Lubricating Oil or Petroleum.—Excess fresh oil should be removed immediately with paper towels or cloth. Avoid wiping. Cover the spot with a dry, powdered, absorbent, inert material (the same used in the poultice) or portland cement, and leave it for 1 day. Repeat until no more oil is absorbed by the powder. If stain persists, or when oil has been allowed to remain for some time and has penetrated the concrete, other methods will be required.

After all oil has been removed from the surface by scrubbing with strong soap, trisodium phosphate, scouring powder, or detergents, a solution of one part trisodium phosphate in six parts water can be applied to the stain. The paste should remain 20 to 24 hours. Remove the paste and scrub surface with clear water. A poultice of 5-percent solution of sodium hydroxide applied for about 20 to 24 hours might be tried. After this period, remove and scrub surface with clear water.

Make a poultice with benzene, apply to stain and allow to remain for 1 hour after the solvent has evaporated. Repeat as necessary, then scrub with clear water, cover stain with ¼-inch-thick layer of asbestos fibers, and saturate with amyl acetate. Apply heat to the slab to draw out dissolved oil.

Paint Stains.—Wet and dried paint films each require different treatment. Wet paint should be carefully soaked up with an absorbent material. Avoid wiping. Immediately scrub the stained area with scouring powder and water. Scrubbing and washing should be continued until no additional improvement is noted. Paint removers or solvents should not be used on wet paint or films less than 3 days old.

Dried paint should be scraped off as much as possible. Then apply a poultice impregnated with a commercial paint remover. Allow to stand for 20 to 30 minutes. Scrub stain gently to loosen the paint film, wash off with water, and any remaining can be scrubbed off with scouring powder. Color that has penetrated can be washed out with dilute hydrochloric or phosphoric acid. This treatment can be applied to dried enamel, lacquer, or oil-based varnish.

Other efficient paint removers are: (1) A mixture of 10 parts methyl acetone, 25 parts benzene, 18 parts denatured alcohol, and 8 parts ethylene dichloride; (2) a solution of 2½ pounds of sodium hydroxide in 1 gallon of hot water; the sodium hydroxide solution can be applied with a poultice or can be brushed onto the surface.

Sandblasting or burning with a blowtorch can be used to remove dried paint films.

Wood Stains.—A wood stain is readily distinguishable by its dark color. The best treatment is that recommended for fire stains.

Miscellaneous Stains.—Stains varying in light intensity from yellow to brown occasionally occur on interior concrete and terrazzo floors. These may have been caused by original finishing and cleaning operations. It is possible to bring the surface back to its original appearance by applying poultices impregnated with an aqueous solution of sodium hypochlorite ($NaOCl$) or by scrubbing the surface with the same solution.

121. Finishing Unformed Surfaces.—Concrete having unformed, exposed surfaces should contain just sufficient mortar to avoid excessive floating. If the mix is wet and oversanded, an excess of water and fine material will be brought to the surface, resulting in a layer of inferior mortar having high water-cement ratio and a tendency to dust, craze, crack, and possibly separate from the mass beneath. Working of the surface in various finishing operations should be the minimum necessary to produce the desired finish. Use of any finishing tool in areas where water has accumulated should be prohibited. Operations on such areas should be delayed until the water has been absorbed, has evaporated, or has been removed by draining, mopping, dragging off with a loop of hose, or other means.

Bureau specifications require unformed surfaces that will be exposed to the weather and that would normally be horizontal to be sloped for drainage. Narrow surfaces such as tops of walls and curbs are usually sloped three-eighths of an inch per foot of width; broader surfaces such as walks, roadways, platforms, and decks are sloped approximately one-quarter of an inch per foot. The classes of finish specified for unformed concrete surfaces are designated as U1, U2, U3, and U4. Surface irregularities allowed in each are shown in table 24. A 10-foot straightedge or template is used for detecting irregularities.

(a) *Finish U1.*—This is a screeded finish used on surfaces that will be covered by fill material or concrete and on surfaces of operating platforms on canal structures. It is also the first stage for finishes U2 and U3. The finishing operations consist of leveling and screeding the concrete to produce an even uniform surface. Surplus concrete should be removed immediately after consolidation by striking it off with a sawing motion of the straightedge or template across wood or metal strips that have been set as guides. Where the surface is curved, as in the invert lining of tunnels, a special screed is used. For long, narrow stretches of invert paving or of flat paving, use of a heavy slip form or of a paving and finishing machine is desirable. The slip form is best for sharply curved inverts; the paving and finishing machine is preferable for flat or long-radius cross sections.

(b) *Finish U2.*—This is a floated finish used on all outdoor unformed surfaces unless other finish is specified. It is used on such surfaces as inverts of siphons and flumes; floors of canal structures, spillways, outlet works, and stilling basins; outside decks of power and pumping plants; floors of service tunnels, galleries, sumps, culverts, and temporary diversion conduits; tops of transmission line and bridge piers and of walls, except tops of parapet walls prominently exposed to view; and surfaces of gutters, sidewalks, and outside entrance slabs. It is also applied to bridge floors and to slabs that will be covered with built-up roofing or membrane waterproofing. Floating may be done by hand or power-driven equipment. It should not be started until some stiffening has taken place and the moisture film or shine has disappeared. The floating should work the concrete no more than necessary to produce a surface that is uniform in texture and free of screed marks. If finish U3 is to be applied, the floating should leave a small amount of mortar without excess water at the surface to permit effective troweling. Any necessary cutting or filling should be done during the floating operations. Joints and edges should be finished with edging tools. Tooled edges are often preferable to formed chamfers.

(c) *Finish U3.*—This is a troweled finish used on inside floor slabs of buildings (except those to receive a bonded concrete or terrazzo finish), on tops of parapet walls prominently exposed to view, on concrete surfaces subject to high-velocity flows, and on interior stair treads and thresholds. Steel troweling should not be started until after the moisture film and shine have disappeared from the floated surface and after the concrete has hardened enough to prevent an excess of fine material and water from being worked to the surface. Excessive troweling tends to produce crazing and lack of durability. Too long a delay in troweling results in a surface too hard for proper finishing. Steel troweling should be performed with a firm pressure that will smooth the sandy surface left by the floating. Troweling should produce a dense, uniform surface free of blemishes, ripples, and trowel marks. Light troweling and slight trowel marks are usually permitted on surfaces to be covered with built-up roofing and membrane waterproofing.

A fine-textured surface that is not slick can be obtained by applying a sweat or light scroll finish immediately after the first regular troweling. This consists of troweling lightly over the surface with a circular motion, keeping the trowel flat on the surface.

Where a "hard, steel-troweled finish" is required as a special finish that will afford added resistance to wear, the regular U3 finish is again troweled after the surface has nearly hardened, using firm pressure and troweling until the surface is hard and has a somewhat glossy appearance.

(d) *Finish U4.*—This finish is specified for canal and lateral linings. The finished surface should be equivalent in evenness, smoothness, and freedom from rock pockets and surface voids to that obtainable by effective use of a long-handled steel trowel. Light surface pitting and light trowel marks are not objectionable. Where the surface produced by a lining machine meets the specified requirements, no further finishing is necessary. If a few rough spots are left by the lining machine, there is no objection to the immediate use of a little mortar to reduce the labor of producing an acceptable finish.

(e) *Preventing Hair Cracks.*—Hair cracks are usually the result of a concentration of water and fines at the surface caused by overmanipulation during finishing operations. Such cracking is aggravated by untimely finishing and by too rapid drying or cooling. When the humidity is so low as to cause checking of the finished surface before it can be covered without damage, the surface should be moistened and kept moist temporarily with a very fine spray of water applied so as not to wash the surface nor form pools on it. As chilling of the green concrete increases its tendency to crack, it is desirable that water used for the preliminary moistening be no colder (preferably warmer) than the concrete. Checking that develops prior to troweling can usually be closed by pounding the concrete with a hand float.

122. Special Requirements for Concrete Surfaces Subject to High-Velocity Flow.—Special requirements for finishing concrete surfaces subjected to high-velocity flow are needed to prevent severe cavitation and destructive erosion. Also, materials and methods that tend to increase the strength of concrete at the surface, or throughout the mass, increase erosion resistance. Bureau laboratory and field tests on small, abrupt, sharp-cornered offsets into and perpendicular to high-velocity waterflow show that, under atmospheric conditions, cavitation and destructive erosion begin downstream from relatively small offsets when water velocities reach about 40 feet per second. When velocities exceed this value, special limitations on offsets are necessary, and special treatment should be accomplished. More stringent requirements than those normally applied for surfaces subjected to low-velocity flow are also needed for other abrupt-type offsets in high-velocity flow.

The severe limitations and treatment required for surfaces subjected to high-velocity flow may be relaxed to some degree when facilities such as aeration devices, designed and constructed in the flow surfaces to introduce and entrain air in the flowing water, damp or cushion damaging forces. In high-velocity-flow tunnel spillways with vertical curves, aeration devices may be designed into the structure to reduce cavitation potential in critical flow areas. An aeration slot was designed and constructed immediately above the point of curvature of the vertical curve of Yellowtail Dam spillway tunnel. Final design, location, and suitability of this slot were determined from hydraulic laboratory model studies. The efficacy of the slot was verified on the project during a 4-day spill at 15,000 cubic feet per second.

Surfaces immediately downstream from gates in outlet works may be aerated by properly designed and located recesses or offsets away from the flow to alleviate damaging cavitation. The use of such recesses or offsets may permit a relaxation of stringent finishing criteria and restrictions on allowable surface irregularities.

The special finishing and treatment requirements for surfaces subjected to high-velocity flow, as discussed in subsections (a), (b), (c), and (d), are applicable to flow surfaces in which designed aeration slots or recesses have not been included.

(a) *Surfaces of Outlet Works Conduits and Tunnels.*—Problems of cavitation resulting from abrupt-type offsets that are into or facing the flow can be especially serious when they occur in the areas immediately downstream from control gates. Also, abrupt-type offsets that are away from the flow may cause cavitation or erosion problems. High-velocity flow passing through a gate opening causes the boundary layer of the flow to be disrupted and a certain length of continuous surface contact, depending upon the velocity, is required for it to be reestablished. Bureau specifications provide for complete elimination of abrupt-type offsets from flow surfaces of outlet works conduits and tunnels for specific distances downstream from the ends of gate frames, by grinding to specified levels as given in table 25.

Precautionary measures are also taken to reduce abrupt offsets into the flow on those surfaces of outlet works conduits and tunnels beyond the areas immediately downstream from gate frames that will be subjected to high-velocity flow. Usually, the precautionary measures are terminated at the first upstream transverse construction joint of the stilling basin. Abrupt offsets into the flow on surfaces upstream from this point should not exceed one-eighth of an inch; if they exceed this limit, they should be completely eliminated by grinding to bevels according to flow velocity as set forth in table 25.

Abrupt-type offsets perpendicular to and away from high-velocity flow and those parallel to such flow that are beyond the critical areas immediately downstream from the control gates are only required to be within the usual allowable irregularities for the specified finish as provided in table 24.

(b) *Surfaces of Tunnel Spillways.*—Surfaces of tunnel spillways subject to flow velocities of 40 feet per second or more are also critical with respect to the need for elimination or reduction of abrupt offsets. Surfaces of spillway tunnel elbows below the centerline or spring line receiving the special stoned finish as provided in section 119(f) must be free of abrupt-type offsets on completion of the stoned finish. Application of the special stoned finish is considered desirable when surfaces of tunnel elbows are to carry water having flow velocities of 75 feet per second or more. Downstream from the elbow at the lower end of a spillway tunnel, abrupt offsets in the surface should be eliminated for a distance of at least five tunnel diameters to prevent cavitation. Grinding to bevels is set forth in table 25.

For surfaces of the spillway tunnel shaft down to the start of the tunnel elbow and surfaces of the spillway tunnel downstream of the 5-diameter section, special precautions should be taken to prevent abrupt offsets not parallel to the flow; if such offsets occur on these surfaces, they should be eliminated by grinding to bevels according to table 25. Abrupt offsets parallel to the direction of flow should not be greater than one-fourth of

Table 25.—Offset and grinding tolerances for high-velocity flow

Velocity range, feet per second	Distance of treatment downstream from gate frame, feet	Grinding bevel, ratio of height to length
40 to 90..................	15	1 to 20
90 to 120.................	30	1 to 50
Over 120.................	50	1 to 100

an inch; the excess should be eliminated by grinding to required bevels according to table 25.

Gradual irregularities, as defined in section 119, on surfaces of spillway tunnels that will be subject to high-velocity flow should not exceed one-fourth of an inch.

(c) *Surfaces of Open-Flow Spillways.*—Precautionary measures should be taken to prevent cavitation or abrasion resulting from abrupt offsets on those surfaces of open-flow spillways that will be subject to high-velocity flow. Abrupt offsets on such surfaces that are not parallel to the direction of flow and that are offset into the flow should not exceed one-eighth of an inch. When this limit is exceeded, the offset should be completely eliminated by grinding to required bevels according to table 25. When flow velocities will be in the 40- to 90-feet-per-second range (or more), the excess should be completely eliminated by grinding to the required bevels. Abrupt offsets away from the flow or parallel to the flow should be required only to meet the maximum allowable limits for the specified finish as provided in table 24.

(d) *Other Treatments of High-Velocity Flow Surfaces.*—Depending on the critical flow conditions that will be involved, grinding to eliminate or reduce irregularities will provide acceptable flow surfaces. This conclusion assumes, of course, that a well-designed concrete mix is used, the work is within permitted tolerances, and the forming and finishing are expertly carried out. However, for extremely critical flow surfaces such as in tunnel spillways, it may be desirable to limit the grinding depth, particularly when flow velocities will exceed 90 feet per second. Grinding irregularities on flow surfaces will reduce many of the aggregate particles at or near the concrete surface and thereby influence mechanical bond capability. Thus, negative pressures on the concrete surfaces can pull out these reduced-size aggregate particles, leaving holes that may cause cavitation.

For repair work below the centerline of Yellowtail Dam spillway tunnel, including the vertical elbow where velocities were in excess of 90 feet per second, grinding irregularities to a depth greater than one-fourth of an inch was prohibited. Instead, irregularities greater than one-fourth of an inch were required to be excavated to a depth below the finished grade and then repaired to that grade. When the maximum width of the area of the irregularity to be eliminated was not more than 1 foot in the minimum dimension, the irregularity was excavated only enough to permit an acceptable filling to finished grade with epoxy-bonded epoxy mortar. When the width of the area of the irregularity was greater than 1 foot in the narrow dimension, the perimeter of the irregularity was saw cut, then excavated to a minimum depth of 1½ inches, and the excavated area repaired to the finished surface with epoxy-bonded concrete.

123. Painting and Dampproofing of Concrete.—(a) *Painting.*—Paints for concrete and their application are described and discussed in the Bureau's Paint Manual.

(b) *Dampproofing.*—Treatment of concrete surfaces during construction with various bituminous and other waterproofing compounds has been required to prevent permeation by moisture from backfill and other ground sources. Experience has shown, however, that the principal penetration of moisture is through cracks, construction joints, or areas of unconsolidated concrete which no ordinary dampproofing treatment could have prevented. The concrete itself, particularly that in heavy walls and conduits, may be expected to be satisfactorily dry on the interior if it is properly constructed in accordance with existing specifications and instructions. For these reasons, there is a decreasing use of dampproofing treatments on Bureau work and more attention is directed to obtaining better construction joints and contraction joints and to placement of well-consolidated, impermeable concrete.

For maintenance and correction of conditions where there is objectionable leakage through cracks, joints, or porous concrete, elastic membranes of waterproofing material are built up on fabrics or sheets. These span the cracks and joints without danger of rupture when there is change in crack width caused by variations in temperature and moisture. A weatherproofing procedure for prolonging the serviceable life of critically exposed portions of concrete structures is described in section 139. This practice is essentially preventive, rather than a repair.

E. Curing

124. Moist Curing.—The water content of fresh concrete is considerably more than enough for hydration of the cement. However, an appreciable loss of this water, by evaporation or otherwise, after initial set has taken place will delay or prevent complete hydration. The object of curing is to prevent or replenish the loss of necessary moisture during the early, relatively rapid stage of hydration. The usual procedure for accomplishing this is to keep the exposed surface continuously moist by spraying or ponding, or by covering with earth, sand, or burlap maintained in a moist condition. Precast concrete and concrete placed in cold weather are often kept moist by steam released within enclosures. These procedures are known as moist curing. Early drying must be prevented or the concrete will not reach full potential quality. In warm, dry, windy weather corners, edges, and surfaces become dry more readily. If these portions are prevented from drying and fully develop hardness and quality, interior portions will have been adequately cured.

Bureau specifications usually require that concrete, which is to be water cured, shall be kept moist at least 14 days. Concrete made with type V cement has a rate of hardening somewhat slower than that of concrete made with type I or II. For this reason, it is especially important that thorough curing be provided when type V cement is used. Tests indicate that a period of drying after completion of moist curing considerably enhances the resistance of concrete to sulfate attack, probably as a result of carbonation.

Moist curing in which the concrete is protected from the sun is less likely to be interrupted by periods of drying and therefore is likely to be more effective than spraying exposed concrete surfaces. Wet burlap in contact with the concrete is excellent for this purpose; it not only shades the concrete, but also holds the moisture needed for good moist curing. Wood forms left in place provide good protection from the sun, but will not keep the concrete sufficiently moist to be acceptable as a method of moist curing for outdoor concrete. However, the surfaces of ceilings and inside walls require no curing other than that resulting from forms being left in place for at least 4 days. The unformed top surfaces of formed concrete, such as tops of walls, piers, and beams, should be moistened by wet burlap or other effective means as soon as the concrete has hardened sufficiently to prevent damage. These surfaces and steeply sloping or vertical formed surfaces should be kept continuously moist, prior to and during form removal, by applying water and allowing it to run down between the forms and the concrete. Soil-soaker hose is particularly suited to such work (see fig. 172). There is no better curing and protection than that provided by well-moistened backfill. Ponding on floors, pavement, and other slabs is effective in reducing crazing, cracking, and wear. Drainage of curing water from the upper surfaces of a structure such as a power or pumping plant, sometimes augmented by drainage from construction activities at higher levels, frequently results in unsightly surfaces, the appearance of which can only be improved by costly cleanup operations.

125. Curing With Sealing Membranes.—Sealing membranes can often be employed to advantage in curing concrete surfaces. The membrane may be an impermeable plastic sheet placed over the surface or a film formed by application of liquid materials (curing compounds) to the surfaces. Acceptable membranes retard evaporation of mixing water so that, under most conditions, sufficient moisture is retained for proper hydration of cement. Laboratory test and field observations indicate that an effective membrane kept intact for 28 days provides the equivalent of 14 days' continuous moist curing.

Membrane curing, especially with curing compounds, is widely practiced because it affords several advantages over water curing. It eliminates the need for supplying water continuously for long periods with the ever-present possibility of interruption or incomplete coverage. The presence of water often hampers other construction activities in the area. Also, the water may stain or disfigure concrete surfaces and require expensive cleaning. Curing compounds tend to prevent deep penetration of stains and even retard somewhat the deposition of stains by rendering the surface impermeable. A coated surface of resin-base compound is paintable after only minor cleanup. On the other hand, wax and clear resin-base compounds must be removed if concrete is to be bonded to the surfaces (construction joints), and the wax-base compound must be sandblasted from the surface if paints are to be applied.

The curing compounds normally used in present Bureau practice include the pigmented wax-base materials (white or gray) and the recently developed essentially clear, resin-base material. These compounds consist of finely ground pigments dispersed in a vehicle of resin, waxes, oils, and/or plasticizers together with solvents. Specifications for these compounds are: for the wax-base compound, "Specifications for Wax-Base Curing Compound," dated May 1, 1973; and for the resin-base compound, "Specifications for Clear Resin-Base Curing Compound, CRC–101," dated May 1, 1973.

The white-pigmented, wax-based compound is generally used for curing canal linings and related structures, interior surfaces such as tunnel and siphon linings, and is sometimes permitted as an alternative to moist curing on diversion dams and other structures. The white compound reflects a considerable amount of sunlight. In hot weather, the decrease in concrete temperature caused by the reflective white compound may be as much as 40° F, and the lower, more uniform temperature minimizes surface cracking caused by thermal expansion and contraction. Tests have shown that use of white-pigmented compounds approximates the effect of shading in maintaining lower concrete temperatures.

Since the white-pigmented compound usually develops a mottled appearance from weathering, it is objectionable where appearance of the cured surface is an important consideration. Therefore, either the gray compound or the clear material is generally used to cure surfaces prominently exposed to public view. The gray compound more nearly resembles concrete in color and, although still imparting a coated or painted appearance on the surface, presents a more pleasing appearance during latter stages of weathering than the white-pigmented type. The clear compound, being essentially transparent, does not conceal the concrete and thus preserves normal appearance. However, since the resin-base compound is costlier than the wax-base materials, both from materials and application standpoints, use is generally restricted to curing exposed surfaces of such structures as power and pumping plants where appearance is considered extremely important. The reflectance of surfaces coated with either of these materials is considered acceptable.

The wax-base compounds can also be used to cure surfaces to which bond of adjacent or subsequent concrete placements is not desired. The paraffin vehicle effectively prevents bond; thus, these compounds are used on contraction joint surfaces in structures of all kinds.

Curing compounds are furnished ready-mixed under Bureau specifications and normally require only thorough mixing prior to use or sampling. (See designation 38 in the appendix.) Diluting by adding thinner or otherwise altering the composition of the compound is not allowed. In cold weather, heating the compound to a maximum temperature of 100° F is permissible to obtain a sprayable viscosity. Heating, if required, should be by use of a hot water bath or heaters specifically designed for this purpose and never over an open flame. The container should be vented and only about three-fourths full to allow for expansion.

Curing compounds are normally applied by spraying. Equipment for spraying wax-base compounds should be of the pressure-tank type with provision for continuous agitation. This equipment may be highly mechanized as in the curing of canal linings where the size of the job justifies a curing jumbo with multiple traveling spray fans synchronized with the movement of the rig. The compounds are more viscous than water but thinner than most paints. The equipment, whether mechanized or portable, must use sufficient pressure and correct nozzles to atomize the material properly. Orchard-type sprayers usually do not meet these requirements and are not adequate.

Standard paint spray equipment of either the conventional pressure-pot or airless types can be used to apply wax-base curing compounds. These types of equipment are considered essential for the clear resin-base compound which is of thin paint consistency and must be applied as a film having extremely uniform thickness. Such equipment includes an agitator, pressure regulators and gages, an oil and water separator, and the correct spray gun tips and nozzles for precise control of the application.

The spraying techniques employed should produce a uniformly thick film. Such uniformity can be observed visually when pigmented compounds are being applied and corrections made as needed. However, the clear compound affords no such visual inspection and requires using the paint technique of multipass application with cross spraying to achieve uniformity. Multipass application consists of two or more spray passes over each point on the surface. Cross spraying is accomplished by first applying about one-half of the compound required for a specific area with parallel passes. After a short interval, the remainder of the compound is then applied at right angles to the first spraying. The clear resin-base compound, being quite thin, will run on vertical surfaces or puddle on horizontal surfaces if applied in one pass. However, it contains a fast-drying solvent so that the cross-spray passes can normally be applied without delay. The delay between passes, if any, should not exceed 30 minutes.

The coverage necessary to ensure effective curing varies considerably with the compound and conditions. A smoothly troweled surface requires less compound than a rough surface. A gallon of the resin compound covers more surface than a gallon of wax-base material. Therefore, it is impracticable to specify coverage rates to apply to all variations in surface conditions. However, the Bureau has established as guides normal maximum coverage requirements of 150 square feet per gallon for the wax-base materials and 200 square feet per gallon for the resin-base compounds. These guides are applicable for reasonably smooth concrete surfaces such as smooth-formed or troweled concrete; rougher surfaces, edges, corners, and other irregularities will require additional compound to obtain the necessary membrane continuity.

Inspection of curing compound applications should confirm that sufficient material has been applied, acceptable uniformity has been attained, and a continuous curing membrane has been formed. Applying the pigmented wax-base materials to a concrete sample at exactly the specified coverage will provide the inspector with a visual guide as to the appearance of that particular compound at its minimum thickness (maximum coverage). This appearance may vary from brand to brand, and even lot to lot within brands, as well as between colors. Coverage of subsequent applications, making allowance for variations in surface roughness, can be judged against this sample. Nonuniformity will be indicated by blotchiness, and the spray technique should be refined until uniformity is attained. The compound on each type of surface should be examined closely to confirm that a continuous membrane has been applied and whether more compound should be added.

Inspection of resin-base compound application is much more difficult since the nearly unpigmented material only darkens the surface slightly. As the clear compound is normally used where appearance is important, extreme care must be exercised in obtaining uniform thickness. Excessive thickness is objectionable as it yields an undesirable gloss. Thus, an inspector should watch closely for correct execution of the multipass, cross spraying technique. Correct coverage should be confirmed by noting the quantity of material applied to a given area.

Proper care of a concrete surface prior to compound application is highly important. Beginning promptly after form removal, formed surfaces should be saturated with a fine spray of water until they will absorb no more water. Thereafter, the surfaces should be moistened frequently to maintain continuous curing through the interval after form loosening and stripping and before the compound is applied. The compound is then applied as soon as the free moisture on the surface has disappeared. On unformed surfaces, the compound should be applied immediately after the bleeding water or shine disappears, leaving a dull appearance. If concrete is allowed to dry before compound application, not only is water unavailable for curing, but the compound will "strike in," resulting in a soft surface layer having poor abrasion resistance. Normally, only small imperfections (those which can be repaired without delaying application of the compound) are repaired prior to compound application. Defective concrete and gross surface imperfections are not repaired until after the compound has been applied.

Under rare circumstances, it may be necessary to augment compound curing with preliminary moist curing. For instance, on one canal lining project, 24 hours of moist curing prior to application of the compound largely eliminated the checking on side slopes experienced previously. If the adequacy of the curing membrane appears to be questionable, preliminary moist curing can be employed as a precautionary measure.

The construction of aesthetically pleasing structures dictates that most exposed concrete surfaces be reasonably free of rust stains, mortar drips, water-deposited scales, and other disfiguring marks. This requires exercising increasing care to prevent traffic, concrete placement, and other construction activities from damaging the appearance. Prevention likely will prove more economical than the cleaning required to restore appearance to an acceptable state.

The continuity of curing membranes must be maintained at least 28 days to be effective. Whenever the film will be subject to damage by traffic or other causes, it should be protected by a layer of sand or earth not less than 1 inch thick placed after the compound has dried 24 hours or by other suitable and effective means. Any damage to the coating occurring during the 28-day curing period should be repaired promptly by application of additional compound.

The Bureau has compiled a list of manufacturers who have demonstrated that they consistently produce good-quality wax-base compounds conforming to specifications. Compounds of manufacturers so listed do not require sampling at the shipping point and may be used on receipt of a compliance certification. However, if such compounds appear to be unusually thin and hiding of surface is inadequate, a sample should be sent to the Denver laboratories for testing. Compounds by manufacturers not on the list should be sampled at the shipping point and tested prior to use. As the clear compound is relatively new in Bureau construction, sampling and testing are always required.

Sheet plastic is particularly adapted to curing slabs and structural shapes. As soon as the concrete has hardened sufficiently to prevent damage, the surfaces are sprayed lightly with water and then completely covered with a white, 4- to 6-mil, plastic sheet. The sheet should be airtight, nonstaining, and vaporproof to effectively prevent loss of moisture by evaporation. Care must be exercised in obtaining an airtight membrane by lapping and sealing all edges of the sheets. Specifications normally require that this type of membrane be maintained a minimum of 14 days.

126. Steam Curing.—Use of steam curing is particularly advantageous under certain conditions, chiefly because of the higher curing temperature and the fact that moisture conditions are favorable. This type of curing is permitted by the Bureau in the manufacture of precast pipe and other precast units. Benefits are also realized in the use of live steam for cold-weather protection of concrete. Steam-cured precast units attain strength so rapidly that forms may be removed and reused very soon after concrete placing.

Data on early strength development of concrete cured with steam at temperatures between 100° and 200° F are presented in figure 173. Greatest acceleration in strength gain and minimum loss in ultimate strength are obtained at temperatures between 130° and 165° F. Higher temperatures produce greater strengths at very early ages, but there are severe losses in strength at ages greater than 2 days. Precast concrete pipe is usually cured at temperatures ranging from 100° to 150° F. Under such conditions, the loss in ultimate strength is relatively small. Use of steam curing in winter to maintain the required initial concrete temperature of 50° F rarely involves an ambient temperature around the concrete over 100° F.

A delay of 2 to 6 hours prior to steam curing will result in higher strength at 24 hours than would be obtained if steam curing were commenced immediately after filling the forms, as was the case in the tests from which the data plotted in figure 173 were derived. If the temperature is between 100° and 165° F, a delay of 2 to 4 hours will give good results; for higher temperatures, the delay should be greater.

It is desirable that the insides and outsides of pipe sections (and all sides of other concrete sections) be simultaneously exposed to the steam curing, especially in cold weather, to avoid stress-producing temperature differences.

The necessary duration of steam curing depends on the mix, the temperature, and the desired results. Pipe is commonly stripped at 12 hours, tipped off the base rings at 36 hours, and considered fully cured at 72 hours. Most pipe mixes are considerably stronger than the mix from which the data in figure 173 were obtained.

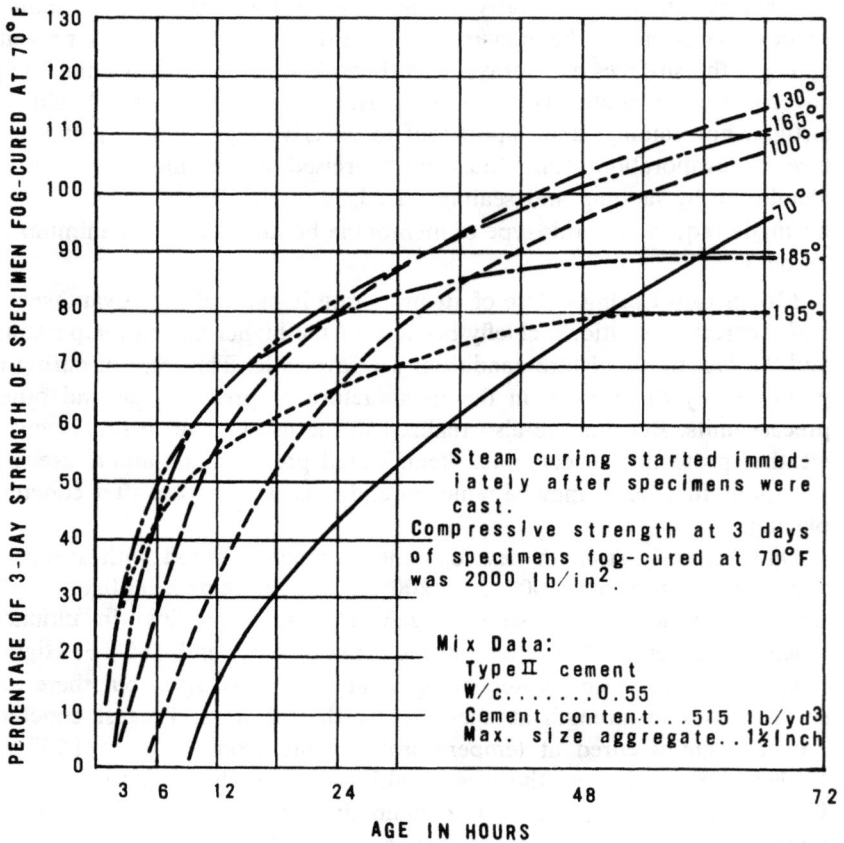

Figure 173.—Effect of steam curing at temperatures below 200° F on the compressive strength of concrete at early ages. 288–D–2659.

F. Concreting Under Severe Weather Conditions

127. Precautions to be Observed During Hot Weather.—Among the various means discussed in section 92 for lowering the temperature of concrete as mixed, there are two which also help in holding the temperature within the specified limit after the concrete leaves the mixer; namely, working at night and shading the operations. Long, exposed pipelines for pumped concrete require special protection during hot weather by preferably being covered with wet burlap or coated with white paint or whitewash. Part of the objection to placing concrete in excessively hot weather is the necessity for richer mixes occasioned by the wetter consistencies required to offset excessive slump loss. The difficulty of securing continuous moist curing for the required period is also increased. In some Southwestern areas the climate is so hot and arid that some specifications prohibit concreting during the summer months.

Although attention to curing requirements is important at all times, it is especially so in hot, dry weather because of the greater danger of crazing and cracking. Higher temperatures with low humidity cause sprinkled surfaces to dry faster and to require more frequent sprinkling; hence, the use of wet burlap and other means of retaining the moisture for longer periods becomes increasingly desirable. Efficiency of curing compounds is reduced in hot weather, and at such times it is particularly important that precautions outlined in section 125 be carefully observed.

128. Precautions to be Observed During Cold Weather.—Figure 10 shows how much more slowly concrete gains strength at low than at average temperatures. For this reason, Bureau specifications require that concrete be protected against freezing temperatures for at least 48 hours after being placed when the mean daily temperature is 40° F or above. They also provide that when the mean daily temperature is below 40° F concrete should have a temperature of not less than 50° F and should be maintained at not less than 50° F for at least 72 hours. Also, according to the specifications, concrete placed in such weather should, to accelerate the set, contain 1 percent calcium chloride by weight of the cement, unless the concrete will be subject to sulfate attack or unless there are other considerations which preclude its use as discussed in section 20(a); these are if galvanized metalwork is embedded or if the concrete will be in contact with prestressed steel. If the concrete is subject to sulfate attack, calcium chloride, because it reduces the resistance of concrete to sulfate attack, should not be used but the maximum water-cement ratio should be reduced to 0.45 ± 0.02.

For concrete placed in moderate climates where freezing and thawing occur less frequently and low temperatures are an exception, a maximum water-cement ratio of 0.60 is usually specified for structures to be covered with fill material, continually submerged, or otherwise protected. In view of this, the maximum water-cement ratio could be reduced to 0.53 instead of 0.45 to accelerate the strength development when calcium chloride is not permitted.

One percent of calcium chloride is required in concrete placed when the mean daily temperature at the worksite is lower than 40° F. When the concrete is cured by membrane curing, no additional protection against freezing is required if the protection at 50° F for 72 hours is obtained by means of adequate insulation in contact with forms or concrete surfaces. If membrane-cured concrete is not protected by insulation, the concrete should be protected against freezing temperatures for an additional 72 hours immediately following the 72 hours of protection at

50° F. Water-cured concrete must be protected against freezing temperatures for 3 days immediately following the 72 hours protection at 50° F. Bureau requirements permit a reduction of water curing to 6 days during periods when the mean daily temperature at the worksite is less than 40° F. However, when temperatures are such that concrete surfaces may freeze, water curing should be temporarily discontinued.

These requirements safeguard the concrete against serious damage from freezing during the early critical period and maintain conditions of temperature and moisture under which hydration can proceed without interruption and the strength can develop to a reasonably satisfactory degree during the period of protection. It is well to be aware that moist surfaces can freeze when the dry-bulb temperature is well above 32° F if conditions are such that the wet-bulb temperature drops to 32° F. At 37° F strong winds at Davis Dam caused ice to form from curing-water spray. The inspector should assure that adequate steps are taken to protect the concrete and that facilities for protection are available when needed.

Before concreting is started, all ice, snow, and frost should be removed from the interior of forms, reinforcement steel, and parts to be embedded. This is best accomplished with steam under canvas covers. Concrete should never be placed on a frozen subgrade. Concrete in contact with frozen subgrade will freeze or its temperature will be considerably below the specified 50° F for the first 72 hours; also, subsequent thawing may cause settlement. Subgrade may be protected from freezing by covering it with straw and tarpaulins or other insulating blankets. At some concrete placements, a short period of covering with an insulating blanket has permitted heat from the earth to eliminate the frost.

Thin reinforced concrete members require much more protection than do massive structures such as piers, abutments, or dam sections. Corners and edges are most vulnerable; methods that will protect these parts will be adequate for other portions of the structure. The approaching need for protection in the fall and the sufficiency of the facilities provided should be determined by taking the temperature of concrete in exposed corners and edges during the coldest periods. Where the surface area is large in relation to the volume, it is important that the forms, reinforcement steel, and embedded parts have a temperature above freezing; otherwise the heat of the concrete may be absorbed by them and the mean temperature fall below freezing. This is particularly true if the forms are metal. Considerable heat may be radiated by reinforcement bars that extend outside the concrete. In massive structures the initial heat within the concrete is not so readily dissipated and is augmented by heat generated by hydration of the cement. However, immediate surface protection is as necessary for massive structures as for others. Less protection is required later.

If the surface is chilled rapidly when forms are removed or when protection against low temperatures is discontinued, cracking and subsequent deterioration may result. For this reason, Bureau specifications require that discontinuance of protection against freezing shall be such that the drop in temperature of any portion of the concrete will be gradual. The surface temperature of mass concrete should not drop faster than 20° F each 24 hours because the probability of cracking from sudden chilling at the surface is great because of the differential between interior and exterior temperatures. Heat of hydration escapes from mass concrete slowly and may raise the initial temperature by 30° F at the surface and as much as 70° F in the interior. In thin sections, on the other hand, surface temperatures may drop gradually as much as 40° F in each 24 hours without damage because the heat of hydration is rapidly dissipated and the probability of excessive differential between the interior and exterior temperatures is not great.

Protection required in cold weather is only that necessary to keep the temperature from falling below specified temperatures during certain initial periods. The most common method of protection is to enclose the structure with an atmosphere warm enough to maintain required temperatures. Another method being used increasingly is insulation. It eliminates the necessity of heating, with its attendant costs and fire hazards. When effectively conserved by insulation, the heat of hydration of hardening concrete is sufficient to maintain early temperatures in most concrete work.

Use of insulation is not new for maintaining the required early temperatures in cold weather. Paving and other nearly horizontal unformed concrete have often been adequately protected by layers of straw, shavings, or dry earth. Wooden forms have afforded considerable protection in less severe exposures. Forms built for repeated use can be insulated, with overall economy, and protection becomes automatic (see fig. 174). As corners and edges are most vulnerable to heat loss, considerable extra insulation is required over them. A double thickness is usually enough. At low temperatures, curing is less urgent. Sealing compound may be applied after insulation or forms are removed.

Tables 26 and 27 show the amount of insulation necessary to maintain various kinds of concrete work at specified temperatures during cold weather of various degrees of severity. The tables are calculated for the stated thickness of blanket-type insulation with an assumed conductivity of 0.25 Btu per hour per square foot for a thermal gradient of 1° F per inch. The values given are for still air conditions and will not be realized where air infiltration caused by wind occurs. Close-packed straw under

Table 26.—Insulation requirements for concrete walls

Concrete placed at 50° F

Wall thickness, feet	Minimum air temperature allowable for these thicknesses of commercial blanket or bat insulation, degrees F			
	0.5 inch	1.0 inch	1.5 inches	2.0 inches
Cement content—300 pounds per cubic yard				
0.5	47	41	33	28
1.0	41	29	17	5
1.5	35	19	0	−17
2.0	34	14	−9	−29
3.0	31	8	−15	−35
4.0	30	6	−18	−39
5.0	30	5	−21	−43
Cement content—400 pounds per cubic yard				
0.5	46	38	28	21
1.0	38	22	6	−11
1.5	31	8	−16	−39
2.0	28	2	−26	−53
3.0	25	−6	−36	....
4.0	23	−8	−41	....
5.0	23	−10	−45	....
Cement content—500 pounds per cubic yard				
0.5	45	35	22	14
1.0	35	15	−5	−26
1.5	27	−3	−33	−65
2.0	23	−10	−50	....
3.0	18	−20	....	....
4.0	17	−23	....	....
5.0	16	−25	....	....
Cement content—600 pounds per cubic yard				
0.5	44	32	16	6
1.0	32	8	−16	−41
1.5	21	−14	−50	−89
2.0	18	−22	....	....
3.0	12	−34	....	....
4.0	11	−38	....	....
5.0	10	−40	....	....

Table 27.—Insulation requirements for concrete slabs and canal linings placed on the ground

Concrete at 50° F placed on ground at 35° F; no ground temperature gradient assumed

Slab thickness, feet	Minimum air temperature allowable for these thicknesses of commercial blanket or bat insulation, degrees F			
	0.5 inch	1.0 inch	1.5 inches	2.0 inches
Cement content—300 pounds per cubic yard				
0.333	(¹)	(¹)	(¹)	(¹)
0.667	(¹)	(¹)	(¹)	(¹)
1.0	47	42	35	29
1.5	37	19	−1	−21
2.0	26	−5	−37	−70
2.5	16	−27	−72	....
3.0	6	−51	....	....
Cement content—400 pounds per cubic yard				
0.333	(¹)	(¹)	(¹)	(¹)
0.667	50	49	47	46
1.0	42	30	17	5
1.5	29	1	−27	−56
2.0	16	−28	−72	−117
2.5	3	−58	....	....
3.0	−10	−86	....	....
Cement content—500 pounds per cubic yard				
0.333	(¹)	(¹)	(¹)	(¹)
0.667	47	42	35	30
1.0	37	19	0	−19
1.5	21	−16	−54	−92
2.0	5	−51	....	....
2.5	−13	....	....	....
3.0	−26	....	....	....
Cement content—600 pounds per cubic yard				
0.333	(¹)	(¹)	(¹)	(¹)
0.667	43	34	24	14
1.0	31	7	−18	−42
1.5	13	−33	−80	−127
2.0	−5	−74	....	....
2.5	−22	....	....	....
3.0	−42	....	....	....

¹ See footnote at end of table.

Table 27.—Insulation requirements for concrete slabs and canal linings placed on the ground—Continued

Concrete at 50° F placed on ground at 40° F; no ground temperature gradient assumed

Slab thickness, feet	Minimum air temperature allowable for these thicknesses of commercial blanket or bat insulation, degrees F			
	0.5 inch	1.0 inch	1.5 inches	2.0 inches
Cement content—300 pounds per cubic yard				
0.333	(¹)	(¹)	(¹)	(¹)
0.667	49	47	44	42
1.0	43	33	22	12
1.5	33	12	−10	−33
2.0	24	−9	−43	−77
2.5	14	−31	−76	. . . .
3.0	5	−52	. . . .	. . . .
Cement content—400 pounds per cubic yard				
0.333	(¹)	(¹)	(¹)	(¹)
0.667	46	40	32	26
1.0	37	22	5	−12
1.5	25	−5	−37	−68
2.0	13	−32	−78	. . . .
2.5	1	−59	. . . .	. . . .
3.0	−11	. . . .	. . . .	. . . .
Cement content—500 pounds per cubic yard				
0.333	(¹)	(¹)	(¹)	(¹)
0.667	42	32	21	10
1.0	32	10	−13	−35
1.5	17	−23	−63	−103
2.0	3	−55	. . . .	. . . .
2.5	−12	. . . .	. . . .	. . . .
3.0	−27	. . . .	. . . .	. . . .
Cement content—600 pounds per cubic yard				
0.333	(¹)	(¹)	48	48
0.667	39	24	9	−5
1.0	27	−1	−31	−59
1.5	10	−40	−90	−139
2.0	−8	−78	. . . .	. . . .
2.5	−25	. . . .	. . . .	. . . .
3.0	−43	. . . .	. . . .	. . . .

[1] Owing to influence of cold subgrade on canal linings and other thin slabs, insulation alone will not maintain the temperature of concrete at the specified 50° F minimum for the first 72 hours after placing in cold weather. At such placements, additional heat is necessary to maintain required temperatures in the concrete by using higher placing temperatures, by preheating the ground, by placing electric resistance wire under the insulation, or by other means, depending on the severity of the prevailing weather. Where it is impracticable to supply such additional heat, insulation mats may prevent the concrete from freezing, but as the concrete temperature will probably fall below 50° F, the period of protection should be increased to obtain concrete strengths at the end of the protection period equivalent to the strength of concrete protected at 50° F for 72 hours.

canvas may be considered a loose-fill type if wind is kept out of the straw. The insulating value of a dead-air space greater than about one-half inch thick does not change greatly with increasing thickness. Textbooks or manufacturers' test data should be consulted for more detailed information on insulations. Insulation equivalents for commonly used materials are as follows:

Insulating material	Equivalent thickness, inch
1 inch of commercial blanket or bat insulation	1.000
1 inch of loose fill insulation of fibrous type	1.000
1 inch of insulating board	.758
1 inch of sawdust	.610
1 inch (nominal) of lumber	.333
1 inch of dead-air space (vertical)	.234
1 inch of damp sand	.023

Heated enclosures should provide sufficient space for circulation of warmed air. As corners and edges are vulnerable to low temperatures, canvas covers or other enclosure material should not rest on them. Enclosures should be tight and windproof. Openings for access should be few and preferably should be self-closing; at least they should be easily closed. Heat may be supplied by live or piped steam, by salamanders or stoves, or by airplane heaters outside the enclosure. Cooling pipe to be embedded in mass concrete has been used temporarily to circulate steam. Salamanders and stoves are easily handled and inexpensive and are convenient for small jobs, but they have the disadvantages of producing dry heat, emitting fumes and smoke which often disfigure the work, and being fire hazards. They often cause fire losses which would more than offset the cost of live steam heating, even on relatively small jobs. On larger centralized work such as a dam, powerhouse, or large canal structure, steam should be no more expensive, considering its advantages, than combustion heating within the enclosure.

Dry heat in cold weather tends to produce rapid drying because warm air will hold much more moisture than cold air. To illustrate, air at 70° F can hold about four times as much moisture as it can at 30° F. Consequently, if air at 30° F, even though saturated, is warmed to 70° F, it will quickly draw moisture from the concrete. It is important, therefore, that the concrete be supplied with adequate moisture when dry heat is used. Live steam is particularly advantageous because it provides moisture as well as heat; however, installations that might rust or be

Figure 174.—Form insulation for protection of concrete during cold weather. This is a reliable method that avoids heating costs and fire hazards. PX–D–33054.

otherwise damaged by steam condensation should be adequately protected. Although drying in the open is not as rapid at temperatures just above freezing as in moderate weather, it is nevertheless important that the concrete be cured and protected from drying as well as from freezing. In fact, the lower temperature necessitates a longer period of curing to yield equivalent results.

Where use of dry heat results in an extremely low humidity, thereby increasing the escape of moisture through the sealing compound, it may be necessary to raise the humidity of the air, use more sealing compound, or employ moist curing for 1 or more days prior to application of the compound. Best results are obtained when live steam is used for early protection and the sealing compound is applied just as the steam is discontinued.

Bureau specifications permit the use of unvented heaters only when unformed surfaces adjacent to the heaters are protected for the first 24 hours from excessive carbon dioxide atmosphere by application of curing compound. If the specifications do not provide for use of curing compound on such surfaces, unvented heaters should not be allowed.

Calcium chloride, salt, or other chemicals in permissible amounts in the mix will not lower the freezing point of concrete to any significant degree. Calcium chloride or additional cement is added during cold weather to assist in maintaining normal rate of hardening of the concrete (see sec. 20(a)) and is not for the purpose of shortening the period or simplifying the type of protection required in cold weather.

Chapter VII

REPAIR AND MAINTENANCE OF CONCRETE

A. Repair of Concrete

129. General Requirements for Quality Repair.—Approved methods and procedures for repairing new and old concrete are described in this chapter. Maintenance is also covered in some detail. The term repairing refers to any replacing, rejuvenating, or renewing of concrete or concrete surfaces after initial placement. The need for repairs can vary from such minor imperfections as she-bolt holes or snap-tie holes to major damages resulting from water energy or structural failure. Although the procedures described may initially appear to be unnecessarily detailed, experience has repeatedly demonstrated that no step in a repair operation can be omitted or carelessly performed without detriment to the serviceability of the work. Inadequate workmanship, procedures, or materials will result in inferior repairs which will ultimately fail.

(a) *Workmanship.*—It is the obligation of the constructor to repair imperfections in his work so that repairs will be serviceable and of a quality and durability comparable to the adjacent portions of the structure. Maintenance personnel have responsibility for making repairs that are inconspicuous, durable, and well bonded to existing surfaces. Since most repair procedures largely involve manual operations, it is particularly important that both foremen and workmen be fully instructed concerning procedural details of repairing concrete and reasons for them. They should also be apprised of the more critical aspects of repairing concrete. Constant vigilance must be exercised by the contractor's and Government's forces to assure maintenance of the necessary standards of workmanship. Employment of dependable and capable workmen is essential. Well-trained, competent workmen are particularly essential when epoxy materials are used in repair of concrete.

(b) *Procedures.*—Serviceable concrete repairs can result only from correct choice of method and careful performance of techniques. Wrong or ineffective repair or construction procedures coupled with poor workmanship lead to inferior quality repair work. Many proven procedures for making high quality repairs are detailed in this chapter; however, not all procedures in repair or maintenance are discussed. Therefore, it is incumbent upon the craftsman doing the work to use procedures that have been successful or that have a high reliability factor.

Repairs made on new or old concrete should be made as soon as possible after such need is realized. On new work the repairs that will develop the best bond and thus are the most likely to be as durable and permanent as the original work are those that are made immediately after early stripping of the forms while the concrete is quite green. For this reason, repairs should be completed within 24 hours after the forms have been removed. Before repairs are commenced, the method and materials proposed for use should be approved by an authorized inspector. Routine curing should be interrupted only in the area of repair operations.

Effective repair of deteriorated portions of structures cannot be assured unless there is complete removal of all affected or possibly affected concrete, careful replacement in strict accordance with an approved procedure, and assurance of secure anchorage and effective drainage when needed. Consequently, work of this type should not be undertaken unless or until ample time and facilities are available. Only as much of this work should be undertaken as can be completed correctly; otherwise the work should be postponed but not so long as to allow further deterioration. Repairs should be made at the earliest possible date.

(c) *Materials*.—Materials to be used in concrete repair must be high quality, relatively fresh, and capable of meeting specifications requirements for the particular application or intended use. Mill reports or testing laboratory reports should be required of the supplier or manufacturer as an indication of quality and suitability. Short of this requirement, certifications stating that the materials meet certain specifications should be required of the supplier. New or unknown materials should never be used in concrete repairing until the inspector or other authorized persons have complete assurance as to quality and suitability.

Materials selected for application must be used in accordance with manufacturers' recommendations or other approved methods. Mixing, proportioning, and handling must be in accordance with the highest standards of workmanship.

130. Methods of Repair.—Five proven methods of repairing concrete are discussed in this chapter. A sixth method, using chemical grout, is discussed briefly in subsection 139(d).

(a) *Dry-Pack Mortar*.—Dry pack should be used for filling holes having a depth equal to, or greater than, the least surface dimension of the repair area; for cone-bolt, she-bolt, and grout-insert holes; for holes left by the removal of form ties; and for narrow slots cut for repair of cracks. Exceptions are use of epoxy-bonded concrete or epoxy-bonded epoxy mortar, and prefabricated plugs for filling of cone-bolt holes or tie-rod recesses. Dry pack should not be used for relatively shallow depressions where lateral restraint cannot be obtained, for filling behind reinforcement, or for filling holes that extend completely through a concrete section.

(b) *Replacement Concrete.*—Concrete repairs made by bonding concrete to repair areas without use of an epoxy bonding agent or mortar grout applied on the prepared surface should be made when the depth of the area exceeds 6 inches and the repair will be of appreciable continuous area, as determined by the contracting officer. Concrete repairs should also be used for holes extending entirely through concrete sections; for holes in which no reinforcement is encountered and which are greater in area than 1 square foot and deeper than 4 inches, except where epoxy-bonded concrete repair is required or permitted as an alternative to concrete repair; and, in reinforced concrete, for holes greater than one-half square foot and extending beyond reinforcement.

(c) *Replacement Mortar.*—Portland cement mortar may be used for repairing defects on surfaces not prominently exposed where the defects are too wide for dry pack filling, the defects are too shallow for concrete filling, and where they are no deeper than the far side of the reinforcement that is nearest the surface. Repairs may be made either by use of shotcrete or by hand methods. For either, the treatment for protection against weathering should be applied.

(d) *Preplaced Aggregate Concrete.*—This method is used advantageously on larger repair jobs, particularly where underwater placement is required or when conventional placing of concrete would be difficult. Information on preplaced aggregate concrete is given in chapter VIII.

(e) *Thermosetting Plastic (Epoxy).*—Epoxies should be used to bond new concrete or mortar to old concrete whenever the depth of repair is between 1½ and 6 inches. Epoxy-bonded epoxy mortar should be used where the depth of repair is less than 1½ inches to featheredges. Epoxies are useful in special applications such as bonding steel anchor bars in old concrete.

131. Prerepair Requirements.—Prerepair considerations are as important as the repairs. Proper materials selection and surface preparation are essential to high quality, durable, functional repair.

(a) *Problem Evaluation and Repair Method Selection.*—The first step in repairing damaged or deteriorated concrete is to ascertain the cause and severity of failure and to determine what repair alternatives are available. The extent of deterioration or damage usually involves more than the eye can detect. Also, the extent of failure generally varies with the cause of the damage. For example, deterioration caused by alkali-aggregate reaction could be considerably more severe than that caused by freezing and thawing even though the two may appear to be somewhat comparable.

Causes of concrete failure can be classified into three general groups: (1) age and natural attrition; (2) unforeseen conditions such as earthquakes, floods, and slides; (3) service conditions such as chemical attack, cavitation, abrasion, subsidence, or freezing and thawing.

Several techniques are available to evaluate the cause and extent of damage or deterioration. In addition to visual examinations, nondestructive testing methods commonly used include impact hammer surveys and sonic velocity studies. Also, it is common practice to extract cores from deteriorated areas to provide a more detailed analysis. These specimens are generally sent to the Denver laboratories where they can be examined visually and subjected to petrographic, chemical, and physical tests. When feasible, it is desirable to conduct an extensive impact hammer and sonic velocity survey and to extract a few select cores for testing. These tests and evaluations provide an excellent analysis of the problem and aid considerably in determining the correct repair method.

Once the extent and cause of the failure have been determined, selection of the repair method can then be made. Other factors must be considered, such as availability of repair materials, relative costs, water seepage, temperatures, accessibility, and future use of the structure.

(b) *Preparation of Concrete for Repair*.—(1) *General requirements*.— Some preparations of the old concrete are required irrespective of the repair method used. Existing concrete surfaces to which new concrete is to be bonded without use of an epoxy bonding agent must be clean, rough, and dry. First, all damaged, loosened, or unbonded portions of existing concrete must be removed by chipping hammers or other approved equipment, after which the surfaces must be prepared by wet sand-blasting, waterblasting with approved waterblasting equipment, bush-hammering, or any other approved method and then cleaned and allowed to dry thoroughly. During the process, care should be taken to prevent undercutting aggregate in the existing concrete. Sometimes concrete in old structures that appear to be sound will slake and soften after a few days' exposure. For this reason, replacement of deteriorated concrete should be delayed several days until reexamination of excavated surfaces confirms the soundness of the remaining concrete. It is far better to remove too much concrete than too little because affected concrete generally continues to disintegrate and, while the work is being done, it costs little more to excavate to ample depth. Cleaning should be done by air-water jets. Surface drying must be complete and may be accomplished by air jet. Compressed air used in cleaning and drying must be free from oil or other contaminating materials.

After surfaces have been prepared and thoroughly cleaned, they must be kept in a clean, dry condition until the placing of concrete has been completed, except for dry-pack repairs. Dry-pack repairs require the application of a mortar bond coat prior to placement of the repair material.

The surface of existing concrete to which concrete and epoxy mortar are to be epoxy bonded must be prepared and maintained in a clean condition in accordance with previous paragraphs, except that wet or dry sandblasting may be used and cleaning may be by water jet or by air jet when approved by the contracting officer.

(2) *Special requirements.*—For the dry-pack method of repair, holes should be sharp and square at the surface edges, but corners within the holes should be rounded, especially when watertightness is a requisite. The interior surfaces of holes left by cone bolts and she bolts should be roughened to develop an effective bond; this can be done with a rough stub of ⅞-inch steel-wire rope, a notched tapered reamer, or a star drill. Other holes should be undercut slightly in several places around the perimeter, as shown in figure 175. Holes for dry pack should have a minimum depth of 1 inch.

To obtain satisfactory results with the replacement concrete method, the conditions should be as follows:

(a) Holes should have a minimum depth of 6 inches in old concrete and 4 inches in new, and the minimum area of the repair should

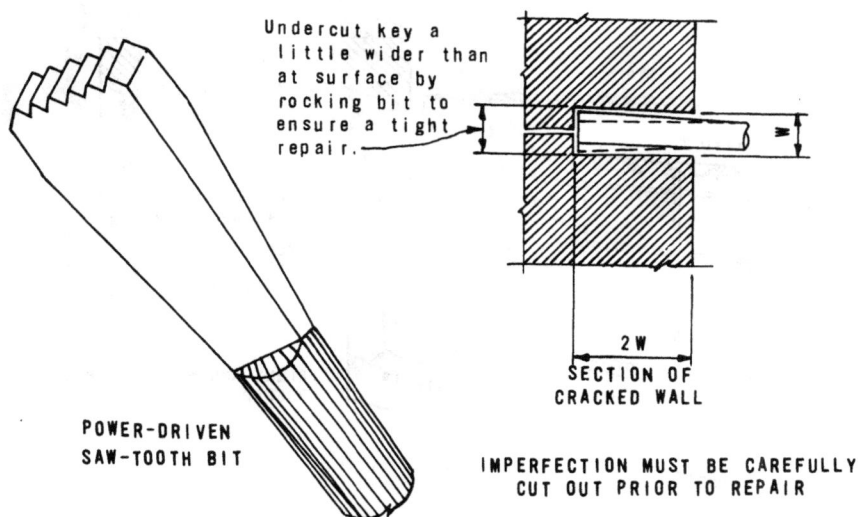

Undercut key a little wider than at surface by rocking bit to ensure a tight repair.

2 W

SECTION OF CRACKED WALL

POWER-DRIVEN SAW-TOOTH BIT

IMPERFECTION MUST BE CAREFULLY CUT OUT PRIOR TO REPAIR

Figure 175.—Saw-tooth bit used to cut a slot for dry packing. 288–D–1547.

be one-half square foot in reinforced and 1 square foot in unreinforced concrete.

(b) Reinforcement bars should not be left partially embedded; there should be a clearance of at least an inch around each exposed bar.

(c) The top edge of the hole at the face should be cut to a fairly horizontal line. If the shape of the defect makes it advisable, the top of the cut may be stepped down and continued on a horizontal line. The top of the hole should be cut on a 1 to 3 upward slope from the back toward the face from which the concrete will be placed (see fig. 176). This is essential to permit vibration of the concrete without leaving air pockets at the top of the repair. In some instances, where a hole extends through a wall or beam, it may be necessary to fill the hole from both sides; the slope of the top of the cut should be modified accordingly.

(d) The bottom and sides of the hole should be cut sharply and approximately square with the face of the wall. When the hole extends through the concrete section, spalling and featheredges may be avoided by having chippers work from both faces. All interior corners should be rounded to a minimum radius of 1 inch.

(e) For repairs on surfaces subject to destructive water action and for other repairs on exposed surfaces, the outlines of areas to be repaired should be saw cut as directed to a depth of 1½ inches

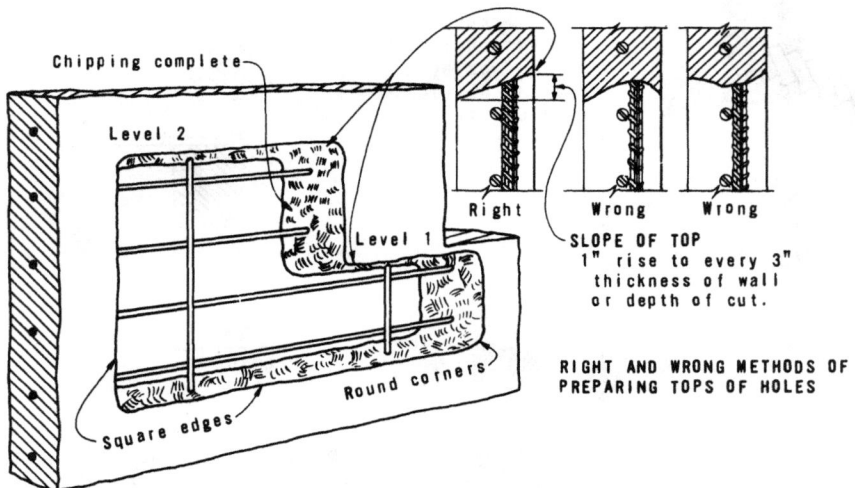

Figure 176.—Excavation of irregular area of defective concrete where top of hole is cut at two levels. 288–D–1548.

before the defective concrete is excavated. The new concrete should be secured by keying or other approved methods.

When a mortar gun is used for the replacement mortar method, comparatively shallow holes should be flared outwardly at about a 1 to 1 slope so that rebound will fall free. Corners within the holes should be rounded. Shallow imperfections in new concrete may be repaired by mortar replacement if the work is done promptly after removal of the forms and while the concrete is still green. For instance, when it is considered necessary to repair the peeled areas resulting from surface material sticking to steel forms, the surface may be filled using the mortar gun without further trimming or cutting. In the repair of old concrete, the importance of removing all traces of disintegrated material cannot be over-emphasized. All areas to be repaired should be chipped to a depth of not less than an inch. Wherever hand-placed mortar replacement is used, the edges of chipped-out areas should be squared with the surface, leaving no featheredges.

Concrete to be repaired with epoxy materials should be heated in sufficient depth, when necessary, so that the surface temperature (as measured by a surface temperature gage) shall not drop below 65° F during the first 4 hours after placement of an epoxy bond coat. This may require several hours of preheating with radiant heaters or other approved means (see fig. 177). If existing conditions prohibit meeting these temperature requirements, suitable modifications should be adopted upon the approval of the inspector or other responsible official. The concrete temperature during preheating should never exceed 200° F, and the final surface temperature at time of placing epoxy materials should never be greater than 105° F.

132. Use of Dry-Pack Mortar.—(a) *Preparation.*—Application of dry-pack mortar should be preceded by a careful inspection to see that the hole is thoroughly cleaned, free from broken or cracked pieces of aggregate mechanically held, and is dry. Occasionally, it may be advantageous to presoak the surfaces prior to application of the dry pack. Care must be taken when using this technique to ensure that no free water is on the surface when the previously mixed mortar bond coat is applied. To apply the bond coat the surface should be thoroughly brushed with a stiff mortar or grout after which the dry-pack material should be immediately packed into place before the bonding grout has dried. The mix for the bonding grout is 1 to 1 cement and fine sand mixed with water to a fluid paste consistency. Under no circumstances should the bonding coat be so wet or applied so heavily that the dry-pack material is more than slightly rubbery.

Where it is not feasible to have the hole prepared as above, the area

Figure 177.—A gas-fired weed burner being used to warm and dry an area prior to the placing of epoxy-bonded concrete in the spillway tunnel at Blue Mesa Dam, Colorado River Storage project. P622A–427–15030.

may be left slightly wet with a small amount of free water on the surfaces. The surfaces should then be dusted lightly and slowly with cement using a small dry brush until all surfaces have been covered and the free water absorbed. Any dry cement in the hole should be removed before packing begins. The holes should not be painted with neat cement grout because it could make the dry-pack material too wet and because high shrinkage would prevent development of the bond that is essential to a good repair.

Dry pack is usually a mix (by dry volume or weight) of 1 part cement to 2½ parts of sand that will pass a No. 16 screen. A mortar patch is usually darker than the surrounding concrete unless special precautions are taken to match the colors. Where uniform color is important, white cement may be used in sufficient amount (as determined by trial) to produce uniform appearance. For packing cone-bolt holes, a leaner mix of 1 to 3 or 1 to 3½ will be sufficiently strong and will blend better with the color of the wall. Sufficient water should be used to produce a mortar that will stick together while being molded into a ball with the hands and will not exude water but will leave the hands damp. The proper amount of water will produce a mix at the point of becoming rubbery when solidly packed. Any less water will not make a sound, solid pack; any more will result in excessive shrinkage and a loose repair. A typical dry-pack mortar repair of cone-bolt holes is shown in figure 178.

(b) *Application.*—Dry-pack material should be placed and packed in layers having a compacted thickness of about three-eighths of an inch. Thicker layers will not be well compacted at the bottom. The surface of each layer should be scratched to facilitate bonding with the next layer. One layer may be placed immediately after another unless an appreciable rubbery quality develops; if this occurs, work on the repair should be delayed 30 to 40 minutes. Under no circumstances should alternate layers of wet and dry materials be used.

Each layer should be solidly compacted over the entire surface by a hardwood stick and a hammer. These sticks are usually 8 to 12 inches long and not over 1 inch in diameter and are used on fresh mortar like a calking tool. Hardwood sticks are used in preference to metal bars because the latter tend to polish the surface of each layer and thus make bond less certain and filling less uniform. Much of the tamping should be directed at a slight angle and toward the sides of the hole to assure maximum compaction in these areas. The holes should not be overfilled; finishing may usually be completed at once by laying the flat side of a hardwood piece against the fill and striking it several good blows. If necessary, a few later light strokes with a rag may improve appearance. Steel finishing tools should not be used and water must not be used to facilitate finishing.

(c) *Curing and Protection.*—Because of the relatively small volume of most repairs and the tendency of old concrete to absorb moisture from new material, water curing is a highly desirable procedure at least during the first 24 hours. When forms are used for repair they can be removed and then reset to hold a few layers of wet burlap in contact with new concrete. In the absence of forms, a wet burlap pad can be supported as shown in figure 179. One of the best methods of water curing is illustrated

Figure 178.—Repairing cone-bolt holes in a bench-flume wall. The holes were packed with wet burlap in the afternoon and the holes filled with dry-pack mortar the next morning. This is a second filling of these holes, necessary because improper procedure caused unsatisfactory results in the first filling. PX–D–33056.

in figure 172, which shows a soil-soaker cotton hose laid along the top of a structure. White curing compound may be used only where its color does not create objectionable contrast in appearance (see sec. 125). When curing compound is used, the best curing combination is an initial water-curing period of 7 days followed, while the surface is still damp, by a coat of the compound. It is always essential that repairs, even dry-packed cone-bolt holes, receive some water curing and be thoroughly damp before the curing compound is applied. If nothing better can be devised for the initial water curing of the dry pack in cone-bolt holes and similar repairs, a reliable workman should be detailed to make the rounds with water and a large brush or a spraying device to keep the repaired surfaces wet for 24 hours prior to application of a curing com-

Figure 179.—Moist curing of surfaces of concrete repairs by supporting wet burlap mats against them. Wetting the burlap twice a day is usually sufficient to keep the surface continuously wet in this excellent method of treatment. PX–D–33059.

pound. Water curing and curing compounds are treated in greater detail in sections 124 and 125.

133. Use of Replacement Concrete.—(a) *Formed Concrete.*—The construction and setting of forms are important steps in the procedure for satisfactory concrete replacement where the concrete must be placed from the side of the structure. Form details for walls are shown in figure 180. To obtain a tight and acceptable repair the following requirements must be observed:

(1) Front forms for wall repairs more than 18 inches high should be constructed in horizontal sections so the concrete can be conveniently placed in lifts not more than 12 inches deep. The back form may be built in one piece. Sections to be set as concreting progresses should be fitted before placement is started.

(2) To exert pressure on the largest area of form sheathing, tie bolts should pass through wooden blocks fitted snugly between the walers and the sheathing.

(3) For irregularly shaped holes, chimneys may be required at more than one level; when beam connections are required, a chimney may be necessary on both sides of the wall or beam. For such construction, the chimney should extend the full width of the hole.

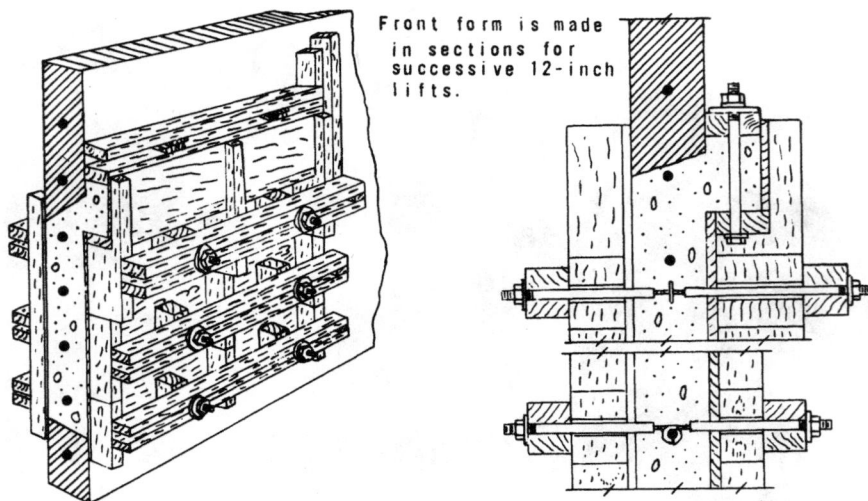

Front form is made
in sections for
successive 12-inch
lifts.

Back form may be
built in one piece.

By use of anchor bolts, these front forms may be used for
replacements in the surfaces of massive concrete structures.

Figure 180.—Detail of forms for concrete replacement in walls. 288–D–1549.

(4) Forms should be substantially constructed so that pressure
may be applied to the chimney cap at the proper time.

(5) Forms must be mortartight at all joints between adjacent
sections and between the forms and concrete and at tie-bolt holes
to prevent the loss of mortar when pressure is applied during the
final stages of placement. Twisted or stranded calking cotton, folded
canvas strips, or similar material should be used as the forms are
assembled.

Surfaces of old concrete to which new concrete is to be bonded must
be clean, rough, and surface dry (see sec. 131). Extraneous materials
on the joint resulting from form construction must be removed prior to
placement. Structural concrete placements should be started with an oversanded mix containing about a ¾-inch-maximum size aggregate; a maximum water-cement ratio of 0.47, by weight; 6 percent air, by volume of
concrete; and a maximum slump of 4 inches. This special mix should be
placed several inches deep on the joint at the bottom of the placement. A
mortar layer should not be used on the construction joints.

Concrete for repair should have the same water-cement ratio as used
for similar new structures but should not exceed 0.47 by weight. As

large a maximum size aggregate and as low a slump as are consistent with proper placing and thorough vibration should be used to minimize water content and consequent shrinkage. The concrete should contain 3 to 5 percent entrained air. Where surface color is important, the cement should be carefully selected, or blended with white cement, to obtain the desired results. To minimize shrinkage, the concrete should be as cool as practicable when placed, preferably at about 70° F or lower. Materials should, therefore, be kept in shaded areas during warm weather. Use of ice in mixing water would be practicable on larger jobs. Batching of materials should preferably be by weight, but batch boxes, if of the exact size needed, may be used. Since batches for this class of work will be small, the uniformity of the materials is important and should receive proper attention. When placing concrete in lifts, placement should not be continuous; a minimum period of 30 minutes should elapse between lifts. When chimneys are required at more than one level, the lower chimney should be filled and allowed to remain for 30 minutes between lifts. When chimneys are required on both faces of a wall or beam, concrete should be placed in one chimney only until it flows to the other.

Best repairs are obtained when the lowest practicable slump is used. This is about 3 inches for the first lift in an ordinary large form. Subsequent lifts can be drier, and the top few inches of concrete in the hole and that in the chimney should be placed at almost zero slump. It is usually best to mix enough concrete at the start for the entire hole. Thus, the concrete will be up to 1½ hours old when the successive lifts are placed. Such premixed concrete, provided it can be vibrated satisfactorily, will have less settlement, less shrinkage, and greater strength than freshly mixed concrete.

The quality of a repair depends not only on use of low-slump concrete but also on the thoroughness of the vibration, during and after depositing the concrete. There is little danger of overvibration. Immersion-type vibrators should be used if accessibility permits. If not, this type of vibrator can be used very effectively on the forms from the outside. Form vibrators can be used to good advantage on forms for large inaccessible repairs, especially on a one-piece back form, or attached to large metal fittings such as hinge-base castings. Immediately after the hole has been completely filled, pressure should be applied to the fill and the form vibrated. This operation should be repeated at 30-minute intervals until the concrete hardens and no longer responds to vibration. Pressure is applied by wedging or by tightening the bolts extending through the pressure cap (fig. 180). In filling the top of the form, concrete to a depth of only 2 or 3 inches should be left in the chimney under the pressure cap. A greater depth tends to dissipate the pressure. After the hole has been filled and the pressure cap placed, the concrete should not be vibrated without a simultaneous application of pressure—to do so may produce a film of water at the top of the repair that will prevent bonding.

Addition of aluminum powder to concrete causes the latter to expand as described in section 182. Under favorable conditions, this procedure has been successfully used to secure tight, well-bonded repairs in locations where the replacement material had to be introduced from the side. Forms similar to those shown in figure 180 should be used. Time should not be allowed for settlement between lifts. When the top lift and the chimney are filled, no pressure need be applied, but the pressure cap should be secured in position so expanding concrete will be confined to and completely fill the hole undergoing repair. There should be no subsequent revibration.

Concrete replacement in open-top forms, as used for reconstruction of the tops of walls, piers, parapets, and curbs, is a comparatively simple operation. Only such materials as will make concrete of proved durability should be used. The water-cement ratio should not exceed 0.47 by weight. For the best durability, the maximum size of aggregate should be the largest practicable and the percentage of sand the minimum practicable. No special features are required in the forms, but they should be mortar-tight when vibrated and should give the new concrete a finish similar to the adjacent areas. The slump should be as low as practicable, and dosage of air-entraining agent should be increased as necessary to secure the maximum permissible percentage of entrained air, despite the low slump. Top surfaces should be sloped so as to provide rapid drainage. Manipulation in finishing should be held to a minimum, and a wood-float finish is preferable to a steel-trowel finish. Edges and corners should be tooled or chamfered. Use of water for finishing is prohibited.

Forms for concrete replacement repairs may usually be removed the day after casting unless form removal would damage the green concrete, in which event stripping should be postponed another day or two. The projections left by the chimneys should normally be removed the second day. If the trimming is done earlier, the concrete tends to break back into the repair. These projections should always be removed by working up from the bottom because working down from the top tends to break concrete out of the repair. The rough area resulting from trimming should be filled and stoned (see sec. 119) to produce a surface comparable to that of surrounding areas. Plastering of these surfaces should never be permitted.

(b) *Unformed Concrete.*—Replacement of damaged or deteriorated paving or canal lining slabs, wherein the full depth of the slab is replaced, involves no different procedures than those described for best results in sections 104 and 108(c). Contact edges at the perimeter should be clean and square with the surface.

Special repair techniques are required for restoration of damaged or eroded surfaces of spillway tunnel inverts and spillway buckets. In addition to the usual forces of deterioration, such repairs often must withstand enormous dynamic and abrasive forces from fast-flowing water and sometimes from suspended solids. Preparation for replacement should follow the instructions in section 131. Depth of repairs should equal the width but should not be less than 6 inches.

If holes are less than 6 inches square, they should be filled using the dry-pack procedure described in section 132 or the epoxy-bonded epoxy mortar method described in section 136. If they are larger, low slump concrete should be used. Slump of the concrete should not exceed 2 inches for slabs that are horizontal or nearly horizontal and 3 inches for all other concrete. The net water-cement ratio (exclusive of water absorbed by the aggregates) should not exceed 0.47 by weight. An air-entraining agent should be used and a water-reducing set-controlling admixture (WRA) may be used. Set-retarding agents should be used only when the interval between mixing and placing is quite long. If practicable, the replacement concrete should be preshrunk by letting it stand as long as possible before it is tamped into the hole.

(c) *Curing and Protection.*—Procedures for curing and protection of concrete are described in subsection 132(c).

134. Use of Replacement Mortar.—Best results with replacement mortar are obtained when the material is pneumatically applied using a small gun. Equipment commonly used for shotcreting is too large to be satisfactory for the ordinarily small-sized repairs of new concrete. Neat work is difficult in the usual small areas, and cleanup costs are high because cleanup is seldom done promptly. However, small-sized equipment such as the air-suction gun shown in figure 124, fitted with a water ring on the nozzle, has been satisfactory for small-scale repair work. After the areas to be repaired have been cleaned, roughened (preferably by sandblasting), and surface dried, the mortar should be applied immediately. No initial application of cement, cement grout, or wet mortar should be made. Small size equipment similar to that shown in figure 124 without the water ring, has been used successfully when the mortar was premixed, including water, to a consistency of dry-pack material.

The dry-mortar mix recommended for the air-suction gun shown in figures 124 and 181 is 1 part cement to 4 parts natural sand by dry volume or weight. Rebound changes these proportions so that the material in place is much richer. Best results are obtained with a well-graded sand passing the No. 16 screen. Cement and sand should be mixed with water to approximately the same consistency as for dry-pack repair. If insufficient water is used, rebound will be high and the applied mortar too rich,

Figure 181.—Application of replacement mortar. The mortar should be applied on dry contact surfaces that are as clean as a freshly broken piece of concrete. PX–D–33057.

but too much water will cause the gun to plug frequently. When the proper consistency is used, the gun will plug occasionally, but it may readily be cleared by holding the nozzle against the ground or the wall, then tapping the gun and suction hose until the congested material is blown out of the suction hose.

If repairs are more than 1 inch deep the mortar should be applied in layers not more than three-quarters of an inch thick to avoid sagging and loss of bond. After completion of each layer there should be a lapse of 30 minutes or more before the next layer is placed. Scratching or otherwise preparing the surface of a layer prior to addition of the next layer is unnecessary, but the mortar must not be allowed to dry.

If a small gun is used in which the water is introduced at the nozzle, as shown in figure 124, care must be exercised to apply mortar of the lowest practicable water content to avoid sagging and later shrinkage cracking. Although the gun should not be used extensively to place mortar around reinforcement bars, good repairs can be made in shallow imperfections where relatively little steel is exposed, if the angle of the gun is varied frequently as this part of the hole is being filled.

In completing the repair, the hole should be filled slightly more than level full. After the material has partially hardened but can still be trimmed off with the edge of a steel trowel, excess material should be shaved off, working from the center toward the edges. Extreme care must be used to avoid impairment of bond. Neither the trowel nor water should be used in finishing. A satisfactory finish may be obtained by lightly rubbing the surface with a soft rag.

For minor restorations, satisfactory mortar replacement may be performed by hand if the repairs are made strictly in accordance with the procedure described in subsection 138(b), followed by the weatherproofing treatment described in section 139. The success of this method depends on complete removal of all defective and affected concrete, good bonding of the mortar to old concrete, elimination of shrinkage of the patch after placement, and thorough curing. (See subsection 132(c) for discussion on curing.)

135. Use of Preplaced Aggregate Concrete.—This concrete placing method, especially adaptable to underwater construction, may be used advantageously on large concrete and masonry repair jobs where placement by conventional means is unusually difficult or where concrete of low volume change is required. Preplaced aggregate concrete has been used in the resurfacing of dams and the repair of tunnel linings, piers, and spillways; it is often particularly well adapted for these types of repair. Preplaced aggregate concrete is discussed in chapter VIII.

136. Use of Thermosetting Plastic (Epoxy)

Note: — **Safety precautions discussed in subsection 136 (i) must be observed.**

(a) *Materials.*—Many proprietary epoxy formulations prepared for bonding old concrete to old concrete, new concrete to old concrete, and epoxy mortar to old concrete are now available. Many of these materials are excellent high-quality products used with reasonable certainty as to the results. However, some of the materials available are unsuitable for most repair applications. Epoxy bonding agents that conform to Federal Specification MMM-B-350B for Binder, Adhesive, Epoxy Resin, Flexible, Type I or Type II are suitable for most concrete bonding applications. Epoxy grout agents conforming to Federal Specification MMM-G-650B for Grout, Adhesive, Epoxy Resin, Flexible, Filled, Type I or Type II are also approved materials. There are many epoxy bonding compounds formulated for specific uses such as floor toppings, patching, crack injection, and underwater use. Type I epoxy should be used only when the temperatures are above 68 °F but less than 104 °F. When concrete temperatures are lower than 68 °F, but above 50 °F, type II should be used. Type I epoxy materials should be stored at 70 °F minimum to 90 °F maximum, and type II epoxy materials should be stored at 65 °F minimum to 80 °F maximum. Epoxies used with sand in concrete repairing should be the two-component, 100-percent solids type, irrespective of whether they meet the Federal specification. When the epoxy is required to conform to Federal Specifications MMM-B-350B and MMM-G-650B, it is common practice to approve use of the material upon receipt of the manufacturer's certification of conformance to those specifications. The certification should identify the specifications number under which the agent is to be used and include the quantity represented, the batch numbers of the resin and hardener, and the manufacturer's results of tests performed on the particular combination of resin and hardener.

In the repair of concrete, the epoxy is generally mixed with sand to make an epoxy mortar. The sand to be used in epoxy mortar must be clean, dry, well graded, and composed of sound particles. For most applications, particularly where featheredging is required, sand passing a No. 16 screen and conforming to the following limits should be used:

Screen No.	Individual percent, by weight, retained on screen
30	26 to 36
50	18 to 28
100	11 to 21
pan	25 to 35 [1]

[1] Range shown is applicable when 60 to 100 percent of pan is retained on No. 200 screen. When 41 to 100 percent of pan passes the No. 200 screen, the percent pan should be within the range of 10 to 20 percent and the individual percentages retained on the Nos. 30, 50, and 100 screens should be adjusted accordingly.

Sand processed for use in concrete rarely contains the required quantity of pan size sand. As a result, problems often arise in obtaining additional pan size material to supplement sand available on the jobsite. A source of silica pan size material may be obtained by contacting the Division of Research, Bureau of Reclamation, Engineering and Research Center, Denver Federal Center, Denver, Colorado 80225. A sand graded as shown above and properly mixed with an epoxy meeting Federal Specification MMM-B-350B will provide a dense, high-strength workable epoxy mortar.

The sand should be maintained in a dry area at not less than 70 °F for 24 hours immediately prior to the time of use. Filler materials other than sand, such as portland cement, can be used. However, for general applications a natural sand is recommended.

No discussion is contained herein relative to materials needed for application where the epoxy is sprayed with a distributor and the aggregate then cast onto it, such as in sealing highways with asphalt emulsions. Such use of epoxies, although sometimes referred to as repairing, is not within the scope of this discussion.

On repair jobs in areas of high-velocity flow or on repairs requiring a considerable quantity of materials, the contractor is required to submit samples of epoxy bonding agent and graded sand to the Denver laboratory for use in mix design determinations. The samples should consist of 1 gallon total quantity of epoxy components and a minimum of 50 pounds of graded sand. Samples should be submitted at least 30 days prior to use in the work and be labeled or otherwise identified with the specifications number under which the material is to be used.

(b) *Preparation of Epoxy Bonding Agent*.—The epoxy-resin bonding agent is a two-component material which requires combination of components and mixing prior to use. Once mixed, the material has a limited pot life and must be used immediately. (Pot life refers to the period of time elapsing between mixing of ingredients and their stiffening to the point where satisfactory placement cannot be achieved.) The bonding agent should be prepared by adding the hardener component to the resin component in proportions recommended by the manufacturer, followed by thorough mixing. Since the working life of the mixture depends on the temperature (longer at low temperature, much shorter at high temperature), the quantity to be mixed at one time should be applied and topped within approximately 30 minutes. The addition of thinners or diluents to the resin mixture is not permitted since it weakens the epoxy.

(c) *Preparation of Epoxy Mortar*.—The epoxy mortar is composed of sand and epoxy bonding agent suitably blended to provide a stiff, workable mix. Epoxy components should be mixed thoroughly prior to addition of sand. Mix proportions should be established, batched and reported on a weight basis, although the dry sand and mixed epoxy may be batched by volume using suitable measuring containers that have been calibrated on a weight basis. Epoxy meeting Federal Specification MMM–B–350B will, using well-graded sand (subsection 136(a)), require approximately 5½ to 6 parts of graded sand to 1 part epoxy, by weight. This is equivalent to a ratio of approximately 4 to 4½ parts sand to 1 part epoxy by volume. If equivalent volume proportions are being used, care must be taken to prevent confusing them with weight proportions. The contracting officer should determine, and adjust where necessary, the mix proportions for the particular epoxy and sand being used. The epoxy bonding agent should be prepared as provided in subsection 136(b). The epoxy mortar should be thoroughly mixed with a slow-speed mechanical device. The mortar should be mixed in small-sized batches so that each batch can be completely mixed and placed within approximately 30 minutes. Addition of thinners or diluents to the mortar mixture is not permitted.

(d) *Application of Epoxy Bonding Agent.*—Immediately after the epoxy-resin is mixed, it must be applied to the prepared, dry existing concrete at a coverage of not more than 80 square feet per gallon, depending on surface conditions. The area of coverage per gallon of agent depends on the roughness of the surface to be covered and may be considerably less than the maximum specified. The epoxy bonding agent may be applied by any convenient, safe method such as squeegee, brushes, or rollers which will yield an effective coverage (see figure 182). Spray-

Figure 182.—Epoxy bond-coat being applied to sandblasted area adjacent to the dentates in Yellowtail Afterbay Dam sluiceway. Following this, epoxy mortar is placed over fluid or tacky epoxy to form epoxy-bonded epoxy mortar. P459–D–68900.

ing of the material is permitted if an efficient airless spray is used and when the concrete surfaces to receive the agent are at a temperature of 70° F or somewhat warmer. Spraying should be demonstrated as providing an adequate job with minimum overspray prior to approval for use. If spray application is used, the operator must wear an operating compressed air-fed hood and no other personnel should be closer than 100 feet if downwind of the operator. Appropriate solvents may be used to clean tools and spray guns, but in no case should the solvents be incorporated in any bonding agent. All tools must be completely dried before reuse.

During application of the epoxy bonding agent, care must be exercised to confine the material to the area being bonded and to avoid contamination of adjacent surfaces. However, the epoxy bond coat should extend slightly beyond the edges of the repair area.

Steel to be embedded in epoxy mortar should be prepared, cleaned, and dried in the same manner as the concrete being repaired. The exposed steel should be completely coated with epoxy bonding agent when the agent is applied to the surfaces of the repair area.

Applied epoxy bonding agent film must be in a fluid condition when the concrete or epoxy mortar is placed. Except when epoxy mortar is placed on steep sloping or vertical surfaces, the agent may be allowed to stiffen to a very tacky condition. Special care must then be taken to thoroughly compact the epoxy mortar against the tacky bond coat. If an applied film cures beyond a fluid condition (or the very tacky condition permitted before epoxy mortar is placed), a second application of bonding agent must be applied while the first coat is still tacky. Epoxy-bonded concrete must be placed while the applied film is still fluid; if the film cures beyond this state but is still tacky, a new bond coat should be applied over the first coat. If any bond coat has cured beyond the tacky state, it must be completely removed by sandblasting, properly cleaned, and a new bond coat applied.

(e) *Application of Epoxy-bonded Concrete.*—Use of epoxy-bonded concrete in repairs requiring forming, such as on steeply sloped or vertical surfaces, can be permitted only when sufficient time has been allowed to place concrete against the bonding agent while it is still fluid. Immediately after application of the epoxy-resin bonding agent while the epoxy is still fluid, unformed epoxy-bonded concrete should be spread evenly to a level slightly above grade and compacted thoroughly by vibrating or tamping (see figure 183). Tampers should be sufficiently heavy for thorough compaction. After being compacted and screeded, the concrete should be given a wood-float or steel-trowel finish as required. Water, cement, or a mixture of dry cement and sand should not be sprinkled on the surface. Troweling, if required, should be performed at the proper

Figure 183.—Epoxy bond-coat being applied to prepared concrete surface before placement of new concrete. New low-slump concrete must be placed while the epoxy bond-coat remains fluid. P622A–427–15035.

time and with heavy pressure to produce a smooth, dense finish free of defects and blemishes. As the concrete continues to harden, the surface should be given additional trowelings.

The final troweling should be performed after the surface has hardened so that no cement paste will adhere to the edge of the trowel, but excessive troweling cannot be permitted. The number of trowelings and time at which trowelings are performed should be approved by the contracting officer.

(f) *Application of Epoxy-bonded Epoxy Mortar.*—Surfaces of existing concrete to which epoxy mortar is to be bonded should be prepared as discussed in subsection 131(b). Epoxy-resin bonding agent should then be applied as outlined in subsection 136(d). The agent should be applied to the areas immediately before placing of epoxy mortar. Special care must be taken to prevent the bond coat from being spread over concrete surfaces not properly cleaned and prepared.

The prepared epoxy mortar should be tamped, flattened, and smoothed into place in all areas while the epoxy bonding agent is still in a fluid condition, except that on steep slopes the bond coat can be brought to a tacky condition as discussed in subsection 136(d). The mortar should be worked to grade and given a steel-trowel finish. Special care must be taken at the edges of the area being repaired, particularly where there are thin featheredges, to assure complete filling and leveling and to prevent the mortar from being spread over surfaces not having the epoxy bond coat application. Steel troweling should best suit prevailing conditions; in general, it should be performed by applying slow, even strokes. Trowels may be heated to facilitate the finishing, but the use of thinner, diluents, water, or other lubricants on placing or finishing tools is not permitted. After leveling of the epoxy mortar to the finished grade where precision surfaces are required, the mortar should be covered with plywood panels smoothly lined with polyethylene sheeting and weighted with sandbags or otherwise braced by suitable means until the possibility of slumping has passed. When polyethylene sheeting is used, no attempt should be made to remove it from the epoxy mortar repair before final hardening. All areas of repair requiring featheredging should be finished with epoxy mortar.

Surfaces of all epoxy-bonded epoxy mortar repairs should be finished to the plane of surfaces adjoining the repair areas. The final finished surfaces should have the same smoothness and texture of surfaces adjoining the repair areas. (See figures 184, 185, and 186).

(g) *Application of Epoxy by Pressure Injection.*—An effective method for repairing cracks in structural members such as walls, piers, floors, ceilings, and pipe is the epoxy pressure injection system. If the cracks to be injected are clean and dry and the epoxy is properly mixed, placed, and cured, the repaired member will be restored to original structural integrity. Although damp or wet cracks can be repaired using this method, development of sufficient bond between epoxy and concrete cannot be assured.

Small repair jobs employing this method can use any system that will successfully deposit the epoxy into the required zones. Such systems could use a prebatch arrangement in which the two components of the epoxy are batched together prior to initiating the injection phase. The relatively short pot life of the epoxy makes this technique rather critical as far as timing is concerned.

Large epoxy injection jobs generally require a single-stage injection technique in which the two epoxy components are pumped independently of one another from the reservoir to the mixing nozzle. At the mixing nozzle, located adjacent to the crack being repaired, the two epoxy components are brought together for mixing and injecting. The epoxy used in this injection technique must have a low initial viscosity and a closely controlled set time. Cracks as small as 0.002 inch in width have been successfully repaired with injected epoxy resin.

Figure 184.—Epoxy-bonded epoxy mortar repairs. In upper photograph epoxy mortar is placed on epoxy bond-coat previously applied to wall of sluiceway at Yellowtail Afterbay Dam. The lower photograph shows a dentate in sluiceway repaired with epoxy-bonded epoxy mortar. The difference in mortar color in the two photographs is caused by use of different colored sands. P459–D–68911 (upper), P459–D–68912 (lower).

Figure 185.—A dentate in overflow weir stilling basin at Yellowtail Afterbay Dam showing typical cavitation damage. P459–D–75010.

Several private companies have proprietary epoxy injection systems (fig. 187). These organizations have developed epoxies and techniques which allow them to make satisfactory repairs under the most adverse conditions. One or more of these companies should be contacted regarding

Figure 186.—A damaged dentate in overflow weir stilling basin at Yellowtail
Afterbay Dam restored to its original condition with epoxy-bonded epoxy
mortar. The concrete color of epoxy mortar was obtained by grinding after
completion of curing. P459–D–68915.

any major repairs requiring the epoxy injection technique. Names and
addresses of these companies can be obtained from the Division of General Research, Bureau of Reclamation, Engineering and Research Center,
Denver Federal Center, Denver, Colorado 80225.

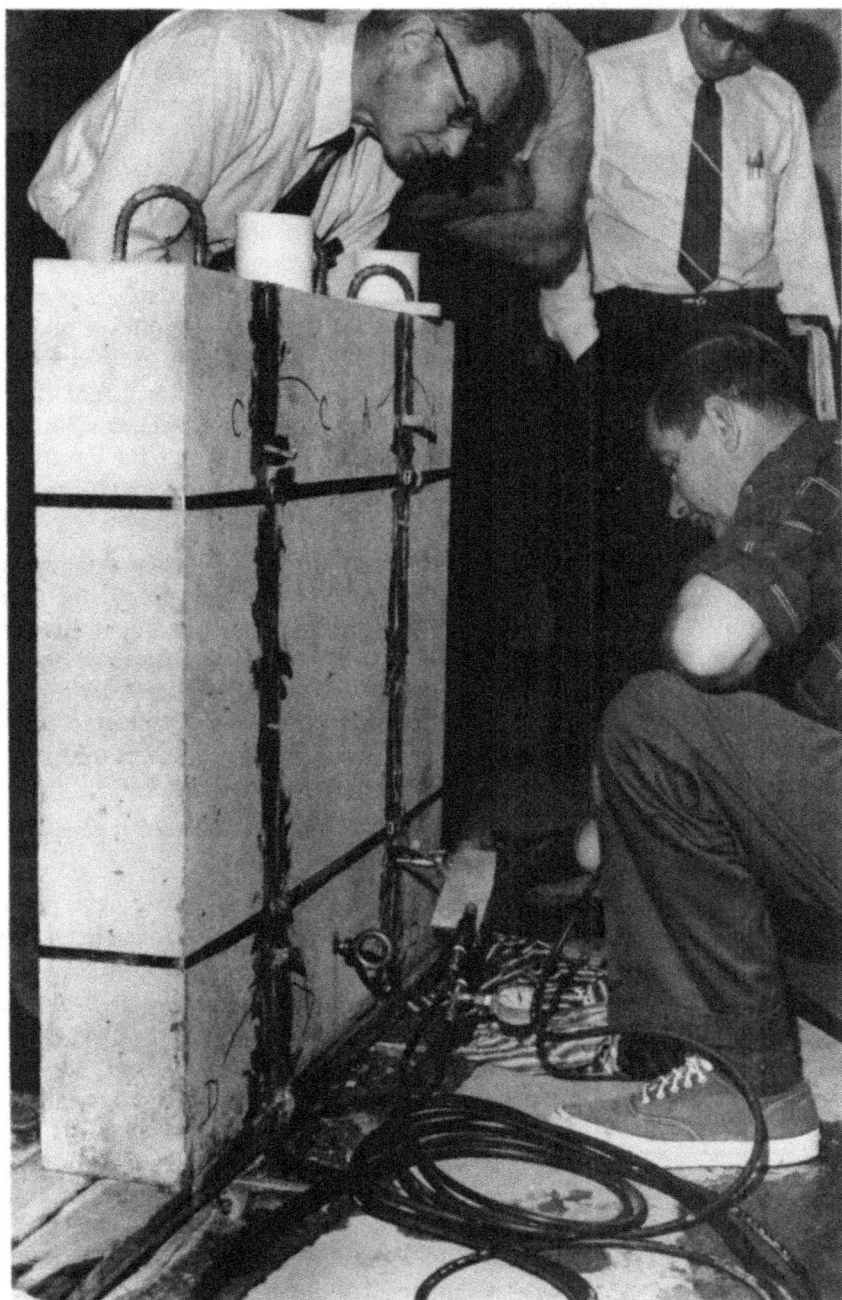

Figure 187.—Demonstration of a proprietary epoxy grout injection system for repairing cracks in concrete structures. P801–D–75011.

(h) *Curing and Protection.*—As soon as the epoxy-bonded concrete has hardened sufficiently to prevent damage, the surface should be moistened by spraying lightly with water and then covering with sheet polyethylene or by coating with an approved curing compound. Curing compound will be used whenever there is any possibility that freezing temperatures will prevail during the curing period. Sheet polyethylene must be an airtight, nonstaining, waterproof covering that will effectively prevent loss of moisture by evaporation. Edges of the polyethylene should be lapped and sealed. The waterproof covering should be left in place for at least 2 weeks. If a waterproof covering is used and the concrete is subjected to any usage during the curing period that might rupture or otherwise damage the covering, the covering must be protected by a suitable layer of clean wet sand or other cushioning material that will not stain concrete. Application of curing compound must be in accordance with the procedures contained in sections 124 and 125. After curing, the covering, except curing compound if used, and all foreign material should be removed as directed.

Epoxy mortar repairs should be cured immediately after completion at not less than 60° F until the mortar is hard. Postcuring should then be initiated at elevated temperatures by heating in depth the epoxy mortar and the concrete beneath the repair. Postcuring should continue for a minimum of 4 hours at a surface temperature of not less than 90° F nor more than 110° F. The heat could be supplied by use of portable propane-fired heaters, batteries of infrared lamp heaters, or other approved sources positioned to attain the required surface temperatures.

In no case should epoxy-bonded epoxy mortar be subjected to moisture until after the specified postcuring has been completed.

(i) *Safety.*—All personnel must be carefully instructed to take every precaution in preventing epoxies and their components from contacting the skin. Protective clothing must be worn, including gloves and goggles, and protective creams for other exposed skin areas should be provided when handling epoxies, as severe allergic reactions and possible permanent health damage can result when these materials are allowed to contact and remain upon the skin. Any deposits acquired through accidental contact of these materials with unprotected skin must be removed immediately by washing with soap and water, never with solvents. Solvents, such as toluene and xylene, may be used only for cleaning epoxy from tools and equipment. Care must also be exercised to avoid contact of cleaning solvents with the skin and to provide adequate ventilation for cleanup operations. All safety equipment used must conform to the requirements of the Occupational Safety and Health Standards of the Occupational Safety and Health Administration.

137. Repairing Concrete Under Unusual Conditions.—(a) *Seepage Conditions.*—Repairs should not be attempted where there is seeping or running water. When the water cannot be diverted, it is often possible, by plugging the outlet with quick-setting mortar, to stop the flow long enough for the repair to be made and the mortar to harden. Mortar for plugging such leaks should consist of 1 part cement and 1 to 2 parts sand, by volume. If mixing water contains 30 to 50 percent of calcium chloride or soda ash equal to about 5 percent of the weight of the cement, the mortar will set in a few minutes while being held tightly in position against the leak. The time of set is determined by the strength of the mixing water solution. Quick-setting proprietary compounds are also available for use in plugging seeps. Plastic sheeting is often used to divert water from areas to be repaired (fig. 188).

Two additional methods of stopping water seepage through cracks are calking and chemical grouting. Sealing by calking requires that the crack be chipped at the surface to form a vee-shaped opening. Lead wool is tamped into this opening to form a dense, tight plug as shown in figure 188. The top surface of the lead wool should be left about one-fourth of an inch below the concrete surface. The repair material can then be placed over the top of the lead wool calking.

Chemical grouting as a means of stopping water seepage through cracks in concrete has been used to a limited but successful extent. Low-viscosity chemical grout (organic monomers are common types) is injected into the seeping cracks through small holes drilled to intersect the crack at some distance below the surface. Grout gel time and injection pressures are controlled in accordance with the requirements of the specific application. Repairing cracks by this method is described in subsection 139 (d).

There are some so-called underwater curing epoxies available on the market. Research conducted by the Bureau indicates that the quality of these epoxies varies widely, and therefore caution must be exercised in selecting and using them. Also, application in wet environments requires knowledge of their limitations.

(b) *Extreme Temperature Conditions.*—As epoxies are thermosetting plastics, they are readily affected by temperature variations. For example, most epoxies will not cure properly at temperatures below about 50° F, but they will cure rapidly at a temperature of 100° F. On the other hand, concrete cures at 50° F with better development of some properties than if cured at higher temperatures. Therefore, extreme temperature conditions should be avoided. If repairs must be made under such conditions, special care should be taken to protect concrete and epoxy.

Epoxies have coefficients of thermal expansion considerably greater than those of concrete. Therefore, particular care must be exercised when

Seeping water is stopped
by chipping the seeping
crack to a vee shape, then
calking with lead wool.
P622–D–56788

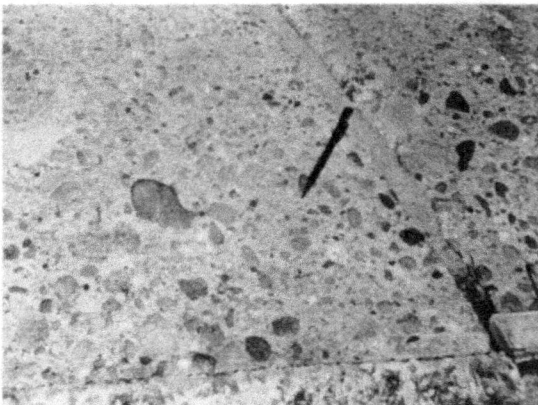

Plastic tents or shelters
provide protection from
rain or other water from
overhead. P459–D–74786

Flowing water is directed
around the repair area by
plastic or dikes made of
quick-setting compounds.
P459–D–74761

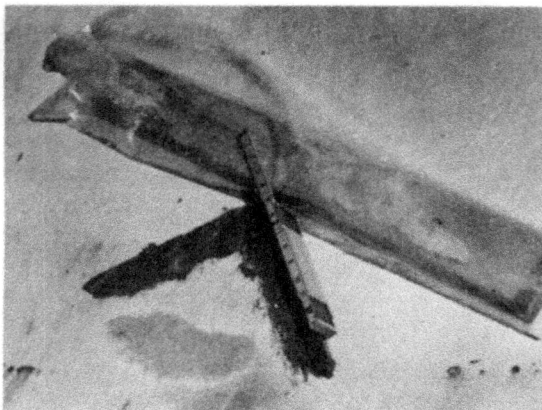

Figure 188.—Typical techniques for maintaining dry work areas during repair
operations.

using epoxy-bonded epoxy mortar in large sunlit areas subject to high temperatures and extreme temperature differentials. For such applications very dark epoxies and sands for mortar should not be used. Excessive heat absorption by the dark materials could cause disbonding and failure of the repair.

(c) *Special Color Considerations.*—A portland cement mortar patch is usually darker than the surrounding concrete unless precautions are taken to match colors. A leaner mix will usually produce a lighter color patch. Also, white cement can be used to produce a patch that will blend with the surrounding concrete. The quantity of white cement to use must be determined by trial.

Epoxy mortars generally produce patches that are darker than the surrounding concrete. Some epoxies available produce a gray-colored mortar resembling concrete. However, these materials will rarely produce an exact color match. Grinding hardened epoxy mortar may lighten its color to about that of the surfaces adjoining the repair areas (fig. 186). Epoxy mortars can be colored by the addition of such materials as iron oxide red, chromium oxide green, lampblack and titanium dioxide white for gray, and ochre yellow. The Bureau rarely uses any materials to color the epoxy other than the sand for the mortar. Use of white silica sand in the mortar will produce a white-looking patch; most natural riverborne sands will produce darker colored mortars (fig. 184). Whenever concrete or epoxy mortar repair materials must be colored to match adjacent concrete, laboratory mixes should be made to ascertain the proper quantities of coloring constituents.

138. Special Cases of Concrete Repair.—(a) *Cracks in Concrete Siphons.*—Transverse cracks sometimes appear in concrete siphons, conduits, and pipelines as a result of shrinkage caused by either a drop in water temperature or drying when the structure is not in service. Because cracks are caused by a strain or movement in concrete, any rigid repair is destined to fail when some later condition causes further opening. For this reason, rigid repairs made with lead wool or portland cement grout or mortar have a poor record of performance and are not recommended. Similarly, a flexible repair from the inside using certain mastic and calking materials has a good record of success when properly installed. The method and materials herein described are a nonproprietary modification of procedure which has been serving with excellent results since 1946 in concrete siphons of the Colorado River Aqueduct.

With reference to figure 189 the method of repair is as follows:

(1) Using a saw-tooth bit as shown in figure 175, cut a trim, narrow, sharp-edged groove about ½-inch wide and 2½-inches deep on the crack and clean it thoroughly as described in section

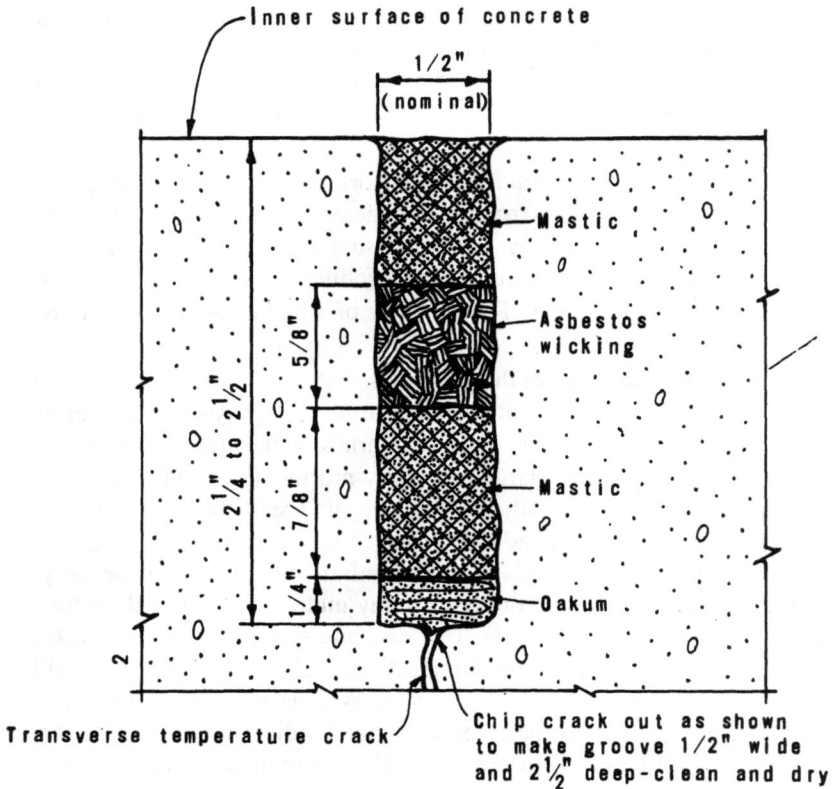

Figure 189.—Calking method used for repair of transverse cracks in concrete siphons. 288–D–1550.

131. The groove should be cleaned frequently during the cutting to make sure the crack is being followed.

(2) After cleaning, the groove may be damp but not wet when the filling treatment begins. If water seeps in from outside the conduit, it may be stopped by lead wool calking before beginning the elastic repair. It may be necessary to excavate outside the siphon or lower the water by pumping until the repair is made.

(3) Tamp oakum tightly into the bottom one-fourth inch of the slot.

(4) Tamp a ¾-inch-diameter rope of the mastic material over the oakum. It should be driven into firm contact with the joint surfaces to establish a satisfactory bond.

(5) Place a section of ⅝-inch, tightly twisted asbestos rope wicking in the groove and calk tightly using hand or pneumatic tools.

(6) Fill the remainder of the groove solidly with a second ¾-inch-diameter rope of the mastic and smooth off at the concrete surface.

The heavy-bodied asphalt mastic intended for use in this repair method is the same as the one often used to seal bell-and-spigot concrete pipe. Federal Specification SS–S–00210 describing this material is titled "Sealing Compound, Preformed Plastic, for Expansion Joints and Pipe Joints." Asbestos rope wicking may be procured under Federal Specification HH–P–41, titled "Packing, Asbestos, Rope and Wick."

(b) *Imperfections in Precast Concrete Pipe.*—(1) *General.*—Allowable repairs of concrete pipe are made in accordance with Bureau "Standard Specifications for Repair of Concrete." If followed closely, the procedures specified will result in acceptable repairs, and pipe not repaired in accordance with these procedures should be rejected.

Imperfections should be detected as early as possible during the manufacturing process, and the cause should be corrected. The occasional imperfections that may still exist should be repaired immediately and properly steam- or water-cured. Damage to precast concrete pipe that may occur after manufacture can sometimes be repaired and the pipe made acceptable.

Pipe that continually has imperfections or damage because of failure to take corrective action in manufacture or handling should be rejected.

(2) *Methods of repair.*—Depending on the severity and location of imperfections or damage, precast concrete pipe repairs may be made with hand-placed mortar, shotcrete, or concrete. Epoxy bonding agents may be used to bond concrete repairs or to make epoxy mortars for repair of shallow imperfections. Epoxy-bonded repairs are sometimes advantageous in that featheredge patches can be produced satisfactorily, and extended curing is not necessary beyond that normally required for adequate concrete strength. Before preparations are started for repair of any pipe, except very minor repairs, the method of repair should be approved by a Government inspector.

Hand-placed mortar should be used only for making superficial repairs on the outside of pipe or for making minor repairs on the inside of pipe that is too small to permit application of shotcrete (usually pipe smaller than 36 inches in diameter). Shotcrete should be used for repair of all other shallow surface imperfections, such as to cover exposed reinforcement steel on the outside of any size pipe and on the interior of pipe 36 inches or more in diameter, and to build up spalled shoulders on spigots for support of rubber gaskets. Shotcrete should not be used where more than one-half square foot extends back of reinforcement steel. Preshrunk concrete should be used for the repair of all other imperfections including areas where more than one-half square foot extends back of reinforcement steel.

(3) *Preparation of imperfections for repair.*—All visibly imperfect concrete should be removed before any type of replacement is made. Where shotcrete is to be used for replacement, unsound materials should be removed to any shape with beveled edges that will not entrap rebound. Where hand-applied mortar is to be used for replacement, the area requiring repair should be chipped to a depth of not less than three-fourths inch; edges of the area should be sharp and squared with the surface, leaving no featheredges. Where concrete is to be used for replacement, the old concrete should be removed to a depth of at least 1 inch back of the first layer of reinforcement steel, even though this involves removal of good concrete. The edges should be sharp and squared with the surface, leaving no featheredges. Keys are not necessary. Where concrete is repaired using epoxy and epoxy mortar, the old concrete should be prepared as described in section 131(b).

As soon as chipping is completed and the area is acceptably shaped, the surface of old concrete should be given a preliminary washing to remove all loose materials and stone dust. Except when epoxy is to be used, surfaces within the trimmed holes should be kept wet for several hours, preferably overnight, before the repair replacement is made. This is best done by packing the holes or covering the areas with several layers of wet burlap as shown in figures 178 and 179. Immediately before new material is applied, all surfaces of trimmed holes or areas to be filled should be thoroughly cleaned with wet sandblasting, followed by washing with an air-water jet to remove all foreign material, dried grout, and any material crushed and embedded in the surfaces by chisels or other tools during trimming. Some equipment for placing shotcrete is effective for wet sandblasting. Other devices such as the air-suction gun shown in figure 124 may be used if they will produce acceptable results. Surfaces to which the replacement concrete mortar is to bond should be damp but not wet when new material is applied. Surfaces to receive epoxy mortar must be dry and warm at the time of application. The prepared surfaces should be inspected before the repair is made.

Individual air holes in gasket-bearing areas of precast concrete pipe may be filled with a hand-placed, stiff, preshrunk 1:1 mortar of cement and fine sand with no other preparation than thorough washing with water. Such fillings should be kept moist under wet burlap for at least 48 hours.

(4) *Hand-placed replacement mortar.*—For application of hand-placed mortar, the pipe should be turned so that new material will rest by gravity on concrete of the pipe. The mortar used for replacement should have the same proportions and air entrainment as mortar used in

the mix of which the pipe was made. Repair mortar should be preshrunk by mixing it to a plastic consistency as long in advance of its use as the cement will permit. Depending on mix, cement, and temperature, the time for preshrinking should range from 1 to 2 hours. Trial mixes should be made and aged to determine the longest period through which the mortar, after reworking, will retain sufficient plasticity to permit application. The mortar should be as stiff as possible and yet permit good workmanship. It is not intended or expected that this relatively stiff, preshrunk mortar should be applied as readily as plaster.

Immediately prior to application of mortar, the damp surface to which the new mortar is to bond should be scrubbed thoroughly with a small quantity of mortar, using a wire brush. Remaining loose sand particles should be swept away immediately before application. The mortar should be compacted into the surface, taking care to secure tight filling around the edges, and shaped and finished to correspond with the undamaged surface of the pipe.

(5) *Shotcrete replacement*.—For shotcrete application, the pipe should be turned so that the repair is in a near vertical position and rebound will fall free and will not be included in the replacement. When shotcrete is used to cover exposed steel on the outside surface of a pipe, the coating should be at least three-fourths inch thick. A similar coating on the inside surface should be between one-half and three-fourths inch thick. The shotcrete coating should extend 1 foot in each direction beyond the limits of the exposed steel.

Shotcrete on the outside surface of a pipe should not be finished other than to sweep off any rebound that would interfere with a good membrane coat of white-pigmented sealing compound. After repair of pipe interiors, bells, and spigots by means of shotcrete, the surfaces should be trimmed to correct shape, care being taken to avoid damaging the bond. Interior surfaces should be finished only by rubbing lightly with a damp rag. Bell-and-spigot surfaces should be tooled and finished to conform to requirements for the joint.

Standard commercial equipment of a size commensurate with the small areas to be treated is available from several manufacturers. Also, the equipment shown in figure 124 is adaptable to such work.

(6) *Concrete replacement*.—For replacement repairs made with concrete, the pipe should be turned so that the area where concrete is to be placed will be on the top of the pipe for an outside repair or on the bottom of the pipe for an inside repair. The pipe should be in the latter position for repair of holes completely through the pipe shell, with the pipe lying in a segment of an outside form. Concrete replacement repairs to bells and spigots should be cast with the pipe in a vertical position and the area to be repaired at the top.

Proportions of concrete used for replacement should be the same as used in the original concrete, including the size and amount of sand and gravel and the amount of cement and air-entraining agent. The slump of the concrete as mixed should be between 2 and 3 inches, but the concrete should not be placed until the slump has dropped to zero. The delay for preshrinking concrete should be as long as the concrete will still respond to vibration and a running vibrator will sink into the concrete of its own weight. Such preshrunk, stiff concrete can be molded by ample vibration into an open, unformed horizontal area with little difficulty and will be much less subject to shrinkage than ordinary concrete.

Immediately prior to placing preshrunk concrete, the prewetted, clean surfaces of old concrete should be thoroughly surface dried and then coated with a thin layer of plastic mortar similar in mix to that in the concrete. The mortar should be worked thoroughly into the old concrete surface by shooting with an air gun, by brushing, or by rubbing with the hand in a rubber glove. Preshrunk repair concrete should then be thoroughly compacted into the repair area while the bonding mortar is still fresh and soft.

(7) *Curing of repairs.*—New concrete or portland cement mortar repairs should be covered with four-ply wet burlap as soon as the burlap can be applied without damage to the surface. The wet burlap should be held in position with boards or forms as shown in figure 179.

Repairs of concrete pipe may be cured using the same procedure as in manufacture of the pipe. When repairs are made early in the manufacture, steam curing is often an efficient and suitable method. When repairs are delayed until they must be made away from steam curing facilities, water curing is also acceptable.

Repairs where bond strength of the patch is critical should be wet cured continually for 28 days with the wet burlap in close contact with the repaired surface. Other repairs, where the serviceability does not depend on bond strength of the patch, may be wet cured for 24 hours, after which the surface may be coated with a membrane coating of an approved white-pigmented curing compound. If the surface of the repair is not moist when the burlap is removed, moist curing should be continued for an additional 24 hours before sealing compound is applied.

Where high bond strength is essential and 28 days' moist curing cannot be assured, epoxy mortar or epoxy-bonded concrete repairs should be used. This repair should be applied in the identical manner to that recommended for epoxy repairs to structural concrete (see sec. 136).

(8) *Testing repaired pipe.*—Each pipe on which major repairs have been made (such as repairs extending through the shell thickness or large repairs to bells) should be tested at the service head to assure that the repair is competent. Pipe having lesser repairs should be tested oc-

casionally to assure the repairs are adequate. Representative units of cracked but unshattered pipe should be tested, and if there is no leakage at 50-foot head other than sweating, the pipe may be accepted for operating heads less than 50 feet. Repairs should be aged at least 1 month after specified water curing, then inspected to determine adequacy of bond before the pipe is tested.

(c) *Concrete Cracking and Other Damage in Canal Linings.*—(1) *Sealing joints and random cracks.*—Since cracks in concrete lining are generally caused by or associated with dimensional changes or movement of the slab, it is essential to seal them with a material that will remain flexible and bond to the concrete. This is mandatory if a canal is to be essentially watertight. Also, a good seal will eliminate entrance of sand and silt into the cracks, preventing excessive stress in the lining when expansion occurs.

Invariably a crack will form between old and new concrete. In removal of the old concrete, it is advisable to cut in straight lines, thus facilitating the forming of the joint as illustrated in figure 190. This type of joint also is suggested where it is known that contraction will occur and crack control is needed.

It is usually good practice to remove the damaged section and replace it with a new panel when linings heave from expansion caused by temperature change. In such linings an expansion joint should be included in the repair so there will not be a recurrence of failure. The type of joint shown in figure 190 can be used to maintain a water-tight lining.

If it is necessary to replace several adjoining panels, crack control grooves, spaced as in the original lining, must also be provided. Figure 191 illustrates a typical groove. To weaken the slab sufficiently for crack control, groove depths must be at least one-third of the lining thickness and grooves should be wide enough so that the combination of extension and compression on the sealant will not exceed 25 percent. To resist hydrostatic pressure and compensate for deterioration, the sealer depth should be at least one-half inch.

Random cracking also may occur in concrete canal linings. Figure 192 shows a random crack and a sealer application. Note that the sealer must not be less than three-eighths inch thick for the repair to be successful. Sandblast cleaning is the best and most economical method of securing a clean and suitable bonding surface.

(2) *Use of polysulfide sealer.*—A coal-tar modified polysulfide (two-component) sealer has superior weathering resistance. Adhesion to clean, dry concrete is excellent but not to wet or damp concrete. Specifications require it to resist extrusion through a one-eighth inch crack under 60 feet of head applied for 7 days; therefore, backup material is not needed except in wider cracks. It can be used for sealing both ran-

Polysulfide sealant

B + 1/8"

B=3/8" To
1/2"

5/8"

Paint to prevent
bond, except
in groove for
sealant.

CONTRACTION JOINT

Polysulfide sealant

New concrete

1/2" Sponge
rubber filler

5/8" Min.

Existing concrete

EXPANSION JOINT

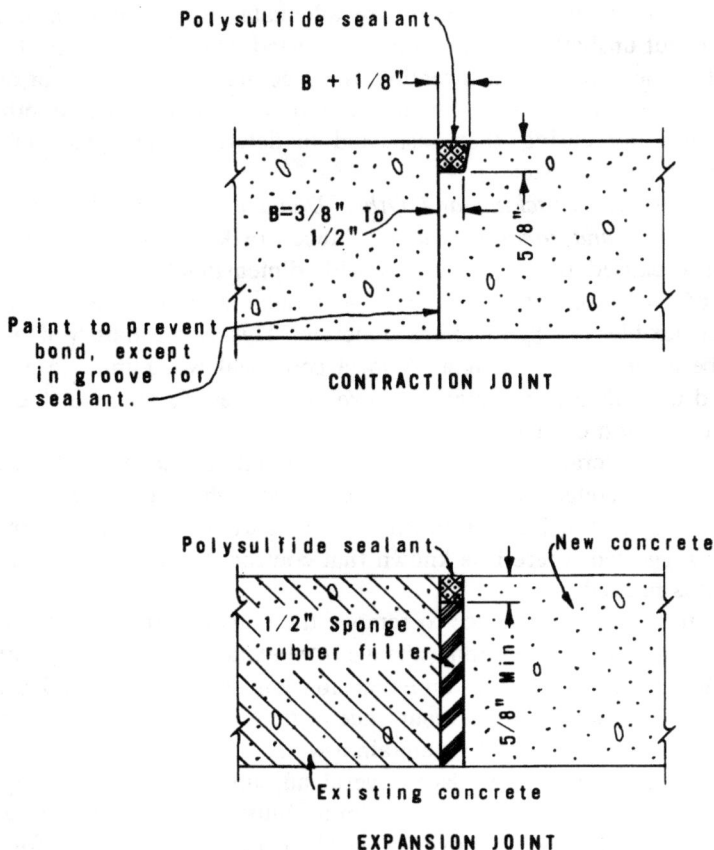

Figure 190.—Contraction and expansion joints for canal lining
repairs. 288–D–3255.

dom cracks and contraction grooves. Experience shows that surfaces must be dry upon application, and air and concrete temperatures must not be less than 50° F.

Economy dictates that a relatively shallow section of sealer be used consistent with hydrostatic pressure but in any case not less than three-eighths inch for cracks and one-half inch for joints. If installed properly, polysulfide sealers should last longer and provide a better seal than rubberized asphalt mastic. Polysulfide canal sealants are available in two types. There is a quick-set material that must be machine applied using a costly mixer-applicator requiring experienced personnel and a slow-set type that can be mixed and applied by hand. The former is more economical on large jobs, and the latter is more adaptable to smaller ones.

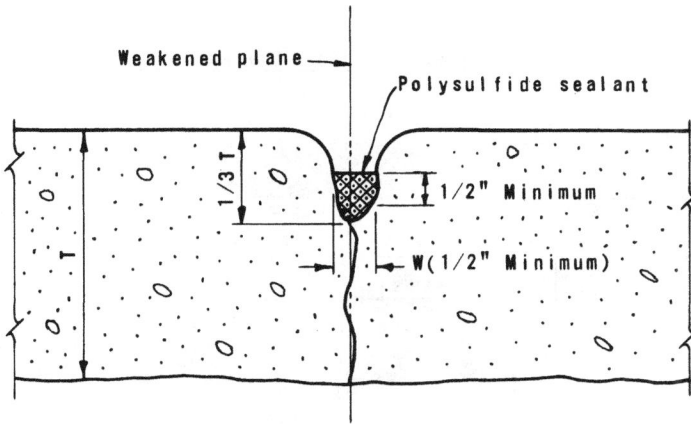

Figure 191.—Crack control groove in canal lining. 288–D–3256.

(3) *Use of preformed reinforced mastic tape.*—Glass cloth reinforced rubberized asphalt mastic tape (fig. 193) has proven satisfactory at several locations for sealing random cracks. The tape must be applied carefully. Both tape and concrete must be warm and dry; also, the tape must be well rolled to assure proper contact of mastic to concrete and to feather the edges of the tape. The concrete surface must be clean before the tape is applied; sandblasting is not always required but is the preferred method of cleaning. Thickness and width of the tape should be determined by judgment. This tape comes in different thicknesses, widths, and lengths and can be purchased commercially.

(4) *Use of other sealing compounds.*—Asphalt mastic sealers have been used by the Bureau for many years for sealing joints and cracks in concrete canal linings. The type used for the past 10 years is low in cost, can be applied as received, is more elastic than previous asphalt mastics, is most durable where continuously submerged or otherwise protected from direct weathering, and depending on service conditions, provides a satisfactory seal for about 5 years. Experience has shown that it will adhere satisfactorily under less than ideal conditions (e.g., damp concrete), less thorough surface preparation, or application over other deteriorated asphalt sealers. It is less weather resistant, however, than polysulfide sealers, and being essentially plastic in nature, it may be extruded through cracks as narrow as one-sixteenth inch under hydrostatic pressures greater than 6 feet of water.

Where the sealer is used in contraction joints under high pressures, a backup material such as butyl sponge rod should be installed at the bottom of the groove, leaving a channel above the backup at least three-fourths inch deep to receive the mastic. It is not recommended for seal-

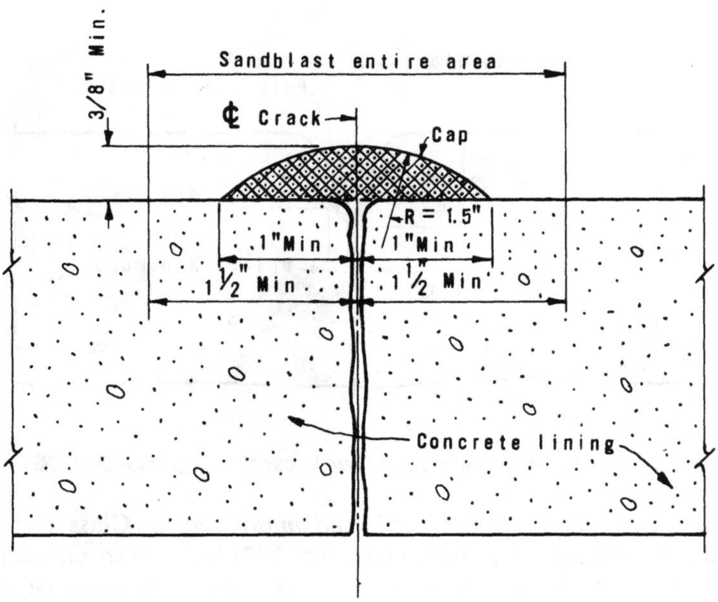

Figure 192.—Random crack repair in canal lining. 288–D–3257.

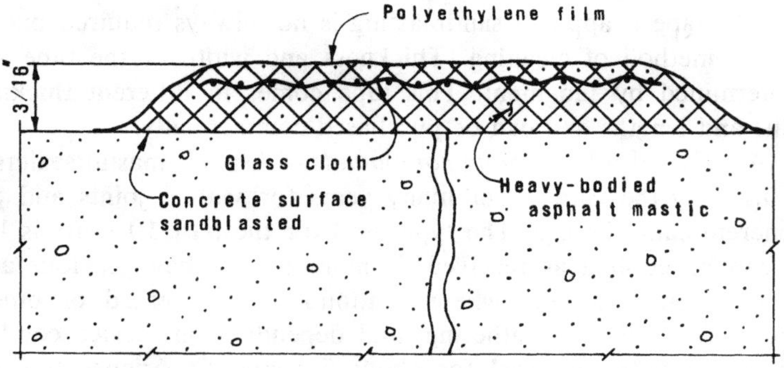

Figure 193.—Section through reinforced asphalt mastic tape used for repair of random cracks. 288–D–3258.

ing cracks under high heads. Under low heads it is vital that the mastic bead be applied over the crack to a depth of at least three-eighths inch to prolong its life. Although this mastic has been used as both joint and crack sealer, it is more satisfactory for sealing contraction joints than for sealing random cracks because cracks tend to reflect through.

The three sealing materials described above can be obtained com-

mercially. Specifications or further information can be obtained from the
Division of General Research, Bureau of Reclamation, Engineering and
Research Center, Denver Federal Center, Denver, Colorado 80225.

(5) *Damage from back pressure or subgrade conditions.*—Where
damage has occurred in concrete canal linings from back pressure or
unstable subgrade conditions, several methods of repair and types of
joints may be necessary. Figure 194 illustrates the replacement of a
section of lining where back pressure has caused damage. Note the 5 to
1 protective slope above the lining and the sandy-clay material placed
near the top to prevent water from getting behind the lining from above.

A flap valve weep near the toe of the slope, shown in figure 194, is
essential to the elimination of hydrostatic back pressure against a concrete lining. Figures 195 and 196 show flap-valve weeps in detail.

B. Maintenance of Concrete

139. Protection of Concrete Against Weathering.—(a) *General.*—As
mentioned in section 128, experience has shown that there are certain
portions of exposed concrete structures more vulnerable than others to
deterioration from weathering in freezing climates. These are exposed
surfaces of the top 2 feet of walls, piers, posts, handrails, and parapets;
all of curbs, sills, ledges, copings, cornices, and corners; and surfaces

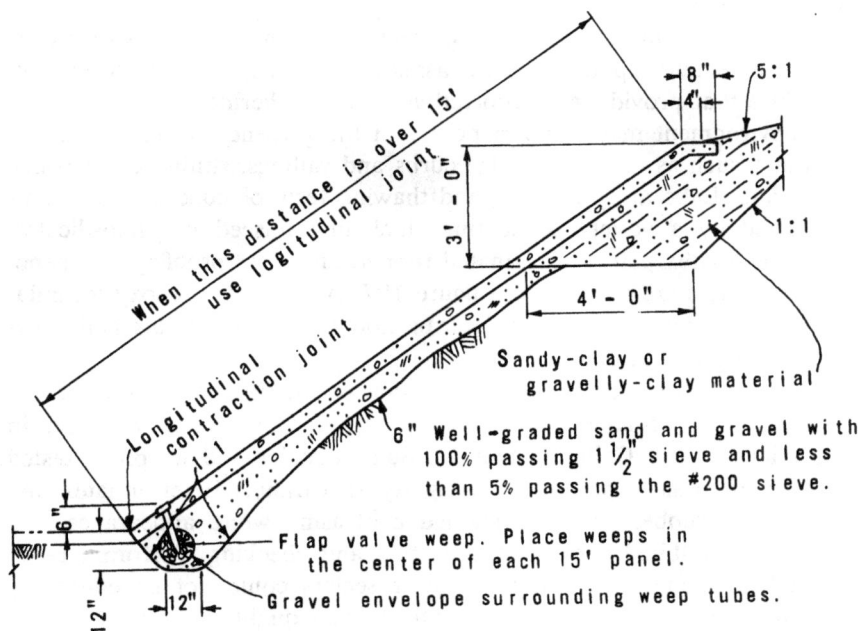

Figure 194.—Canal lining repair where back pressure exists. 288-D-3259.

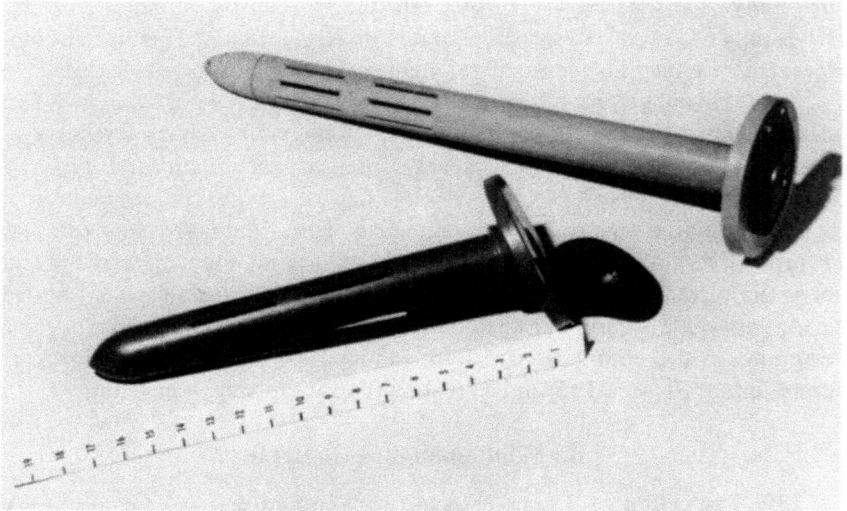

Figure 195.—Flap-valve weeps. P801–D–74764.

in contact with spray or water at frequently changing levels during freezing weather. The durability of these surfaces can be considerably improved and serviceability greatly prolonged by preventive maintenance such as weatherproofing treatment.

Selecting the most satisfactory protective treatment depends to a considerable extent upon correctly assessing the exposure environment. Coatings that provide good protection from weathering in an essentially dry environment may perform poorly in the presence of an abundance of water such as on some bridge curbs and railings, stilling basin walls, and piers. Laboratory freezing and thawing tests of concrete specimens protected by a variety of coatings including linseed oil, fluosilicates, epoxy and latex paints, chlorinated rubber, and waterproofing and penetrating sealers are compared in figure 197. As indicated, epoxy formulations clearly excel in resisting deterioration caused by freezing and thawing in the presence of water.

In observations of effects of outdoor weathering, only four of the listed coatings remained intact on concrete during 6 years' exposure in the relatively dry Colorado climate. Some epoxies and neoprene tested weathered so as to present an unsightly appearance. Exterior latex and chlorinated rubber coatings retained a pleasing white appearance, but the latter exhibited considerable dusting and checking. Performance of linseed oil alone and clear penetrating sealers could not be effectively evaluated as these specimens resembled good quality untreated control concrete after only 6 years of exterior exposure.

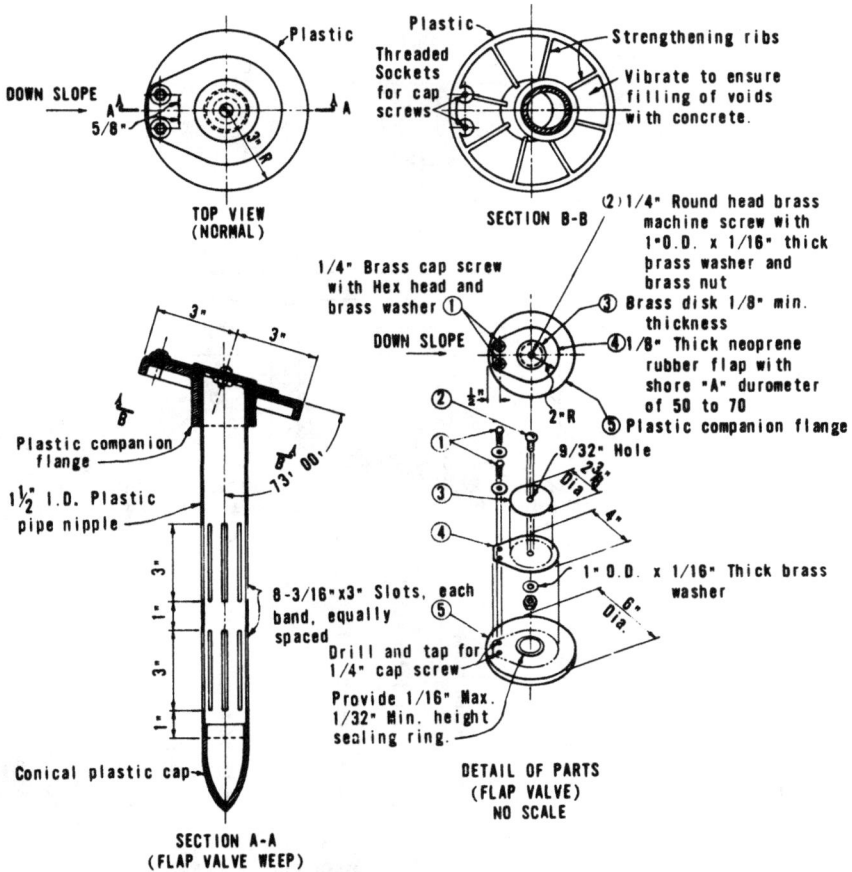

Figure 196.—Details of flap-valve weeps. 288-D-3260.

If epoxy material is used, the project should contact the Denver office for methods of mixing, application, curing, and precautions to be exercised during placement. Although initially more expensive, epoxy probably will not require replacement as frequently as linseed oil-turpentine paint applications.

Except for hand-placed mortar restorations of deteriorated concrete (see sec. 134), weatherproofing treatment is ordinarily not applied on new concrete construction. The treatment is most advantageously used on older surfaces when the earliest visible evidence of weathering appears; that is, the treatment is best used before deterioration advances to a stage where it cannot be arrested. Such early evidence consists primarily of fine surface cracking close and parallel to edges and corners. The need for protection may be indicated by pattern cracking. By treatment of

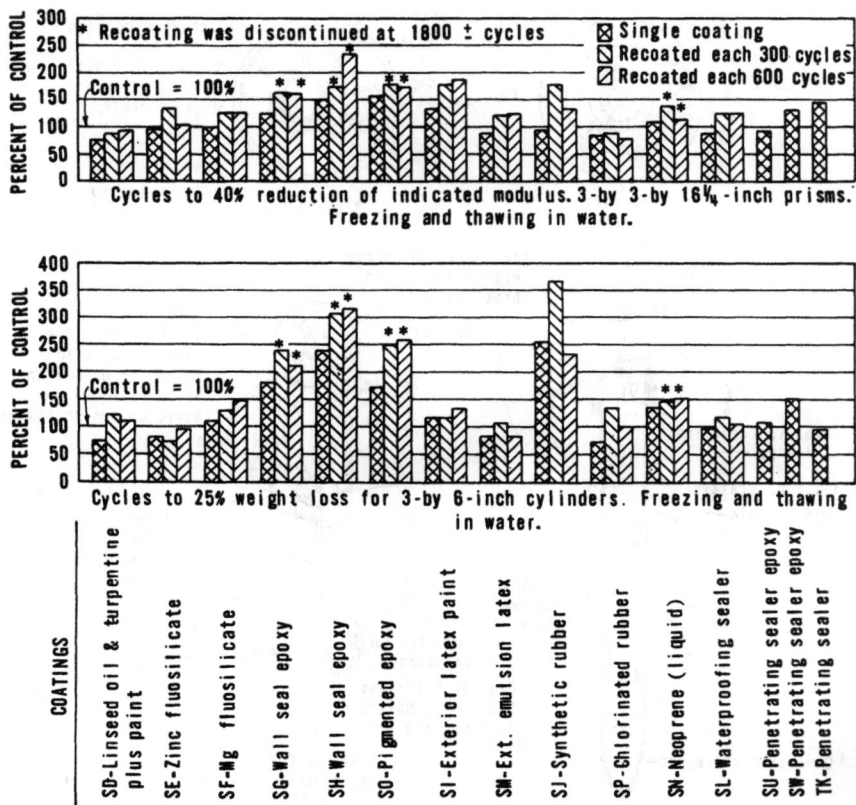

Figure 197.—Comparison of coatings to protect concrete against weathering. Several types of coatings for concrete increase resistance to deterioration caused by freezing and thawing. 288–D–3261.

these vulnerable surfaces in the early stage of weathering, later repairs may be avoided or at least postponed for a long time.

Linseed oil turpentine paint applications have been widely and successfully used by the Bureau to retard deterioration caused by weathering. This treatment is described below:

(b) *Preparation of Surfaces.*—After completion of a curing period, a repair should be allowed to dry 2 or 3 weeks before the waterproofing is applied. New mortar and concrete patches, aged less than 1 year, should be given a neutralizing wash to prevent saponification of linseed oil when used in the waterproofing treatment. A 3-percent phosphoric acid solution brushed over the surface and allowed to dry 48 hours is effective. This application is not necessary on old concrete. Rinsing or brushing after the neutralizing wash has dried is unnecessary. Before applying the waterproofing, the repair must be clean and dry. Dust and loose material should be brushed off. Efflorescence may be removed by scrubbing with a 10-percent solution of hydrochloric acid.

(c) *Treatment of Surfaces.*—After the surface is clean and dry, two coats of linseed oil are applied. The first coat consists of a mixture of 50-percent raw linseed oil and 50-percent turpentine, heated to a temperature of 175° F and applied with an ordinary paint brush. Better results are obtained when the atmospheric temperature is above 65° F. After the first coat has set 24 hours, spots will be evident where the concrete is more porous than the remaining surface. Such areas should be spot treated with the hot mixture and allowed to set 24 hours before the second coat is applied. The second coat consists of undiluted raw linseed oil heated to 175° F and applied in the same manner.

After the second waterproofing coat is thoroughly dry, the entire treated surface should be given two coats of any standard outside white lead and oil paint. The first paint coat should be thinned by the addition of 2 quarts of turpentine and 2 quarts of boiled linseed oil to a gallon of paint so that it will not produce a heavy pigment coat susceptible to scaling but will be heavy enough to brush out uniformly and evenly. The final paint coat should be applied at package consistency without thinning or diluting. Without the protection of this pigmented paint, the oil treatment is subject to rapid deterioration, and its potential value will be seriously impaired. If desired, the top coat can be obtained in a color resembling that of concrete. The paint should be formulated in conformance with Federal Specification TT–P–102, "Paint, Oil Alkyd (Modified), Exterior, Fume-resistant, Ready-mixed, White and Tints."

When there are open cracks in the surface being repaired, a more effective waterproofing may be obtained by filling the cracks. This system lacks flexibility to suitably cover working cracks subject to movement.

(d) *Sealing Cracks in Concrete.*—Small hairline cracks in concrete can be sealed by using the previously described linseed oil treatment or other materials such as modified epoxies. However, sealing of larger cracks probably would require a different technique. Chemical grouting is a method that has been used successfully. Small grout holes (± one-half inch in diameter) are drilled at points located away from the crack to intersect the crack 10 to 15 inches below the concrete surface. A low initial viscosity two-part grouting solution is injected through a mixing head into the grout holes. Pressures required to inject the grout depend upon several variables including grout hole diameter, width of the crack, and grout viscosity. The grouting materials are formulated to set into a rigid or semirigid mass at a predesignated time interval. As the rate of grout setting is influenced by temperature, chemicals in mixing water, and the media being grouted, it is recommended that preliminary field tests of the material be made before repairs are begun.

Many different chemical grouts are available commercially. Some set to form a very hard material, and others set to form a semihard gelatinous material. Materials of the latter type ordinarily have lower initial viscosities and are pumped more readily into small cracks. When maintained in a damp condition, most grouting materials are stable for indefinite periods. Drying produces shrinkage and subsequent loss of repair effectiveness. Specific information concerning chemical grouting may be obtained from the Division of General Research, Bureau of Reclamation, Engineering and Research Center, Denver Federal Center, Denver, Colorado 80225.

Repairing cracks in concrete by the epoxy pressure injection system is described in subsection 136(g).

Chapter VIII

SPECIAL TYPES OF CONCRETE AND MORTAR

A. Lightweight Concrete

140. Definition and Uses.—Lightweight concrete has been used in this country for more than 50 years. Its strength is roughly proportional to weight, and resistance to weathering is about the same as that of ordinary concrete. As compared with the usual sand and gravel concrete, it has certain advantages and disadvantages. Among the former are the savings in structural steel supports and decreased foundation sizes because of decreased loads, and better fire resistance and insulation against heat and sound. Disadvantages include greater cost (30 to 50 percent), need for more care in placing, greater porosity, and greater drying shrinkage.

The principal use of lightweight concrete in Bureau work is in construction of underbeds for floors and roof slabs, where substantial savings can be realized by decreasing dead load. It is also used in some insulated sections of floors and walls.

Lightweight concrete may be obtained through use of lightweight aggregates, as discussed in the following sections, or by special methods of production. These methods include the use of foaming agents, such as aluminum powder, that produce concrete of low unit weight through generation of gas while the concrete is still plastic. Lightweight concrete may weigh from 35 to 115 pounds per cubic foot, depending on the type of lightweight aggregate used or the method of production. In Bureau construction, lightweight concretes have been limited to those whose lightness depends on inorganic aggregates.

141. Types of Lightweight Aggregate.—Lightweight aggregates are produced by expanding clay, shale, slate, diatomaceous shale, perlite, obsidian, and vermiculite through application of heat; by expanding blast-furnace slag through special cooling processes; from natural deposits of pumice, scoria, volcanic cinders, tuff, and diatomite; and from industrial cinders. Lightweight aggregates are sold under various trade names.

(a) *Cinders.*—Cinders are residues from high-temperature combustion of coal or coke in industrial furnaces. Cinders from other sources are not considered suitable. The Underwriters Laboratories limit the average combustible content of mixed fine and coarse cinders for manufacturing precast blocks to not more than 35 percent by weight of the dry, mixed aggregates. Sulfides in the cinders should be less than 0.45 percent and sulfate should be less than 1 percent. Stockpiling of cinders to permit washing away undesirable sulphur compounds is recommended. Cinder concrete weighs about 85 pounds per cubic foot, but when natural sand is used to increase workability in monolithic construction the weight increases to 110 to 115 pounds per cubic foot.

(b) *Expanded Slag.*—Expanded slag aggregates are produced by treating blast-furnace slag with water. The molten slag is run into pits containing controlled quantities of water or is broken by mechanical devices and subjected to sprays or streams of water. The products are fragments that have been vesiculated by steam. The amount of water used has a pronounced influence on the products, which may vary over wide ranges in strength and weight. Concrete in which the aggregate is expanded slag only has unit weights ranging from 75 to 110 pounds per cubic foot.

(c) *Expanded Shale and Clay.*—All expanded shale and clay aggregates are made by heating prepared materials to the fusion point where they become soft and expand because of entrapped expanding gases. With the exception of one product made from shale, the raw material is processed to the desired size before it is heated. The particles may occasionally be coated with a material of higher fusion point to prevent agglomeration during heating.

In general, concrete made with expanded shale or clay aggregates ranges in weight from 90 to 110 pounds per cubic foot.

(d) *Natural Aggregate.*—Pumice, scoria, volcanic cinders, tuff, and diatomite are rocks that are light and strong enough to be used as lightweight aggregate without processing other than crushing and screening to size. Of these, diatomite is the only one not of volcanic origin.

Pumice is the most widely used of the natural lightweight aggregates. It is a porous, froth-like volcanic glass usually white-gray to yellow in color, but may be red, brown, or even black. It is found in large beds in the Western United States and is produced as a lightweight aggregate in several States, among which are California, Oregon, and New Mexico. Concrete made with sound pumice aggregate weighs from 90 to 100 pounds per cubic foot. Structurally weak pumice having high absorption characteristics may be improved in quality by calcining at temperatures near the point of fusion.

Scoria is a vesicular glassy volcanic rock. Deposits are found in New Mexico, Idaho, and other Western States. Scoria resembles industrial cinders and is usually red to black in color. Very satisfactory lightweight concrete, weighing from 90 to 110 pounds per cubic foot, can be made from scoria.

When obsidian is heated to the temperature of fusion, gases are released which expand the material. The interiors of the expanded particles are vesicular and the surfaces are smooth and quite impervious. Expanded obsidian has been produced experimentally. The raw material was crushed and screened to size and coated with a fine material of higher melting point to prevent agglomeration.

The rock from which perlite lightweight aggregate is manufactured has a structure resembling tiny pearls compacted and bound together. When perlite is heated quickly, it expands with disruptive force and breaks into small expanded particles. Usually, expanded perlite is produced only in sand sizes. Concrete made with expanded perlite has a unit weight ranging from 50 to 80 pounds per cubic foot. It is a very good insulating material.

Vermiculite is an alteration product of biotite and other micas. It is found in California, Colorado, Montana, and North and South Carolina. The color is yellowish to brown. On calcination, vermiculite expands at right angles to the cleavage and becomes a fluffy mass, the volume of which is as much as 30 times that of the material before heating. It is a very good insulating material and is used extensively for that purpose. Concrete made with expanded vermiculite aggregate weighs from 35 to 75 pounds per cubic foot; strengths range from 50 to 600 pounds per square inch.

142. Properties of Lightweight Aggregates.—Properties of various lightweight aggregates, as reflected by those of the resulting concrete, vary greatly. For example, the strength of concrete made with expanded shale and clay is relatively high and compares favorably with that of ordinary concrete. Pumice, scoria, and some expanded slags produce a concrete of intermediate strength; perlite, vermiculite, and diatomite produce a concrete of very low strength.

Insulation properties of low-strength concretes, however, are better than those of the heavier, stronger concretes. The insulation value of the heaviest material (crushed shale and clay concrete) is about four times that of ordinary concrete.

All lightweight aggregates, with the exception of expanded shales and clays and scoria, produce concretes subject to high shrinkage. Most of the lightweight concretes have better nailing and sawing properties than do the heavier and stronger conventional concretes. (For information on nailing concrete, see part C of this chapter.) However, nails, although easily driven, fail to hold in some of these lighter concretes.

143. Construction Control of Lightweight Concrete.—Commercially available lightweight aggregate is usually supplied in three principal sizes depending upon its application. These are fine, medium, and coarse and range in size to ¾-inch maximum. Production of uniform concrete with lightweight aggregate involves all the procedures and precautions that have been discussed elsewhere in this manual in connection with ordinary concrete. However, the problem is more difficult where lightweight aggregates are used because of greater variations in absorption, specific gravity, moisture content, and amount and grading of undersize. If unit weight and slump tests are made frequently and the cement and water content of the mix are adjusted as necessary to compensate for variations in the aggregate properties and condition, reasonably uniform results can be obtained.

Concretes made with many lightweight aggregates are difficult to place and finish because of the porosity and angularity of the aggregates. In some of these mixes the cement mortar may separate from the aggregate and the aggregate float toward the surface. When this occurs, the condition can generally be improved by adjusting aggregate grading. This can be done by crushing the larger particles, adding natural sand, or adding filler materials. The placeability can also be improved by adding an air-entraining agent. The amount of fines to be used is governed by the richness of the mix; as sand content is increased, the optimum amount of fines is reached when the concrete no longer appears harsh at the selected air content. From 4- to 6-percent air is best for adequate workability, and the slump should not exceed 6 inches. Pumping of lightweight aggregate concrete is discussed in section 102.

To ensure material of uniform moisture content at the mixer, lightweight aggregate should be saturated 24 hours before use. This wetting will also reduce segregation during stockpiling and transportation. Dry lightweight aggregate should not be fed into the mixer; although this will produce a concrete which can be readily placed immediately after being discharged, continuing absorption by the aggregate will cause the concrete to segregate and stiffen before placement is completed.

It is generally necessary to mix lightweight concrete for longer periods than conventional concrete to assure proper mixing. Workability of lightweight concrete with the same slump as conventional concrete may vary more widely because of differences in type, porosity, and specific gravity of the materials. For the same reason, the amount of air-entraining agent required to produce a certain amount of air may also vary widely. Continuous water curing, by covering with damp sand or use of a soil-soaker hose, is particularly advantageous where concrete is made with lightweight aggregate.

B.   Heavyweight Concrete

144. Definition and Uses.—Wide use and application of heavy concrete is relatively new in the construction industry, coming of age coincidentally with the development and practical application of atomic energy. Heavy concrete may vary in unit weight between approximately 150 pounds per cubic foot for concrete using conventional aggregate and 290 pounds per cubic foot for concrete containing steel shot as fine aggregate and steel punchings as coarse aggregate. It is used principally to shield personnel and to provide protection from nuclear radiation. With such a limited use, it has no application to date on Bureau projects.

145. Types of Heavy Aggregate.—Heavy aggregates, or materials utilized as such, are both naturally occurring and manufactured. Heavier aggregates are of the latter category and are in the form of smelted metal; however, various quarry materials or ores, which of course are less expensive, have been utilized for nuclear shielding with satisfactory results.

(a) *Barite*.—Barite is a quarry rock composed prinicipally of 90 to 95 percent of barium sulfate, $BaSO_4$, and small percentages of iron oxide, chalcedony, clay, quartz, and zeolites and having an apparent specific gravity ranging from 4.1 to 4.3. This rock is amenable to use in both conventional and preplaced aggregate concrete. These concretes develop an optimum density of 232 pounds per cubic foot and optimum compressive strength of about 6,000 and 5,000 pounds per square inch, respectively. In general, the same gradings and mix proportions can be used with barite that are employed with conventional concrete aggregate.

(b) *Mineral Ores (Limonite, Magnetite)*.—Limonite and magnetite are iron ores of high density, ranging in specific gravity between 3.6 and 4.7. These types are readily available at lower cost than barite, and are amenable to processing and use as fine and coarse aggregate in preplaced aggregate and conventional concretes. Densities from 210 to 224 pounds per cubic foot and compressive strengths of 3,200 to 5,700 pounds per square inch can be obtained. The same gradings and mix proportions are generally used with these materials as with barite and iron products.

(c) *Iron Products*.—Iron products in the form of ferrophosphorous, steel punchings, and sheared bars for coarse aggregate and steel shot for fine aggregate, having specific gravities between 6.2 and 7.7, produce the heaviest concrete (250 to 290 pounds per cubic foot). This concrete compares favorably with concrete containing conventional aggregates, with strengths of about 5,600 pounds per square inch at 28 days being obtained.

None of the heavy aggregate concrete has proven to be suitable for exposure to weathering or abrasion, but when protected from these forces such concrete should provide good service. All procedures, methods of handling materials, and precautions discussed elsewhere in connection with production and control of conventional concrete should be followed in producing heavy concrete of optimum quality. Because of limited application of heavy concrete to Bureau construction, the elastic, physical, and thermal behaviors are not discussed.

C. Nailing Concrete

146. Definition, Use, and Types.—Concrete into which nails can be readily driven and which will hold the nails firmly is called nailing concrete. In Bureau work such concrete is used for constructing cants to which roofing material and flashing can be nailed. Among the aggregates that produce good nailing concrete are sawdust, expanded slag, natural pumice, perlites, and volcanic scoria. Because of the widespread availability of modern mechanical fasteners suitable for attachment to conventional concrete, the Bureau no longer specifies nailing concrete for new construction.

147. Sawdust Concrete.—Good nailing concrete can be made by mixing equal parts, by volume, of portland cement, sand, and pine sawdust with sufficient water to give a slump of 1 to 2 inches. Nailing is easier if the sand passes a No. 16 or No. 8 screen. Rigid adherence to the stated mix proportions is not essential. If the concrete is too hard, the amount of sawdust may be increased as much as 100 percent while keeping the quantities of cement and sand the same. Concrete proportioned on this basis is very workable and bonds well with the base concrete. After sawdust concrete is 3 days old, nails can easily be driven into it and have excellent holding power.

The concrete should be mixed thoroughly, preferably in a mixing machine unless the quantity is very small. It should be moist cured for 2 days and then allowed to dry for a day or more before any nailing is done.

148. Types and Grading of Sawdust.—Sawdust should be clean, free from chips and lumps that will not pass a ¼-inch screen and not so fine that all will pass a No. 16 screen. Concrete made with coarse sawdust requires about 24 hours to harden, whereas that made with fine sawdust requires about 48 hours. An increase in the fineness of sawdust (greater surface area of the wood particles) may result in extraction of a larger percentage of organic acids and consequently in retarded set and reduced strength.

The following tabulation gives results of tests of different types of sawdust. The tests show that some types are entirely unsuitable for use.

Material	*Notes*
Sugar pine	Set hard at 1 day. Good nailability.
Pine	Set hard at 2 days. Good nailability.
Pine and fir mixture	Set hard at 3 days. Good nailability.
Hickory, oak, or birch	No set at 3 days, some at 14 days. Never satisfactory.
Oregon fir	The sawdust is very fine. Partial set at 28 days.
Cedar	No set at any time.

Analyses of pine and cedar sawdust for tannin (tannic acid) showed the cedar to contain 2-percent tannin and the pine to contain none. Appreciable amounts of bark in the sawdust retarded setting and weakened the concrete.

In view of the variable behavior of different kinds of sawdust, it is advisable to try a sample of the material before procuring the quantity required.

D. Porous Concrete

149. Definition and Use.—Porous concrete is a special type that is commonly used either where free drainage is required or where lighter weight and low conductivity are to be provided without the use of lightweight aggregates. (Sometimes the use of lightweight aggregates is not practicable or desirable.) Porous concrete is ordinarily produced by gap grading or single-size aggregate grading. In special draintile, a No. 4 to $\frac{3}{8}$-inch or $\frac{3}{8}$- to $\frac{1}{2}$-inch aggregate is frequently used alone; a low water-cement ratio and the minimum amount of cement are required to merely cover and cement the aggregate particles together into a mass much resembling that obtained in a popcorn ball. Occasionally, inserts of porous concrete may be installed as weep holes or drains in hydraulic structures such as canal linings to prevent back pressure or uplift from breaking the lining upon dewatering. Such concrete may require type V cement, especially for drainage structures or special draintile where soluble sulfate conditions exist.

Occasionally, porous concrete is placed upon rock foundations under split sewer pipes to drain ground water. Specifications call for 7-day strengths, as determined by 6- by 12-inch cylinders, of not less than 1,000 pounds per square inch and porosity such that water will pass through a slab 12 inches thick at the rate of not less than 10 gallons per minute per square foot of slab with a constant 4-inch depth of water on the slab.

E. Preplaced Aggregate Concrete

150. Definition and Use.—Preplaced aggregate concrete, sometimes referred to as prepacked concrete, is made by forcing grout into the voids of a compacted mass of clean, graded coarse aggregate. The preplaced aggregate is washed and screened to remove fines immediately before placing in the form. As the grout is pumped into the forms it displaces any water and fills the voids, thus creating a dense concrete having a high content of aggregate. The advantage is the ease with which preplaced aggregate concrete can be placed in certain locations where placement of conventional concrete would be extremely difficult. Preplaced aggregate concrete is especially adaptable to underwater construction, to concrete and masonry repairs, and, in general, to certain types of new structures. It has been used in construction of bridge piers, atomic reactor shielding, plugs for outlet works in dams and tunnels, in mine workings, and for embedment of penstocks and turbine scrollcases, as well as a great variety of repair work. Recently, it has been used for architectural treatment applications. Since preplaced aggregate concrete is most adaptable to special types of construction, it is essential that the work be undertaken by well-qualified personnel experienced in this method of concrete construction.

151. Properties of Preplaced Aggregate Concrete.—Although preplaced aggregate concrete develops strength somewhat more slowly than regular concrete, the strengths of both concretes containing 1½-inch-maximum size aggregate are about equal after 90 days. Under ordinary drying conditions, following proper curing, the drying shrinkage of preplaced aggregate concrete containing 1½-inch-maximum size aggregate is within the range of 200 to 400 millionths; the drying shrinkage of ordinary concrete containing the same maximum size aggregate is from 400 to 600 millionths. Preplaced aggregate concrete has shown excellent bond to many old concrete structures where it has been used for repair and has superior resistance to alternate cycles of freezing and thawing with the proper amount of entrained air.

152. Grout Materials and Consistency.—Grout for preplaced aggregate concrete may consist of sand of specified gradation for concrete, cement, and water mixed at high speeds to a creamy consistency. Alternatively, it may contain fine sand, portland cement, pozzolanic filler, and an agent designed to increase the penetration and pumpability of the mortar. In the first method a minimum size aggregate of 1½ inches is used. The second method uses a ½-inch-minimum size aggregate. The maximum sizes for either method may be as large as are available, provided the aggregate can be readily handled and placed.

In general, the gradation of fine sand described in the latter method should be such that all the sand will pass a No. 8 screen and at least 95 percent will pass a No. 16 screen. The fineness modulus will usually range between 1.2 and 2 for best pumping characteristics. Quality of sand should be equal to that of good concrete sand. Natural sand is preferable because of its particle shape.

The pozzolanic filler reacts with lime liberated during hydration of cement to form insoluble strength-producing compounds. This finely divided material also increases the flowability of the grout and tends to decrease bleeding and separation.

A water-reducing, set-controlling agent is added to inhibit early stiffening of the grout; also, it enhances the fluidity and holds the solid constituents in suspension. It contains a small amount of aluminum powder which causes the grout to expand slightly before initial set, thus reducing settlement shrinkage.

The consistency of grout for preplaced aggregate concrete should be uniform from batch to batch and should be such that it may readily be pumped, under reasonably low pressure, into the voids of the preplaced mass of aggregate. Consistency is affected by water content, sand grading, type of cement, and type and amount of agent. For each mix there are optimum amounts of filler and agent which produce best pumpability or consistency, and tests are necessary for each job. Consistency of grout may be determined by any one of several methods. One method is based on the time required for a cone filled with the grout to be emptied by gravity. Another method is by use of a torsion pendulum consistency meter.

153. Coarse Aggregate.—The coarse aggregate should meet all requirements applying to coarse aggregate for ordinary concrete. It is important that the aggregate be clean. It should be well graded from ½- or ¾-inch-minimum size to the largest maximum size practicable and, after compaction in the forms, should have a void content of from 35 to 40 percent. For preplaced aggregate concrete, in which a sand of conventional grading is used, the minimum nominal size of coarse aggregate should be not less than 1½ inches.

154. Construction Procedures.—Forms for preplaced aggregate concrete may be of wood, steel, or other materials suitable for conventional concrete. The form workmanship should be of better quality than is normally suitable for conventional concrete. This is important to minimize grout leakage. Also, since the grout is fluid longer than concrete is in a plastic condition, forms must be constructed to resist more lateral pressure than occurs with conventional concrete placed at normal rates. Bolts should be tightly fit through the sheathing. Possible points of grout leakage should be calked. This usually can be done from the outside during grout injection. Leaking forms, besides affecting appearance, can be a source of trouble and should be avoided wherever possible by careful form construction.

The grout pipe system is used to deliver and inject grout into the pre-placed aggregate, to provide means for determining grout elevations within the aggregate mass, and to provide vents in enclosed forms for water and air escape. Proper design and arrangement of the pipe system is essential for a successful placement. The simplest and most reliable system consists of a single pipeline connected to insert pipes positioned during placing of the aggregate. The pumping system should have a bypass for returning grout to an agitating tank.

The length of delivery line should be kept to a practicable minimum. Pipe sizes should be such that during operation under normal conditions the grout velocity ranges between 2 to 4 feet per second, or at a pumping rate of about 1 cubic foot of grout per minute through a 1-inch-diameter pipe. Higher velocities require excessive pumping pressure. The recommended velocity range is for delivery pipes up to 300 feet long. From 300 to a maximum of 1,000 feet, the diameter will need to be increased about one pipe size to avoid excessive pumping pressure.

Grout insert pipes for intruding grout into an aggregate mass are normally ¾ to 1 inch in diameter and may be placed vertically or at various angles to inject grout at the proper point. The pipes should be in sections about 5½ feet long for easy withdrawal. For depths below 15 feet they should be flush coupled. For shallower depths standard pipe couplings may be used.

Connections between the grout delivery line and insert pipes should be quick-opening fittings. Quick-disconnect pneumatic fittings are not suitable because of the reduction in cross section of the flow area. Valves should be quick-opening, plug type which can readily be cleaned.

Spacing of insert pipes depends on aggregate gradation, void content, depth, and area of work and location of embedded items. Spacing of insert pipes may range from 4 to 12 feet; 5 or 6 feet spacing is commonly used. For the purpose of insert pipe layout it is assumed that the grout surface will be on about a 1 to 4 slope in a dry placement and 1 to 6 underwater, although actual grout surfaces may be considerably flatter. It is helpful to color code or number and record location of each insert pipe so there is no question where the outlet end is reaching.

Through sounding wells, usually slotted pipes, the level of the grout can be determined with reasonable accuracy. The ratio of sounding wells to insert pipes ranges from 1:4 to about 1:10. These sounding wells, through which a sounding line (equipped with a 1-inch-diameter float weighted so that it will sink in water and float on grout) may be lowered, are usually 2 inches in diameter. There should be no burrs or obtrusions inside the sounding well on which the float will catch.

The pump should be of a positive displacement type, such as a piston pump or a progressive cavity type. Although a well-proportioned grout mix will retain solids in suspension within a piping system, pumps shut down for prolonged periods will permit the sand to settle within the pump and lines with attendant difficulties. As pumps normally require a period of maintenance on each shift, one or more standby pumps should be provided for quick changeover to maintain continuous operations. The pump should have a pressure gage on the outlet line to indicate any incipient line blockage.

Fundamentally, grout injection should start at the lowest point within the form and be continued until the placement is completed. Usually a sufficient quantity of grout is pumped through the insert pipe to raise the grout level from 6 to 12 inches. The insert pipe outlet is set initially 6 inches from the bottom and progressively raised as grouting proceeds. The lower end should remain embedded 12 inches below the grout surface. The grout surface should be kept relatively level, although often a gentle slope is maintained. Care should be exercised not to permit grout to cascade on a steep slope through the aggregate, causing separation of sand from grout. Adequate venting should be provided to ensure complete filling and prevent entrapment of air or water in enclosed spaces.

Internal vibration cannot be employed with this method of placing, but external vibration of the forms can be and is beneficial in improving surface appearance. If it is not done, a spotty appearance will develop where coarse aggregate particles have been in contact with the form.

Quality control of the preplaced aggregate concrete is maintained by controlling consistency or thickness of the grout. For this purpose a flow cone as shown in figure 198 is used. The cone is filled to the level indicated by the pointer with the outlet end closed by thumb pressure. The efflux is timed by a stopwatch. The consistency or efflux for fine sand grout with a fluidifier and fly ash should range between 18 to 22 seconds.

F. Prestressed Concrete

155. Definition and Use.—Prestressed concrete is based upon the principle of using high tensile steel alloys to produce a permanent precompression in areas of a concrete structure that would normally be subjected to tension, for which concrete has very little strength. By introducing compression into such members before normal loading is applied, a portion of potential tensile stress can be counteracted, thereby reducing the total cross-sectional area of steel reinforcement required.

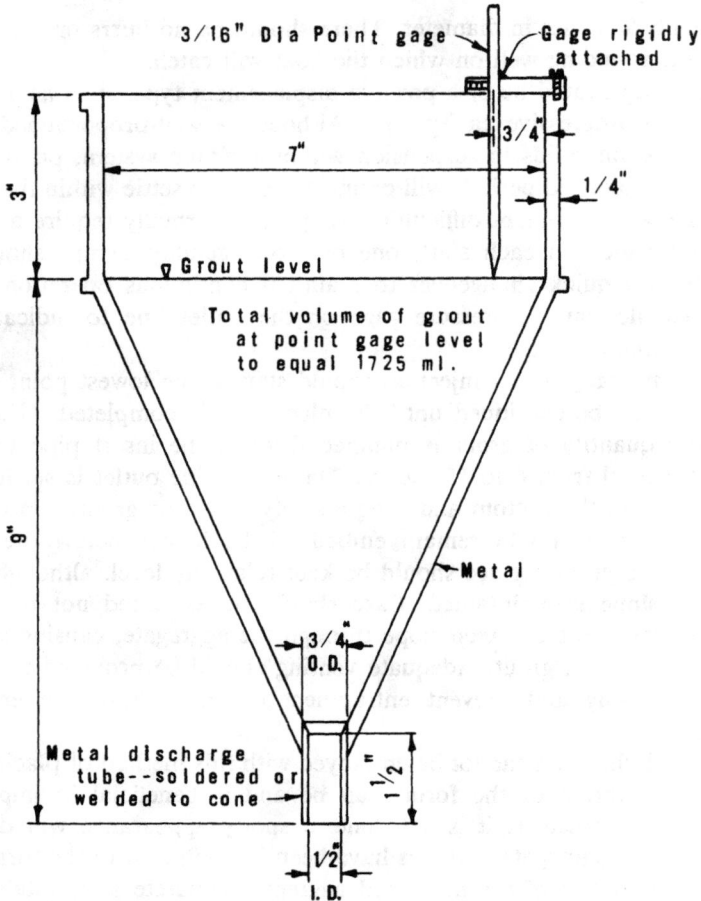

Figure 198.—Cone for measuring consistency of grout for preplaced
aggregate concrete. 288–D–3262.

There are two modern methods of prestressing concrete. One method,
known as pretensioning, consists of placing concrete around reinforcement
tendons that have been properly placed in the form and stressed under
tension to the desired degree. Concrete is carefully placed, consolidated,
and cured to assure adequate bond. After the concrete has developed
the necessary minimum strength, the tensile anchorage of reinforcement
is released; and through bond between the steel and concrete, the initial
tension in the steel produces the required compression in the concrete.

The other method, called posttensioning, involves preforming voids
or ducts throughout the length of the concrete structural member or in-
corporating tubes or sheathing in the member and placing the reinforcement tendons within the channels or tubes in such a manner that they

will be free to move throughout their entire length after the concrete has hardened. After adequate curing of the concrete, the reinforcement tendons are stretched to the required tension and anchored to the concrete at the ends to retain tension in the steel and compression in the concrete. If the steel tendons are in open channels or conduits, these spaces are then usually pumped full of grout to bond the steel tightly to the concrete throughout the length of the tendon, thus aiding the uniform transfer of stresses during live loading and protecting the stressed steel tendons from possible corrosion by moisture, gases, or other corrosive materials. Occasionally, the stressed tendons of single span and continuous unbonded posttensioned beams are not grouted, and the beams are designed for inclusion of additional unprestressed tendons. The tendons to be prestressed are coated with a corrosion-preventive lubricant and covered with a moisture-resistant jacket. The tendons are held by endplates and grippers sealed with a suitable drypack or mortar. Such beams will act together as a flexural member after cracking occurs and not as a shallow tied arch.

Prestressed concrete design is currently being applied to many types of concrete construction. It is especially applicable to structural members such as beams, girders, or bridge-deck panels which can be precast in a central casting yard and incorporated in a continuous structure. It has also been highly successful in the construction of roof tees and slabs and in circular concrete tanks and pipe where cracking must be eliminated.

Advantages of prestressed concrete construction include the following:

(1) There is a high degree of crack reduction. The areas where cracking normally occurs because of tensile stresses are placed under compressive load to largely offset this tendency. Cracking caused by drying shrinkage can also be largely eliminated by proper prestress design.

(2) The freezing and thawing durability of prestressed concrete is slightly higher than that of similar unstressed concrete. This is partially because of the reduction in cracking and the fact that compressive stresses keep shrinkage cracks tightly closed.

(3) Precasting of structural units provides distinct advantages in many building projects. High quality of the finished product, made possible by excellent concrete control in a permanently established plant having controlled curing, provides superior structures. Precast parts may be delivered to a construction site of limited area and rapidly erected without extensive formwork and shoring. Modern methods enable the expeditious bonding and incorporation of these units into continuous structures with a minimum of skilled help and construction equipment. The precise, many-use forms employed in precasting provide a greater degree of uniformity and better surface

appearance than is usually possible for concrete produced on the job.

(4) Efficient utilization of weight in fabrication of prestressed concrete members is also an advantage. The high precision of placement and high tensile strength of steels normally used in prestressing, along with use of concrete under compressive stress to carry tensile loads, give maximum efficiency in size and weight of structural members, thus providing space economy and transportation economy in building modern structures.

(5) Applications of various prestressed techniques enable quick assembly of standard units such as repetitive bridge members, building frames, and roof and bridge decks to provide important construction time economies. It is possible that the structure can even be largely fabricated elsewhere while the site is being prepared.

Conditions in the use of prestressed concrete are:

(1) Good quality materials should be used and quality control maintained. For most construction uses, types I, II, or III cements will be suitable. However, if the prestressed members are to be in contact with sulfate-bearing soils or water, use of a moderate sulfate-resisting cement (type II) or sulfate-resisting cement (type V), depending on the sulfate conditions, would be necessary. When known or suspected reactive aggregates are to be used in concrete, the cement, regardless of type used, should also have a low-alkali content to provide positive protection against potential disruptive expansion caused by alkali-aggregate reaction.

(2) Use of a water-reducing admixture (WRA) will improve strength and reduce the amount of water required for the same slump. Some WRA's also entrain air; if the desired air content is obtained with the WRA, use of an air-entraining agent (AEA) as well would not be necessary. In some construction, however, the use of an AEA might be desirable.

(3) Calcium chloride or admixtures containing calcium chloride should not be used in concrete for prestressing, as the stressed condition of the steel reinforcement makes it more subject to corrosion in the presence of chlorides.

(4) Mixes for prestressed concrete must necessarily be highly workable and produce high strength concrete, usually 4,000 or more pounds per square inch at 28 days. Workability is essential, as placement must be accurate and thorough in relatively close clearance areas, and high bond must be produced. High strength is necessary in effective use of prestressed concrete; consequently, relatively rich, low water-cement ratio mixes, usually with ¾- or 1½-inch-maximum size aggregate, are customarily used. Fre-

quently, high-early-strength cement is employed, and curing, at least during the early ages, is by low-pressure steam so that valuable forms and casting space may be reused at the earliest possible moment. High-early-strength cement also accelerates the necessary development of minimum strength for application of the compressive load. Steam or warm-moist curing is preferable. Such curing should be continuous until the concrete has attained a high percentage of the required strength. Usually the concrete is cured until companion test cylinders, made and cured with prestressed members, have attained compressive strengths meeting the required strength for stress release of the tendons. When the concrete is heat cured, the detensioning should be accomplished while the concrete is warm and moist to prevent cracking or undesirable stresses caused by dimensional changes that may occur.

(5) The concrete should be so designed that the shrinkage and creep in prestressed applications are minimized so that the steel may retain as much initial tension and produce the maximum feasible compression possible.

(6) The proper eccentricity for tensile steel in prestressed concrete is an important design consideration to be carefully maintained so that the structure will perform efficiently. This eccentricity is an important part of the efficient application of steel and concrete for minimum weight and for maximum space saving. Careful workmanship to accurately maintain this important design relationship and to provide adequate bond of concrete to steel is of utmost importance.

G. Vacuum-Processed Concrete

156. Definition, Characteristics, and Uses.—Vacuum-processed concrete is produced by applying a vacuum to formed or unformed surfaces of ordinary concrete immediately or very soon after the concrete is placed. This patented treatment removes water from concrete adjacent to the surface and removes air bubbles which would otherwise appear as holes in the surface. (See fig. 199.) Air bubbles, being noncontinuous, are removed from the surface but not from the interior. The depth of water extraction and the amount of water removed depend on coarseness of the mix, mix proportions, and the number of surfaces to which vacuum is applied. Water content can be reduced to a depth of 6 to 12 inches, and in amounts up to one-third of the mixing water a few inches from the surface. Removal of an average of 20 percent of the water from a 6-inch surface layer is common. Experience has demonstrated that best results are obtained from vacuum treatment when (1) the mix contains the practicable minimum of fines, (2) newly placed concrete fully covers

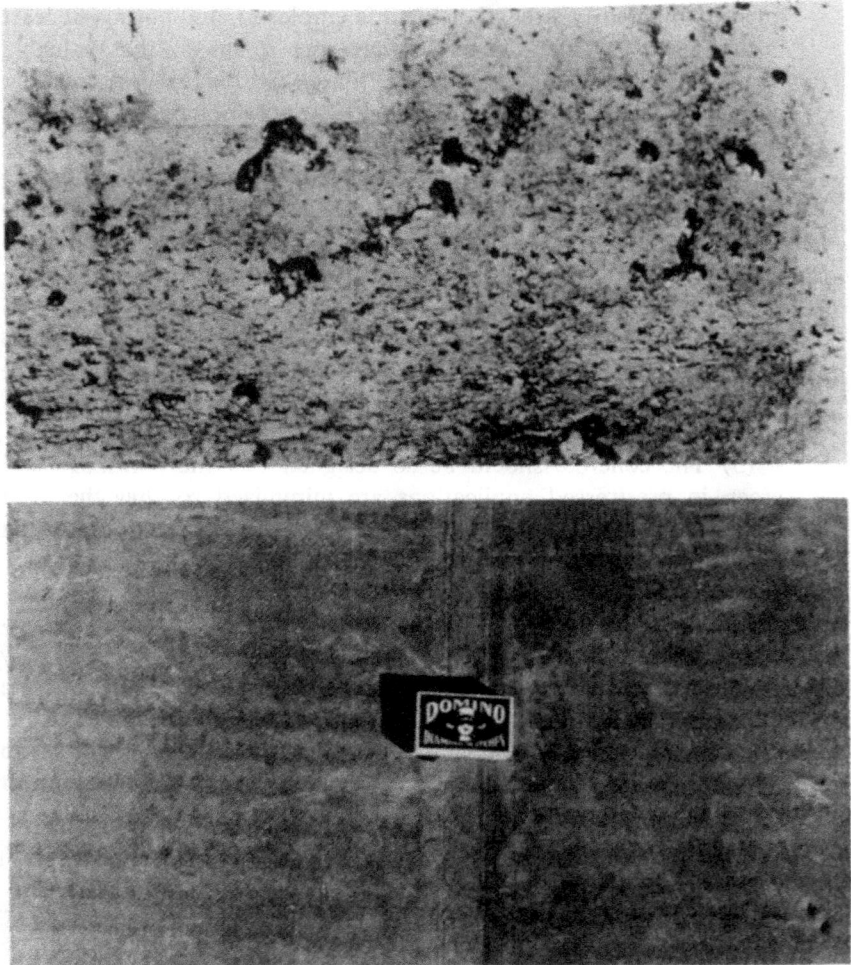

Figure 199.—Surfaces of concrete formed on an 0.8 to 1 slope at Shasta Dam, Central Valley project, California. Upper and lower pictures show surfaces produced by wood forms and vacuum forms, respectively. PX–D–33060.

the vacuum panel so that the vacuum can be applied promptly while the concrete is still plastic, and (3) concrete near the panel is vibrated during the first few minutes of the vacuum treatment.

The marked reduction in water content with vacuum treatment results in higher strength and greater durability. Vacuum treatment increased the 3-day strength of one concrete from 800 to 1,800 pounds per square inch. The earlier strength is of great advantage in precast concrete work because it permits immediate removal of forms and early release of base plates and thus appreciably lessens the investment in such equipment.

Elimination of surface pits on precast concrete pipe is especially desirable.

Vacuum treatment for the sole purpose of improving appearance is not justifiable; neither can it ordinarily compete with the much less expensive air entrainment as a means for enhancing durability. Vacuum treatment, like the use of absorptive form lining, has been specified to produce a surface that would have increased resistance to the action of water flowing at high velocity. It appears, however, that a close approach to perfect alinement of flow-line surfaces is much more important in this respect than increased strength or elimination of surface pitting.

157. Vacuum Forms and Panels.—Vacuum is applied to concrete surfaces through vacuum hose attached to special vacuum mats or form panels. (See figs. 200 and 201.) The mats for unformed surfaces are generally reinforced plywood faced with two layers of screen wire covered by muslin. For unformed curved surfaces, such as the buckets of dams, flexible steel which will adapt to the curved surface is used instead of plywood. Similar vacuum liners are clamped to the inside face of forms. A fiberglass cloth has been used without screen backing for the vacuum lining of steel forms for concrete pipe. The perimeter of each mat or area of form served by a vacuum outlet is sealed against loss of vacuum by a narrow strip of calking compound in the screen wire. The seal of the vacuum mats for unformed surfaces is further ensured by a 1-inch

Figure 200.—Placing concrete in spillway bucket at Shasta Dam. Vacuum processing mats are shown at the right. PX–D–33061.

Figure 201.—Vacuum panels being constructed at Angostura Dam, in South Dakota. View shows rubber strips and hardware cloth before covering with muslin. PX–D–33514.

flap of rubberized cloth attached to the perimeter of the mat and pressed lightly onto the concrete. Areas of mats are usually about 12 square feet, and a variety of shapes can be used. A 3- by 4-foot mat is a common size. The individual vacuum areas on the forms are usually several feet long horizontally and 12 to 18 inches high so that each area may be quickly covered with concrete and the vacuum applied while the concrete is quite fresh. The muslin can be treated with a liquid product to permit it to be stripped cleanly from the concrete at any convenient time.

158. Processing Procedure.—Improvements in processing procedures are made from time to time, and the latest information should be secured before using vacuum treatment on any project.

On one job a stationary vacuum pump with a capacity of 800 cubic feet per minute and located a maximum of 1,300 feet from the forms maintained an average vacuum of 20 inches of mercury on 120 square feet of surface. A pump having a capacity of 1,100 cubic feet per minute maintained an average vacuum of 15 inches on 400 square feet of un-formed surface. For smaller jobs special vacuum pumps are not required since the necessary vacuum can be obtained by connecting the intake of an air compressor to the vacuum tank and exhausting the compressor discharge.

Vibration of formed concrete during the first few minutes the vacuum is applied is most important in securing best quality and watertightness. By use of such vibration, the small openings and channels created as the vacuum draws out the water are taken up and closed as rapidly as they are formed. For best consolidation a proper balance must be secured between the duration and intensity of the vibration and the vacuum treatment. Too weak or too short a period of vibration may not fully close all the new voids (left by extraction of water) in concrete which is rapidly stiffening under the vacuum. Too strong a vacuum, especially at the start, may stiffen the concrete too rapidly for effective vibration.

Sticky mixes with an excess of fines do not respond well to vacuum treatment. The treatment is more effective at lower temperatures and with coarser sands, minimum practicable percentages of sand, and lower cement contents.

H. Concrete Floor Finish

159. Requirements for a Satisfactory Finish.—A good concrete or mortar floor should have a surface that is durable, nonabsorptive, of suitable texture, and free from cracks, crazing, or other defects. The floor should satisfactorily withstand wear from traffic, the purpose for which it is intended; it should be sufficiently impervious to prevent staining or ready passage of water, oils, or other liquids; and it should possess a texture in keeping with requirements for appearance, easy cleaning, and safety against slipping. It should be structurally sound and, for separately placed floor topping, it should be well bonded to the subfloor.

160. Concrete Floors Placed as a Monolith.—From an economic standpoint, the top surface of a structure or portion of a structure placed as a monolith often can be finished to serve as a floor surface. Although not as durable as the surface of bonded floors, which are usually placed with a net water-cement ratio of not greater than 0.36, this type of floor surface will serve adequately for many purposes. Some improvement can be obtained by using a lower slump concrete and by increasing the richness of the mix used. Also, a slight reduction in entrained air content may facilitate finishing the surface. Care should be taken, however, that the mix proportions of the course to be finished are not enough different from the structural concrete to cause cracking from differential shrinkage or other differential properties of the two mixes. For this type of floor surface, the use of good quality material and the same good workmanship as for finishing surfaces of bonded floors are required.

161. Bonded Concrete and Mortar Floors.—Since an accurately proportioned concrete topping is much superior to a mortar topping, there is no reason for construction of poor concrete floor surfaces, provided an aggregate of suitable quality is available, proper procedures are followed, and the work is done by experienced workmen. Some inherent weaknesses of mortar topping are: (1) The large percentage of fine sand brought to the surface by floating and troweling forms a skin that wears poorly, dusts, sometimes scales, and has a strong tendency to craze (if these fines are completely removed from the sand, the mortar may bleed and be too harsh to permit satisfactory finishing); (2) the topping has high porosity, a well-known characteristic of mortars; (3) there is a deficiency of wear-resistant aggregate particles at the surface, as illustrated by figure 202 which shows sections through wearing courses of 1:2 mortar and of 1:1:2 concrete, bonded to concrete bases; and (4) exceptional care is required to obtain a good bond.

162. **Aggregate.**—A desirable grading of sand for floor topping is one that conforms with Bureau specifications for regular concrete sand, except with respect to content of fines. Best results are obtained if the percentage passing a No. 50 screen does not exceed 10 percent and the percentage passing a No. 100 screen does not exceed 5 percent. However, use of regular concrete sand is permitted if material that meets the foregoing requirements is not available. It is usually required that the gravel shall pass a ½-inch screen with not more than 10 percent passing a $\frac{3}{16}$-inch screen. If the concrete floor is to be highly resistant to wear, the aggregate must be tough, hard, and dense. Relatively small changes in grading are not important except as they affect consistency.

163. **Proportioning and Mixing.**—Mix proportions for concrete floor topping are usually 1 part cement, 1 part sand, and from 1¾ to 2¼ parts gravel, by weight, based on dry materials. At the Tracy Pumping Plant it was found that with 2 percent of entrained air, the topping mix worked and finished better at zero slump than non-air-entrained concrete at 1-inch slump. Use of this limited amount of entrained air is, therefore, advisable. When the floor is to be hand finished, it is generally required that the concrete be the driest consistency that can be worked with a sawing motion of the strike-off board and that the net water-cement ratio be not greater than 0.40.

For a power-floated finish involving no time interval between placing and finishing, it is necessary that the mix be considerably stiffer than for hand finishing; otherwise the machine will gouge the surface and satisfactory results will be difficult to obtain. The mix should be stiff enough to prevent excess mortar from working to the surface when the material is tamped and trimmed to grade prior to power floating. Such concrete, with the water-cement ratio frequently limited to a maximum of 0.36 by weight, will usually have no slump and can be efficiently mixed in a paddle-type mixer. The mix should respond to the power float sufficiently to fill and seal irregularities of the surface and yet be stiff enough to permit power floating immediately after trimming. Concrete that will stick together on being molded into a ball by a slight pressure of the

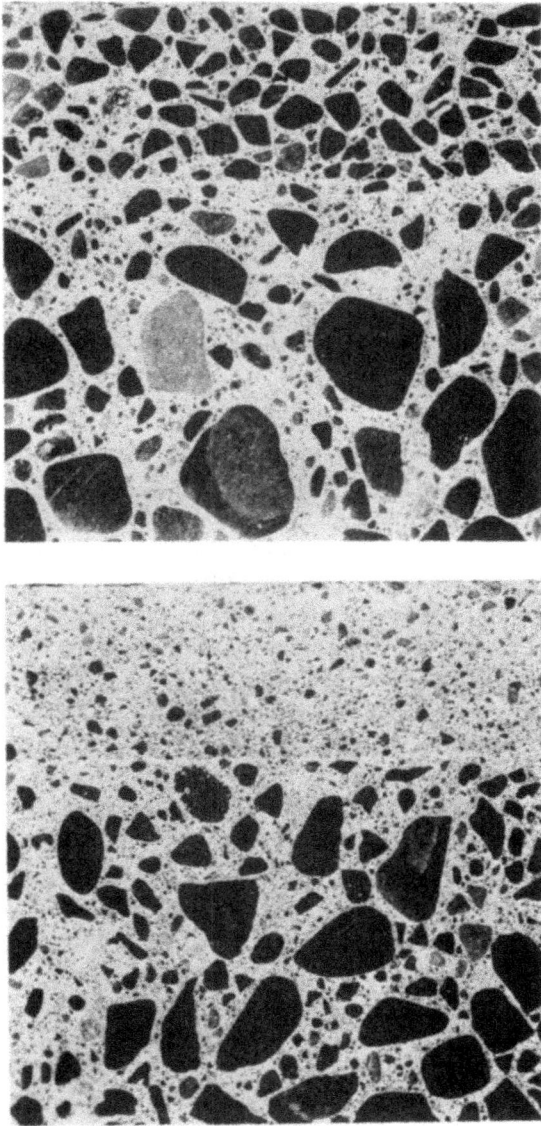

Figure 202.—Comparison of concrete and mortar floor toppings. Concrete topping (upper photo) is superior to mortar topping (lower photo) because it contains less water and fewer fines, is less porous, and has more wear-resistant aggregate at the surface. PX–D–33515.

hands and will not exude free water when so pressed but will leave the hands damp should meet these requirements.

For best results with either hand-finished or power-floated topping, the consistency of the concrete must be uniform. For this reason the aggregate should be reasonably uniform in grading and in moisture content, and the facilities for adding mixing water and controlling consistency should be the best. Small changes in water content produce marked differences in workability of the topping, and a surprisingly small increase in water will make the topping too wet for power floating.

Because of the dry mix used, precautions should be taken to prevent cement from accumulating on the mixer blades and shell. Mixing time for topping should not be less than 2 minutes regardless of the method of finishing.

164. Preparation of the Base.—Concrete floor toppings used on Bureau projects are usually bonded to a hardened concrete base which must be cleaned so that a surface free of all laitance and foreign material will be exposed at the time the topping is applied. Any oil, grease, or other contaminants on the surface should first be removed. If wet sandblasting or vacuum blasting for preparation is permitted, such blasting may eliminate the contaminants from the surface. However, if the surface is to be acid etched (or if sandblasting is not effective in removing contaminants) it will be necessary to remove oil and grease by solvent washing and other contaminants by use of strong detergents, with final surface cleanup by a trisodium phosphate wash followed by a water rinse. (See section 120 for methods for removal of stains and other foreign materials from concrete surfaces.)

One of the better treatments for thoroughly removing laitance and providing a suitable bonding capability to subgrade surfaces utilizes acid etching with muriatic acid (commercial or technical grade HC1—approximately 30 percent HC1), accompanied by vigorous scrubbing with a stiff wire or fiber-bristled brush or broom and followed by complete and thorough scrubbing with clean water to remove all traces of acid and reaction products.

When acid cleanup is not feasible or safe, wet sandblasting shortly before the topping is placed or the use of any type of equipment which will effectively remove all laitance and foreign matter from the surface of the base concrete, followed by washing with water under pressure, gives good assurance of adequate bond between the topping and the base. However, this method of cleanup is objectionable around operating machinery as the air will become laden with moisture and particles of floor cuttings which will be deposited on the machinery.

When topping is placed after installation of equipment, the base concrete surfaces may be cleaned and roughened by mechanical routers or with blasting equipment provided with a vacuum system for collecting the cutting medium and refuse from the surfaces. If the vacuum system is properly operated and is provided with adequate dust collectors, very little dust will escape into the atmosphere. The cutting medium for this equipment must be steel grit or shot, aluminum oxide, silicon carbide, or other effective abrasives. Washing with water under pressure is not required with this equipment.

Bond between topping and base course is improved by thoroughly water curing the base course for the prescribed period or preferably until the surface is cleaned in preparation for placement of topping. All cleaned surfaces to be topped should be completely dry, and care should be exercised to prevent recontamination from any source. No traffic should be allowed upon the prepared surfaces prior to concrete placement, and the necessary steps should be taken to provide that the temperature of the base course approximates that of the topping mix. The topping should be bonded by (1) an epoxy bonding agent applied to the surface of the base course or (2) by scrubbing a mortar thoroughly into the surface of the base course.

Epoxy resin bonding agents of the thermosetting plastic type and conforming to Federal Specifications MMM–B–350 are required for Bureau work when epoxy-bonded concrete floor finish is specified. Use of type I or type II epoxy conforming to this specification depends on the temperature of the concrete to receive the epoxy, as discussed in section 136. Preparation and application of the epoxy resin bonding agent to prepared, dry surfaces of the base course are discussed in paragraphs (b) and (d) of section 136. The epoxy bonding agent should be applied to the base course immediately prior to placing of the topping and only over a small area at a time to assure that the applied film is fluid when the topping is placed.

When mortar is to be used as the bonding media, a 1:1 mortar-sand mix should be scrubbed into the surface just prior to placing of the topping. This mortar should be composed of cement and fine well-washed sand (preferably passing a No. 16 screen), should have a medium consistency, and should not exceed one-sixteenth inch in average thickness. Such mortar is more satisfactory for this purpose than neat cement grout as its properties more closely conform with those of the base and topping courses.

165. Screeds.—Screeds are set as guides for a straightedge to bring the surface of floor concrete to the required elevation. They must be sufficiently rigid to resist distortion during spreading and leveling of the floor topping. Metal strips or pipe spaced not more than 10 feet apart (preferably 4 to 6 feet) make effective screeds. At Canyon Ferry Power-plant, screeds were made of ¾-inch-diameter pipe spaced 4 to 5 feet apart and supported at intervals of 3 to 4 feet. Supports consisted of a 1½- by 4- by ⅜-inch steel plate tapped and threaded near each end and welded at the center to the steel pipe. The screeds were leveled to grade by bolts through threaded holes in the supports, the ends of the bolts resting on the concrete base. Locknuts held the bolts in place. The screeds, after being leveled, were held in position by wire ties attached to rivets shot into the concrete base using a tool powered by an explosive charge. After the topping had been leveled, wire ties were cut, screeds and supports removed, and recesses filled with topping material.

Wooden blocks approximately 2 inches square and of suitable thick-ness have been used as screeds, but most installations now make use of screed strips. When blocks are used, they are usually spaced 10 feet apart in each direction. Each block is supported by and slightly embedded in a small amount of mortar, with the top level and accurately set to finish grade. After the blocks are in position, dry cement dusted over the mortar will cause it to harden rapidly and hold the blocks in place. After the floor topping has been compacted and leveled, the screed blocks are removed and recesses filled with topping material.

166. Depositing, Compacting, and Screeding.—When all preparations for placing have been completed as described in section 164, the epoxy resin bonding agent is applied or the thin coat of mortar described in section 164 is brushed thoroughly into the surface of the base for a short distance ahead of the topping. The topping should be applied im-mediately while the epoxy bonding agent is still fluid or before the mortar coat has stiffened.

The finish course should be spread evenly with rakes to a level slightly above grade and compacted thoroughly by tamping. Tamping should be sufficiently heavy for thorough compaction. After being compacted the topping is trimmed to grade with a steel-faced straightedge or scraper. The screeding is followed by power or hand floating, as discussed in the following section. Power floating results in a sounder, more durable top-ping—sounder because a stiffer mix having less tendency toward volume change can be used, and more durable because the power float will compact a mix containing a high percentage of coarser aggregate, thus increasing resistance to surface wear.

167. Finishing.—Floor finishing should never be performed by inex-perienced operators. It is a critical task that, for satisfactory results, requires the best efforts of skilled workmen. Two operations are usually required in producing a finished floor surface: First, a compaction and truing (to a rather rough texture) of the trimmed or screeded surface by use of power-driven floats or by hand floating with wood floats; and second, a final compaction and smoothing (to a much finer texture) by steel troweling. If, for economy or appearance, a coarse texture is desired, the troweling operation may be omitted. A fine, even-grained, or scroll finish can be attained by light troweling; a very smooth finish can be attained by hard troweling.

(a) *Floating.*—Preliminary finishing, or floating, should be performed with power-driven revolving disks equipped with vibrating devices. Power floating is begun as soon as the screeded topping has hardened sufficiently to bear the weight of a man without leaving an indentation—usually within 30 minutes after scraping—and is continued until the hollows and humps are removed or, if the surface is to be troweled, until a small amount of mortar is brought to the surface. The floated surface should be checked with a straightedge to see that it is accurately on grade.

Hand floating is used when the floor area is too small to justify power floating. The screeded surface is compacted and smoothed with a wood float and tested with a straightedge to make sure that high spots and depressions are eliminated. Floating should be continued just long enough to produce a true and smooth surface, and if a trowel finish is required, to bring up a small amount of mortar. Excessive floating may produce a floor that will dust or craze.

(b) *Troweling.*—Finishing with steel trowels may be commenced as soon as the floated topping has hardened enough to prevent excess fine material from working to the surface. This operation, which should be performed with heavy pressure, should produce a dense, smooth, watertight surface free from blemishes. Sprinkling cement or a mixture of sand and cement on the surface to absorb excess moisture or to facilitate troweling should be prohibited. Troweling too soon, or excessive troweling in one operation, produces an unsound, nondurable finish. If an extra hard, durable finish is desired, a second troweling should be done after the floor has nearly hardened.

Power-driven troweling machines are suitable for use on large floor areas.

If a ground finish is required, the surface is lightly troweled, no attempt being made to remove all trowel marks.

(c) *Grinding.*—When properly constructed of materials of good quality, ground floors are dustless, dense, easily cleaned, and attractive in appearance. When grinding is specified, it should be started after the surface has hardened sufficiently to prevent dislodgment of aggregate particles and should be continued until the coarse aggregate is exposed. The machines used should be an approved type with stones that cut freely and rapidly. The floor is kept wet during the grinding process, and the cuttings are removed with a squeegee and then flushed with water.

After the surface is ground, airholes, pits, and other blemishes are filled with a thin grout composed of 1 part No. 80-grain carborundum grit and 1 part portland cement. This grout is spread over the floor and worked into the pits with a straightedge, after which it is rubbed into the floor with the grinding machine. When the fillings have hardened for 7 days, the floor receives a final grinding to remove the film and to give the finish a polish. All surplus material is then removed by washing thoroughly.

168. Protection and Curing.—The finished floor surface should be adequately protected from damage that might be caused by building operations, weather conditions, or other causes. For curing, it is usually required that the floor be completely covered with airtight, nonstaining, vaporproof, plastic waterproof membrane covering which will effectively prevent loss of moisture from the concrete by evaporation. The covering should be applied as soon as can be without damaging the surface. The edges of the covering should be lapped and sealed and the covering left in place for not less than 14 days. A light fog spray applied just before the waterproof covering is laid will improve the curing action.

Coverings of nonstaining sand or cotton or burlap mats are also effective if kept continuously and completely moist. Other means sometimes used for curing floor surfaces are not as satisfactory as the plastic waterproof membrane or moist coverings.

When the floor is to be subjected during the curing period to any usage that might rupture the covering or damage the finish, it should be protected by a suitable layer of cushioning material.

Protection of concrete floor finishes during cold weather is of particular importance as the sections involved are usually thin and the effects of low temperature are correspondingly intensified. The space both below and above the floor should be enclosed and maintained at an appropriate temperature throughout the curing period. Heaters should be insulated from the floor by a heavy layer of sand to prevent excessive drying in their immediate vicinity.

169. Liquid Hardener Treatments.—A well-constructed concrete floor surface in which first-class materials have been used will give satisfactory service under most conditions without special treatment. Any concrete surface will dust to some extent and may be benefited to a degree by proper treatment with solutions of certain chemicals. Included in these chemicals are fluosilicates of magnesium and zinc, sodium silicate, gums and waxes. When the compounds penetrate the pores in the topping, they form crystalline or gummy deposits and thus tend to make the floor less pervious and reduce dusting either by acting as plastic binders or by making the surface harder.

Application of such chemicals will have little effect on improving wearing or abrasion qualities of a high quality concrete surface. Surface hardener treatments will temporarily improve the wearing or abrasion resistance of poorer quality floors, but to remain effective they must be reapplied periodically.

The Bureau requires that concrete floor hardeners consist of magnesium or zinc fluosilicate crystals, or a combination of both, dissolved in water. Two coats of hardener are normally applied after the floors have been cured thoroughly and the concrete is at least 28 days old. At time of application the surfaces should be thoroughly clean of all dirt, grease, laitance, and other foreign matter and should be dry. The first coat of hardener should consist of one-half pound of crystals per gallon of water; for the second coat, 2 pounds of crystals per gallon of water should be used. The solution should be applied liberally by floor mops, spreading each coat uniformly over the entire surface, avoiding pools of the hardener solution. The first coat should be allowed to dry thoroughly before the second coat is applied. The coverage rate for each coat should not be more than 100 square feet per gallon. After the second coat has dried the floor surfaces should be brushed and washed with water to remove any crystals which may have formed on the surface.

170. Nonslip Finish.—Surfaces of ramps and other surfaces required to retain a highly nonslip texture under traffic are sometimes treated with an abrasive grit incorporated in the surface during the floating operation. The grit is sprinkled uniformly over the surface at the rate of one-fourth to one-half pound per square foot.

For the Grand Coulee Third Powerplant an epoxy-bonded, carborundum-grit, nonslip finish was required for stair treads and landings. After curing, surfaces of the treads and landings were lightly sandblasted, cleaned, and brought to a completely dry condition. Epoxy-resin-base grout conforming to Federal Specification MMM–G–650A, properly mixed, was applied to surfaces at a coverage rate of approximately 60 to 80 square feet per gallon. While the applied epoxy was still fluid, No. 50 carborundum-grit was sprinkled over the epoxy coat to obtain an excess of grit over the surface. After the grit had been rolled sufficiently into the epoxy and after the epoxy had hardened, the excess grit was brushed from the surface.

171. Colored Finish.—The principal materials used for coloring concrete floors are (1) pigment admixtures, (2) chemical stains, and (3) paints. Pigment admixtures may either be added integrally to the topping mix by blending with dry cement or by dusting onto the topping immediately after it has been screeded.

Where resistance to wear is of prime importance and for floors subjected to outdoor exposure, the use of pigment admixtures added integrally to the topping mix is much better than surface treatments. Of the pigment admixtures, the synthetic mineral pigments are preferable to relatively impure materials. Because of their color intensity, the amount needed is smaller and the quality of the concrete is enhanced by avoiding excessive inert fines. For indoor floors subjected to only light traffic, the depth of colored concrete obtained from integral mixing is not needed and dust-on coloring may be used. With careful application, this type of coloring material provides a colored layer one-thirty-second to one-sixteenth inch thick.

Regardless of the type of material selected, the quantity of pigment to be used will depend not only on the depth of colored layer desired but also on the color itself. The correct quantity should be determined from test panels made with the materials to be used in the work. The pigment procured for the job should be thoroughly blended to assure a constant color and shade. Where a pigment is to be mixed integrally with topping mix, it should be accurately weighed for each batch and thoroughly blended with cement in a separate mixing device before it enters the concrete mixer. It is essential that each pigmented batch be thoroughly mixed. If successive batches are not similar in all respects, a uniform color will not be produced. It is also essential that the finishing and curing procedures be the same over all portions of the floor area.

Colored surfaces may be cleaned and brightened and thin films of efflorescence obscured by rubbing with a mixture of equal parts of paraffin oil and benzine or naphtha. Waxing colored floors adds an attractive luster, gives them a more uniform appearance, and reduces marring from scratches and stains.

Recommended pigments are as follows:

Reds and pinks ....Red oxide of iron.
Yellows and buffs ....Yellow oxide of iron.
Brown ....Brown oxide of iron.
Blacks and grays ....Black oxide of iron.
Greens ....Chromium oxide, 98 percent pure.
Blues ....Cobalt blue, 98 percent pure and free from sulfates. (Ultramarine may not be dependable.)

Chemical stains are primarily applicable to inside floors where some variation in tone is preferred to the flat colors produced by pigmented admixtures or painting and where the surfaces may be kept varnished and waxed to prevent wear. Proprietary compounds should be used in strict accordance with the manufacturer's directions.

Painting is the least desirable of the three decorative floor treatments, as rapid and uneven wear from traffic necessitates frequent repainting.

When concrete floors are to be painted, the Bureau's Paint Manual should be consulted for information with respect to preparation of surfaces and selection of type of paint and its application.

172. Terrazzo Finish.—Terrazzo floor finish is occasionally used on floors of Bureau buildings. Portions of the floors in several powerplants were finished in terrazzo.

This type of construction is highly specialized and should be performed by experienced workmen. Complete details concerning the materials and procedures involved are contained in the construction specifications.

I. Shotcrete

173. Definition and Use.—Shotcrete is mortar or concrete shot into place by means of compressed air. There are two processes for producing shotcrete. In the *dry-mix process,* the dry materials are thoroughly mixed with enough moisture to prevent dusting. This dry mixture is forced through the delivery hose by compressed air and water is added at the nozzle. In the *wet-mix process,* all materials and water are mixed to produce mortar or concrete. The product is then forced through the delivery hose to the nozzle where air is injected to increase velocity.

In past years manufacturers of equipment for shotcrete application used several names to promote their products, although basically they were the same. The word "shotcrete" is a nonproprietary term adopted by technical societies to describe pneumatically placed mortar or concrete which by high-velocity application adheres to the surface on which it is projected. When coarse aggregate is used in the dry process, a set accelerator can be used which aids in holding the coarse aggregate within the mass. The accelerator also produces high early strength necessary in tunnel support. Until recently, the use of an accelerator has been limited to the dry-mix process, but the wet-mix process has reportedly been modified to allow use of accelerators. Because the Bureau has made little use of the wet-mix process, the information that follows is directed toward the dry-mix process.

Shotcrete containing ¾-inch aggregate is increasingly gaining acceptance for use in lieu of steel sets for tunnel support, where adaptable. With the accelerator it will adhere to wet surfaces and in most instances seal off water seepage sometimes encountered in tunnel excavation. Either sand or coarse aggregate shotcrete can be applied readily on surfaces of various materials, regardless of shape or inclination. Shotcrete is used extensively for repairing and strengthening buildings; for protective coatings for structural steel, masonry, and rocks; and for various kinds of relatively thin linings. Shotcrete has also been used for special ground support in tunnel construction, coating steel pipe, canal lining, and certain types of repairs.

Failures of shotcrete applied to the surface of concrete are often attributable to defects of the base on which coatings were applied rather than to weakness of the coatings. A heavy base that is subject to structural cracking would not be restrained from further cracking by a thin layer of shotcrete, regardless of its quality; and it cannot be expected that such a coating will not break its bond with underlying concrete when the two are subjected to different volume changes occasioned by variations in temperature and moisture.

174. Preparation of Surfaces to be Treated.—Surfaces to be covered with shotcrete should be cleaned thoroughly of all loose material and all dirt, grease, oil, scale, and other contaminations. When reinforcement is to be covered, it should be held in place by expansion bolts or dowels anchored firmly. When shotcrete is used for tunnel support, it should be applied as soon as possible after the round is shot and close to the face. With a mole-driven tunnel, shotcrete should be applied as soon as possible behind the cutterhead. In a successful operation, space for positioning the nozzleman was designed between the cutterhead and the machine. It is believed that the immediate application prevents the losing of fines from the rock joints, thus maintaining a keyed rock support about the opening.

175. Sand.—Sand for shotcrete should be uniformly graded. Hard particles are desirable because soft grains crumble as they pass through the discharge hose and form fine powder which may reduce the bonding value of cement. Such pulverization increases with increase in the hose length. Specifications require that the sand grading conform with requirements for concrete sand. For coarse aggregate shotcrete the quantity of sand passing the No. 100 screen may be substantially increased if needed for added plasticity and adhering qualities, provided that quality and strength are not detrimentally affected.

For shotcrete containing no coarse aggregate, rebound, as defined in section 176, will be less, and a smoother surface texture will be obtained when the sand contains an excess of fines (material retained on the No. 50 and No. 100 screens) and less coarse material (retained on the No. 8 and No. 16 screens). However, shotcrete made with finer sand will have a higher water requirement and a correspondingly increased drying shrinkage; it will also have greater tendency to slug in the machine. If a sand is deficient in fines, the addition of diatomaceous earth (not more than 3 percent of the cement, by weight) will improve plasticity of the mix and decrease the amount of rebound.

Sand should contain 3- to 6-percent moisture for efficient operation of equipment for application of both sand and coarse aggregate shotcrete. If the sand is too dry, there is difficulty in maintaining uniform feeding and also increased rebound because of a greater tendency for aggregate and cement to separate. If the aggregate is too wet, there will be frequent plugging of the equipment. Use of moist sand avoids discomfort to the operator from discharge of static electricity.

176. Rebound.—Because of the velocity at impact, a portion of the mixture bounces from the surface on which it is being applied. This material is known as rebound. When applying dry-mix shotcrete to overhanging surfaces or squaring off corners, the rebound averages about 30 percent; for vertical surfaces about 25 percent; and on nearly level surfaces, it is close to 20 percent. The amount of rebound tends to increase with increased nozzle velocity. Within the range of ordinary consistencies, when other factors remain the same, the amount of rebound for sand shotcrete varies inversely with the water-cement ratio. As the percentage of water is increased, the mortar becomes more plastic and sticky and has greater tendency to adhere to the surface.

Unless the accelerator in coarse aggregate shotcrete causes the cement to set instantaneously, very little coarse aggregate will be incorporated in place. If the accelerator is working properly, the quantity of rebound should not exceed that of the sand-type shotcrete. The amount of coarse aggregate rebound is believed to be related to particle shape to some degree with more angular particles rebounding to a lesser degree.

177. The Optimum Mix.—Rebound has a greater percentage of coarse sand particles and a much smaller cement content than the shotcrete as it leaves the nozzle. The cement content of materials as mixed should, therefore, be less than that desired for shotcrete in place.

Although increasing the water content decreases the amount of rebound, water content must be limited as overwetness of the shotcrete causes it to slough from its initial position on the structure. The optimum mix contains a little less water than that which will cause sloughing and just enough cement for the desired water-cement ratio. On one large job the optimum mix for sand shotcrete (as discharged at the nozzle) was 1 part cement to 4.5 parts sand and coarse aggregate by weight; this gave proportions in place of 1:3.2 to 1:3.8. The water-cement ratio of the fresh shotcrete in place was about 0.57 for sloping and 0.54 for overhanging surfaces; these were approximately the maximum ratios that could be used without causing sloughing. Diatomaceous earth equal to 3 percent of the weight of cement was added to make the mix more plastic.

The optimum mix proportions for coarse aggregate shotcrete are estab-listed through test panels for each job. Although the ratio of 1:4.5 is commonly used as a trial mix, this ratio is sometimes modified to meet strength criteria. A minimum cement content per cubic yard as discharged from the nozzle is usually specified; the proportion of fine to coarse aggregate ranges between 0.55 and 0.60 percent. The proportions are adjusted during early tests to provide minimum rebound.

178. Mixing.—Thorough mixing of all ingredients, especially any coarse aggregate or accelerator if used, is essential to good quality shotcrete. The materials tend to cake on mixer blades and inner surface of the shell, and the mixer requires frequent cleaning to maintain mixing efficiency. The mixing period should be not less than 1½ minutes. Unused cement and aggregate mixed material that stands longer than 1 hour should be wasted. If an accelerator is used, it should be well dispersed into the shotcrete mix immediately before entering the shooting equipment.

179. Equipment.—One type of machine for placement of dry-mix shotcrete consists of two compression chambers, one above the other. The sand-cement mixture is introduced into the upper chamber, which is alternately under pressure and free from pressure. When the upper chamber is closed and the pressure becomes equal to that in the lower chamber, a valve separating the two opens. The material drops into the lower chamber in which a constant pressure is maintained. In the bottom of the lower chamber a feed wheel, driven by an air motor, takes the material to the outlet where air, introduced through a gooseneck, forces it through the outlet valve and hose to the nozzle. Water under pressure is conducted by a separate hose to the nozzle where it enters the water ring and is sprayed radially into the stream of mixed sand and cement.

Several makes of gunning equipment are available for placing dry-process shotcrete containing ¾-inch aggregate. Each machine has a basically horizontal chamber into which the material falls and which then is rotated through an airlock to drop the mixed material into the airstream. The material is then conveyed by air to the nozzle where water is injected into the material stream. The wet process gun has two chambers. One is used for mixing while the mixture from the other is being placed. Thus continuous placing is achieved.

Another type of machine feeds the mixture of sand and cement to the material hose by screws. The dry mixture is forced by air through the hose to the nozzle where water is added as described in the preceding paragraphs.

Use of elevators or conveyors and gravity feed greatly increases output of most units and adds materially to the quality of the product through increased efficiency of equipment. The mobile unit shown in figure 203 is an example of such an assembly.

180. Placing and Curing.—For proper application, the nozzle should be held normal to and about 3 feet from the surface to be coated, as shown in figure 204. The most favorable velocity for material leaving the nozzle depends on size of nozzle. For a 1¼-inch nozzle, the velocity should average about 475 feet per second. In finishing off corners, or in confined spaces, lower velocities (hence lower pressures) are more satisfactory.

Average compressor capacities for various nozzle sizes and required air pressures are shown in the following tabulation:

Compressor capacity, ft^3/m	Hose diameter, in	Maximum size of nozzle tip, in	Operating air pressure available, lb/in^2
250	1	¾	40
315	1¼	1	45
365	1½	1¼	55
500	1⅝	1½	65
600	1¾	1⅝	75
750	2	1¾	85

In the dry process it is essential that water pressure be greater than air pressure to ensure complete wetting of the materials at the nozzle and to give the nozzleman a quicker, more positive control. Maximum, minimum, and average air pressures, water pressures, and hose lengths are given in the following tabulation:

	Maximum	Minimum	Average
Air pressure, pounds per square inch..........	70	35	50
Water pressure, pounds per square inch........	130	50	70
Hose length, linear feet...............................	350	50	200

When coatings 1 inch thick or more are to be applied to vertical or overhanging surfaces, shotcrete without coarse aggregate should be applied in several layers to prevent sloughing of freshly placed material. For level or slightly sloping surfaces the thickness of a single layer may vary up to a maximum of 3½ inches. When more than one layer is applied, a delay of 30 minutes to 1 hour between applications is usually sufficient to prevent sloughing. For shotcrete containing coarse aggregate and an accelerator, no delay is necessary since initial set takes place almost immediately and it gains strength rapidly. Layers should be applied before the previously placed shotcrete has set completely, otherwise a glaze coating will form on the surface of the previous layer. There is no apparent difference between finished placements started at the top and those placed from the bottom upward. It is essential that the surface to be coated be free of rebound.

Shotcrete is ordinarily placed by a crew of three: a nozzleman, a machine operator, and a person to clear the rebound. Only experienced persons should be employed as the quality of coating depends in large part on the skill of those who place it. The nozzleman places the shotcrete to line and grade, adds the correct amount of water at the nozzle, applies

CONCRETE MANUAL

Figure 203.—Shotcrete mixer and drum elevator used on the Gila project, Arizona. PX–D–33516.

Figure 204.—Shotcrete being applied to canal prism with wire mesh reinforcement installed in the Auburn-Folsom South Canal, Central Valley Project, California. P859–245–5383.

the shotcrete systematically so that rebound can be kept cleared away from the work, and minimizes rebound by holding the nozzle in proper position. The machine operator regulates the air and water pressures and the rate of feed to produce a uniform flow of the proper velocity at the nozzle. This enables the nozzleman to place a coating of good quality. The third person clears away rebound so that it will not become incorporated in the shotcrete and also helps the nozzleman move the hose when changing positions.

Operations should be suspended when wind blows spray from the nozzle and prevents proper control of consistency.

Contrary to some belief, shotcrete has no special virtue because of the method of application. Density and other properties are not materially different from or superior to those of other mortars or concrete of similar mix and water-cement ratio. It can be screeded and troweled the same as other mortar or concrete without impairment of properties. However, where the security of bond with underlying materials is important, as in repair of structural work, shotcrete should be screeded and troweled with extreme caution. This special care in finishing is not necessary where the shotcrete is used for canal or reservoir lining on an earthen subgrade.

To gain sufficient strength shotcrete must receive proper curing. When shotcrete is used as protective coating, curing can be minimal; but when it is to be used as part of a structure or as permanent structural tunnel support, some curing is usually necessary. If the ambient relative humidity is above 85 percent or the shotcrete is applied to wet rock, sufficient curing water should be present. In dry tunnels with low humidity, bulkheads and an atomized moisture environment of at least 85-percent relative humidity should be maintained. If the shotcrete is used as tunnel support and later covered with concrete, the shotcrete needs only to achieve a specified strength. Depending on moisture conditions in the tunnel, it may or may not need additional curing.

If shotcrete is placed above ground, it should be water cured and protected from direct rays of the sun for 3 days, unless it will be flooded as in canals. Much of the completed area of shotcrete becomes coated with rebound. In some applications, such as in canal lining where sand shotcrete is placed, it may be advantageous to allow the coating of rebound to remain in place because of its ability to retain water and thus enhance the effectiveness of water curing. For some canals, shotcrete canal lining has been treated by curing compound, and in these instances the coverage rate should be more than the usual 1 gallon per 150 square feet. Where membrane curing is used and rebound has not been removed, an excessive amount of curing compound is needed for effective sealing of the rough, porous surface. Pneumatically applied canal lining should be swept or the rebound troweled to a surface that can be effectively covered with curing compound. Rebound on shotcrete coatings applied to steel pipe should be swept off where membrane curing is used. This should be done as the work progresses and before rebound becomes too hard.

Because of the thinness of shotcrete coatings on steel pipe, good curing of these coatings is of special importance. Recommended procedures are discussed in sections 184 through 187 and in section 125.

Test cylinders (6- by 12-inch) for shotcrete containing sand can be made by shooting the mortar vertically into cylindrical cages of ½-inch mesh hardware cloth mounted on a board. The mortar outside the mold should be removed immediately after shooting the specimen so that the wire mesh can be detached before testing. Because the above method may not give good representative samples, it is now common practice to cut cubes or core cylinders from panels made of shotcrete for compression strength tests. If cubes are made, a correction factor should be applied to relate the cubes to cylinders having a height-to-diameter (H/D) ratio of two. Cubes of comparable concrete average about 15 percent higher in compressive strength than do cylinders with a height-to-diameter (H/D) ratio of two. Cores can also be drilled from in-place shotcrete at various ages to evaluate compressive strength where the shotcrete is of sufficient thickness.

J. Grouting Mortar

181. Uses and Essential Properties.—The term "grouting mortar" as used in the following discussion has particular reference to special sand-cement mortars for sealing joints of precast pipe, seating machinery and structural steel members on foundations, and filling reglets for roof flashing. Neat cement grout for pressure grouting of contraction joints and rock foundations and sand-cement grout for pressure filling of cavities behind tunnel linings are not within the scope of this discussion.

Grouting mortar must readily and completely fill the space to be grouted and, insofar as practicable, must permanently retain original volume. Ordinary plastic and fluid mortars are unsatisfactory in these respects because of the inherent tendency of solid constituents to settle and leave a layer of water at the top surface. A second but less objectionable characteristic is shrinkage that occurs when such hardened mortar dries. Settlement can be practically eliminated by using special ingredients or treatments, but drying shrinkage can be reduced for a given mix only by use of stiffer mortar. Fortunately, drying shrinkage of the grout sections usually is so small that it may be disregarded.

Factors influencing the amount of settlement for a given mix are (a) consistency of mix, which, in turn, depends on unit water content, (b) grading of sand, (c) fineness of cement, (d) time that elapses between placement and initial set, and (e) length of time interval before placing during which the mortar is maintained in a plastic condition by continuous or intermittent mixing. In the following section mortars are described which are so fluid that they readily flow into and thoroughly fill small spaces but have negligible settlement.

182. Types of Nonsettling Mortars.—In general, nonsettling mortar is prepared by a prolonged or delayed mixing of ordinary mortar, by adding a special ingredient to ordinary mortar, or by using a special cement. In all preparations the sand should preferably contain approximately 25 percent of material that will pass a No. 50 screen. The mortar should be no wetter than necessary for satisfactory placement.

(a) *Prolonged or Delayed Mixing*.—Reduction of the interval between time of placement and initial set, by extending the mixing period or by delaying final mixing, results in material reduction of settlement. A mix of 1 part cement to 2.5 parts sand with a water-cement ratio of 0.50 and a 6-inch slump after about 10 minutes of mixing in a mixer and 1 hour of mixing in a mortar box with a hoe has been used by the Bureau for grouting reinforcement bars in holes drilled in rock. This method, termed premixing, has also been successfully used for several years in minor repairs of disintegrated concrete by the Oregon State Highway Department. The reduction in settlement that may be expected from prolonged mixing is indicated in table 28.

(b) *Addition of Aluminum Powder*.—Aluminum powder added to concrete reacts chemically with alkaline constituents of the cement and generates hydrogen gas. Expansion of the mortar, which results from generation of the gas, causes the mortar to fit snugly in the space which confines it. Such mortar is, therefore, useful where tight grout fillings are required. The ground aluminum powder should contain no polishing agents such as stearates, palmitates, and fatty acids and may be of any variety that produces the desired expansion.

Some brands of aluminum powder do not react as expected; consequently, tests should be performed with the materials prior to their being used in construction work to establish the required amount and effectiveness of the variety. Extremely small amounts are required. Laboratory

Table 28.—Effect of prolonged mixing of grouting mortars

Mix, cement to sand	W/C by weight	Mixing time, minutes	Slump, inches	Unit 24-hour settlement	Mixing time, minutes	Slump, inches	Unit 24-hour settlement
1:1	0.40	15	10¼	0.0011	105	9½	0.0005
1:2	0.50	15	10	0.0037	135	9½	0.0005
1:3	0.65	15	9¾	0.0073	150	9½	0.0005

Type I portland cement and concrete sand within Bureau specifications were used. The F.M. of the sand was 2.67. Each value of settlement represents the average of three specimens stored in laboratory air for 24 hours. After 4 hours of mixing, the 1:1 mix had a 1-inch slump.

tests have demonstrated that a mortar suitable for use under machine bases may be produced by adding to a 1:1.5 mortar mix having a water-cement ratio of 0.50, a quantity of aluminum powder equal to 50 to 60 millionths of the weight of cement used (about a teaspoonful per bag of cement). With well-graded sand such a mix will have a slump of about 11 inches. A 1:2 mix with a slump of 1 inch and containing the same proportion of aluminum to cement is satisfactory as a filling for roof flashing reglets.

It is important that the dosage for each batch be very carefully prepared and weighed. The aluminum powder should first be blended in proportions of 1 part powder to 50 parts cement or pozzolan by weight. The blend is then added by sprinkling over the batch. Dosage of the blended material will be governed by the amount and chemical composition of the cement used, placing temperatures, and whether the aluminum admixture is used in a grout, sand-cement mortar, or concrete. The amount to be used should be adjusted as necessary to obtain effective expansion. To assist in establishing proper amounts of blended material for the particular work involved, the following dosages are suggested for preliminary trial mixes:

Concrete or grout	Blended aluminum powder, ounces per bag of cement		
	70° F placing temperature		40° F placing temperature
Concrete	6.5	to	10.0
Sand-cement grout	5.5	to	8.5
Neat-cement grout	4.5	to	7.0

It is advisable to mix the aluminum thoroughly with the cement and sand before water is added because aluminum powder has a tendency to float on water. Batches should be small enough to allow placement of freshly prepared mortar as the action of the aluminum becomes very weak about 45 minutes after mixing. After all ingredients are added, the batch should be mixed for 3 minutes.

(c) *Use of Special Expansive Cements and Mortars.*—These are proprietary products designed to expand sufficiently during initial hardening and curing processes to offset subsequent shrinkage and assure complete filling of the grouted space. Confinement of the grout is essential to produce the small compressive stresses necessary.

Expansive cements are essentially portland cement with small amounts of expansive components introduced during manufacture or subsequently interground with cement clinker. The expansion is caused by formation of a solid compound rather than gas.

Mortars are prepared mixtures of cement and fine aggregate. Sometimes an expansive cement provides expansion, and in other cases, a component acting on portland cement forms gas bubbles similar to the action of aluminum powder.

Since many of these products are relatively new, assurance of suitability should be ascertained by performance records or preliminary tests.

183. Machine Base Grouting Procedure.—The effectiveness of a hardened cement mortar in firmly securing a machine to a base depends to a great extent on the procedure used in placing the material. In practical terms, grout or mortar is a plastic material introduced between a piece of machinery and the foundation. The method of introducing the mortar may vary, but certain fundamental steps are required for assurance that the space is completely filled and that the mortar will remain in intimate contact with base and machinery.

The preparation of the foundation should be accomplished before the machinery is set. It is important that the concrete base be thoroughly cleaned and wet before grouting begins. The surface may be prepared through use of either a pneumatic or electrically driven chipping hammer equipped with a bull point or spade point chisel or with a hand bush hammer where air or electric tools are not available. Oil or grease should be removed as described in section 120 and thoroughly flushed or removed by chipping to a sufficient depth.

The machine base or soleplate should be cleaned of rust, mill scale, paint, oil, or grease before it is set into place. When a soleplate is used and it is necessary to lubricate between the soleplate and machine base in the final alinement, either a light coating of paraffin or flake graphite or other special lubricant should be used. The metal surfaces should be wet before grouting to facilitate the flow of grout around and under foundation bolts and machine parts.

The forms around bases should be built of lumber not less than 1½ inches thick and should be braced securely to minimize bending and slipping during grouting operations. To assure that the space to be grouted remains full during grouting, the grouting should be done under pressure. This may be accomplished (1) by using an expanding agent such as aluminum powder as discussed in section 182(b), or (2) by providing a static head pressure by extending a part of the form at least 6 inches above the machine base or soleplate. Where bond between metal parts and grout is not desired, flake graphite or paraffin should be applied to the metal parts.

In proportioning the grout mixture, use of too much water should be avoided. A low water-cement ratio will aid in reducing shrinkage and also in developing strength. The water-cement ratio should never be greater than 0.50. A mix proportion of 1 part cement and 2 parts sand with a slump of approximately 4 or 5 inches should be used for machinery set with light loading. For heavy loading, the mix should be 1 part cement to 1½ parts sand with a slump of not more than 3 inches. When greater flowability of grout is needed, the water content may be increased provided the water-cement ratio is 0.50 or less. When the vertical grout space exceeds 3 inches, additions of 3½ parts of clean coarse aggregate up to ½-inch size can be included.

After the grouting mortar has been allowed to settle in place for 30 minutes, surplus air and water can be eliminated by rodding. This can be facilitated by having previously laid a length of chain or hoop steel under the machine and extending it from the forms so that it can be grasped and drawn back and forth as additions of grout are made.

K. Mortar Lining and Coating of Steel Pipe

184. Definition and Uses.—Cement-mortar as used for the protection of steel pipe against corrosion is basically a mixture of portland cement, sand, and water. Generally, mortar for these applications contain, in addition, pozzolan or natural cement. Normally the mortar is applied at a thickness of five-sixteenths to one-half inch for interior linings and one-half to three-fourths inch for exterior coatings in Bureau construction. Other uses may require different thicknesses. The coating owes its ability to mitigate corrosion largely to the fact that as portland cement hydrates, calcium hydroxide is liberated and, being a strongly basic compound, stifles rusting. Thus, it makes little difference whether the coating becomes saturated, and hairline cracking can be tolerated.

Cement-mortar coatings may be applied by pneumatic placement, extrusion, brush coating, or any other method that will give equivalent results. Mortar linings are commonly placed by the centrifugal method. In rehabilitation work they have also been applied in place on waterlines over 4 inches in diameter by means of special pipe-cleaning and mortar-application machines. Cement-mortar linings are best adapted to pipe continuously filled with water; they may not serve well where the lining will dry, as in exposed steel siphons. Pneumatically placed, steel-reinforced mortar has been used by the Bureau for some time as an exterior coating for buried steel pipe. More recently cement mortar has come into wide use on interior surfaces of steel pipe.

185. Inplace Mortar Linings.—Field application of mortar linings is very often employed in rehabilitating old, scaled, or tuberculated water-carrying steel pipelines to stop internal corrosion and increase carrying capacity. The thick, rigid lining seals small undetected holes in the steel, and the alkaline environment it creates next to the metal effectively stifles further corrosion which, with a pipe in near failing condition, might soon lead to extensive leakage. Steel pipelines installed unprotected or with temporary protective coatings, and which consequently tuberculated severely, have been restored to nearly full capacity by application of a smooth, continuous mortar lining. Inplace mortar linings may also be used with new piping, especially large-size piping which cannot be shop lined because of size or need for pressure testing prior to the lining application.

An important advantage of a mortar application is that an elaborate metal cleaning process, such as sandblasting, is not required. All loose rust, scale, and deteriorated paint coatings should be removed; various machines have been developed for this purpose. Usually, these consist of scraping tools drawn by winches or driven by hydraulic pressure through the piping. The materials so loosened are then pulled from the pipe with rubber swabs. Pools of water must also be removed; however, a thoroughly dry metal surface is not required.

Two types of special inplace lining equipment are suitable for small-diameter piping, that is, piping with a diameter of 4 to 16 inches. In one process, cement mortar previously fed into the pipe is distributed by pulling a conical-ended, cylindrical form through the pipe. The form, having a diameter selected to produce the desired coating thickness, drives the mortar supply ahead, leaving behind it a smooth layer over the pipe surface. Water is squeezed from the mortar through perforations in the cylinder, drying and densifying the lining, and the pressure of the form forces the mortar into intimate contact with the pipe. A second process for lining small pipe employs a centrifugal machine which distributes the mortar by spinning it from a rapidly revolving head, thickness being controlled by rate of travel. This process does not produce a troweled finish. For large piping, up to 12 feet or more in diameter, the mortar is also distributed by a centrifugal machine; however, it is further smoothed by rotating mechanical trowels. (See figs. 205 and 206.)

Inplace mortar application to small-diameter pipe ordinarily necessitates that access be obtained at short intervals, say every 250 feet. With larger piping, the interval may be increased. Also, certain machines will line around bends of larger sized pipe. Otherwise, these bends must be hand coated, and the result should be generally equivalent to that obtained by machine. Operation of lining equipment and correct proportioning of mortar mix for proper consistency require special skills, and experienced contractors employing trained men are best suited to perform such applications.

Curing a newly applied, inplace mortar lining consists essentially of keeping it moist. Thus, immediately following mortar placement, the lined section should be closed to prevent air circulation. Water may be intro-

Figure 205.—Workmen shoveling mortar into a lining machine which will distribute it over the interior surface of a steel pipe. Reinforcement has been fastened to the surface to strengthen the lining. Normally, the lining is placed without reinforcement. P126–100–19.

duced 24 hours after placement to continue the cure, although high velocities which might erode the mortar should be avoided for an additional 2 days. Moist curing should be continued for at least 7 days. Most pipelines lined by inplace methods are buried, hence are not subject to drying or expansion and contraction caused by temperature extremes. Further, it is often desired to place these pipelines back in service as soon as possible. They are, therefore, likely to remain in a damp condition; however, it is not desirable to allow the mortar to become thoroughly dry either during the curing period or thereafter.

186. Shop-Applied Mortar Linings and Coatings.—By far the majority of mortar linings and coatings for steel pipe used by the Bureau are applied in the shop prior to installation, and procedures for the applications are set forth in Bureau construction specifications. The General Services Administration has issued Federal Specification SS–P–385 titled "Pipe, Steel (Cement-mortar Lining and Reinforced Cement-mortar Coating)."

The following paragraphs review procedures and provisions reflecting present requirements.

(a) *Surface Preparation.*—As with inplace mortar applications, surface preparation of exterior and interior pipe surfaces for shop-applied mortar

Figure 206.—Trowels smoothing mortar which has been spun to the pipe surface by the rotary head in the lining machine. This produces a surface which has good hydraulic flow characteristics. P126–100–18.

coatings is intended to remove greasy or loosely adhering materials from the surface but not necessarily to expose the base metal. Although intimate mortar contact with the pipe is desired, adhesion in the usual sense is not depended upon to hold the mortar in place as with other coatings.

(b) *Materials.*—Materials for coating and lining consist of water, cement, and a graded sand. Water should be free of objectionable quantities of silt, organic matter, acids, alkalies, salts, and other impurities.

The type of cement to be used in coating or lining a pipe is usually required to be the same as used in concrete structures. Where the lining or coating will be in contact with sulfate-bearing waters or with soils containing soluble sulfate, it is often necessary to require use of a cement that will provide positive resistance to sulfate attack. Where the soluble sulfate concentration in soil or ground water is moderately high, use of type II cement is specified. Type V sulfate-resisting cement is specified when the soil or water contains soluble sulfates in such concentrations as would cause serious deterioration if other types of cements were used.

In addition to providing protection from sulfate attack for the lining or coating, it is necessary to provide protection from alkali-aggregate re-action. This is usually accomplished by requiring use of a low-alkali cement, which contains not more than 0.60 percent alkalies (percentage of sodium oxide (Na_2O) plus 0.658 times the percentage of potassium oxide (K_2O)). If the aggregates to be used in a coating or lining are known to be deleteriously reactive with high-alkali cement, or if the re-activity of the aggregates is not known, use of low-alkali cement is re-quired to assure adequate protection against potential alkali-aggregate reaction.

Sand for the mortar should be hard, dense, durable, uncoated, and otherwise conform to quality requirements for sand used in concrete. The following sand grading is suitable for mortar:

U.S. standard sieve No.	Individual percent, by weight, retained on sieve
4	0
8	0-5
16	10-20
30	20-30
50	25-40
100	15-20
Pan	3-7

This grading permits some latitude to allow the advantageous use of materials at hand and, particularly when a lining is to be applied, the most suitable selection for the thickness specified.

(c) *Exterior Coatings.*—For coating the pipe exterior, mortar mix may consist of 1 part cement to about 4 parts dry sand, by weight, together with sufficient water to produce suitable consistency, quality, and uni-formity. Some modification of these proportions may be required to obtain best results with the particular materials at hand. The cement and sand should be machine mixed for at least 1½ minutes prior to placement by pneumatic, steam, or other method. Accurate measurement of quan-tities is required to assure uniformity of resulting mortar and hence ap-plication characteristics. Rebound may be used to replace sand in amounts up to 50 percent by weight of the amount of sand required in the cement-mortar mixture.

Exterior mortar coatings require reinforcement. The reinforcement may consist either of helically wound, cold-drawn steel wire of a cage of welded wire fabric, 2- by 2-inch 14-gage or 2- by 4-inch 13-gage, con-forming to Federal Specification RR–W–375; or 1-inch hexagonal twisted wire fabric, 18- or 20-gage, ribbon mesh with both edges selvaged, conforming to ASTM Designation A 390. The netting may be either galvanized or not galvanized. Since reinforcement must be approximately centered in the mortar layer for maximum effect, welded cages or the wire netting should be self-positioning.

The helically wound wire, not less than No. 14 gage, should be carefully placed during application, with the spacing not to exceed 1¼ inches. The reinforcement should be embedded at the approximate center of the cement-mortar coating, and tension on the reinforcement must be sufficient to prevent sagging because of the weight of the mortar as it is placed. Ends of the netting should be securely fastened at laps. The reinforcement is intended to cover uniformly the entire surface; hence, adjoining circumferential strips of netting should overlap each other. When mortar is placed, the reinforcement should be free of dirt or grease.

In applying exterior mortar by means of pneumatic or steam pressure, the sand and cement are thoroughly mixed, then combined with a carefully controlled amount of water in a nozzle or spray head and propelled a distance of 3 to 5 feet onto the pipe surface by air or steam pressure. The water should be maintained at a uniform pressure of at least 15 pounds per square inch greater than the pressure in the placing machine, which should be at least 35 pounds per square inch. The mortar stream should be directed normal to the pipe surface.

Premixed plastic mortar may also be applied as an exterior coating by revolving belts or brushes, or other mechanical means such as extruding. In the first two methods, straight sections of piping, rotating on rollers, move past the mortar jet at a rate sufficient to produce the desired coating thickness, which can be controlled within narrow tolerances by careful adjustment of the rate of travel. In the extrusion method, the pipe moves longitudinally through a device which places the coating and immediately covers it with a spirally wrapped polyethylene membrane. This membrane supports the coating and provides an airtight covering.

Hand-held equipment is used for coating irregularly shaped sections not suitable for rotating. Close control of the mortar consistency is essential to secure high strength, adhesion, and a dense coating. Excessive moisture tends to cause sloughing, while a very dry mix will result in excessive rebound. Where sand pockets or signs of sloughing or separation from the surface are found, the coating should be stripped away promptly and a new mortar placed.

The following thicknesses are often used for the sizes of pipe shown:

Inside pipe diameter, inches	Exterior mortar thickness, inches
4 to 12	½
14 to 18	⅝
Over 18	¾

Extreme importance is attached to securing good curing of mortar coatings and preventing drying in the early stages. Accordingly, the coated piping (except for extruded coatings) is carefully moved to a curing area where, as soon as the mortar has set, it is kept continuously moist by sprinkling for at least 7 days. To expedite the coating process, accelerated curing can be accomplished by storing the pipe in a steam environment at elevated temperatures. Steam curing at temperatures between 130° and 150° F may be substituted for wet curing on a time ratio of 1 hour of steam curing to 4 hours of wet curing. The pipe should be brought to curing temperature at a rate of not more than 30° F per hour, but in no case should the temperature exceed 100° F within 2 hours after the mortar is applied. After steam curing, the pipe should be protected from rapid drops in temperature which may damage the lining or coating. Cement mortar coatings applied by extruding are cured by airtight wrapping of polyethylene membrane placed on the pipe immediately after coating. Failure to achieve a satisfactory cure may result in shrinkage cracking of the coating, lowered mortar strengths, and decreased durability.

(d) *Interior Linings.*—For the pipe lining, a richer mortar mix is usually specified and consists of 1 part cement to about 2½ parts sand, by weight. After a 3-minute mixing period, the mortar should be applied promptly.

The pipe to be lined, having been cleaned as previously described, is rotated on rolls as the mortar is introduced by a feed line or trough. The quantity of mortar must be accurately gaged to produce a finished lining of the specified thickness, and the initial speed of rotation is selected so that the mortar will flow and distribute itself uniformly over the surface. Thereafter, more rapid rotation compacts and densifies the mortar and smooths the surface. Water is forced out of the mortar by centrifugal force of high-speed spinning, and the mortar develops sufficient density and bond to the metal so that the pipe can be gently moved from the spinning rig to a curing area. Before moving the pipe, excess water is removed and the lining inspected for uniformity of thickness and any defects. When rotated at high speed, pipe sections should exhibit no eccentricity which would cause vibration and result in undesirable nonuniformity of lining thickness. Any pipe sections that fail to rotate smoothly should be trued or stiffener rings installed before being lined. Lining thicknesses commonly used for the pipe sizes shown are as follows:

Inside pipe diameter, inches	Mortar lining thickness, inches
4 to 12	5/16
14 to 18	3/8
Over 18	1/2

Curing of linings, as with coatings, consists of preserving a moist atmosphere. This is done by capping the pipe with a waterproof cover immediately following the mortar application. This procedure prevents drying air from circulating during the first stages of mortar curing and eliminates maintaining a sprinkler system where piping will be lined but not coated. If additional moisture is required to maintain a moist condition, water is introduced inside the pipe. If steam curing is used, it is necessary to cap the pipe ends when steam curing is interrupted or completed. A minimum of 7 days of water curing is required and, as with mortar coatings, steam curing may be substituted on the basis of 1 hour of steam curing for 4 hours of water curing.

Proper curing of mortar linings assumes even greater importance than curing of coatings on the exterior since the lining is usually unreinforced and will have no backfill to assist in holding it in place. Even with proper curing, if followed by thorough drying, the mortar may develop sizable longitudinal and circumferential cracks which penetrate to the pipe steel. These cracks at least partially close upon rewetting and, so far as is known, have little effect on serviceability of the lining. However, in larger pipe sizes the cracks may be proportionately larger, and since mortar thickness compared to pipe diameter is smaller, the arch effect holding the lining in place is weaker. During transportation and handling, more severe flexing of the less rigid large piping can be expected and the potentiality is increased for damage to a cracked lining, particularly when very high temperatures prevail.

If lining is improperly cured, not only will strength be decreased, but cracks of excessive width may develop. Ideally, of course, the lining should remain moist from the time of application through installation, and a practical approach to this ideal may be justified. Transparent plastic membrane covers used on pipe ends retain moisture on the pipe interior for long periods. Droplets of water on the plastic give visual evidence of the effectiveness of the seal obtained and permit rapid inspection for curing adequacy of large quantities of piping. Obtainable at minor cost, the covers prevent high-velocity dry air from passing through the pipes during shipment from lining plant to jobsite and will retain moisture in the mortar through a storage period.

(e) *Specials.*—Bends, transitions, manifolds, and other irregular shapes require the same mortar linings and coatings as adjoining straight pipe sections. On the exterior, with reinforcement installed as previously described, the mortar may be applied by means of a hand-held nozzle. This method is used also on the interior, or the mortar may be carefully troweled on. The quality of coating should always equal machine-applied coating.

187. Field Coating of Joints.—Mortar-lined or mortar-coated pipe sections may be connected by means of a variety of joints, including bell and spigot, welded, and rubber-gasketed joints and mechanical couplings, and the coating protection must be extended to exposed metal when the joint is complete in the field. On piping too small to be entered, ends or couplings are coated before joining with a mortar consisting of 1 part cement to 1 part sand. As the sections are thrust together, some of this mortar squeezes out, projecting into the path of waterflow. A smooth lining, flush with adjacent mortar, can be secured by drawing an inflated rubber ball through the joint. Joints in large piping are troweled smooth by hand. The exterior surfaces of the joint must likewise be protected. Reinforcement comparable to that embedded in the machine-placed mortar is first secured around the joint; then mortar is placed to the same thickness as required on piping. The mortar may be applied either pneumatically or troweled on, and it should be coated with white-pigmented curing compound until backfill is placed.

L. Concrete Polymer Materials

188. Concrete Polymer Materials.—New materials for construction.— (a) *General.*—A major advance in concrete technology in recent years has been the development of polymer-impregnated concrete—concrete that is impregnated with a liquid monomer or resin system that is subsequently converted to a solid polymer. Research carried out cooperatively since 1966 by the Bureau of Reclamation and the former Office of Saline Water, the Brookhaven National Laboratory, and the Atomic Energy Commission has indicated that impregnation and polymerization processes have induced substantially improved structural and durability properties in polymer-impregnated concrete formed as compared with conventional concrete. Maximum increases in strength up to almost four times have been obtained. Water permeability has been reduced to negligible values, and water absorption has been decreased by as much as 95 percent. Abrasion and cavitation resistance has shown significant improvement. Freeze-thaw resistance has been greatly improved, as has corrosion resistance to distilled water, sulfate solution, and acid. These and other improvements in physical and chemical resistance properties of polymer-impregnated concrete as compared with those of unimpregnated portland cement concrete are shown in table 29.

This section defines concrete polymer materials and briefly summarizes the properties of polymer-impregnated concrete. Users of concrete polymer materials will find that continued reading of the technical literature describing on-going research in this new technology of concrete to be essential in utilizing the benefits in a variety of applications in construction.

(b) *Definition.*—Concrete polymer materials may be defined as con-

Table 29.—Typical properties of polymer-impregnated concrete

Property	Unimpregnated Concrete	Polymer-Impregnated Concrete [1]
Compressive strength, lb/in^2	5,300	18,200
Modulus of elasticity, 10^6 lb/in^2	3.5	6.2
Tensile strength, lb/in^2	400	1,500
Modulus of rupture, lb/in^2	700	2,300
Flexural elasticity modulus, 10^6 lb/in^2	4.3	7.1
Hardness, "L"-type impact hammer	32	52
Abrasion loss,		
inches	0.050	0.015
weight loss, grams	14	4
Cavitation loss, inches	0.32	0.02
Water absorption, percent	6.4	0.3
Water permeability, 10^3 ft/yr	53	14
Thermal conductivity, 73° F Btu/ft-hr $-°$ F	1.33	1.27
Coefficient of expansion, 10^6 in/in	4.02	5.25
Diffusivity at 73° F, ft^2/hr	0.0387	0.0385
Specific heat, 73° F, Btu/lb/°F	0.241	0.220
Specific gravity	2.317	2.386
Freeze-thaw durability,		
cycles	490	3,650
percent weight loss	>25	2
Sulfate attack resistance (accelerated test),		
days exposure	480	1,436
percent expansion	0.50	0.017
Resistance to 15 percent HC1,		
days exposure	105	1,395
percent weight loss	>25	10
Resistance to 5 percent H_2SO_4,		
days exposure	49	133
percent weight loss	>25	25

[1] Conventional concrete impregnated with methyl methacrylate
Thermal-catalytic polymerization
Specimens contain 4.6 to 6.7 percent polymer by weight

crete either with or without portland cement to which has been added, either in the hardened state or to fresh concrete during mixing, a chemical system subsequently polymerized. The chemical system is generally a liquid and may be composed of monomers, resins, polymer solutions, or polymer emulsions, which are converted to a solid plastic by polymerization. The polymerization may be initiated by irradiation, by chemical action in conjunction with heat, or through the use of a promoter in conjunction with a chemical initiator under ambient conditions. There are three types of concrete polymer materials:

(1) Polymer-impregnated concrete (PIC) consists of preformed, hardened, and cured concrete that is impregnated with a liquid monomer or resin system and subsequently polymerized. The process generally includes oven drying of the concrete, vacuuming, and pressure soaking to ensure complete impregnation. The concrete may be either completely impregnated or partially impregnated from the surface to a desired depth, depending on the intended use of the material and required properties.

(2) Polymer-cement concrete (PCC) consists of a portland cement concrete to which a monomer or polymer resin system has been added during the mixing period and subsequently polymerized either in the forms or after stripping. PCC is in the early stages of development, but it appears that it may not be possible to develop the exceptional strength properties obtained with PIC. However, PCC may be useful in applications where impermeability or durability is important.

(3) Polymer concrete (PC) consists of a monomer or resin system that is mixed with graded aggregate and subsequently polymerized. PC contains no portland cement. Preliminary tests have shown that PC can be fabricated using conventional concrete mixing and placing equipment and that full strength is developed 3 hours after placing. Mixes have been designed to produce a material having approximately the same polymer content and compressive strength as PIC.

The production of polymer-impregnated concrete is not diffcult, but it requires some special equipment and strict adherence to safety regulations. Complete impregnation, which increases resultant strength and durability, is more easily achieved by drying the hardened concrete and placing it under vacuum to remove moisture and air. While still under vacuum, the specimen is immersed in monomer and pressure applied to thoroughly fill the voids and capillaries. With volatile monomers the concrete should then be wrapped or enclosed in an appropriate medium to minimize evaporation and drainage loss of monomer during the period between impregnation and polymerization. This can be accomplished by wrapping in aluminum foil, by impregnating and polymerizing in the mold, or by immersing and polymerizing in a liquid that will not react with or dissolve the monomer. Polymerization is then accomplished using heat or radiation, and the concrete is removed ready for use.

(c) *Investigations*.—Investigations are progressing to develop methods, processes, new materials, and design criteria to efficiently and safely utilize concrete polymer materials. Investigations are also in progress to provide design criteria for PIC structural members. In addition to the greatly improved properties, PIC shows promise of having economic advantages over plain portland cement concrete in applications that require exceptional strength or durability and in situations where maintenance or repair is expensive or impracticable. Some of these are concrete pipe used for conveyance of irrigation water, sewage, and municipal and industrial wastewater; housing applications, including prefabricated beams, wall and floor panels, and load-bearing columns; materials resistant to chemicals and distilled, mineral-free water encountered in sewers and desalination plants; and bridge decks, curbs, and sidewalks with high resistance to abrasion and effects of deicing chemicals.

Appendix

Designation 1

1. Sampling Aggregate.—The task of obtaining a truly representative sample of aggregate is complicated because of segregation that takes place when the aggregate is handled or moved. The following paragraphs describe methods which, if carefully adhered to, will generally compensate for segregation.

2. Aggregate Samples from a Deposit.—The first step in obtaining a truly representative sample of a deposit is the systematic placing of drill holes and test pits over the entire area. Such procedure provides the opportunity for securing accurate samples. The next step, equal in importance, is the method of taking the sample itself. In sampling test pits the objective is to obtain a continuous sample from top to bottom. If the walls of the pit do not require timbering, collection of such a sample is simple. One sampler holds a box against the wall while another with pick or shovel cuts a vertical channel in the wall 6 inches to 1 foot wide and 2 to 6 inches deep. Collection of material in this manner should be stopped at regular vertical intervals and a new box used. The sample taken should be carefully labeled with the number of the hole and depths from which materials were removed. Vertical intervals vary with different deposits. Where the gravel contains thin sand or clay beds, different boxes should be used for each material. Where the deposit is uniform, a 5-foot interval is sufficient. The sampler should keep a record as to any unusual occurrence in the deposit such as single boulders or nests of boulders, water seams, changes in bedding, etc. His notes should also show from which wall of the pit the sample was taken. It is often advisable to take a similar sample from the opposite wall for verification.

Where the walls of a pit will not stand without shoring, the sampling must be done as the pit is sunk. It is seldom that gravel pits will not stand in place for a vertical distance of 2 feet. Therefore, the timber should be kept within 2 feet of the bottom. Usually 4 to 5 feet are ample

491

protection. In such pits the wall channel samples are taken just prior to placing the timber. The procedure is the same except that each 5 feet is sampled separately instead of sampling the whole depth at one time.

Where timber must be kept close to the bottom or in advance of the digging, as in loose dry sand or quicksand, a different method must be employed. Such deposits are by nature more homogeneous, and vertical accuracy is not so essential except for notation of the depth at which radical changes occur. At these deposits it is often sufficient to take a channel sample across the floor of the pit at each vertical foot in depth.

3. Sampling with Augers.—Sampling by earth augers is, of course, automatic. The material removed by the auger should be placed along a line on the ground surface. It is then bagged in predetermined vertical intervals, tagged, and recorded.

4. Sampling with Bailer or Drill.—Sampling of the material obtained by bailer or churn drill or wherever water is used or encountered in the hole presents other difficulties in obtaining accuracy. Accurate samples are obtainable by churn drill or bailer only where it is possible to drive the casing ahead of the drill bit or sand pump. Where the bit precedes the casing, contamination is always present from material falling in from walls of the hole. The presence of water tends to segregate fine sands and clay from coarser material.

These fine materials are maintained in suspension by churning action of the drill or bailer. Accurate sampling requires that for each vertical interval cut by the drill all material, coarse or fine, be saved. This means that the material in suspension must be saved along with the coarse material. To accomplish this, the simplest procedure is to drive the casing 2 feet ahead of the bit or bailer, add the necessary water, and drill or bail to the bottom of the casing. All material, including the water, as removed from the hole should be placed in a box from which no water is allowed to escape before the fines settle out. When completely settled the clear water is poured off and the material in the box bagged and labeled for hole number and depth.

When using the small clamshell bucket within casing, the pipe should be driven ahead of the digging a predetermined distance and the coarse material removed and placed in a continuous pile or ridge. When the bucket has dug to the bottom of the pipe, the water and suspended material should be removed by a bailer and allowed to settle. This fine material is then added to the coarser material and thoroughly mixed before bagging the sample.

When drilling below water level, it is difficult to obtain samples of both fine and coarse material because water enters from the bottom of the casing. This water may bring with it fine material from surrounding areas beyond the casing. At best, it provides an excess of water which so dilutes the suspended material that it is almost impossible to obtain the fine and the coarse material from the same vertical section. Many times the incoming water enters so fast that it is impossible to bail the hole dry. This greatly dilutes the fine material, and much of it is consequently lost. Therefore, samples taken below the waterline must be considered to have been subjected to loss of fines.

When boulders are uncovered and casing cannot be driven, a new hole should be started as close as possible or the boulder blasted. In deposits containing many loose or clustered boulders, drilling holes may be more expensive than test pitting because of the loss of holes and consequent extra work.

Requirements for extra holes can be determined from the sampler's notes.

5. Aggregate Samples from Conveyor Belts or Chutes.—To secure a representative sample of aggregate from a belt or chute, a complete cross section of the stream should be taken over a short period, rather than just a portion of the stream over a longer period. Samples should be taken at regular intervals until the entire supply has been sampled. The number and size of such samples will depend on quantity and uniformity of the aggregate.

6. Aggregate Samples from Railroad Cars.—Samples from a railroad car are best taken at points equally spaced on straight lines along the sides and center of the car. The size of samples will depend on the size of the car, the number of points from which samples are taken, and the maximum size of aggregate particles. At each point the sample should be dug down to obtain aggregate well beneath the surface. A standard tube sampler should be used for sand and, when possible, for coarse aggregate. The tube sampler is usually a steel pipe about 2 inches in diameter and 6 feet long, pointed at the lower end, and having a handle at the top. A series of openings is punched along the pipe so that a line of ears projects from one side of the openings. The tube is forced into the aggregate as far as possible, turned until the ears have scooped sufficient material into the tube for a sample, and then withdrawn, keeping the openings on top.

As it is sometimes difficult to obtain a representative sample from a railroad car by the above method, especially with large coarse aggregate, samples should be taken so far as practicable while the material is being loaded or unloaded. While a car is being unloaded, a fairly representative sample may be obtained by taking a shovelful at regular intervals, provided care is taken that the larger pieces do not roll off the shovel.

•

Whatever method is used to obtain the sample, it should be one that will assure representative material.

7. Aggregate Samples from Stockpiles.—The entire sand stockpile from top to bottom should be sampled. Sampling should start at equally spaced points along the bottom of the pile and proceed upward, at equal intervals, over the sides and top. If only part of the pile is to be used for a portion of the job, just the part to be used should be sampled.

Where practicable, gravel samples are taken with a standard tube sampler. If this cannot be done, samples should be obtained with a shovel and consist of material from well beneath the surface. By holding a short piece of board against the pile just above the point of sampling, the inclusion of unwanted surface material may be avoided.

8. Treatment of Sample.—The sample should be reduced to a test sample by quartering or splitting as described in sections 9 and 10. It may be required that a separate analysis be made on each sample rather than one analysis of a composite sample so that the variations in the material may be ascertained.

9. Sand.—Samples of sand should be reduced to test size by the quartering method or use of a sample splitter.

(a) *Quartering Method.*—The sample is placed on a hard, clean surface where there will be neither loss of material nor accidental addition of foreign matter. The sample is mixed thoroughly by turning the entire lot over three times with a shovel. This can best be accomplished by two men, one on each side of the sample, beginning at one end and taking alternate shovels of the material as they advance the length of the pile. With the third or last turning, the entire sample is shoveled into a conical pile by depositing each shovelful on top of the preceding one. The conical pile is carefully flattened to a uniform thickness and diameter so that the material will not be transposed from one quarter to another. The flattened mass is then marked into quarters by two lines that intersect at right angles at the center of the pile. Two diagonally opposite quarters are removed and the cleared spaces brushed clean. The remaining material is mixed and quartered successively until the sample is reduced to 50 pounds or less. This quantity is passed through the sample splitter, one-half being discarded and the other half split again. This procedure is repeated until the sample is reduced to the desired size.

(b) *Sample Splitter.*—The entire sample is passed through the splitter, one-half set aside and the other half split again. The procedure is repeated until the sample is reduced to the desired size. (The sample splitter should be similar or equal to the Jones sample splitter.)

10. Coarse Aggregate.—Samples of coarse aggregate should be reduced to test size by the quartering method or use of a suitable sample splitter.

(a) *Quartering Method*.—Same as section 9(a).

(b) *Sample Splitter*.—A representative sample is passed through the splitter, one-half set aside and the other half split again. This procedure is repeated until the sample is reduced to the desired size. (The sample splitter should be similar or equal to the Gilson sample splitter.)

With large size aggregate it may be more desirable or convenient to handpick the sample. When this is done, extreme care should be taken to obtain a representative sample.

SAMPLING CONCRETE

Designation 2

1. Sampling Fresh Concrete.—It is important that the samples of concrete be representative of the concrete being placed. The batch sampled should be a typical batch, and if practicable, samples of the sand and gravel fractions that enter the batch should be taken so that moisture determination and grading analyses may be made to correspond with the concrete specimens. The exact amount of water added at the mixer should be noted. This, together with the moisture in the aggregates, is used in computing water-cement ratio.

Sampling from chutes and conveyors is usually unsatisfactory; but if it is necessary, care should be taken to obtain a full cross section of the flow. Sampling from a transporting container is also undesirable as the concrete is often segregated. If the concrete is dumped from the mixer into a loading hopper, fairly representative samples may be taken from a slide gate opening about halfway down the side of the hopper.

It is difficult to obtain representative samples of concrete discharged directly from the mixer, and this should not be attempted unless arrangements have been made to catch the entire flow for a short period. It is possible to obtain reasonably representative samples from certain types of mixers, after the drum has been stopped, by using a long-handled shovel. When this method is used, small quantities of concrete are taken from several points and mixed. From revolving-drum truck mixers or agitators, samples are taken at three or more regular intervals during the discharge of the batch. In obtaining the samples, the rate of discharge of the batch should be regulated by the speed of the drum and not by the size of the gate opening. When the method of sampling has been selected that best meets the conditions involved and results in representative samples, this procedure should be used throughout the job so that the results of tests will be comparable.

The size of a sample will depend largely on the intended use. Separate portions of the sample should be used for slump and unit weight tests and for casting the cylinders. For concrete in which the maximum size of aggregate is 1½ inches, an ordinary water bucket about two-thirds full will usually suffice for a slump test or for one 6- by 12-inch test specimen. When the concrete contains aggregate larger than 1½ inches, it will be necessary to remove such oversize material from the concrete to be used for slump tests or for 6- by 12-inch test specimens. Oversize material is removed by wet screening through a screen having 1½-inch-square openings. The maximum size of aggregate in the sample should not be greater than one-fourth the minimum dimension of the test specimen because results of compressive-strength tests will be unreliable if such oversize is present. For wet screening, the screen should be conveniently mounted for easy shaking and rapid removal of the oversize. A screen similar to that shown in figures 46 and 47 has been satisfactorily used. The screened concrete should be remixed with a shovel into a uniform mass before making slump tests or test specimens. When concrete containing large aggregate is sampled, it will be necessary to take a larger sample.

Slump tests and cylinder specimens should be made without delay after sampling. A delay of 15 minutes may decrease the slump as much as 50 percent.

2. Sampling Hardened Concrete.—Tests on control cylinders made during construction are not a direct index of the quality of concrete in place. More accurate and more informative data may be obtained from specimens taken from the completed work. Effectiveness of placing methods in avoiding rock pockets and other forms of separation is revealed, in many cases, by core-drill investigations. Such investigations have disclosed relative strengths and the extent of the surface conditions attributable to different curing procedures and exposures; comparative strengths of concrete at various depths from the surface; and bonding efficiencies resulting from various cleanup methods used on horizontal construction joints. Some information has been gained concerning different procedures for compacting concrete on canal slopes and in arch sections of tunnels. Drill cores have also been helpful in the analysis of data from crack surveys and in determining the extent of damage or deterioration in structures showing evidence of alkali-aggregate reaction.

Concrete core-drilling equipment is illustrated in figure 207. Concrete cores from 2 to 10 inches in diameter and of any reasonable length can be obtained with these machines. Bases of some drilling machines may not be of sufficient size to accommodate a 10-inch-diameter core barrel, in which event cribbing will be required as shown in figure 208. Bureau

Figure 207.—Drilling a 10-inch core in concrete floor. PX–D–33518.

machines for cores less than 18 inches in diameter are frequently furnished with diamond (bort) bits. (See fig. 209.) A gasoline engine or an electric motor may be used for power. A water-pump attachment permits use of these machines where water in a pressure line is not available. Only persons experienced in diamond-drill work should operate

Figure 208.—Concrete core drilling equipment. Cribbing is used to give clearance beneath drilling machine for 10-inch-diameter core barrel. PX–D–33519.

Figure 209.—A commercial 6-inch bit. PX–D–33520.

core drills.[1] Drilling cores 18 inches and greater in diameter is usually performed with a calyx drill (see fig. 210). This method employs steel shot as the cutting medium. Although calyx drilling is slower than diamond drilling, it is justifiable from the standpoint of economy.

Concrete to be sampled should be hard enough to prevent damage by drilling to the bond between the mortar and coarse aggregate in the

---

[1] Extreme care should be taken not to keep the bort bit in service beyond a point where the stones begin to loosen as they would soon break out, thus causing serious damage to the bit and loss of bort. For concrete drilling, the majority of the bits are set with white bort diamonds which are less expensive than black diamonds. Borts must be used with care as they will not stand up under forced drilling. It is customary to set many of the white borts in the face of the bit and to use the more expensive black diamonds for reaming the sides of the hole for clearance. The black diamond, being more resistant to shock is more extensively used where the material being drilled is fragmented or loose. It should not be permitted to become excessively heated as this decreases the wearing quality. Another precaution to be taken in drilling with white stones is to use the correct amount of water. Grooving of metal on the cutting face of the bit may sometimes be prevented or decreased by reducing the quantity of water that reaches the bit. Care should be taken when drilling through steel, particularly in a vertical "down" hole, to use enough water to expel metal cuttings from the hole and thus reduce wear on the metal of the bit. If there is any doubt as to continued service of a diamond-drill bit, it should be repaired rather than risk costly damage.

Figure 210.—Calyx drill extracting 22-inch-diameter core. PX–D–33521.

specimen. For compression tests, the core specimen should be as nearly as possible a cylinder with length twice the diameter; it should have a diameter at least three times that of the largest aggregate and should not have any weak joints or bedding planes not perpendicular to the axis. The location for cores taken horizontally from walls should be selected carefully and described accurately in the drilling record because the lower portion of a deep concrete lift may be more dense, and the upper portion less dense, than the middle portion.

Where cores are taken for examination of joints or for study of effectiveness of compaction methods (as evidenced by the amount of segregation at reinforcement intersections, or at joints or water stops), the holes should be drilled as carefully as possible, without forcing, so as to minimize core breakage. Examination of holes drilled for these purposes will often reveal conditions that cannot be detected by observing the cores.

Bort bits will normally not be damaged by drilling through steel if the steel is firmly embedded in the concrete. Embedded steel can, however, be located by means of an astatic compass or an electronic metal detector and metal-free cores obtained. An astatic compass consists of two parallel magnetic needles of equal, but opposite, magnetic moments; it is fairly reliable for locating steel close to the surface and spaced about a foot apart, such as reinforcement bars in canal lining.

For concrete walls up to 12 inches thick that are accessible on both sides reinforcement can be located by use of radioisotopic techniques. This involves placing a radioisotope that emits gamma rays on one side of the wall and a photographic plate on the other side. A radiograph similar to an X-ray is obtained that shows the location of steel. For very thin walls a Geiger counter can be used in lieu of a photographic plate.

Drill-core specimens are cut to proper lengths for test in the Denver laboratories, where suitable cutting equipment is available. (For testing procedure see designations 32 and 33, also ASTM Designation C 42.) The cores should be painted with some legible identification mark and packed so they will not be broken in transit. Extra care should be taken in wrapping the cores in waterproof paper, sealed tightly with tape and well surrounded with damp sawdust in tight substantial boxes. (See figs. 211 and 212.) Data sheets should be transmitted with each shipment. The exact location of the hole, together with information indicated in figure 213, is needed for the proper classification and study of each core. Cost of drilling varies considerably with type of aggregate, amount of steel reinforcement, and accessibility of the drill setting, and is affected by many other details. Ordinarily, 1 to 2 hours are required for preliminary operations at each hole. Under average conditions 1 foot of 6-inch-diameter core can be cut per hour.

SAMPLING SOIL AND WATER FOR CHEMICAL ANALYSES

Designation 3

1. General.—Prior to construction, definite knowledge concerning the soluble sulfate content of both soil and water is essential to permit selection of the most suitable type of cement to provide for adequate

Figure 211.—Well-identified concrete cores from Clear Creek Tunnel, Central Valley project, California. Cores show number, location marks, and drilling directions. P416–D–26421.

Figure 212.—Proper packaging of cores for shipment. Cores are well wrapped in waterproof paper and solidly packed in damp sawdust in substantially built core boxes. PX–D–33522.

7-1579
(8-65)

UNITED STATES
DEPARTMENT OF THE INTERIOR
BUREAU OF RECLAMATION

CONCRETE CORE DATA SHEET

PROJECT	FEATURE	
CORE NO.	CORE DIA.	LENGTH
DATE PLACED	DATE DRILLED	
DIRECTION DRILLED		
LOCATION IN STRUCTURE		
PRESENT CONDITION OF STRUCTURE		
RATE DRILLED PER HOUR		
REMARKS		

THE FOLLOWING INFORMATION IS DESIRED
CHARACTER OF CONCRETE

CEMENT: BRAND	TYPE	
SAND SOURCE	GRAVEL SOURCE	
MAX. SIZE AGGREGATE	TOTAL AGGREGATE LBS/CU YD	PERCENT SAND
TYPE OF CONCRETE (exterior - interior)	TYPE OF CURE	
FIELD MIX, BY WEIGHT 1:	W/C	SLUMP	AIR %
CEMENT CONTENT, LB PER CU YD	POZZOLAN CONTENT, LB PER CU YD	
WATER CONTENT, LB PER CU YD	UNIT WEIGHT, LB PER CU FT	
TEMPERATURE: AIR, MAX. F	MIN. F	MIX WATER F	CONCRETE F
TYPE MOLD USED FOR TEST CYLINDERS	TEST CYLINDER NO.	
CONTROL CYLINDER COMP. STR. PSI 7-DAY	28-DAY	90-DAY
REMARKS		

Figure 213.—Record of core drilling. (Form 7–1579)

protection of the concrete against chemical attack. It is also important to know the quality of water to be used for mixing and curing concrete. This information can be obtained by chemical analyses made in the Denver laboratory of carefully collected samples of soil and water.

The following considerations should be borne in mind during sample collection in order that the results of the laboratory analyses will not be rendered valueless because of ineffective methods of sampling:

(a) The concentration of soluble sulfate salts in any particular soil is affected appreciably by weather conditions, such as rainfall and temperature, by fluctuations in the water table, by topographic features, by type of soil, and by vegetable growth.

(b) During or immediately following a rainy season it is desirable to gather samples of soil and water at depths from 2 to 12 feet below the surface. The depth is dependent on the permeability of the soil, the elevation of the normal water level, and the amount and distribution of rainfall. Where the water surface is near or at the ground surface, as at an excavation site, surface samples are valuable, although they may frequently represent a concentration or even a dilution.

(c) Too much reliance should not be placed on samples of seepage water from test holes as such samples may not be representative because of a concentration by evaporation or dilution by runoff water or rainfall. Such samples of water are frequently of value when accompanied by samples of representative soil from the same point. Samples of underground water are always of value.

(d) It is good practice to obtain samples of soil from surface depressions. The sample report should include mention of any surface encrustations or other evidence of alkali on the surrounding area.

(e) In general, soil samples should not be taken where there is extensive vegetation. The possible exception to this rule is the collection of samples near such plants as saltbush and greasewood which grow in saline soil. Such samples should be obtained about 2 feet below the plant roots.

(f) In a prolonged dry season it is permissible to take soil samples at the surface or from piles of material already excavated, such as diggings from test pits. When this procedure is used, avoid sampling soil from which chemical salts have been leached by rainwater.

(g) Samples of water proposed for use in the manufacture or curing of concrete should be taken directly from the source. The water should be allowed to flow freely for at least 10 minutes and the containers should be rinsed with the water before the sample is collected.

2. Sampling, Shipment of Samples.—Soil samples should weigh a minimum of 1 pound and may be shipped either in fruit jars or paper cartons. Water samples should consist of at least 1 pint and should be shipped in clean pint fruit jars tightly capped or in polyethylene plastic bottles. Glass sample containers should be well packed to avoid breakage during shipment. Samples should be shipped to the Chief, Division of General Research, Bureau of Reclamation, P.O. Box 25007, Engineering and Research Center, Denver Federal Center, Denver, Colorado 80225. Table 30 illustrates the type of information that should be transmitted with samples.

Table 30.—Data on soil and water samples

Project: *Moon River—Wyoming*
Feature: *West Canal, Lateral A2*
Date: *March 25, 1962*

Sample No.	Location (Station No.)	Structure site	Eleva-tion at surface	Eleva-tion of sample	Soil classification and description [1]
1	1225+52 ..	Siphon	4916.0 .	4912.5 .	Shale, fractured (CL).
1A	1230+00 ..	Siphon	4865.5 .	4862.5 .	Clayey sand (SC).
2	1360+25 ..	Wasteway ..	2 feet below natural surface on center line.		Silty gravel (GM).
3	1380+00 ..	Canal	Excavated cut		Seepage water.

[1] See the Bureau's Earth Manual.
Remarks: *Seepage water collected from pool as it filled after being drained.*

SCREEN ANALYSIS OF SAND

Designation 4

1. General.—This test covers the procedure for determining the particle size distribution of fine aggregate by screening.

2. Apparatus.—(a) A balance or scale accurate within 0.1 percent of the test at any point within the range of use.

(b) Sample splitter (similar or equal to Jones type).

(c) Oven of appropriate size capable of maintaining a temperature of not less than 212° nor more than 230° F.

(d) Motor-driven mechanical screen shaker (preferred, but not required).

(e) Standard square-mesh screens designated by the following numbers: 4, 8, 16, 30, 50, and 100. The wire sizes and screen openings should conform to the requirements given in table 31.

(f) Miscellaneous sample pans, brass wire, and soft hair bristle brushes for cleaning screens.

3. Test sample.—The sample of sand to be tested for screen analysis

Table 31.—United States standard screen openings and wire diameters [1]

Screen number or size in inches	Average screen opening		Average wire diameter	
	Millimeter	Inch	Millimeter	Inch
200	0.074	0.0029	0.053	0.0021
100	0.149	0.0059	0.110	.0043
50	0.297	0.0117	0.215	.0085
30	0.595	0.0234	0.390	.0154
16	1.19	0.0469	0.650	.0256
8	2.38	0.0937	1.00	.0394
4	4.76	0.187	1.54	.0606
3/8	9.51	0.375	2.27	.0894
1/2	12.7	0.50	2.67	.1051
3/4	19.0	0.75	3.30	.1299
1	25.4	1.00	3.80	.1496
1 1/2	38.1	1.50	4.59	.1807
2	50.8	2.00	5.05	.1988
3	76.1	3.00	5.80	.2283

[1] The dimensions conform to the requirements of ASTM Designation E 11, "Standard Specification for Wire-Cloth Sieves for Testing Purposes." (Screen openings given are square openings.)

shall be thoroughly mixed and reduced by use of a sample splitter or by the quartering method (designation 1, sec. 9) to an amount suitable for testing. The entire portion from each side of the final sand split or opposite quarters of the last quartering should be used in this test for two samples. The sand shall be moistened before reduction to minimize segregation and loss of dust. Samples for testing shall be approximately the weight desired when dry and shall be the end result of the reduction method. Reduction to an exact predetermined weight shall not be permitted.

For concrete sands, the weight of the test sample should be determined by the fineness of the sand. Accurate results on the fine screens will not be obtained with test samples much larger than the following:

Kind of sand	Net weight of sample, grams
Coarse sand (F.M. 2.50-3.50)	400-1,000
Fine sand (F.M. 1.50-2.50)	200-400
Very fine sand (F.M. 0.50-1.50)	100-200

In no case shall the fraction retained on any 8-inch-diameter screen at the completion of the screening exceed 200 grams.

4. Procedure.—The screen analysis of sand is obtained by screening a representative sample of the size indicated through 8-inch screens, using a mechanical shaker or by hand shaking. Screens larger than 8 inches in diameter should not be used for sand analyses.

(a) Prior to making a screen analysis, test samples should be dried to substantially constant weight at a temperature of 212° to 230° F.

(b) Each sample is placed on a set of screens (conforming to the requirements of table 31) nested in order of decreasing size of opening from top to bottom. The nested screens are placed in the mechanical shaker and screened until not more than 1 percent of the residue passes any screen during a 1-minute period. Generally, the shaker should be run 10 minutes.

(c) The weighing is cumulative, starting with the material retained on the No. 4 screen and continuing in order of decreasing screen size until the material in the pan has been weighed. The weighing should be to the nearest gram. The material on each screen is not poured directly into the balance pan but transferred to an auxiliary pan of convenient size, or onto a paper, and then placed in the balance pan. This will prevent spilling.

(d) Each screen is carefully cleaned with a brush provided for that purpose. A brass wire brush may be used for cleaning the No. 4, 8, 16, 30, and 50 size screens. Only a soft hair bristle brush should be used for cleaning the No. 100 size screen. Frequent inspection is necessary to guard against use of broken or stretched screens.

(e) Where power facilities are not available, screen analysis may be made by hand shaking a representative sample through a set of 8-inch-diameter screens. Screening should be continued until not more than 1 percent, by weight, of the residue passes any screen during 1 minute. Otherwise, the general procedure is the same as previously outlined.

(f) The gradation samples should be retained for determining the percentage of material passing the No. 200 screen and, if necessary, for screen analysis of washed sand.

5. Calculations.—(a) Cumulative weights of each sample are converted to percentages of total weights and an average grading is expressed in terms of the nearest whole percent.

(b) Individual percentage is that percentage of material which is retained between consecutive screens shown in the following tabulation.

(c) The fineness modulus of sand is computed by adding the cumulative percentages retained on the six standard screens, from the No. 4 to the No. 100, inclusive, and dividing the sum by 100. The following tabulation shows typical sand-analysis results with computation for fineness modulus:

Screen No.	Cumulative percentages retained	Individual percentages retained
4	1	1
8	19	18
16	39	20
30	58	19
50	76	18
100	92	16
Pan		8
	285	100

$$285 \div 100 = 2.85 \text{ F.M.}$$

6. Report.—The report shall include the following:

(a) Total (i.e., cumulative) percentage of material retained on each screen.

(b) Individual percentage of material retained on each screen.

(c) Fineness Modulus (F.M.).

SCREEN ANALYSIS OF COARSE AGGREGATE

Designation 5

1. General.—This test covers the procedure for determining the particle size distribution of coarse aggregate by screening.

2. Apparatus.—(a) Scale accurate within 0.1 percent of the test load at any point within the range of use.

(b) Motor-driven mechanical or manual screening equipment.

(c) Standard square-mesh screens having openings in sizes 6 inch, 3 inch, 1½ inch, ¾ inch, ⅜ inch, and No. 4 (³⁄₁₆ inch), and other sizes as required. The wire sizes and screen openings should conform to the requirements given in table 31 (designation 4). It is recommended that screens be mounted on substantial frames, the smallest dimension of which shall be 18 inches, and constructed in such a manner that will prevent loss of material during screening.

(d) Miscellaneous sample pans, splitters, brushes for cleaning screens, etc.

3. Test Sample.—The test sample should not be less than 100 pounds of material for 1½-inch-maximum size aggregate and preferably not less than 500 pounds for aggregate of larger maximum size. The sample shall be obtained by a sample splitter or the quartering method (designation 1, sec. 10).

4. Procedure.—The coarse aggregate sample is separated into the various sizes and the individual sizes weighed. The amount of coarse aggregate separated at one time shall be of such quantity that it will not overload any of the screens. Unless there are appreciable quantities of adherent fine material, which should be determined and analyzed, it is not necessary to dry wet samples for this test. Screening should continue until not more than 1 percent of the coarse aggregate passes any screen during a 1-minute period.

5. Calculations.—(a) Weights of each coarse aggregate size fraction are converted to individual percentages retained, based on total sample weights, expressed in terms of the nearest whole percent.

(b) Individual percentages retained of each coarse aggregate size fraction are converted to cumulative percentages retained.

(c) The fineness modulus is computed by adding the cumulative percentages retained on the specified screens, from the largest to the No. 4 size inclusive, dividing the sum by 100 and adding 5.00 to the result. The 5.00 represents the sum of the percentages retained on specified sand screens, finer than No. 4, divided by 100.

6. Report.—The report shall include the following:

(a) Total (i.e., cumulative) percentage of material retained on each screen.

(b) Individual percentage of material retained on each screen.

(c) Fineness Modulus (F.M.).

SCREEN ANALYSIS OF COMBINED SAND AND COARSE AGGREGATE (COMPUTED)

Designation 6

An approximate analysis, sufficiently accurate for use in plotting combined grading curves, may be computed from the routine grading tests made at the batching plant. The screen analysis of the combined sand and coarse aggregate for a given mix may be obtained from the screen analysis of the sand (designation 4) and the complete screen analysis of each fraction of coarse aggregate (designation 5) through the range of undersize and oversize. Figure 214 illustrates the sizing nomenclature for the ¾- to 1½-inch coarse aggregate size fraction and shows the

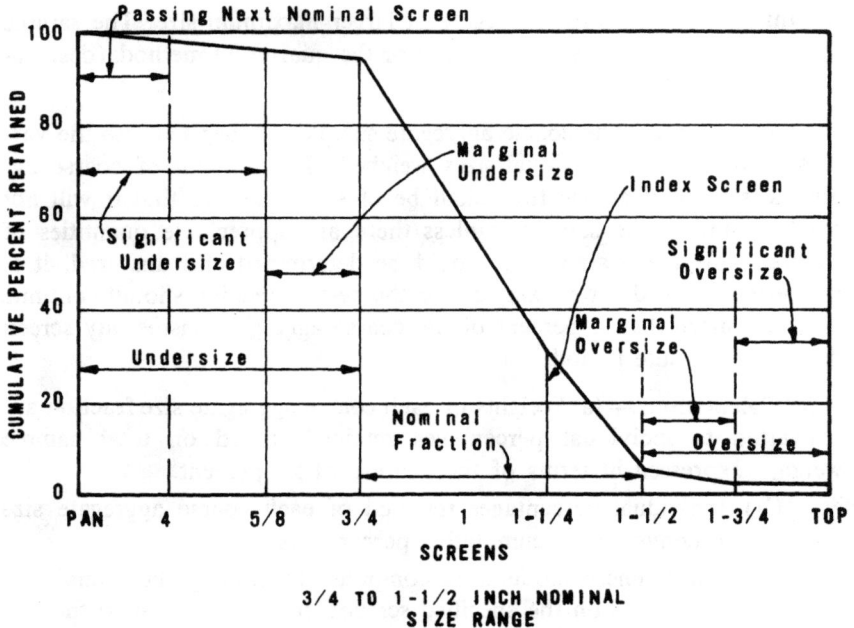

Figure 214.—Sizing nomenclature for concrete aggregate. 288–D–3263.

significant and marginal undersize and oversize material as well as the index screen size for this nominal size range. The computation is similar to that shown in figure 69 with complete analysis of the sand included as its proportional part of the total aggregate.

PETROGRAPHIC EXAMINATION OF AGGREGATES

Designation 7

1. General.—This test involves visual inspection and a segregation of the constituents of coarse and fine aggregates according to petrographic, chemical, and physical differences.

2. Apparatus.—(a) Screens conforming to those given in table 31, designation 4.

(b) An anvil and hammer suitable for breaking pebbles.

(c) Hand lens, stereoscopic microscope, petrographic microscope, and auxiliary equipment necessary for adequate petrographic examination and identification of rocks and minerals.

3. Procedure.—(a) Sufficient aggregate should be roughly screened to provide at least the following quantities of the various sizes:

Size of aggregate	Weight, grams
1½ inch to ¾ inch	9,000
¾ inch to ⅜ inch	1,000
⅜ inch to $\frac{3}{16}$ inch	200
$\frac{3}{16}$ inch to No. 8	100
No. 8 to No. 16	100
No. 16 to No. 30	100
No. 30 to No. 50	100
No. 50 to No. 100	50
Minus No. 100 to pan	50

(b) Examination of the aggregate should be performed by a qualified petrographer who is familiar with the problems of concrete. Each of the above size fractions should be examined individually.

Each fraction is examined to establish whether the particles are coated with mineral substances (such as opal, calcium carbonate, or gypsum), silt, or clay. Organic coatings may also occur. Where such coatings are present, they will be identified and evaluated by petrographic or chemical techniques. The potential physical and chemical effects of the coating on quality and durability of concrete are recorded. Where coatings are present, the aggregate may be washed, after preliminary examination, to remove the coating. The sample is dried in an oven at 220° to 230° F; observed are the degree of agitation and extent to which the coating substances or soft particles are disintegrated and removed. The sample is further examined to determine particle shape, flatness and angularity, and other pertinent properties.

The coarse aggregate fractions are examined particle by particle (each particle being broken on the anvil with a hammer) and identified by a hand lens or microscope. The individual particles are segregated according to three criteria: (1) petrographic identity, (2) physical condition, and (3) anticipated chemical stability in concrete. The percentage in each category is determined by particle count on the basis of 300 particles for each size fraction. Each category is classified by name (for example, granite, sandstone, limestone, and weathered basalt) and a designation of quality determined according to the following scale:

(1) Physical quality:

Satisfactory aggregate particles which will contribute to high or moderate strength, abrasive resistance, and durability of concrete under any climatic condition.

Fair aggregate particles which will contribute to moderate strength, durability, and abrasive resistance under ideal conditions but might contribute to physical breakdown of concrete under rigorous conditions.

Poor aggregate particles which will contribute to low strength and poor durability of concrete under any climatic condition and cause physical breakdown of concrete under rigorous climatic conditions.

(2) Chemical quality:

Deleterious aggregate particles (alkali reactive rock types) which will produce adverse effects on concrete through chemical reactions between particles and cement alkalies. If significant proportions of deleterious particles are present in an aggregate, the use of low-alkali cement or an effective combination of portland cement and pozzolan is recommended.

After analysis of the gravel is completed, the particles in each category are counted and their proportions calculated as particle percentages of the whole sample.

The fine sand is washed and the wash water tested chemically for dissolved salts of chlorides and sulfates. Each fraction of fine aggregate is examined under a binocular microscope. Ordinarily, complete segregation of fine aggregate fractions is not required, but physically unsound or chemically deleterious particles should be segregated and petrographically identified, and appropriate designation of quality should be indicated. The fractions passing the No. 30 screen are generally examined in immersion oils under the petrographic microscope to determine the presence and approximate quantity of any unsound or deleterious substances.

4. Reporting the Petrographic Examination.—Following completion of the examination and analysis, the petrographer prepares a report summarizing observations and conclusions regarding the suitability of the aggregate under the anticipated conditions of service and indicating any necessary qualifications as to its use or the need for special tests to elucidate the significance of particular properties.

The petrographic analysis of the aggregate is reported in tables (see illustrative tables 32, 33, and 34) which indicate for each constituent its petrographic identity, its proportion, its fundamental characteristics, and the petrographer's estimate of its physical and chemical quality.

FALSE SET IN CEMENT

Designation 8

1. General.—This test is for determining whether portland cement has false set. It is performed in accordance with Method 2501.1, "Early Stiffening of Hydraulic Cement Paste," Federal Test Method Standard No. 158a.

2. Apparatus.—(a) Balance having a capacity of 2 kilograms and sensitive to 0.1 gram.

(b) Vicat apparatus (see fig. 215).

Table 32.—Illustrative general petrographic analysis of coarse aggregate

Rock types	Description of rock types	Physical quality	Percentage by particle count	
			1½–¾ inch	¾–⅜ inch
Sandstone	Hard, compact, dense, fine to medium grained, white to red colored. Includes quartzite and shale.	Satisfactory	7	10
	Somewhat softened, fractured, slightly weathered or coated.	Fair	10	8
	Soft, porous, absorptive, friable.	Poor	2	3
Limestone	Hard, compact, dense, massive to crystalline, gray to buff colored. Contains some dolomite.	Satisfactory	7	7
	Somewhat softened, fractured, slightly weathered or coated.	Fair	3	12
Granite	Hard, compact, dense, medium to coarse grained. Includes monzonite and gneiss.	Satisfactory	7	4
	Somewhat softened, fractured, slightly weathered or coated.	Fair	3	6
	Soft, porous, absorptive, friable.	Poor	1	1
Schist	Hard, compact, dense, fine grained. Includes amphibolite and amphibole schist.	Satisfactory	2	3
	Somewhat softened, fractured, slightly weathered or coated.	Fair	1	—
Basalt	Hard, compact, dense, massive to vesicular. Vugs contain some opal and nontronite.	Satisfactory	7	4
	Somewhat softened, fractured, slightly weathered or coated.	Fair	3	1
Ferruginous particles	Soft, porous, absorptive, friable, sandstone and concretions.	Poor	1	2
Altered volcanics	Hard, compact, dense, massive, altered groundmass. Includes andesite, dacite, and rhyolite.	Satisfactory	22	18
	Somewhat softened, fractured, slightly weathered or coated.	Fair	12	8

Table 32.—Illustrative general petrographic analysis of coarse aggregate—Continued

Rock types	Description of rock types	Physical quality	Percentage by particle count	
			1½–¾ inch	¾–⅜ inch
Glassy volcanics [1]	Hard, compact, dense, glassy groundmass. Includes andesite and rhyolite.	Satisfactory	4	5
	Somewhat softened, fractured, slightly weathered or coated.	Fair	3	2
Chert [1]	Hard, compact, dense, microcrystalline to chalcedonic, gray to black colored.	Satisfactory	4	4
	Somewhat softened, fractured, slightly weathered or coated.	Fair	1	2

[1] Alkali-reactive rock types.

Table 33.—Illustrative summary of petrographic analysis for quality of coarse aggregate

Quality		Percentage by particle count	
		1½–¾ inch	¾–⅜ inch
Physical quality	Satisfactory	60	55
	Fair	36	39
	Poor	4	6
Chemical quality	Alkali-reactive	12	13

Particle shape: Essentially subround with about 30 percent subangular material. About 15 percent of the particles are flattened in shape.

Encrustation: About 10 percent of the particles are moderately to heavily covered by a calcium carbonate and sand grain coating which is moderately well-bonded to the particle surface.

Table 34.—Illustrative general petrographic analysis of sand

Rock and mineral types	Percentage by particle count		
	No. 8	No. 16	No. 30
Sandstone	16	10	—
Limestone	8	5	2
Granite	9	4	—
Basalt	3	2	—
Altered volcanics	21	12	—
Quartz	20	39	72
Feldspar	8	13	16
Amphibole	1	3	5
Mica	—	3	2
Opal	—	trace	1
Garnet	—	—	1
Ferruginous particles	4	2	trace
Chert	7	5	1
Glassy volcanics	3	2	—
Percent unsound	10	5	3
Percent alkali reactive	10	7	1
Percent coated	8	3	1
Percent flat	7	3	—

Remarks: The coarse sand (+ No. 30 sieve size) is subangular to angular in shape. The fine sand (— No. 30 sieve size) is almost subangular to angular and contains decreasing amounts of the fine-grained rock types found in the coarse sand and increasing amounts of monomineralic grains of quartz, feldspar, amphibole, mica, garnet, magnetite, gold, and a few miscellaneous detrital minerals.

The fine sand contains about 2 percent physically unsound material and a trace of alkali-reactive rock types.

The material removed by washing (about 5 percent) contains silt with quartz, calcite, mica, and a few miscellaneous detrital minerals, a trace of carbonaceous material, clay, and some soluble sulfate (gypsum).

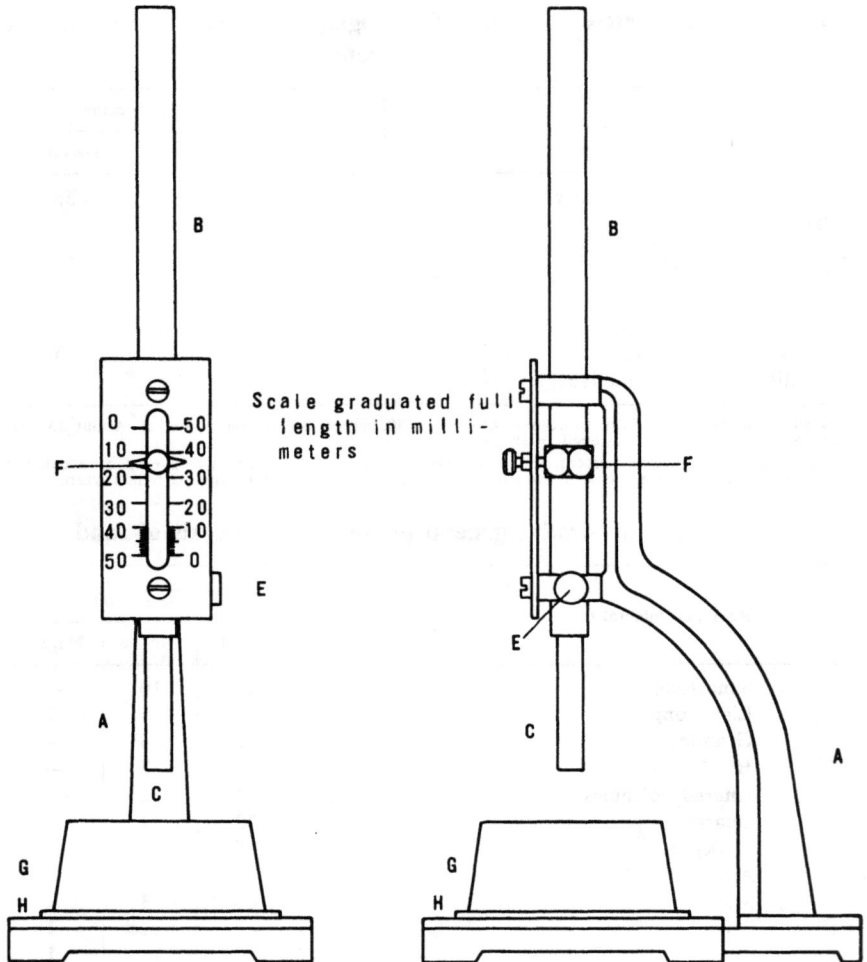

Figure 215.—Vicat apparatus used in test for false set in cement. 288–D–1551.

    (c) Glass plates 4 inches square.
    (d) Rubber gloves.
    (e) Trowel.
    (f) Stopwatch or clock with second hand.
    (g) Mechanical mixer (similar or equal to Hobart N–50).

    3. Preparation of Cement Paste.—Mix 500 grams of cement with sufficient water to produce a paste with an initial penetration of 32 ± 4 millimeters using the following procedure:
    (a) Place the dry paddle and the dry bowl in the mixing position in the mixer.

(b) Introduce the materials for a batch into the bowl and mix in the following manner:

(1) Place all the mixing water in the bowl.

(2) Add cement to the water and allow 30 seconds for the absorption of the water.

(3) Start the mixer and mix at slow speed (140 ± 5 revolutions per minute) for 30 seconds.

(4) Stop the mixer for 15 seconds and during this time scrape down into the batch any paste that may have collected on the sides of the bowl.

(5) Start the mixer at medium speed (285 ± 10 revolutions per minute) and mix for 2 minutes and 30 seconds.

4. Determination of Initial Penetration.—After completion of mixing, the paste is quickly formed into a ball with gloved hands. The ball, resting in the palm of one hand, is pressed into the larger end of the conical ring **G** (fig. 215) held in the other hand, completely filling the ring with paste. Excess paste at the larger end is then removed by a single movement of the palm of the hand. The ring is then placed with larger end down on glass plate **H**, excess paste at the smaller end sliced off at the top of the ring by a single oblique stroke of a sharp-edged trowel held at a slight angle with the top of the ring, and the top smoothed, where necessary, with a few light touches of the pointed end of the trowel. During these operations care should be taken not to compress the paste. The paste, confined in the ring and resting on glass plate **H**, is set under rod **B** about one-third of the diameter from the ring edge. Plunger end **C** is brought in contact with the surface of the paste and setscrew **E** tightened. The movable indicator **F** is set to the upper zero mark of the scale, and rod **B** released exactly 20 seconds after completion of the mixing period. The apparatus should be free from all vibration during the test. The paste is considered to have proper consistency when the rod settles 32 ± 4 millimeters below the original surface in 30 seconds after being released. Trial pastes should be made with differing percentages of water until this consistency is obtained. This consistency is the initial penetration.

5. Determination of Final Penetration.—After the initial penetration requirement is satisfied, the plunger is removed from the paste, cleaned, and the ring and plate reset to a new position under rod **B** about one-third of the diameter from the ring edge and about the same distance from the initial penetration. This operation should be performed with as little disturbance as possible to the paste confined in the Vicat ring. The plunger is then again brought into contact with the surface of the paste, the setscrew tightened, and the movable indicator set to the upper zero mark of the scale. Five minutes after completion of the mixing period the plunger is released and allowed to settle for 30 seconds before reading the millimeters of penetration.

6. Test Results.—The cement is considered to have false set if the ratio of the final penetration to initial penetration is less than 50 percent.

SPECIFIC GRAVITY AND ABSORPTION OF SAND [2]

Designation 9

PYCNOMETER METHOD

1. General.—This method describes the "pycnometer method" of determining the specific gravity and absorption of sand [3] (saturated surface-dry basis). Specific gravity and absorption may also be determined by direct measurement in accordance with method C of designation 11.

2. Apparatus.—(a) A pycnometer consisting of a 1-quart glass fruit jar with top surface ground level and fitted with a plate glass disk.

(b) A balance or scale accurate within 0.1 percent of test load at any point within the range of use.

(c) A water storage jar of about 5-gallon capacity for maintaining water at room temperature.

(d) A conical metal mold 1½ inches in diameter at the top, 3½ inches in diameter at the bottom, and 2.9 inches high.

(e) A vacuum attachment, capable of producing 20 inches of mercury vacuum, with the necessary pipe, connections, and gages for removing the entrapped air (preferred, but not required).

(f) Sand splitter, drying cloths, pans, thermometer, large spoon or small trowel, drying oven of appropriate size and capable of maintaining a temperature between 212° and 230° F, electric fans, and tamping rods.

(g) Gas or electric hot plate.

3. Sample Preparation.—(a) A 1,200- to 1,500-gram sample of sand is obtained from the large sample by means of a sand splitter or the quartering method (designation 1, sec. 9) to ensure an average or representative sample. The sample is oven-dried to remove accumulated moisture. Determinations are based on saturated surface-dry conditions. The sample is placed in a pan and covered with water for a period of 24 hours, after which time the water is drained off and the sample is placed

[2] Based on ASTM Designation C 128.
[3] Generally, in Bureau work that material passing the No. 4 screen is considered as "sand."

on a hot plate to speed up the drying action, with care being exercised to prevent loss of the fine particles and to guard against overdrying of the sample. The sample is then placed on a drying cloth and brought to a surface-dry condition. Drying of the sample may be accelerated by a current of air circulated over the sample by an electric fan. The complete sample must be stirred at all times to facilitate even drying.[4]

(b) The surface-dry condition should be determined by one of the following means. The cone method is preferred as it is considered to be the most consistent.

Cone method.—The sample of wet sand is placed loosely in the conical metal mold and the surface tamped lightly 25 times with a 12-ounce metal tamping rod having a flat circular tamping surface 1 inch in diameter. Afterwards the mold is lifted vertically. If free moisture is present, the cone of sand will retain its shape. The drying should be continued and this test made at frequent intervals until the cone of sand slumps upon removal of the mold, indicating that a surface-dry condition has been reached.

Cutting method.—In this method the dried sand is formed into a pile and is cut vertically by means of a large spoon or small trowel. If the cut portion of the pile remains vertical, further drying is necessary. When the surface-dry condition has been reached, the pile should slump off.

Visual inspection.—This method is dependent on the experiences of the operator and is based on the principle that most types of sand will change color upon drying. The surface-dry condition is reached immediately after the sample is changed from a dark (wet) to a light (dry) color. It must be noted that in wet sand fine particles will often adhere to the coarser particles or to themselves. When the sample has reached the surface-dry condition, the particles should no longer possess this power to adhere and should be lighter in color.

(c) When the sample has reached a surface-dry condition, it should be split into two parts by use of the sand splitter. One part will be used for the absorption test and the other for the specific gravity test.

4. Procedure.—(a) *Absorption.*—Weigh the absorption sample and record as saturated surface-dry weight B. This surface-dry sample is then placed in the oven and dried at a temperature of 212° to 230° F to a constant weight. The oven-dry weight A is determined, and the percent absorption is calculated.

---

[4] The above procedure is based on a 24-hour soaking period. Specific gravity and absorption of saturated surface-dry sand are also determined for 30-minute periods, with the procedure being identical in all other respects. The specific gravity and absorption of sand for 30 minutes are used in laboratory concrete mix investigations.

(b) *Specific Gravity.*—Weigh the specific gravity sample, record as saturated surface-dry weight *B*. The pycnometer is filled about three-quarters full of water of known temperature, and the saturated surface-dry sand sample *B* is added. Entrapped air is removed either by a vacuum applied to the top of the jar, as illustrated in figure 216, or by rolling the jar or otherwise agitating the sand. The jar is then filled with water and

Figure 216.—Aspirator installation for removing entrapped air from pycnometer.
PX–D–33524.

covered with a glass disk by sliding the disk across the top of the jar. The jar is shaken vigorously to remove all remaining entrapped air, after which the disk is removed and the jar carefully refilled with water. When no more air bubbles appear after repeating this operation, the jar is wiped thoroughly dry and weighed (weight W) with the glass disk in place.

5. Pycnometer Calibration.—The pycnometer jar should be calibrated for various water temperatures ranging from 40° to 90° F by filling it with water of known temperature, applying the vacuum for 15 minutes or agitating the water to remove entrapped air, and weighing with the disk in place. A curve is then plotted through the points obtained using the weights as the ordinate and temperature as the abscissa. The weight (W_c) of the pycnometer, with water at a specific temperature, can then be selected from the curve.

6. Calculations.—(a) *Specific Gravity.*—The bulk specific gravity is computed by one of the following formulas:

Dry basis:

$$\text{Specific gravity} = \frac{A}{W_c + B - W}$$

Saturated surface-dry basis:

$$\text{Specific gravity} = \frac{B}{W_c + B - W}$$

where:

A = weight of oven-dry sample,
B = weight of saturated surface-dry sample,
W_c = weight of pycnometer filled with water at same temperature as water used in test, and
W = weight of pycnometer with water and sand sample.

(b) *Absorption.*—The absorption (dry basis) is computed by the following formula:

$$\text{Absorption, } \% = \frac{B - A}{A} \times 100$$

where A and B are the same as in the specific gravity formulas.

SPECIFIC GRAVITY AND ABSORPTION OF COARSE AGGREGATE [5]

Designation 10

SUSPENSION METHOD

1. General.—This method is intended for use in determining specific gravity and absorption (after 24 hours in water at room temperature) of

---

[5] Based on ASTM Designation C 127.

coarse aggregate.[6] Specific gravity and absorption may also be determined by direct measurement in accordance with method C of designation 11.

2. Apparatus.—(a) Balance or scale accurate within 0.1 percent of the test load at any point within the range of use.

(b) A wire basket of No. 8 mesh, approximately 8 inches in diameter and 2½ inches in height. If it is in continual use, the basket should be constructed of copper or galvanized material.

(c) A suitable container for immersing the wire basket in water and suitable apparatus for suspending the wire basket from center of scale pan of balance.

(d) Oven of appropriate size capable of maintaining temperature of not less than 212° nor more than 230° F.

(e) Electric fans and miscellaneous drying cloths, pans, etc.

3. Sample.—A representative sample is obtained by a sample splitter or the quartering method (designation 1, sec. 10) and the following weights are used in the tests for the various sizes of aggregate:

Size of aggregate	Weight, grams
³⁄₁₆ to ⅜ inch	1,000 to 1,500
⅜ to ¾ inch	1,500 to 2,000
¾ to 1½ inches	2,500 to 3,500
1½ to 3 inches	4,000 to 6,000
Larger than 3 inches	A sufficient amount to make a representative sample.

4. Procedure.[7]—(a) The sample is immersed in water at room temperature and the mixture thoroughly agitated to remove dust or other coatings from the surface of the particles; the sample is then dried to a constant weight at a temperature of 212° to 230° F, cooled in air and immersed in water at room temperature for 24 hours. The sample is then removed from the water and rolled in a large cloth until all visible films of water are removed, although the surfaces of the particles still appear to be damp. The larger fragments may be individually wiped. Care should be taken to avoid evaporation from surface-dry particles. The weight of the sample in the saturated surface-dry condition is then obtained.[8] This and all sub-

---

[6] Generally, in Bureau work that material retained on a No. 4 (4.76 mm) screen is considered as "coarse aggregate."

[7] The specific gravity and absorption of saturated surface-dry coarse aggregate are also determined for 30-minute periods, the procedure being identical in all other respects. The specific gravity and absorption of coarse aggregate for 30 minutes are used in laboratory concrete mix investigations.

[8] The balance should be tared to compensate for the weight of the empty basket immersed to a depth sufficient to cover it and the test sample during weighing.

sequent weights should be determined to the nearest 1.0 gram.

(b) After it is weighed, the saturated surface-dry sample is placed immediately in the wire basket and its weight in water determined.

(c) The sample is then dried to constant weight at a temperature of 212° to 230° F, cooled to room temperature, and weighed.

5. Calculations.—(a) *Specific gravity.*—The bulk specific gravity is calculated by one of the following formulas:

Dry basis:

$$\text{Specific gravity} = \frac{A}{B-C}$$

Saturated surface-dry basis:

$$\text{Specific gravity} = \frac{B}{B-C}$$

where:

A = weight of oven-dry sample in air,
B = weight of saturated surface-dry sample in air, and
C = weight of sample in water.

(b) *Absorption.*—The percentage of absorption (dry basis) is calculated by the following formula:

$$\text{Absorption, } \% = \frac{B-A}{A} \times 100$$

where A and B are the same as in the specific gravity formulas.

SURFACE MOISTURE OF AGGREGATE
(ALSO SPECIFIC GRAVITY AND ABSORPTION)

Designation 11

A. HOT PLATE METHOD

1. General.—This is an approximate method for determining the surface moisture of sand and coarse aggregate.

2. Apparatus.—(a) A balance or scale accurate within 0.1 percent of the test load at any point within the range of use.

(b) Drying oven of appropriate size capable of maintaining a temperature between 212° and 230° F.

(c) A small shallow pan.

(d) A stirring rod or spoon.

(e) A hot plate or stove.

3. Procedure.—(a) A representative sample (designation 1, sec. 9 or sec. 10) of the aggregate (about 2,000 grams for sand and 4,000 grams or more for gravel) is weighed and spread in a thin layer in the pan.

(b) The sample is heated slowly and stirred continuously. As the material approaches a saturated surface-dry condition, extreme care is necessary to avoid driving off more than the surface moisture. The pan containing the sample should be removed from the heat while a small amount of free moisture is still present. The sample is stirred as it is allowed to cool while the saturated surface-dry condition (designation 9, sec. 3b or designation 10, sec. 4a) is approached and obtained, after which it is weighed.

4. Calculations.—The amount of surface moisture is computed by one of the following formulas:

Saturated surface-dry basis:

$$\text{Surface moisture, \%} = \frac{S-B}{B} \times 100$$

Dry basis:

$$\text{Total moisture, \%} = \frac{S-A}{A} \times 100$$

where:

S = weight of wet aggregate tested,
B = saturated surface-dry weight of aggregate tested, and
A = weight of oven-dry sample.

B. FLASK METHOD

1. General.—This method is used for rapid determination of surface moisture in fine aggregate.

2. Apparatus.—(a) A standardized volumetric flask of 500-milliliter capacity, calibrated to 0.15 milliliter.

(b) A balance or scale accurate within 0.1 percent of the test load at any point within the range of use.

(c) A pipette, ¼-inch-diameter glass tube of sufficient length to be used to adjust the water level in the flask.

3. Procedure.—The flask is filled with water at room temperature to just above the calibrated 200-milliliter mark. The level of water is then lowered by means of the pipette until the lower part of the water surface coincides with the 200-milliliter mark.

From a representative sample (designation 1, sec. 9) of approximately 1,000 grams obtained at the mixing plant, 500 grams are introduced into the flask. The flask is rolled and shaken vigorously to remove entrapped air bubbles. When the entrapped air bubbles are removed, read the combined volume of water and fine aggregate directly from the graduated scale of the flask.

4. Calculations.—The percentage of the surface moisture in the fine aggregate can be taken directly from the nomograph (fig. 217) for the

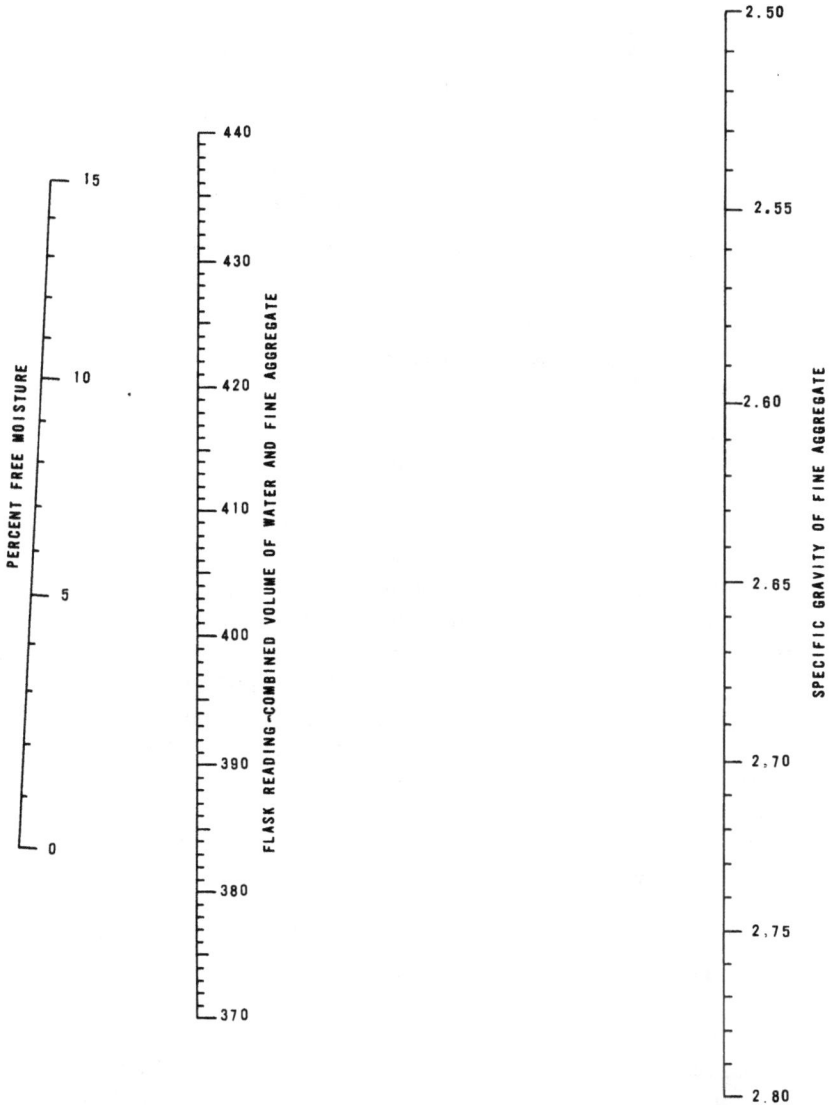

Figure 217.—Nomograph for determining free moisture in fine aggregate with a calibrated flask. 288–D–1552.

corresponding flask reading and bulk specific gravity of the material or may be computed from the following formula:

$$\text{Free moisture, } \% = \frac{V - \dfrac{500}{s.g.} - 200}{200 + 500 - V}$$

where:

V = flask reading = combined volume of the fine aggregate and water in the flask, and

$s.g.$ = bulk specific gravity of the saturated, surface-dry fine aggregate.

C. DIRECT MEASUREMENT

1. General.—This method provides rapid measurement of surface moisture, absorption, and specific gravity without computations or charts.

2. Apparatus.—(a) Moisture apparatus illustrated in figure 218.

(b) A balance or scale accurate within 0.1 percent of the test load at any point within the range of use.

3. Procedure.—A representative sample (designation 1, sec. 9 or sec. 10) of the aggregate (sample weight B) is obtained. Sample weight B is a constant, determined for each direct measurement on the moisture apparatus at the time the specific gravity scale is calibrated. The siphon can is filled with water and allowed to discharge to the siphon cutoff point. Discharge water is drained from the glass discharge reservoir, discarded, and the petcock closed. The specific gravity of the aggregate must be known to measure surface moisture or absorption and may be determined by designation 9, designation 10, or section 3(c) below. For determining surface moisture or absorption the zero percent mark of the movable moisture scale is alined with the aggregate specific gravity on the fixed specific gravity scale and the value read. Several aggregate samples may be tested without emptying the siphon can. However, care should be exercised to see that the siphon can is filled to the siphon cutoff point before placing each sample in the can.

(a) *Surface moisture.*—B grams of wet aggregate are placed in a full siphon can. When the siphon stops discharging, the percent of moisture is read on the moisture scale.

(b) *Absorption.*—B grams of dry aggregate are placed in a full siphon can. The aggregate is left submerged for 30 minutes, after which the glass reservoir is drained, the petcock closed, and all discharged water is returned to the siphon can. After the second discharge is complete, the percent of absorption is read on the moisture scale.

(c) *Specific gravity.*—B grams of saturated surface-dry aggregate (designation 9, sec. 3 or designation 10, sec. 4) are placed in a full siphon can. The specific gravity is read on the specific gravity scale after the siphon has discharged.

4. Calibration.—The moisture scale is movable and can be adjusted for different specific gravities. The zero mark of the moisture scale should be moved opposite the appropriate specific gravity. Weight B is a constant

Figure 218.—Apparatus for direct measurement of surface moisture, absorption, and specific gravity of aggregate. PX–D–33523.

determined for each apparatus by the manufacturer at the time the specific gravity scale is calibrated.

If moisture, absorption, and specific gravity are to be determined on the dry basis, the scales on the equipment must be calibrated on that basis.

UNIT WEIGHT OF AGGREGATE FOR CONCRETE [9]

Designation 12

1. General.—This test is intended to cover the determination of unit weight of sand, coarse aggregate, and mixed aggregate for concrete.

2. Apparatus.—The apparatus required consists of a cylindrical metal measure, a tamping rod, and scale.

(a) *Measure.*—The measure should be cylindrical in form, with top and bottom in planes normal to the cylinder axis. It should be metal, preferably machined to accurate dimensions on the inside, watertight, of sufficient rigidity to retain its form under rough usage, and preferably provided with handles.

The measure should be $\frac{1}{10}$-, $\frac{1}{2}$-, 1-, or 3-cubic-foot capacity, depending on the maximum size of the aggregate, and of the following dimensions:

Capacity, cubic feet	Inside diameter, inches	Inside height, inches	Minimum thickness of metal U.S. gage	Maximum size of aggregate, inches
0.10	6.00	6.10	No. 11	$\frac{1}{2}$
0.50	10.00	11.00	No. 8	$1\frac{1}{2}$
1.00	14.00	11.23	No. 5	3
[1] 3.02	17.00	23.00	$\frac{1}{16}$ inch	6

[1] Garbage can, reinforced outside, top, and bottom with $\frac{1}{4}$- by $1\frac{1}{4}$-inch bands.

(b) *Tamping Rod.*—The tamping rod is a straight rod $\frac{5}{8}$ of an inch in diameter and 24 inches in length, hemispherical-tipped at the lower end.

(c) *Balance.*—A balance or scale accurate within 0.3 percent of the test load at any point within the range of use with the measure empty or full.

3. Procedure.—(a) The volume of the measure is determined by dividing the weight of water at 70° F required to fill it by 62.3, the unit weight of water at that temperature.

---

[9] Based on ASTM Designation C 29.

(b) The sample of aggregate should be room dry and thoroughly mixed.

(c) Aggregate is placed in the measure until it is one-third full; the surface of the aggregate is leveled with the fingers. The mass is rodded with the tamping rod, using 25 strokes evenly distributed over the surface. Aggregate is then added until the measure is two-thirds full and leveled and rodded as before. The measure is then filled to overflowing, rodded, and the surplus aggregate struck off, using the tamping rod as a straight-edge. In rodding the first layer the rod should not be permitted to strike forcibly against the bottom of the measure. In rodding the other layers only enough force should be used to cause the tamping rod to barely penetrate the surface of the preceding layer.

For aggregate larger than 2 inches, jigging rather than rodding is preferable. The container is jigged by raising alternate sides of the container about 2 inches from a firm foundation such as a concrete floor and allowing it to drop 25 times on each side, 50 times in all, for each one-third filling. Aggregate smaller than 2 inches may also be compacted by jigging or vibration, but the unit weights thus obtained are appreciably greater than those obtained by dry rodding, the increase being larger for finer materials. This greater unit weight should be taken into consideration when results obtained in this manner are correlated with data based on unit weight obtained on dry-rodded samples.

(d) The net weight of the aggregate in the measure is determined.

(e) Three determinations should be made.

4. Calculations.—(a) The unit weight of the aggregate is obtained by dividing the net weight of the aggregate by the volume of the measure found as described in section 3(a).

(b) If the spread of the three tests exceeds 1 percent of the average net weight, additional tests should be made. The average of the net weights for all tests is used in determining the unit weight.

FRIABLE PARTICLES IN AGGREGATE [10]

Designation 13

1. General.—This method of test covers the approximate determination of friable particles in natural aggregates.

2. Apparatus.—(a) A balance or scale accurate within 0.1 percent of the test load at any point within the range of use.

(b) Containers of such size and shape as will permit spreading of the sample in a thin layer and of sufficient depth to permit submergence of the largest particle.

---

[10] Based on ASTM Designation C 142.

(c) Screens conforming to the requirements of table 31.

(d) Drying oven of appropriate size capable of maintaining a temperature between 212° and 230° F.

3. Samples.—(a) Aggregates for this test are obtained by means of the quartering method or a sample splitter (designation 1, sec. 9 or sec. 10) from representative material to be tested. They should be handled so as to avoid breaking friable particles which may be present.

(b) Sand samples should consist of the particles coarser than a No. 16 screen and should weigh not less than 100 grams.

(c) Coarse-aggregate samples are separated into sizes, using the following screens: No. 4, ⅜ inch, ¾ inch, and 1½ inches. The weights of samples for the different sizes should be not less than those indicated in the following tabulation:

Size of particles	Minimum weight, grams
³⁄₁₆ to ⅜ inch	1,000
⅜ to ¾ inch	2,000
¾ to 1½ inches	3,000
Over 1½ inches	5,000

(d) The samples for this test shall consist of the material remaining after washing the individual size fractions of aggregate over the smaller specified screen size in paragraphs (b) and (c).

(e) Mixtures of sand and coarse aggregate should be separated on the No. 4 screen and the separated samples prepared in accordance with paragraphs (b) and (c).

4. Procedure.—Dry the sample to a constant weight at a temperature between 212° and 230° F and obtain the weight. Spread the sample in a thin layer on the bottom of the container and cover with water. After soaking for 24 hours, the water is decanted and the sample examined for friable particles. Particles are considered as friable when they have softened so that they may be reduced sufficiently with the fingers to pass the specified washing screen. After all discernible friable particles have been broken, the sample is washed over the screens indicated in the following tabulation:

Size of particles	Size of washing screen
Sand (retained on No. 16 screen)	No. 30
³⁄₁₆ to ⅜ inch	No. 8
⅜ to ¾ inch	No. 4
¾ to 1½ inches	No. 4
Over 1½ inches	No. 4

The material retained on the screen is dried to a constant weight at a temperature between 212° and 230° F and weighed.

5. Calculations.—The percentage of friable particles is calculated to the nearest 0.1 percent in accordance with the following formula:

$$\text{Friable particles, } \% = \frac{W - W_w}{W} \times 100$$

where:

W = weight prior to 24-hour soaking, and

W_w = weight after washing.

ORGANIC IMPURITIES IN SAND [11]
Designation 14

1. General.—This test is an approximate method for determining the presence of injurious organic compounds in natural sands which are to be used in cement mortar or concrete. A dark color, obtained in the test, is not necessarily conclusive evidence that the sand is unfit for use, as certain relatively harmless organic and even inorganic substances produce such a color. Some coals and manganese minerals are examples. The principal value of the test is to indicate whether further tests should be made to determine the durability and strength of concrete in which the sand is to be used.

2. Apparatus and Supplies.—(a) Twelve-ounce graduated bottles of colorless glass equipped with watertight caps or stoppers.

(b) A 3-percent solution of sodium hydroxide and water. The 3-percent sodium hydroxide solution may be made by dissolving 1 ounce of sodium hydroxide in enough water to make 32 fluid ounces.

(c) Standard color solution for reference with sample (sec. 3).

3. Standard Color Solution.—(a) A temporary standard color solution is prepared by adding 2.5 milliliters of a 2-percent solution of tannic acid to 97.5 milliliters of 3-percent sodium hydroxide solution. The 2-percent tannic acid solution is made by adding 10 milliliters of 190-proof alcohol and 2 grams of tannic acid powder to 90 milliliters of water. The color solution is placed in a 12-ounce bottle, stoppered, shaken vigorously, and allowed to stand for 24 hours. The color of this solution is not permanent, and a new standard should be prepared for each set of tests.

(b) A permanent standard color may be prepared by dissolving 9 grams of chemically pure (c.p.) ferric chloride ($FeCl_3 \cdot 6H_2O$) and 1 gram of c.p. cobalt chloride ($CoCl_2 \cdot 6H_2O$) in 100 milliliters of water to which one-third milliliter of hydrochloric acid (HCl) has been added. This mixture when hermetically sealed in a glass bottle will remain stable indefinitely.

---

[11] Based on ASTM Designation C 40.

4. Procedure.—(a) A representative test sample of sand weighing about 1 pound is obtained by the quartering method or by use of a sample splitter (designation 1, sec. 9).

(b) A 12-ounce graduated clear glass bottle is filled to the 4½-ounce mark with the sand to be tested.

(c) A 3-percent solution of sodium hydroxide in water is added until the volume of the sand and liquid after shaking is 7 liquid ounces.

(d) The bottle is stoppered, shaken vigorously, and allowed to stand for 24 hours.

(e) If a temporary solution (sec. 3) is to be used as the standard, it should be made up immediately after the sample is shaken vigorously.

(f) After the solution has stood 24 hours, the color of the clear liquid above the sand is compared with that of the standard color solution or with a glass of similar color.

(g) Solutions darker than the standard color indicate the presence of more than 500 parts per million of tannic acid. If a solution darker than the standard color is obtained, the test should be repeated on a representative sample of sand which has been washed in water (in accordance with designation 16) to determine whether this washing removes contaminating organic compounds. If the solution color remains darker than the standard color, further tests are required to determine the nature of the material and its effect on the time of set and strength of cement.

(h) If a color equal to or lighter than standard is obtained, the sand is considered suitable (from an organic standpoint) without further testing.

(i) A sand failing in this test may be used if, when tested in accordance with designation 20 ("Effect of Organic Impurities in Fine Aggregate on Strength of Mortar"), the relative mortar strength at 7 and 28 days is not less than 95 percent.

SEDIMENTATION TEST FOR APPROXIMATE QUANTITY OF CLAY AND SILT IN SAND

Designation 15

1. General.—This test is a rapid method for determining the approximate quantity of clay or silt in sand.

2. Apparatus.—32-ounce graduated bottles of colorless glass equipped with watertight caps or stoppers.

3. Procedure.—(a) A representative test sample of sand weighing about 3 pounds is obtained by the quartering method or by use of a sample splitter (designation 1, sec. 9).

(b) A 32-ounce graduated clear glass bottle is filled to the 14-ounce mark with sand and clear water added to the 28-ounce mark. The bottle is stoppered, shaken vigorously, and the contents allowed to settle for 1 hour. One ounce of sediment above the sand is roughly equivalent to 3 percent by weight of clay or silt. The sand may be washed, samples again tested, and the sand approved for use if requirements are fulfilled.

(c) When the test is made for organic impurities in sand (designation 14), it is unnecessary to make an additional test for the approximate quantity of clay and silt. In the test for organic impurities, ½ ounce of sediment at the top of the 7 ounces of sand corresponds roughly with 3 percent by weight of clay or silt. Sedimentation results are obtained in the 3-percent solution of sodium hydroxide much more rapidly than in water.

PERCENTAGE OF AGGREGATE PASSING NO. 200 SCREEN [12]

Designation 16

1. General.—This test outlines the procedure for determining the total quantity of material, contained in the aggregate, passing a No. 200 (74-micrometer) screen.

2. Apparatus.—(a) Two nested screens, the upper screen a No. 16 and the lower a No. 200, both conforming to the requirements of ASTM Designation E 11, "Standard Specification for Wire-Cloth Sieves for Testing Purposes."

(b) A pan or vessel of a size sufficient to contain the sample covered with water and to permit vigorous agitation without loss of any part of the sample or water.

(c) A drying oven of appropriate size capable of maintaining temperature between 212° and 230° F.

(d) A balance accurate within 0.1 percent of the test load at any point within the range of use.

3. Test Sample.—A representative test sample is obtained by a sample splitter or the quartering method (designation 1, sec. 9 or sec. 10) from material which has been mixed thoroughly and which contains sufficient moisture to prevent segregation. The size of the sample should be sufficient to yield not less than the appropriate weight of dried material, as shown in the following tabulation:

Size of aggregate	Weight, grams
0 to 3/16 inch (sand)	500 to 1,000
3/16 to 3/8 inch	1,000 to 1,500
3/8 to 3/4 inch	1,500 to 2,000
3/4 to 1½ inches	2,500 to 3,500
Larger than 1½ inches	A sufficient amount to make a representative sample.

[12] Based on ASTM Designation C 117.

4. Procedure.—The test sample is dried to constant weight at a temperature not exceeding 230° F and weighed to the nearest 0.1 gram for each 500 grams.

The test sample, after being dried and weighed, is placed in the container and sufficient water added to cover it. The contents of the container are agitated vigorously.

The agitation should be sufficiently vigorous to result in complete separation from the coarse particles of all particles finer than the No. 200 screen and to bring the fine material into suspension so that it will be removed with the wash water. Immediately after agitation the wash water is decanted and poured over the nested screens, arranged with the coarser screen on top. (Do not pour gravel or sand over screens because the screens will not withstand the weight.) The operation is repeated until the wash water is clear.

The test sample may be washed on the nested screens by a water spray until the wash water is clear. To prevent overloading of the screens when this procedure is used, small portions of the test sample should be washed individually until the entire sample has been washed. Care should be exercised in washing the sample or transferring washed material to drying pans to avoid loss of any part of the sample.

All material retained on the nested screens is returned to the washed sample, which is then dried to constant weight at a temperature not exceeding 230° F and weighed to the nearest 0.1 gram for each 500 grams.

5. Calculations.—The results are calculated from the following formula:

$$\text{Material finer than No. 200 screen, } \% = \frac{A-E}{A} \times 100$$

where:

A = original dry weight, and
E = dry weight after washing.

6. Check Determinations.—When check determinations are desired, the wash water is either evaporated to dryness, or filtered through tared filter paper which is subsequently dried. The following formula is used:

$$\text{Material finer than No. 200 screen, } \% = \frac{W_r}{A} \times 100$$

where:

A = original dry weight, and
W_r = weight of residue.

PERCENTAGE OF PARTICLES LESS THAN 1.95 SPECIFIC GRAVITY IN SAND

Designation 17

1. General.—This test outlines a procedure for the approximate determination of lightweight material in sand. Material having a specific gravity less than 1.95, including most shale and coal, will be removed in this test.

2. Apparatus.—(a) A balance or scale accurate within 0.1 percent of the test load at any point within the range of use.

(b) A No. 30 screen.

(c) Two or more small strainers having openings smaller than those of a No. 30 screen.

(d) Three 1,000-milliliter glass beakers.

(e) Hydrometers accurate to 0.01 and suitable for determining specific gravity of zinc chloride solution and hydrometer flasks of sufficient size for hydrometer measurements.

3. Sample.—A representative sample of sand of sufficient size to yield 100 to 200 grams of material coarser than the No. 30 screen is selected by use of a sample splitter or the quartering method (designation 1, sec. 9). The quantity will vary depending on the coarseness or fineness of the sand grading. Material selected is oven dried to constant weight at a temperature between 212° and 230° F.

The sand is then screened over a No. 30 screen and that portion retained on the screen is weighed to the nearest 0.1 gram.

4. Procedure.—A solution of zinc chloride ($ZnCl_2$) is prepared having a specific gravity of 1.95 ± 0.02 at 70° to 80° F.

(**NOTE.**—*Zinc chloride of this concentration should be handled with caution as it may cause irritation or burns on the skin. Immediate application of copious amounts of water is a satisfactory antidote.*)

Approximately 600 milliliters of solution is placed in a 1,000-milliliter glass beaker and the sample poured into the solution. As the sample is added, the solution is stirred vigorously. When the sample is all in suspension, stirring is stopped and the sample allowed to settle (approximately 30 seconds) until there is a definite cleavage plane between the rising lightweight material and the settling sand. The solution is then decanted over a strainer until the sand appears near the lip of the beaker. Care should be taken that only floating particles are poured off with the liquid. Samples containing a substantial amount of lightweight material may require additional separation, which may be accomplished by adding more solution to the sample, agitating and decanting. The samples should not be in contact with the zinc chloride solution for more than 2½ minutes during the test.

Material retained on the strainer is thoroughly washed to remove the zinc chloride and then oven dried to constant weight at a temperature between 212° and 230° F and weighed to the nearest 0.1 gram.

5. Calculations.—The percentage of lightweight material is calculated from the following formula:

$$\text{Lightweight material, } \% = \frac{W_d}{W} \times 100$$

where:

W = weight of portion originally retained on No. 30 screen, and

W_d = weight of decanted particles.

It should be noted that only lightweight material in the fraction retained on the No. 30 screen is separated from the original sample.

PERCENTAGE OF PARTICLES LESS THAN 1.95 SPECIFIC GRAVITY IN COARSE AGGREGATE [13]

Designation 18

1. General.—This test outlines a procedure for the approximate determination of lightweight material in coarse aggregate. Materials that have a specific gravity less than 1.95, including most shale and coal, will be removed in this test.

2. Apparatus.—(a) A balance or scale accurate within 0.1 percent of the test load at any point within the range of use.

(b) A metal tank and wire-screen basket of such capacity that the sample may be submerged with not less than 2 inches of solution over the sample.

(c) Two or more small strainers with openings not larger than No. 8 mesh (0.0937 inch).

(d) Hydrometers accurate to 0.01 and suitable for determining specific gravity of zinc chloride solution and hydrometer flasks of sufficient size for hydrometer measurements.

3. Sample.—A representative sample is obtained by a sample splitter or the quartering method (designation 1, sec. 10) and the following weights are used in the tests for the various sizes of aggregate:

---

[13] Based on AASHTO Designation T 150.

Size of aggregate	*Weight, grams*
$\frac{3}{16}$ to $\frac{3}{8}$ inch	1,000 to 1,500
$\frac{3}{8}$ to $\frac{3}{4}$ inch	1,500 to 2,000
$\frac{3}{4}$ to $1\frac{1}{2}$ inches	2,500 to 3,500
$1\frac{1}{2}$ to 3 inches	4,000 to 6,000
Larger than 3 inches	A sufficient amount to make a representative sample.

The sample is oven dried to constant weight at a temperature between 212° and 230° F and weighed to the nearest 0.1 gram.

4. Procedure.—A solution of zinc chloride ($ZnCl_2$) is prepared having a specific gravity of 1.95 ± 0.02 at 70° to 80° F.

(**NOTE.**—*Zinc chloride of this concentration should be handled with caution as it may cause irritation or burns on the skin. Immediate application of copious amounts of water is a satisfactory antidote.*)

The sample of coarse aggregate is placed in the wire basket and lowered into the tank. The sample and solution should be stirred vigorously with a large mixing spoon for 1 minute. Floating particles should be skimmed off within 1 minute after stirring has stopped.

The removed particles are thoroughly washed to remove the zinc chloride and then oven dried to constant weight at a temperature between 212° and 230° F and weighed to the nearest 0.1 gram.

5. Calculations.—The approximate percentage of lightweight material is calculated from the following:

$$\text{Lightweight material, } \% = \frac{W_d}{W} \times 100$$

where:

W = weight of original dry sample, and
W_d = weight of decanted particles.

SOUNDNESS OF AGGREGATE (SODIUM SULFATE METHOD)[14]

Designation 19

1. General.—The results of this test are used as an indication of the ability of aggregate to resist weathering.

2. Supplies and Apparatus.—(a) A supply of saturated solution of sodium sulfate prepared in accordance with section 3.

---

[14] Based in general on ASTM Designation C 88.

(b) Screens conforming to the requirements of table 31.

(c) Suitable containers for immersing the samples of aggregate in the solution. No. 3 porcelain evaporating dishes are suitable for sands and larger porcelain bowls for the gravel. The containers should be equipped with covers that will prevent dilution or contamination of the solution.

(d) A constant-temperature room.

(e) A balance or scale accurate within 0.1 percent of the test load at any point within the range of use.

(f) A drying oven constructed to provide free circulation of air through the oven and capable of maintaining a temperature between 221° and 230° F. The rate of evaporation at this range of temperature must average at least 25 grams per hour for 4 hours. The rate of evaporation is determined by the loss of water from 1-liter, Griffin low-form beakers, each initially containing 500 grams of water at a temperature of 70° ±3° F placed at each corner and the center of each shelf of the oven. This evaporation requirement applies when the oven is empty, except for the beakers of water, and the oven doors are closed. Heat in a gas oven may be increased when the samples are first put in the oven to maintain a nearly constant temperature.

(g) Hydrometers accurate to ±0.001 and suitable for determining specific gravity of sodium sulfate solution and hydrometer flasks of sufficient size for hydrometer measurements.

3. Sodium Sulfate Solution.—Only c.p. anhydrous salt should be used. The salt should be dissolved slowly, accompanied by vigorous stirring, in clean water at a temperature not exceeding 86° F and not less than 77° F. It is recommended that not less than 350 grams of the anhydrous salt be used for each liter of water. The solution should be kept at a temperature of 73.4° ±2° F at all times in a covered vitreous earthenware crock or similar container not affected by the solution. Before it is used, a newly mixed solution should be kept at constant temperature for 48 hours to allow saturation. During this period it should be stirred frequently. If, at any time, the temperature of the saturated solution should vary more than several degrees and then return to normal, the solution should be stirred at least 30 minutes before using. It is essential that a considerable excess quantity of crystals be present on the bottom of the container at all times to assure complete saturation of the solution. The solution should have a specific gravity of not less than 1.151 and not greater than 1.174 and should be vigorously stirred immediately before using to eliminate layering. It is recommended that one batch of sodium sulfate solution be used for not more than 10 cycles of test.

4. Procedure.—(a) Sufficient aggregate is roughly screened to provide the following approximate quantities of the various sizes. (For sands, 150 grams will usually be sufficient to allow for washing and screening.)

Size of aggregate	Weight, grams
1½ to ¾ inch	2,000
¾ to ⅜ inch	500
⅜ to $\frac{3}{16}$ inch	100
$\frac{3}{16}$ inch to No. 8	100
No. 8 to No. 16	100
No. 16 to No. 30	100
No. 30 to No. 50	100

For coarse aggregate sizes larger than 1½ inches, use approximately 6,000 grams of material and determine the loss through the screen on which the sample was retained before testing.

(b) The sand fractions are recombined and thoroughly washed on a 100-mesh screen. The coarse material may be washed in any suitable container, and both sand and gravel should be dried to constant weight between 221° and 230° F. After cooling to room temperature both fine and coarse aggregate are screened for 20 minutes on a mechanical screen shaker (material larger than ¾-inch may be screened by hand). Particles of any size fraction sticking in the screen should be wasted. The fractions of the sample are weighed and placed in separate containers resistant to the action of the solution.

(c) Sufficient sodium sulfate solution is poured into the containers to cover the fractions to a depth of not less than one-half of an inch. The solution is siphoned from the lower parts of earthenware storage vessels after it has been stirred for several minutes. The fractions are then allowed to soak for 18 hours, during which period they are maintained at 73.4° ± 2° F. The containers should be covered to prevent contamination.

(d) After the 18-hour immersion period the samples are removed from the solution (or excess solution decanted), placed in an oven,[15] and dried to constant weight at a temperature between 221° and 230° F. Care should be taken to see that samples do not remain in the oven longer than is necessary for complete drying since excessive disruptive forces may result from overdrying. Thorough drying should take place under the above temperatures in approximately 4 hours, but every operator should make tests to determine the minimum length of time necessary to dry samples to a constant weight with his particular equipment. After drying, the sample fractions are cooled to approximately 73.4° F, and alternate immersions and dryings should proceed as outlined in paragraphs (c) and (d).

(e) At the end of each five cycles, the test samples should be inspected and record made of observations. Each fraction is then washed

---

[15] It has been found necessary to decant the sands again after approximately 30 minutes in the oven to prevent explosion of entrapped sulfate solution and resultant showering of particles from their containers.

thoroughly.[16] The sands should be washed, as previously, on a 100-mesh screen with tapwater, and the gravel should be washed in some suitable container.

(f) After removal of the sodium sulfate from the fractions, they are dried thoroughly and cooled as described in paragraph (d). Each fraction is screened for 15 minutes in a mechanical screen shaker, and the quantity of material retained on each screen is weighed and recorded. All material remaining in the screens should be carefully brushed into the part of the sample to be weighed. If the ¾-inch material is screened by hand, all pieces appearing unsound should be tried with a slight pressure of the fingers; pieces that crumble easily, together with the material passing the screen during the hand-screening operation, should be considered as loss. If it is desired to continue the test to more complete disintegration, the test should proceed as outlined.

5. Data Reported.—The following data should be reported. Table 35 is suggested as a suitable report form.

(a) Weight of each fraction before test.

(b) Material from each fraction, finer than the screen on which the fraction was retained before the test, expressed as a percentage by weight of the fraction.

(c) Weighted average loss for coarse aggregate, computed from the percentage of loss for each fraction, and weighted according to the following:

Size of material	Grading, percent	
	¾-inch-maximum sample	1½-inch-maximum sample
1½ to ¾ inch	....	50
¾ to ⅜ inch	60	30
⅜ to 3/16 inch	40	20

(d) The weighted average loss for sand, computed from the percentage of loss for each fraction and weighted according to the typical sand grading given in table 35.

(e) Where sufficient material is unavailable for testing in any size fraction, that size shall not be tested. For the purpose of calculating the test results, it shall be considered to have the same loss in sodium sulfate

---

[16] To assure complete removal of the sodium sulfate during the washing, a small quantity of the wash water is obtained and a few drops of barium chloride ($BaCl_2$) solution added. The presence of sodium sulfate is indicated by a formation of white precipitate of barium sulfate ($BaSO_4$).

Table 35.—Illustrative sodium sulfate test results

Weighted percentage of loss in 5 cycles of sodium sulfate tests, based on typical grading and 1½-inch-maximum size aggregate.

Screen size		Typical grading, percent	Weight of test fractions before test, grams	Percentage passing finer screen after test (actual percent loss by difference)	Weighted averages (corrected percent loss)
Passing	Retained on				
Coarse Aggregate					
1½ inch	¾ inch	50	2,000	2.6	1.3
¾ inch	⅜ inch	30	500	2.0	0.6
⅜ inch	³⁄₁₆ inch ...	20	100	6.8	1.4
Totals	....	100	2,600		3.3
Sand					
No. 4	No. 8	20	100	7.1	1.4
No. 8	No. 16	20	100	5.6	1.1
No. 16	No. 30	30	100	9.6	2.9
No. 30	No. 50	30	100	6.7	2.0
Totals	....	100	400		7.4

treatment as the average of the next smaller and the next larger size; or where one of these sizes is absent, it shall be considered to have the same loss as the next larger or next smaller size, whichever is present.

EFFECT OF ORGANIC IMPURITIES IN FINE AGGREGATE ON STRENGTH OF MORTAR [17]

Designation 20

1. General.—This method covers the procedure for measuring the mortar-making properties of fine aggregate for concrete by a compression test on specimens made from a mortar of plastic consistency and gaged to a definite water-cement ratio. It covers the determination of the effect on mortar strength of organic impurities determined in accordance with the Method of Test for Organic Impurities in Sand (designation 14).

2. Apparatus.—(a) Flow table, flow mold, and caliper, conforming to the requirements of ASTM Specifications C230, for Flow Table for Use in Tests of Hydraulic Cement.

---

[17] Based on ASTM Designation C 87.

(b) Tamper, made of a nonabsorptive, nonabrasive, nonbrittle material such as rubber compound having a Shore A durometer hardness of 80 ± 10 or seasoned oak wood rendered nonabsorptive by immersion for 15 minutes in paraffin at approximately 400° F and having a cross section of ½ by 1 inch and a convenient length of about 6 inches. The tamping face shall be flat and at right angles to the length of the temper.

(c) A round, straight steel tamping rod, three-eighths inch in diameter and approximately 12 inches in length, having one end rounded to a hemispherical tip of the same diameter as the rod.

(d) Trowel, having a steel blade 4 to 6 inches in length, with straight edges.

(e) Cube molds 2 inches in dimension conforming to the requirements of ASTM Method C 109, Test for Compressive Strength of Hydraulic Cement Mortars (Using 2-inch Cube Specimens); or cylindrical molds 2.00 ± 0.02 inch in diameter by 4.0 ± 0.1 inch in height, of nonabsorbent material, watertight, and substantial enough to hold their form during molding of test specimens.

3. Procedure.—(a) The fine aggregate shall be compared in mortar, as described in this method, with a sample of the same aggregate that has been soaked and washed in a 3-percent solution of sodium hydroxide followed by thorough rinsing in water. The treatment shall be repeated a sufficient number of times to produce a washed material having a color lighter than the standard of designation 14. Sufficient fine aggregate for the untreated and treated samples shall be washed over a 200-mesh screen to remove excess silt or organic impurities removable by washing. Dry and split the material, using one part for the untreated and the other part for the treated sample. The soaking and washing in sodium hydroxide shall be performed in such a way as to minimize loss of fines. The washed and rinsed aggregate shall be checked with a suitable indicator such as phenolphtalein or litmus to assure that all sodium hydroxide has been removed prior to preparation of the mortar. The samples as tested shall have a standard grading (2.74 F.M.) as shown below:

Screen Size		Grading, percent
Passing	Retained on	
No. 4	No. 8	15
No. 8	No. 16	15
No. 16	No. 30	25
No. 30	No. 50	24
No. 50	No. 100	16
No. 100	No. 200	5

(b) Unless otherwise specified or permitted, strength comparisons shall be made at 7 and 28 days.

(c) Mold batches of mortar with the aggregate treated in sodium hydroxide and batches with the untreated aggregate on the same day.

(d) Mold six 2-inch cubes or 2-inch by 4-inch cylinders from each batch.

(e) Test three cubes or cylinders from each batch at the ages of 7 and 28 days.

4. Temperature.—The temperature of the mixing water, moist closet, and storage tank shall be maintained at 73.4° ± 3° F.

5. Preparation of Mortar.—Place cement and water in quantities [18] that will give a water-cement ratio of 0.6 by weight, in an appropriate vessel,[19] and permit the cement to absorb water for 1 minute. Mix the materials into a smooth paste with a spoon. Beat into the mixture a known weight of the sample of sand under test that has been brought to a saturated surface-dry condition [20] as described in section 3b, Method of Test for Specific Gravity and Absorption of Sand (designation 9). Mix until the material appears to be of the desired consistency (flow 100 ± 5). Continue the mixing for 30 seconds, and make a determination of the flow in accordance with section 6.

6. Flow Test.—Carefully wipe the flow-table top clean and dry and place the flow mold at the center. Immediately after completing the mixing operation place a layer of mortar about 1 inch in thickness in the mold and tamp 20 times with the tamper. The tamping pressure shall be just sufficient to ensure uniform filling of the mold. Fill the mold with mortar and tamp as specified for the first layer. Cut off the mortar to a plane surface, flush with the top of the mold, by drawing the straight edge of the trowel (held nearly perpendicular to the mold) with a sawing motion across the top of the mold. Wipe the table top clean and dry being especially careful to remove any water from around the edge of the flow mold. Lift the mold away from the mortar 1 minute after completing the mixing operation. Immediately drop the table through a

---

[18] For six cubes, 600 g of cement and 360 ml of water will usually give sufficient mortar. The quantity of sand used with this amount of cement may vary from 1,200 g for fine sand to 2,000 g or more for coarse sand.

[19] An ordinary saucepan of 4-quart capacity, about 9½ inches in diameter at the top and about 4½ inches in height, and a large iron kitchen spoon are suitable for use in mixing the mortar.

[20] If the absorption as determined in accordance with designation 9 is known, sand may be prepared for test by adding to known weight of dry sand the amount of water it will absorb, mixing thoroughly, and permitting the sand to stand in a covered pan for 30 minutes before use.

height of one-half inch 10 times in 6 seconds. The flow is the resulting increase in average diameter of the mortar mass, measured on at least four diameters at approximately equal angles, expressed as a percentage of the original diameter. Should the flow be too great, return the mortar to the mixing vessel, add additional sand, and make another determination of the flow. If more than two trials must be made to obtain a flow of 100 ± 5, consider the mortar as a trial mortar and prepare test specimens from a new batch. If the mortar is too dry, discard the batch. Determine the quantity of sand used by subtracting the weight of the portion remaining after mixing from the weight of the initial sample.

7. Molding Test Specimens.—Immediately following completion of the flow test, place the mortar in cylinder molds in three layers or in cube molds in two layers. Rod each layer in place with 25 strokes of the tamping rod. After the rodding has been completed, fill the molds to overflowing. Place the specimens in a moist closet (sec. 4) for curing. Three to four hours after molding, strike off the specimens to a smooth surface and cover with a glass plate or plastic sheet. Twenty or twenty-four hours after molding, remove the specimens from the molds and, except for those to be tested at that time, store in water until tested.

8. Capping Specimens.—If the specimens are cylindrical, before testing, cap them in such a manner that the ends are plane within 0.002 inch and perpendicular to the axis of the cylinder within 1°. The material used for capping and the thickness of the cap shall be such that it will not flow or fracture under the load. Cubes made in suitable molds need not be capped and shall be tested at right angles to the direction of molding.[21]

9. Testing Specimens.—(a) Test the specimens immediately after their removal from the moist closet for 24-hour specimens and from storage water for all other specimens. When more than one specimen at a time is removed from the moist closet for the 24-hour tests, keep these specimens covered with a damp cloth until the time of testing. When more than one specimen at a time is removed from the storage water for testing, keep these specimens in water at a temperature of 73.4° ± 3° F and of sufficient depth to immerse completely each specimen until the time of testing.

(b) Surface dry each specimen and remove any loose sand grains or incrustations from the faces that will be in contact with the bearing blocks of the testing machine. Check these faces by applying a straight-

---

[21] It is desirable that the capping material have a modulus of elasticity equal to or greater than that of the mortar.

edge and inserting a 0.002-inch thick feeler gage.[22] If the bearing surface departs from plane by more than the 0.002 inch, grind the face or faces to plane surfaces or discard the specimen.

(c) Apply the load continuously and without shock to plane faces of the specimen. Carefully place the specimen in the testing machine below the center of the upper bearing block. Use no cushioning or bedding materials. In testing machines of the screw type the moving head shall travel at a rate of about 0.05 inch/min when the machine is idling. In hydraulically operated machines apply the load at a constant rate within the range of 20 to 50 lb/in²/sec. During the application of the first half of the maximum load, a higher rate of loading will be permitted. No adjustment shall be made in the controls of the testing machine while a specimen is yielding rapidly immediately before failure.

10. Calculation.—Calculate the compressive strength of each specimen by dividing its maximum load carried during the test by the cross-sectional area. As a measure of the effect of the organic impurities, express the relative strength of mortar made with untreated fine aggregate as a percentage of that of mortar made with the same aggregate washed in sodium hydroxide. Sand contaminated by organic impurities may be rejected if it produces a 7 and 28 day strength less than 95 percent of a neutralized sand from the same source.

ABRASION OF COARSE AGGREGATE BY USE OF THE LOS ANGELES MACHINE [23]

Designation 21

1. General.—This test is intended for determining the abrasion resistance of crushed rock, crushed slag, uncrushed gravel, and crushed gravel.[24]

---

[22] Results much lower than the true strength will be obtained by loading faces of the specimen that are not truly plane surfaces. Therefore, it is essential that specimen molds be kept scrupulously clean; otherwise, large irregularities in the surfaces will occur. Instruments for cleaning of the molds should always be softer than the metal in the molds to prevent wear. When grinding of specimen faces is necessary, it can be accomplished best by rubbing the specimen on a sheet of fine emery paper or cloth glued to a plane surface; paper or cloth glued to a plane surface and applied only with a moderate pressure should be used. Such grinding is tedious for more than a few thousandths of an inch; where more than this is found necessary, it is recommended that the specimen be discarded.

[23] Based on ASTM Designations C 131 and C 535.

[24] Ledge rock, hand broken in approximately cubic fragments of the different sizes shown, when tested by this method, has been found to have a loss of approximately 85 percent of that for crushed rock of the same quality.

2. Apparatus.—(a) The Los Angeles Machine consists of a hollow steel cylinder having an inside diameter of 28 inches and an inside length of 20 inches and closed at both ends. The cylinder is mounted on stub shafts attached to the ends of the cylinder, but not entering it, and is mounted so that it may be rotated on its axis in a horizontal position. An opening is provided in the cylinder for introduction of the test sample. The opening should be closed with a removable dusttight cover bolted in place. The cover should be designed to maintain the cylindrical contour of the interior surface. A removable steel shelf, projected radially 3½ inches into the cylinder and extending its full length, should be mounted along one element of the interior surface of the cylinder. The position of the shelf should be such that the distance from the shelf to the opening, measured along the circumference of the cylinder in the direction of rotation, is not less than 50 inches. A shelf of wear-resistant steel, rectangular in cross section and mounted independently of the cover, is preferred. However, a shelf consisting of a section of rolled angle, properly mounted on the inside of the cover plate, may be used provided the direction of rotation is such that the charge will be caught on the outside face of the angle. The shelf should be of such thickness and so mounted, by bolts or other approved means, as to be firm and rigid. The machine shall be so driven and so counterbalanced as to maintain a substantially uniform peripheral speed of from 30 to 33 revolutions per minute. Backlash or slip in the driving mechanism is very likely to furnish test results which are not duplicated by other Los Angeles abrasion machines producing constant peripheral speed.

(b) A balance or scale accurate within 0.1 percent of the test load at any point within the range of use.

(c) Standard square-mesh screens designated by the following sizes or numbers: 1½ inch, 1 inch, ¾ inch, ½ inch, ⅜ inch, No. 3 (¼ inch), No. 4, and No. 12 conforming to the requirements of ASTM Designation E 11, "Standard Specification for Wire-Cloth Sieves for Testing Purposes."

(d) Miscellaneous pans, wire brushes for cleaning screens.

3. Abrasive Charge.—(a) The abrasive charge consists of cast iron or steel spheres approximately 1⅞ inches in diameter and weighing between 390 and 445 grams each.

The weights of the cast iron or steel spheres should be checked periodically; those weighing less than 390 grams should be discarded. The total weight of the selected combination of metal spheres used in the abrasive charges for "A", "B", and "C" test sample gradings must conform to the requirements of paragraph (b).

(b) The charges used with the grading described in section 4 are as follows:

Grading	Number of spheres	Weight of charge, grams
A	12	5,000±25
B	11	4,584±25
C	8	3,330±20

4. Test Sample.—The test sample consists of 5,000 grams of clean, dry, representative aggregate conforming to one of the gradings given in the following tabulation. The grading used should be that most nearly representing the aggregate furnished for the work.

Screen size (square openings)		Weight in grams		
Passing	Retained on	Grading A	Grading B	Grading C
1½ inch	1 inch	1,250		
1 inch	¾ inch	1,250		
¾ inch	½ inch	1,250	2,500	
½ inch	⅜ inch	1,250	2,500	
⅜ inch	No. 3 (¼ inch)			2,500
No. 3 (¼ inch)	No. 4			2,500

5. Procedure.—The test sample and the abrasive charge are placed in the Los Angeles abrasion-testing machine and the machine rotated for 100 revolutions. The material is removed from the machine and screened on a No. 12 screen conforming to the requirements of ASTM Designation E 11, "Specification for Wire-Cloth Sieves for Testing Purposes." The material retained on the screen is accurately weighed to the nearest gram. Care should be taken to avoid loss of any part of the sample. The entire sample, including the dust of abrasion, is returned to the testing machine and the machine rotated 400 revolutions. The sample is then removed, screened on a No. 12 screen, and weighed to the nearest gram.

6. Calculation.—The differences between the original weight of the test sample and the weights of the material retained on the screen at 100 revolutions and 500 revolutions are expressed as percentages of the original weight of the test sample. These values are reported as percentages of wear.

SLUMP [25]

Designation 22

1. General.—The consistency of concrete usually is measured by the slump test in accordance with the following procedure.

2. Mold.—The test specimen is formed as the frustum of a cone with a base 8 inches in diameter, the upper surface 4 inches in diameter, and the height 12 inches. The base and the top of the mold are open and parallel to each other and at right angles to the axis of the cone. The mold is provided with foot pieces and handles, as shown in figure 219.

3. Sample.—The sample should be representative of the batch. For samples of concrete having aggregate exceeding 1½ inches in size, the pieces of aggregate larger than 1½ inches are removed by wet screening.

4. Procedure.—The mold should be dampened and placed on a flat, moist, nonabsorptive surface, where the operator holds it firmly in place by standing on the foot pieces while it is being filled. The mold is filled in three layers, each approximately one-third the volume of the mold. The vertical dimensions from the bottom of the mold to the tops of the layers are approximately 2½, 6, and 12 inches. In placing each scoopful of concrete, the scoop is moved around the top edge of the mold as the concrete slides from it to ensure symmetrical distribution of concrete within the mold. Each layer is rodded with 25 strokes of a ⅝-inch rod, 24 inches long and hemispherically tipped at the lower end. The strokes are distributed uniformly over the cross section of the mold and should just penetrate into the underlying layer. The bottom layer should be rodded throughout its depth. After the top layer has been rodded, the surface of the concrete is struck off so that the mold is exactly filled and the spilled concrete cleaned from the base. The mold is immediately removed from the concrete by raising it slowly and carefully in a vertical direction. The slump is measured to the nearest one-fourth of an inch immediately thereafter by determining the difference between the height of the mold and the average height of the top surface of the concrete after subsidence.

Slump specimens which break or slough off laterally give incorrect results and should be remade with a fresh sample. After the slump measurement is completed, the side of the concrete frustum should be tapped gently with the tamping rod. The behavior of the concrete under this treatment is a valuable indication of its cohesiveness, workability, and placeability (see fig. 2). A well-proportioned, workable mix will

---

[25] Based on ASTM Designation C 143.

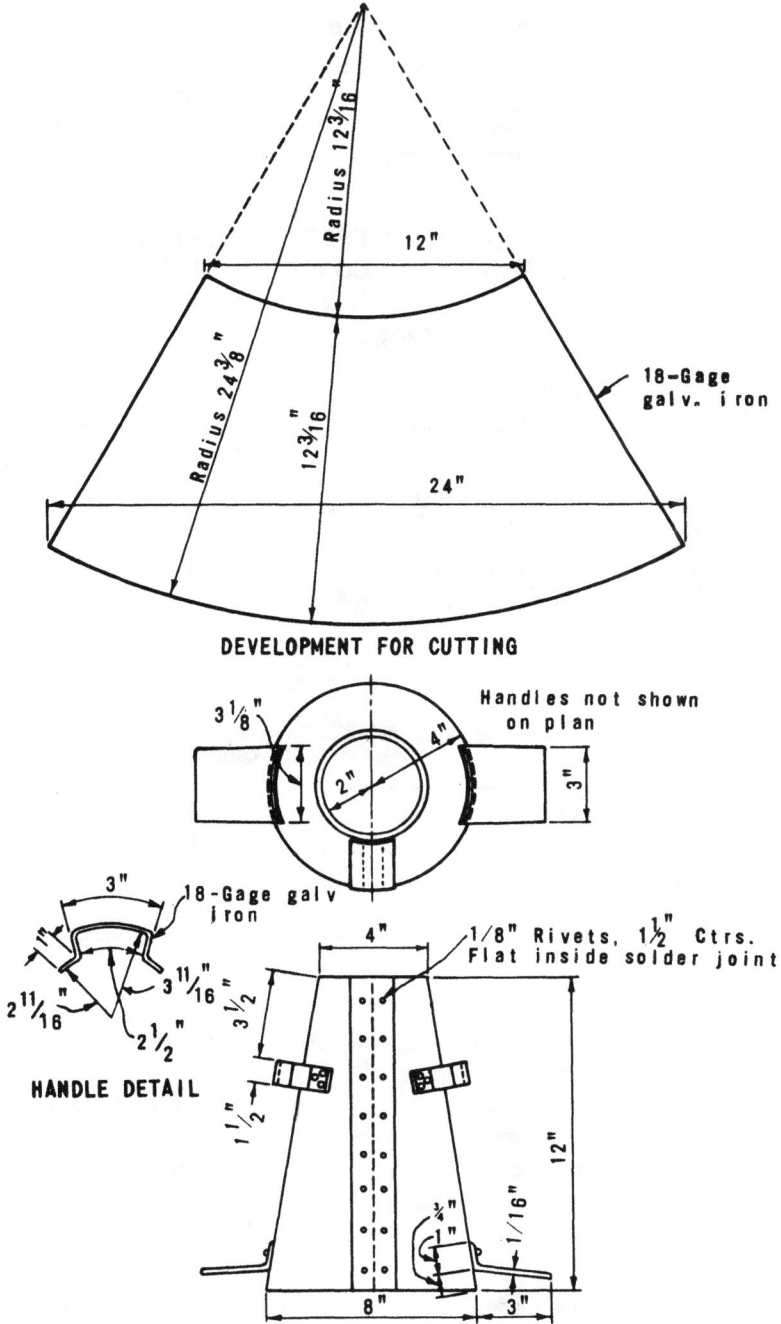

DEVELOPMENT FOR CUTTING

Radius 12 3/16"
12"
Radius 24 3/8"
12 3/16"
24"
18-Gage galv. iron

Handles not shown on plan
3 1/8"
4"
2"
3"

3"
18-Gage galv iron
3 11/16"
2 11/16"
2 1/2"

HANDLE DETAIL

1/8" Rivets, 1 1/2" Ctrs. Flat inside solder joint
4"
3 1/2"
1 1/2"
12"
3/4"
1/16"
8"
3"

Figure 219.—Mold for slump test. 288–D–2662.

slump gradually to lower elevations and retain its original identity; however, a poor mix will crumble, segregate, and fall apart. Operations involved in making the slump test are illustrated in figure 220.

When practicable, duplicate slump tests should be made and the average of the two slumps reported.

UNIT WEIGHT AND VOLUME OF FRESH CONCRETE; AND CEMENT, WATER, AIR, AND AGGREGATE CONTENTS OF FRESH CONCRETE

Designation 23

1. General.—This procedure is for determining the unit weight and volume of fresh concrete at the mixer.[26] Simultaneously, determinations are made of cement, water, air, and aggregate contents of the concrete.

2. Unit Weight Test Apparatus.—Tests for the unit weight of fresh concrete are made in a cylindrical measure of $\frac{1}{10}$-, $\frac{1}{4}$-, $\frac{1}{2}$-, $\frac{3}{4}$-, or 1-cubic-foot capacity, the size depending on the maximum size of aggregate in the concrete to be tested. The requirements are as follows:

Maximum size of aggregate, inches	Minimum capacity, cubic feet	Required accuracy of scales, pounds
1½ and smaller	0.25	±0.02
3	.50	±.05
4	.75	±.10
6	1.00	±.20

3. Procedure for Determining Unit Weight.—(a) The test is made immediately after mixing is completed.

(b) The measure is dampened and excess moisture removed with a damp cloth just before starting the test.

(c) The measure is filled in two layers, each approximately one-half the volume of the measure. Care should be taken to ensure that only representative concrete is used for the test. If the maximum size of ag-

[26] Usually this test is made at the mixer; but in any situation where it is indicated that more than normal amounts of air are lost as a result of special conditions of handling, transportation, or vibration, it will be desirable to make a check test for air content on a sample of concrete taken after the treatment in question.

Rodding and filling cone using three separate layers of concrete

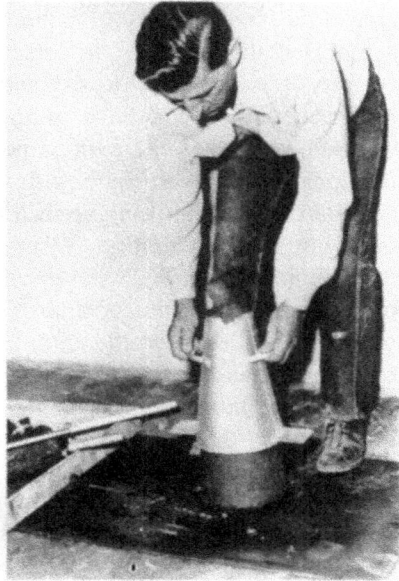

Slow, steady, vertical removal of the mold

Measuring the slump after subsidence

Tapping the concrete to observe the plasticity

Figure 220.—Making a slump test. A good indication of concrete workability is obtained by tapping the slump test specimen with the tamping rod. PX–D–1747, 1745, 1752, and 1753.

gregate is larger than 1½ inches, special care must be taken to obtain a representative sample; in mixes containing 6-inch-maximum size aggregate, it is advisable to estimate and include the number of cobbles which should be contained in the test concrete. Each layer should be vibrated with the minimum amount of vibration necessary to obtain freedom from large surface voids and rock pockets, the sufficiency of this amount of vibration being verified by the appearance of concrete cylinders receiving similar vibration. Where a laboratory vibrator is not available for internal vibration, external vibration may be applied to the outside of the container by use of a large vibrator. Because of the small quantity of concrete in the test, the amount of vibration is often greater than may be apparent and, if so, may cause excessive loss of air. Average concrete of medium slump may require no more than 1 or 2 seconds of in-and-out immersion of the vibrator. No attempt should be made to vibrate so as to simulate the loss of air handling, placing, and vibration. (See footnote 26.) Vibration of the second layer should not penetrate more than 1 inch into the lower layer. After vibration, the concrete is made level with the top of the container with a steel or hardwood straightedge. Overflow mortar should be thoroughly cleaned from the sides of the measure. Weighing should be done on approved scales of the accuracy indicated in section 2. The unit weight is determined by dividing the net weight of the concrete by the volume of the container as determined by calibration.

4. Computations.—The solid unit weight of each ingredient of a mix is equal to its specific gravity multiplied by the weight of water per cubic foot. The solid volume occupied by an ingredient in a batch is determined by dividing the batch weight of the ingredient by its solid unit weight. Computations for a typical batch of concrete are illustrated in the following tabulation:

Ingredient	Weight, pounds per batch	Specific gravity	Solid unit weight, pounds per cubic foot	Solid volume, cubic feet
Water	168.3	1.00	62.30	2.70
Pozzolan	70	2.50	155.75	0.45
Cement	282	3.15	196.25	1.44
Sand	765	2.65	165.10	4.63
Gravel	1,327	2.70	168.21	7.89
Air				V_a
Batch total	2,612			$17.11 + V_a$

The volume of air is the difference between the volume of the batch of concrete and the solid volume of the ingredients.

$$\text{Measured unit weight} = 143.5 \text{ pounds per cubic foot}$$

$$\text{Volume of batch} = \frac{\text{Total weight of batch}}{\text{Measured unit weight}} = \frac{2{,}612}{143.5}$$

or

$$= 18.20 \text{ cubic feet}$$

$$\frac{18.20}{27} = 0.674 \text{ cubic yard}$$

$$\text{Volume of air} = \text{Batch volume} - \text{solid volume of ingredients,}$$

$$\text{or } 18.20 - 17.11 = 1.09 \text{ cubic feet}$$

$$\text{Percent air} = \frac{\text{Volume of air (cubic feet)}}{\text{Volume of batch (cubic feet)}} \times 100$$

$$= \frac{1.09}{18.20} \times 100 = 6 \text{ percent}$$

The quantity of each ingredient in a cubic yard of concrete equals the batch weight divided by the batch volume in cubic yards:

$$\text{Water} = \frac{168.3}{0.674} = 250 \text{ lb. per cubic yard}$$

$$\text{Pozzolan} = \frac{70}{0.674} = 104 \text{ lb. per cubic yard}$$

$$\text{Cement} = \frac{282}{0.674} = 418 \text{ lb. per cubic yard}$$

$$\text{Sand} = \frac{765}{0.674} = 1{,}135 \text{ lb. per cubic yard}$$

$$\text{Gravel} = \frac{1{,}327}{0.674} = 1{,}970 \text{ lb. per cubic yard}$$

5. Volume of Hardened Concrete.—The volume of hardened, air-entrained concrete in place may be about 98 percent of that of the concrete as mixed and occasionally may be even less. The loss of volume may be attributed to the following:

(1) Decrease in the combined volume of cement and mixing water during hydration.

(2) Compression of entrained air in the bottom of the lift.

(3) Loss of part of entrained air during handling, placing, and vibration.

(4) Losses caused by settlement and bleeding.

(5) Small losses caused by drying shrinkage.

Computation of volume of hardened concrete must also take into account any concrete wasted or otherwise lost. It should be remembered that any discrepancy between the actual and nominal dimensions of the formed concrete, such as might be occasioned by spreading of forms during placement or inaccurate form construction, will result in incorrect volume measurements.

When more accurate figures are not available, the volume of hardened concrete may be estimated as 2 percent less than the net volume of the fresh concrete. On this basis, the cement content of hardened concrete would be equal to the cement content of the fresh concrete at the mixer, as determined above in section 4, divided by 0.98.

AIR CONTENT OF FRESH CONCRETE BY PRESSURE METHODS

Designation 24

1. General.—The Bureau of Reclamation has adopted the pressure-type air meter for use on its projects (see fig. 221). The operation of these meters is based on Boyle's law of gases. A volume of air at a certain initial pressure is allowed to expand into a container of fresh concrete, compressing the entrained air voids. The reduction in pressure indicates the percentage of air voids in the concrete, the percentage being read directly on a gage. This type of air meter is particularly advantageous in certain types of Bureau work because only a small quantity of water is required for operation and the calibration is not affected by changes in barometric pressure.

This designation does not cover other types of air meters. For their operation consult the appropriate ASTM standards.

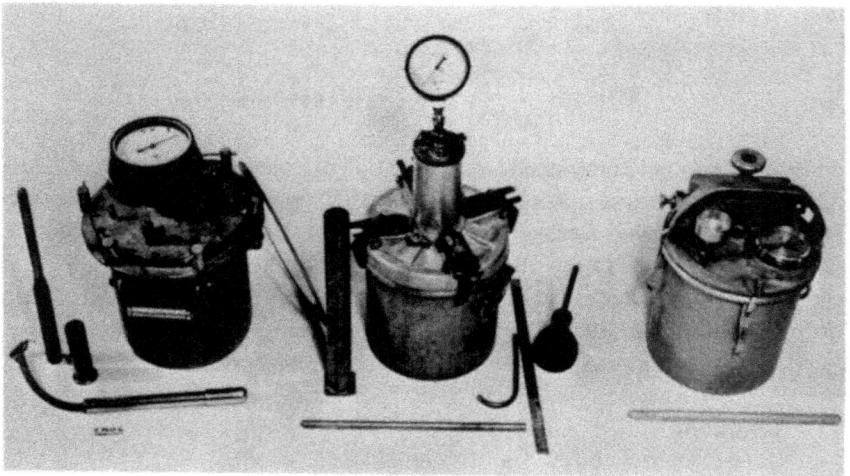

Figure 221.—Pressure-type air meters. PX–D–33527.

2. Apparatus.—(a) A base container consisting of a flanged cylindrical container, preferably of steel or other hard metal not readily attacked by cement paste and sufficiently rigid to limit its expansion. The upper surface of the container is machined smooth where it fits the cover so as to be pressure tight. The minimum size of container will depend upon the size of coarse aggregate in the concrete sample (designation 23).

(b) A cover assembly of steel or other hard metal not readily attacked by cement paste. The rim is machined so as to provide a pressure-tight fit with the base container when the two are clamped together. The cover is fitted with an air chamber; a pump, either built in or separate, for developing pressure in the air chamber; a valve for bleeding the air chamber to atmospheric pressure; an operating valve for allowing air in the air chamber to enter the container; a bypass valve which will release air in the container directly to atmosphere without passing through the air chamber; and a pressure gage with a suitable range.

(c) A calibration cylinder having a volume equal to approximately 1 percent of the volume of the base container. Glass plates suitable for covering the calibration cylinder and base container.

(d) Miscellaneous equipment, including trowel, strike-off bar, tamping rod, vibrators, syringes, funnel, or other equipment which may be necessary for operation of the equipment.

3. Calibration of Apparatus.—The calibration procedure will vary depending on the design of apparatus used. Manufacturer's instructions should be followed for calibrating the particular design of meter being used. A typical calibration would consist of the following essential steps: The base container is filled with water and the cover clamped on. If equipment is of the type which uses a small amount of water in the operation, this water is added. The valves to the air chamber are closed and air is pumped into the chamber to slightly more than the initial pressure mark. The air is allowed to cool for a few seconds and the chamber is bled to the initial operating pressure mark while the gage is tapped lightly. The operating valve which allows air to go from the air chamber to the container is then opened, and the air is allowed to warm for a few seconds while the gage is tapped lightly. The indicated air content should be 0 percent. The above procedure is repeated, drawing off successive increments of water into the calibration cylinder or measure before each measurement in accordance with instructions. The indicated air content should be that represented by the amount of water removed from the meter.

The air that replaces the water removed from the meter is at the top of the container and is not under an initial hydrostatic pressure. Air in the concrete sample is under an initial pressure varying from zero at the top to a value at the bottom equal to the pressure exerted by weight of the concrete. Although the calibration as described does not allow for this difference in initial hydrostatic pressure, the error is so small that it is within the accuracy of the test and may be neglected. If the indicated air content does not equal the correct amount, the equipment should be adjusted by changing the initial line, re-marking the dial of the pressure gage, shifting the dial of the pressure gage, or adjusting the pressure gage, as appropriate.

4. Procedure for Determining Air Content of Fresh Concrete.—(a) The base container is filled with a representative sample of fresh concrete, following the procedure for determining unit weight of concrete given in designation 23. If the fresh concrete contains aggregate larger than 1½-inch nominal size and a ¼-cubic-foot container is used, the sample should be wet screened over a screen having 1½-inch square openings.

(b) The flange of the container and the cover are cleaned and the cover clamped in place with the operating and bleeding valves open. If the equipment uses water, water is added in accordance with manufacturer's instructions. (For the equipment shown on the right in figure 221, the valve between the air chamber and the concrete must be closed before water is added; otherwise water will enter the air chamber.) Air chamber valves are then closed, and pressure in the air chamber is increased to slightly more than the initial pressure mark. After waiting a few seconds for the air to cool, the needle is brought to the initial mark by opening the bleeder valve while the gage is tapped lightly. The valve between the air chamber and the concrete is then opened rapidly.

The gage is tapped lightly and, after the gage has stabilized, the percent of air is read on the dial.

METHOD OF TEST FOR BLEEDING OF CONCRETE

Designation 25

1. Scope.—This test covers the procedure for determining the relative rate of bleeding and the quantity of mixing water that will bleed from a sample of freshly mixed concrete under the conditions of test. The apparatus described may be used with samples of concrete containing any size of aggregate graded up to and including 2½-inch maximum size (nominal maximum size of 2 inches).

2. Apparatus.—(a) A cylindrical container of ½-cubic-foot capacity.

(b) A platform scale of at least 100-pound capacity, accurate to 0.1 pound.

(c) A pipette or similar instrument for drawing off free water from the surface of the test specimen.

(d) A glass graduate of 100-milliliter capacity for collecting and measuring the quantity of water withdrawn.

(e) A tamping rod, consisting of a ⅝-inch-round steel rod approximately 24 inches in length and hemispherically tipped at the lower end.

3. Test Specimen.—Freshly mixed concrete should be used. The container should be filled with concrete to a height of 10 inches ± ⅛ inch, as described in designation 23. The top surface of the concrete should be leveled to a reasonably smooth surface by a minimum amount of troweling.

4. Procedure.—During the test, the ambient temperature should be maintained between 65° and 75° F. Immediately after troweling the surface of the specimen, the container and contents are weighed and the time recorded. The specimen and container are placed on a level platform or floor free from noticeable vibration, and the container is covered with a suitable lid, which should be kept in place throughout the test except when drawing off water. The water is drawn off with a pipette or similar instrument at 10-minute intervals until cessation of bleeding. To facilitate collection of "bleeding" water, the specimen is tilted carefully by placing a block approximately 2 inches thick under one side of the container 2 minutes prior to each time water is withdrawn. After the water is removed, the container is returned to a level position without jarring. After each withdrawal the water is transferred to a 100-milliliter graduate. The accumulated quantity of water is recorded after each transfer.

5. Calculations.—(a) Calculate accumulated "bleeding" water expressed as a percentage of the net mixing water contained within the test specimen, as follows:

$$\text{Bleeding, } \% = \frac{B}{C \times 453.6} \times 100$$

where:

$$C = \frac{w}{W} \times A = \text{weight of water in the test specimen in pounds,}$$

W = total weight of batch in pounds,

w = net weight of water in the batch in pounds,

A = weight of sample in pounds,

B = total quantity of water withdrawn from the test specimen in milliliters, and

$C \times 453.6$ = quantity of water in the test specimen expressed in milliliters.

(b) Bleeding per unit area of surface can be obtained as follows:

$$\text{Bleeding, ml/cm}^2 \text{ of surface} = \frac{V_1}{A_1}$$

where:

V_1 = volume of bleeding water, in milliliters, and

A_1 = area of exposed concrete in square centimeters.

VARIABILITY OF CONSTITUENTS IN CONCRETE

(A Test of Mixer Performance)

Designation 26

1. General.—The mixer performance test is used to check the ability of a mixer to mix concrete that will be within prescribed limits of variability. The uniformity of fresh concrete is evaluated by comparing variations in quantity of coarse aggregate and unit weight of air-free mortar of two samples, one taken at the front and one at the back of the mixer or from the first and last portions of the batch as discharged from the mixer. The test is applicable to all types and sizes of mixers.

2. Apparatus.—(a) An air meter of suitable capacity to meet the requirements outlined in designation 23 for unit weight containers or a 1/4-cubic-foot air meter with equipment required for wet screening aggregate larger than 1½-inch size from test samples.

(b) Scales of required capacity with accuracy of 0.1 percent of net weight of sample.

(c) A No. 4 brass screen of suitable capacity (approximately 18 by 18 inch).

(d) Four 24- by 24-inch pans.

(e) Equipment required to weigh, suspended in water, material retained on the No. 4 screen.

3. Samples.—Two samples of freshly mixed concrete of sufficient quantity are taken to make a test of air content by an air meter of adequate capacity. (If a 1/4-cubic-foot air meter is used for concrete with larger than 1½-inch-maximum size aggregate, the size of sample should correspond with that required for unit weight test for each specific maximum size of aggregate in the concrete as shown in designation 23.)

4. Procedure.—Samples are taken either from the front and rear of the mixer or from the first and last portions of the batch while being discharged from the mixer. In the use of the 1/4-cubic-foot air meter for testing concrete with larger than 1½-inch-maximum size aggregate, the samples are carefully weighed and wet screened through the 1½-inch screen. Material retained on the screen is then washed and dried to a saturated surface-dry condition before weighing.

Immediately after the samples are taken and wet screened if required, weights and air contents of the samples are determined by the air meter. The portion of sample used in each test is then washed on a No. 4 screen, removing all mortar from the sample. To establish the weight and volume occupied by the portion of aggregate between the No. 4 and 1½-inch sizes in each sample, the material retained on the No. 4 screen may be dried to a saturated surface-dry condition and weighed, or the wet material may be weighed suspended in water and the saturated surface-dry weight and volume computed from the specific gravity of the material. The latter method requires less time for performance of the test.

5. Calculations.—The air-free unit weight of the mortar and the weight of coarse aggregate per cubic foot of concrete (table 36) are calculated from the following formulas:

For air-free unit weight of the mortar:

$$M = \frac{b-c}{V - \left(A + \dfrac{c}{G \times 62.3}\right)}$$

For weight of coarse aggregate per cubic foot of concrete (using air meter of sufficient capacity),

$$W = \frac{c}{V}$$

where:

M = unit weight of air-free mortar in pounds per cubic foot,
b = weight (in air) of portion of sample in air meter,
c = weight (in air) of material retained on No. 4 screen,
V = volume of sample, cubic feet,
A = volume of air computed by multiplying the volume of container, V, by percent of air divided by 100,
G = specific gravity of coarse aggregate, and
W = weight of coarse aggregate per cubic foot of concrete.

For W computed from a submerged weight,

$$W = (\text{Submerged weight}) \times \frac{G}{G-1}$$

In testing a wet-screened sample to remove larger than 1½-inch coarse aggregate, the weight of the total coarse aggregate per cubic foot of concrete is:

$$W = \frac{c + \left(\dfrac{S-T}{T}\right) \times b}{V + \left[b\left(\dfrac{S-T}{T}\right) \times \dfrac{1}{G_1 \times 62.3}\right]}$$

where:

c = weight (in air) of material retained on No. 4 screen,
b = weight (in air) of portion of sample in air meter,
S = weight of sample before wet screening,
V = volume of sample in air meter,
T = weight of portion of sample passing through 1½-inch screen, and
G_1 = specific gravity of the plus 1½-inch material.

Table 36.—Example of computations for mixer performance test

Plant __East__   Mixer __No. 8__   Date __7-5-70__   Shift __Graveyard__   Time __0130 hours__   Mix No. __17__

Batch No. __195__   Mixing time __2 minutes__   Max size of agg __6 inches__   Slump __3.5 inches__

Specific gravity __2.70__   Inspector __J.C.R.__

Line		Sample from front of mixer		Sample from back of mixer	
		Wt in lb	Vol. in ft³	Wt in lb	Vol. in ft³
1	Weight of sample (Test result)	297.3		303.0	
2	Weight of material retained on 1½-inch screen (Test result)	48.72		57.06	
3	Difference in weights (1-2)	248.58		245.94	
4	Percent of material retained on 1½-inch screen (2 ÷ 3 × 100)	19.6%		23.2%	
5	Weight and volume of samples in air meter (Test result)	36.45	0.2500	36.96	0.2500
6	Percent air by air meter (Test result)	4.2%		3.6%	
7	Volume of air (6 × 5 vol.)		0.0105		0.0090
8	Volume of air-free sample (5 vol. − 7)	36.45	0.2395	36.96	0.2410
9	Submerged weight of material retained on No. 4 screen (Test result)	12.14		12.87	
10	SSD weight of material retained on No. 4 screen (9 × sp. gr.) ÷ (sp. gr. − 1)	19.28		20.44	
11	Solid volume of material retained on No. 4 screen (10 ÷ sp. gr. × 62.3)		0.1146		0.1215
12	Weight representing mortar in sample (5 wt − 10)	17.17		16.52	

13	Volume representing mortar in sample (8-11)	0.1249		0.1195	
14	Unit weight of air-free mortar (12 ÷ 13)		137.5 lb/ft³		138.2 lb/ft³
15	Weight of material retained on 1½-inch screen for air meter sample to represent full mix (4 × 5 wt)		7.14		8.57
16	Solid volume of material retained on 1½-inch screen for air meter sample to represent full mix (15 ÷ sp. gr. × 62.3)	0.0424		0.0509	
17	Unit wt of full mix (5 wt + 15) ÷ (5 vol. + 16)		149.08		151.31
18	Total coarse agg in air meter sample plus 1½-inch material to represent full mix (10 + 15)		26.42		29.01
19	Total coarse agg per cubic foot of concrete (18 ÷ 5 vol. + 16)		90.36		96.41

Efficiency: Variation in unit wt of air-free mortar from average $= \dfrac{\text{Av} - 14 \text{ front or } 14 \text{ back}}{\text{Av}} \times 100$

Maximum variation allowed $= 0.8$ percent

Variation of coarse agg per cubic foot of concrete from average $= \dfrac{\text{Av} - 19 \text{ front or back}}{\text{Av}} \times 100$

Maximum variation allowed $= 5.0$ percent

MIXER EFFICIENCY EXAMPLE

Average unit weight of air-free mortar $= \dfrac{137.5 + 138.2}{2} = 137.85$ lbs

Variation in unit weight from average
Maximum allowed 0.80% $= \dfrac{137.85 - 137.50}{137.85} \times 100 = 0.25\%$

Average weight of coarse aggregate ft³ $= \dfrac{90.36 + 96.41}{2} = 93.38$

Variation of coarse aggregate per ft³
Maximum allowed 5.0% $= \dfrac{93.38 - 90.36}{93.38} \times 100 = 3.23\%$

NOTE.—The term $\left(\dfrac{S-T}{T}\right) b$ gives the weight of coarse aggregate larger than 1½-inch required in the air meter samples to represent the original concrete. The unit weight of the full mix can be computed by dividing the total weight of the air meter sample and weight of the plus 1½-inch material required to represent the original sample, by the volume of the air meter sample increased by the volume of plus 1½-inch material computed from its weight and specific gravity.

6. Mixing Time.—In analyzing several hundred mortar tests of samples taken during the construction of several dams, it is apparent from figure 222 that variation in the unit weight of the mortars and variations in the water-cement, sand-cement, and water-fines ratios reflect the adequacy of mixing.

BATCHING FOR MACHINE-MIXED LABORATORY CONCRETE

Designation 27

1. Measurements.—The exact quantities of cement and aggregate are determined by weight; the water may be weighed or be measured in a graduated container.

2. Record.—A data sheet showing the batch weights for the proposed mix is furnished to the weigher. The weights of the aggregate are recorded cumulatively, beginning with the smallest size and progressing through successive larger sizes to the maximum used.

3. Procedure.—(a) The aggregate is weighed on an indicating dial scale accurate to one-quarter pound. A container or aggregate bucket, large enough to hold a half batch or full batch, is placed on the scale platform and the tare set on the balancing beam at the weight required to bring the pointer exactly to zero.

(b) All weights should be checked before the next fraction of aggregate is added.

(c) The cement is weighed carefully on a scale accurate to one-hundredth pound, and a record of the weight and the mix number is placed on the container for identification.

(d) The aggregate should be room dry when weighed and should be brought to room temperature (65° to 75° F) before being mixed.

(e) As the mix proportions are based on saturated surface-dry aggregate, the weight of aggregate in the batch must be reduced by the amount of water absorbed by the aggregate during mixing. The amount of absorption is generally considered to be that absorbed in 30 minutes of immersion in water by the room-dry aggregate.

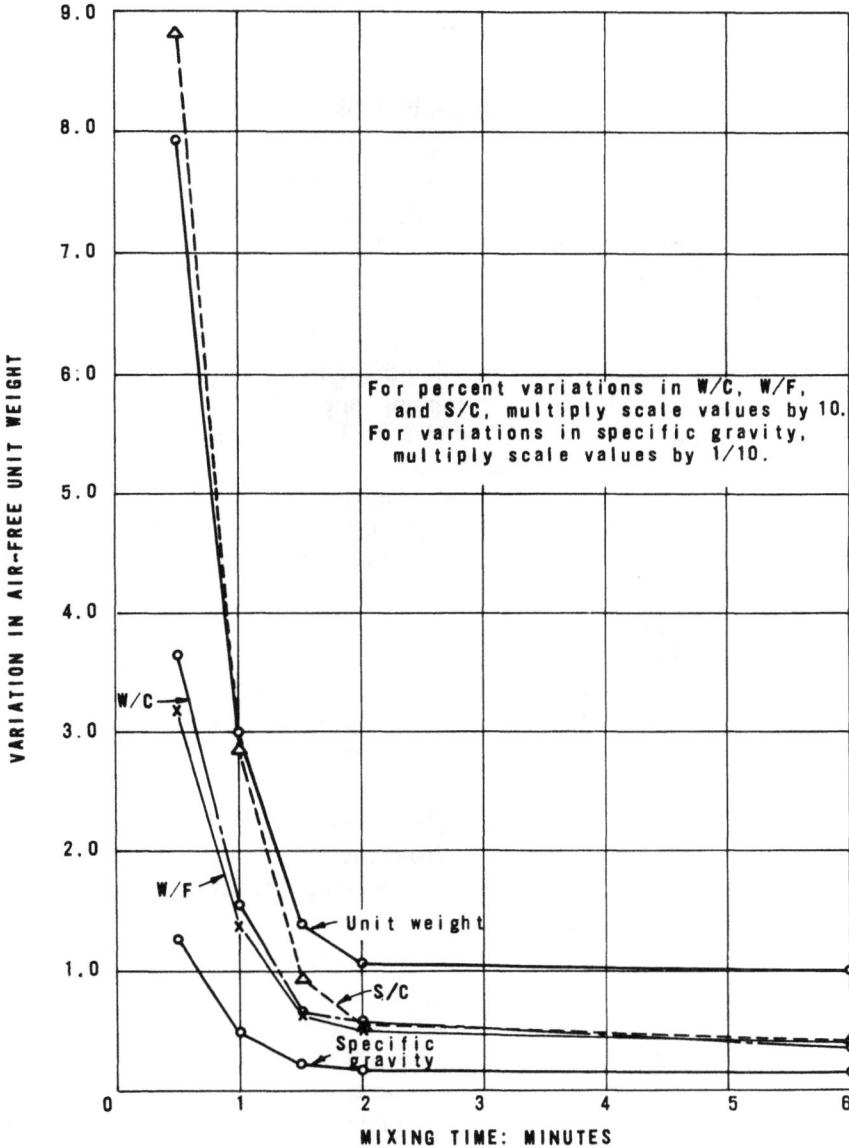

Figure 222.—Results of mixing time tests on 4-cubic-yard mixers. Adequate mixing time of concrete is indicated at about 2 minutes by specific gravity and unit weight as well as by the water-cement, water-fines, and sand-cement ratios. 288–D–1117.

LABORATORY CONCRETE MIXING

Designation 28

1. Machine Mixing.—(a) Before the trial batch is placed in the mixer, the mixer should be primed by mixing a partial batch (about one-half the normal size) having approximately the composition of the trial batch. The priming batch is discharged and wasted, leaving a normal coating of mortar on the inside of the mixer drum.

(b) When the mixer is equipped with a loading skip, the charge of aggregate is placed in the loading skip first and the cement placed on top of the aggregate. With the mixer running, approximately four-fifths of the estimated required water is placed in the mixer, and then the total charge of aggregate and cement is introduced. The mixing time should start immediately after all aggregates are in the mixer. The remainder of the water is added slowly, all being added in approximately one-half minute. Occasionally, it will be necessary to add a small amount of water as mixing progresses to obtain the desired consistency.

(c) When the mixer is not equipped with a loading skip, the aggregate is placed in the mixer and the cement introduced on top of the aggregate. Approximately four-fifths of the estimated required water is added and the mixer started. The remainder of the water is added immediately. It may occasionally be necessary to add a small amount of water as mixing progresses to obtain the desired consistency.

(d) The mixing time should be not less than 3 minutes.

(e) The concrete is dumped from the mixer into a watertight and nonabsorptive receptacle of such size and shape that the concrete can be turned over with a shovel to eliminate segregation.

(f) Tests should be made immediately after the concrete is emptied from the mixer.

2. Hand Mixing.—Hand-mixed concrete is mixed in batches of such size as to leave a small quantity of concrete after the test specimens are molded. The batch should preferably be mixed in a shallow metal pan with a brick-mason's 10-inch trowel which has been blunted by cutting off about 2½ inches of the point. Mixing should be performed as follows:

(a) The cement and aggregate are thoroughly mixed dry.

(b) Sufficient water is added to produce concrete of the required consistency.

(c) The mass is mixed thoroughly until the resulting concrete is homogeneous in appearance.

CASTING CYLINDERS IN CAST IRON MOLDS OR DISPOSABLE MOLDS

Designation 29

1. Mold.—The general design of a 6- by 12-inch cast iron mold in general use by the Bureau is shown in figure 223. (If fabrication details are desired, request drawing No. 288–D–2663 from Chief, Division of General Research, Engineering and Research Center, Denver Federal Center, Denver, Colorado 80225.) All molds should be provided with machined base plates securely fastened to the mold with machine screws or clamps. Particular care should be taken to obtain tight molds so that mixing water will not escape during molding. The inside surfaces of cast iron molds should be greased with soft (No. 2) graphite grease before using. It is not necessary to grease plastic or metal disposable molds.

2. Molding Specimens, Using Vibration.—For 6- by 12-inch and 8- by 16-inch cylinders, the fresh concrete is placed in the mold in two layers, each approximately one-half of the volume of the mold. Each layer is consolidated by vibrating with an immersion-type vibrator having a vibrating element approximately 1 inch in diameter and a frequency of 10,000 vibrations per minute when immersed. In consolidating the bottom layer, the vibrator should not be allowed to rest on the bottom of the mold. While vibrating the top layer, the vibrating element should penetrate approximately 1 inch into the bottom layer.

The period of vibration will depend on the slump of the concrete and the effectiveness of the vibrator. (Three insertions of 3 or 4 seconds each are usually sufficient.) The vibration can be judged adequate when the top surface of the concrete exhibits a shiny wet appearance, at which time the vibrator is slowly withdrawn. In placing each scoopful of concrete, the scoop should be moved around the top edge of the mold as the concrete slides from it to ensure a symmetrical distribution of concrete. The mold should not be filled so full that mortar runs over the top when the vibrator is inserted. After vibration of the second layer, enough concrete should be added and worked into the underlying concrete with a trowel to bring the level about one-eighth inch above the top of the mold. The specimen is then moved to the curing room, struck off flush, and smoothed with a trowel.

3. Molding Specimens, Using Hand Compaction.—Where concrete is compacted by hand rodding in 6- by 12-inch and 8- by 16-inch cylinders, the molds are filled in three layers, each approximately one-third the volume of the mold. Each layer is rodded 25 strokes with a ⅝-inch-diameter rod 24 inches in length, hemispherically tipped at the lower end. The strokes are distributed in a uniform manner over the cross section of the mold and should just penetrate into the underlying layer. The bottom layer

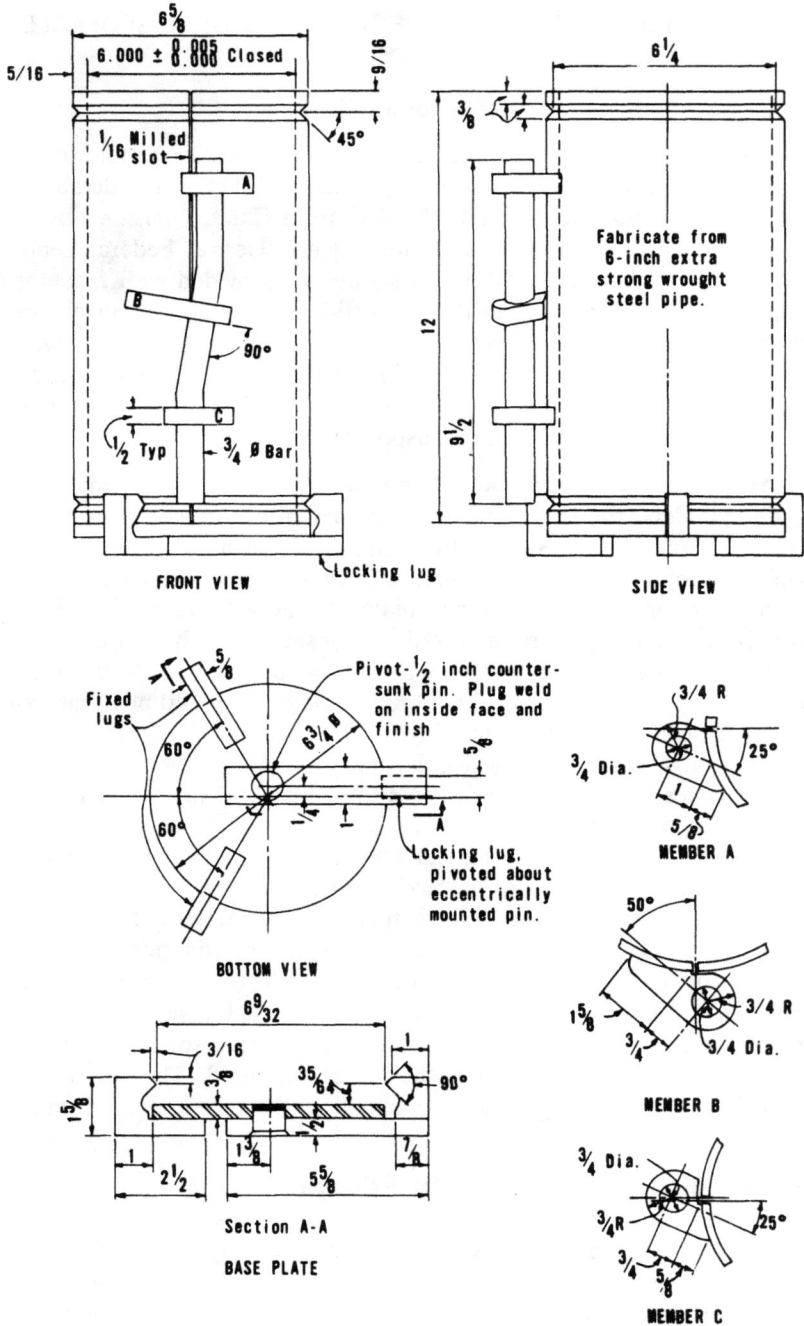

Figure 223.—Test cylinder mold. 288–D–3270.

should be rodded throughout its depth. Care should be taken that the rod does not strike the bottom of disposable molds. After the top layer has been rodded, the specimen is removed to the curing room and the excess concrete struck off with a trowel.

4. Storage and Handling of Specimens.—(a) Specimens should be removed from the molds 18 to 24 hours after casting, weighed, and returned to the curing room.

(b) Specimens made in the field at the site of placement should be kept, as nearly as practicable, at 73.4° F and protected from the sun and from moisture loss.

CASTING CYLINDERS IN CANS

Designation 30

1. Mold.—The can should be 6 inches ± $\frac{1}{16}$ inch in diameter and 12½ inches ± ⅛ inch high with single friction lids. The lids should be tight fitting. The cans should be round and true to shape. Attached handles are not required. Bottoms of cans should be completely flat on the inside

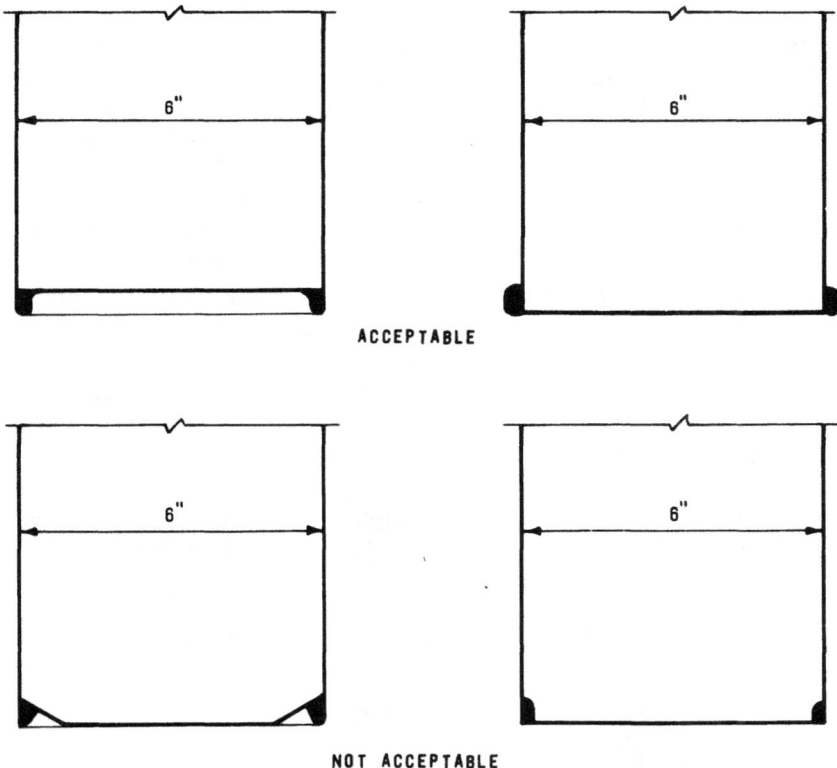

ACCEPTABLE

NOT ACCEPTABLE

Figure 224.—Fabrication details of can bottoms. 288–D–2660.

without beads or other profile (see fig. 224). All seams should be double seamed and should be flat on the inside of the can, but slightly protruding longitudinal seams on the inside of the can are acceptable. The cans should be watertight and may be made so by coating with a suitable thin, non-reactive coating or by soldering. Minimum grade of material acceptable for the cans should be 107-pound base weight, 0.50-pound, electrolytic tinplate.

2. Procedure.—(a) The fresh sample is mixed thoroughly before concrete is placed in the can.

(b) Before casting the specimen, the can should be placed on a smooth, level surface that is free from vibration. If the cylinder is to be transported immediately after fabrication, it must be done by hand and in a vertical position. The specimen should then be allowed to set on a level surface for at least 6 hours and preferably 24 hours, during which time it must be protected from extreme temperatures.

(c) The can is filled in three layers, each approximately one-third the volume of the can. Each layer is compacted by hand rodding 25 strokes with a ⅝-inch-diameter rod 24 inches in length, hemispherically tipped at the lower end. The strokes are distributed in a uniform manner over the cross section of the can and should penetrate into the underlying layer. The bottom layer should be rodded throughout its depth, care being taken not to strike the bottom of the can hard enough to dent or mar it. Convex ends on cylinders and bottoms damaged by rodding will be avoided if each can, during filling and until the concrete has hardened, is rested symmetrically on a wooden or metal disc 5½ inches in diameter. After the top layer has been rodded, its surface is smoothed off not less than one-half nor more than three-fourths of an inch below the top of the can. A small trowel with a rounded point should be used for this purpose. Particular care shall be taken to secure smooth cylinder ends that are perpendicular to the axis of the cylinder.

(d) The friction cover is pressed into place immediately after filling, care being taken to assure that the rim of the can is not bent and that the cover does not touch the top concrete surface.

(e) Specimens should be stored where the temperature of the concrete will be maintained at approximately 73.4° F. Before testing, cylinders should be capped on both ends in the manner prescribed in designation 32.

FIELD-LABORATORY CURING, PACKING, AND SHIPPING OF TEST CYLINDERS

Designation 31

1. Removal of Molds.—All concrete test specimens, except those cast in cans, should be removed from the molds not sooner than 18 nor later than 48 hours after casting. They should then be marked, weighed, and stored in water, damp sand, or a fog room.

2. Water Curing.—The specimens should be completely immersed in water, preferably maintained at 73.4°±3° F. (See sec. 60, chapter IV.) When such temperature control is impracticable, the actual average temperature conditions should be reported. The water should be removed completely at intervals not to exceed 30 days.

3. Damp-Sand Curing.—The specimens should be completely buried in sand. The sand should be thoroughly saturated with water when the specimens are placed in it, and sufficient water should be added each day to maintain a saturated condition. Requirements for temperature control are stated in section 2.

4. Fog-Room Curing.—The specimens should be stored in a fog room (see sec. 60, chapter IV) maintained at 73.4°±3° F and at a relative humidity of 100 percent. The specimens should be placed in the fog room immediately after casting and should remain there until tested, except for the brief periods necessary for the removal of molds and for capping.

5. Temperature Control of Cylinders Cast in Cans.—When a room maintained at a temperature of approximately 73.4° F is not available, several simple expedients may be used to obtain the desired storage temperature for cylinders cast in cans. In cold weather an insulated box may be heated with an electric light controlled by a simple thermostat of the type used in incubators. In hot weather the cans may be stored in a cellar, in which temperature may be controlled by ventilation or by sprinkling. Storing the cans in damp sand or under moist burlap in a shaded location may also provide nearly constant temperature curing.

6. Packing and Shipping.—Where it is necessary to ship test cylinders to an adjacent project or to Denver for testing, the specimens should be wrapped in moistureproof paper and crated in damp sawdust to prevent moisture loss or damage in transit.

CAPPING CONCRETE CYLINDERS

Designation 32

1. General.—The ends of all concrete specimens 8 inches in diameter or less which have not been ground smooth or which have not been cast against a machined base plate should be capped with a melted mixture of sulfur and finely ground, screened material. Milled fire clay has been found to be the best material to use with sulfur for capping. Commercial capping materials specially compounded are suitable, provided they develop compressive strength equal to or greater than the anticipated strength of the specimen at time of test.

2. Capping.—(a) Before capping, all loose particles are removed from the end of the cylinder with a wire brush; the end should be free from laitance and moisture. All specimens cured in cans and having smooth, glazed ends should be roughened with a sharp tool or by sandblasting.

(b) The capping material, if a mixture of 3 parts of sulfur by weight to 1 part of finely ground fire clay, is heated at a temperature between 350° and 400° F until the mixture is thick and viscous. The right temperature of the mixture for capping is determined through observation and experience with the mixture used. If a commercial preparation is used, the manufacturer's recommendation should be followed as to temperature. Overheating or reheating of the mixture should be avoided as these practices tend to make the cap rubbery rather than brittle as desired. Enough material to make the cap is poured in a lightly greased capping mold (see fig. 225) and the specimen pressed firmly into the melted material, care being taken to keep the cylinder plumb by holding it against the angle-iron guide. This operation should be done quickly before solidification begins so that the mixture will adhere to the specimen. Caps should be as thin as practicable and should not flow or fracture when the specimen is tested.[27]

After the cap has hardened, the plate may be loosened by striking the edge of the mold. If any holes in the end of the cylinder are more than 0.25 inch deep, they should be partially filled by pouring capping material into them and allowing it to harden prior to forming the full cap. When the ends of cylinders are square with the longitudinal axis, capping can be done on a machined steel plate. This plate must be flat to 0.002 inch tolerance in 6 inches. The bearing surface of the capped specimen in contact with the lower bearing block of the testing machine should not depart from perpendicularity by more than 0.25° (approximately equal to one-sixteenth inch in 12 inches) and the combined departure of the two bearing surfaces from perpendicularity to the axis should not exceed 3°.

(c) Before testing the cylinder, the cap should be tested for air pockets by tapping it lightly with the handle of a putty knife or other suitable instrument. A cap that sounds hollow in places should be removed and replaced by a good, solid cap. Solid caps are easily obtained when the capping surface is dry and clean, the mixture is at the right temperature, and the cylinder is placed immediately in the melted mixture on the capping plate.

(d) Evidence indicates that sulfur-mixture caps thicker than three-sixteenths inch cause strength reductions that may be as much as 20 percent, particularly in testing high-strength concrete. Sulfur-mixture caps should not be used when:

(1) Ends of cylinders are convex by more than 0.20 inch;

---

[27] Based on ASTM Designation C 617.

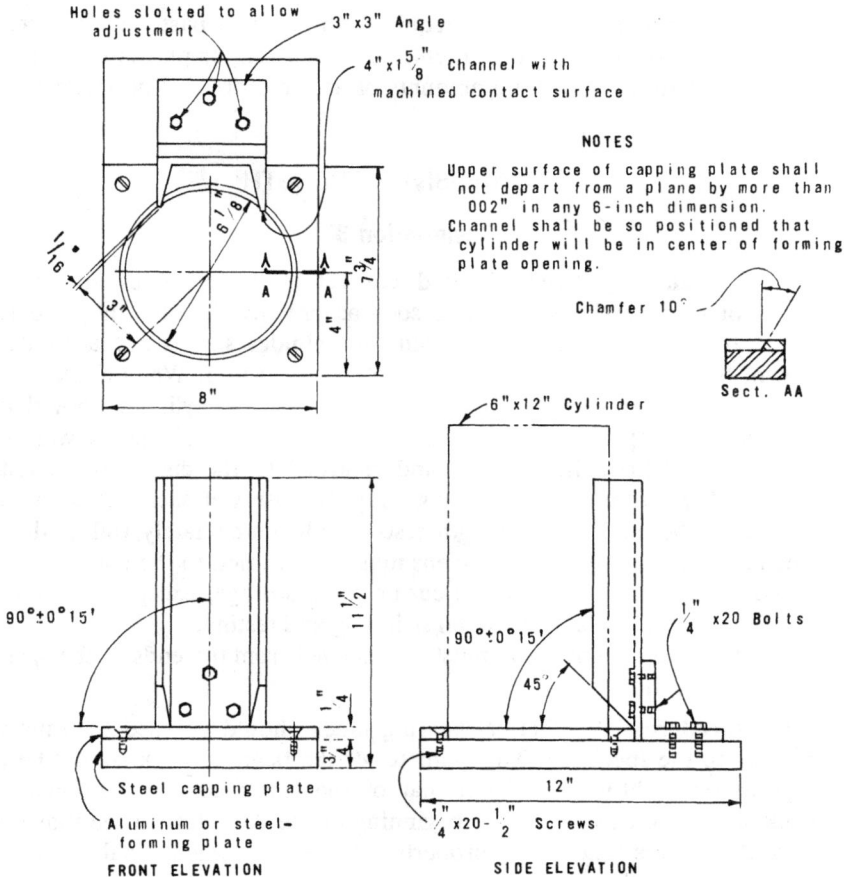

Figure 225.—Cylinder capping mold and alining jig. 288–D–1554.

(2) Aggregate or other bulges protrude from the ends by more than 0.20 inch;

(3) These or other irregularities result in caps which are more than 0.25 inch thick over most of the end area; or

(4) Ends are not at right angles to the axis by more than 0.3 inch in the 6-inch diameter of the specimen.

Cylinders with any of the above defects should be discarded, or they should be capped with neat cement paste. Before capping, the ends of the cylinders should be prewetted, wire brushed, and cleaned. Plastic, pre-shrunk, neat cement paste should be applied and molded in a true plane normal to the axis, and of minimum practicable thickness, with a piece of plate glass lightly greased with a 50–50 mixture of lard oil and paraffin. Caps should be moist cured for at least several days with the cylinders.

Where a rapid-hardening cap is required, the cement paste can be accelerated as necessary with calcium chloride, soda ash, or plaster of paris. It is preferable to use moderate amounts of accelerators and permit caps to harden at least 24 hours.

COMPRESSIVE STRENGTH

Designation 33

1. General.—(a) Standard-cured concrete test specimens should be tested for compressive strength as soon as practicable after they are removed from the curing room. When the cylinder size is not accurately known, the cylinder should first be carefully measured. When a check on the cement content of the concrete is desired, the cylinder should be weighed. If the specimens are to be sulfur capped, they should be weighed and measured first, then capped and returned to the curing room until tested or kept from drying by being wrapped with wet sacks. If elasticity tests are to be made, the strength tests should immediately follow them. When there is a delay, the specimens must be returned to the curing room. Sealed specimens do not require curing-room storage, except when more than 6 hours are to elapse between stripping and testing.

(b) All foreign material must be removed from the ends of the specimen before testing.

2. Procedure.—A spherical bearing block should be used to transmit the load to the specimen. The diameter of the bearing block should be as large as or slightly larger than that of the test specimen. The bearing block should be checked at the beginning of the day's test to see that it is lubricated and is functioning properly. The test specimen should be carefully centered with respect to the bearing block.

The load should be applied uniformly and without shock at a rate of 2,000 pounds per square inch per minute. It is important that no adjustment be made in the controls of the testing machine while a specimen is yielding rapidly, immediately before failure.

3. Test Record.—The total load indicated by the testing machine at failure of the test specimen should be recorded immediately. Exceptional variations in the type of break and angle of fracture should be noted.

4. Factors Affecting Test Results.—(a) *Effect of Wet Screening and Type of Curing.*—Cylinders of concrete used for compressive strength tests are generally 6 inches in diameter and 12 inches long. The strength of mass concrete in mass-cured larger cylinders, (18- by 36-inch or 12- by 24-inch) can be estimated using figure 226. If the strength of fog-cured 6- by 12-inch cylinders made with 1½-inch maximum aggregate wet screened from the full mass mix is known, the strength of the mass concrete at various ages can be predicted with a reasonable degree of accuracy. Also, if the desired strength of the mass concrete is known, figure 226 may be used to determine the required strength of wet screened 6- by 12-inch cylinders.

(b) *Effect of Cylinder Size.*—The test data shown in figure 227 are typical of the relative strengths to be expected when the same concrete is tested in cylinders varying in size from 2 by 4 inches to 36 by 72 inches, except that concretes containing ⅜-inch and ¾-inch maximum

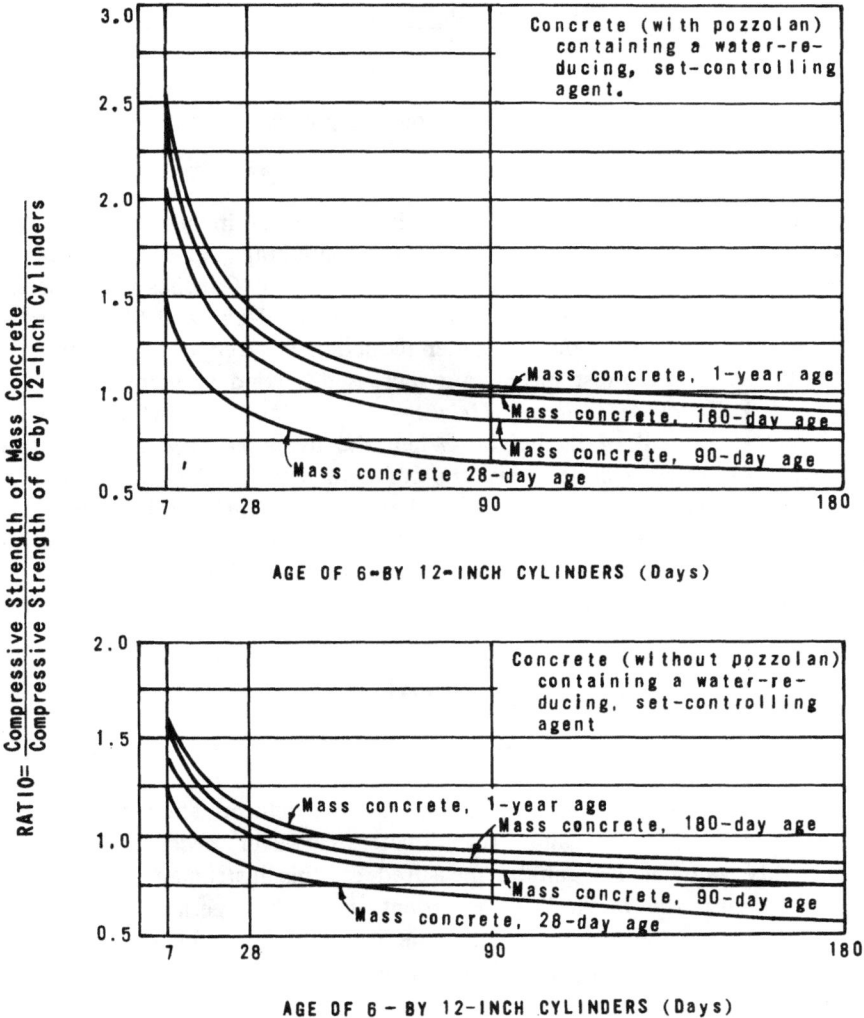

Figure 226.—Ratios of mass concrete compressive strengths in seal-cured cylinders to compressive strengths of 6- by 12-inch fog-cured cylinders fabricated from minus 1½-inch MSA wet-screened concrete. 288–D–3264.

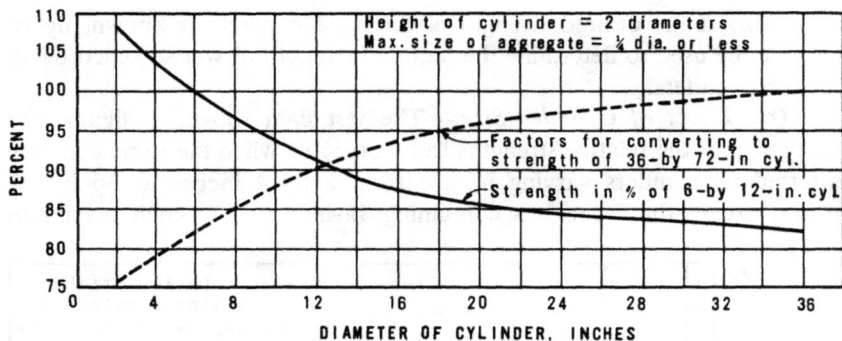

Figure 227.—Effect of cylinder size on compressive strength of concrete. 288–D–869.

aggregate were not tested in cylinders larger than 18 inches and 24 inches in diameter, respectively, and concretes containing 1½-inch maximum aggregate were not tested in cylinders smaller than 6 inches in diameter. The values shown are based on the average of 28-day and 90-day tests. It is of interest that a much smaller reduction in strength is indicated as the diameter of the test specimen is increased beyond 18 inches.

(c) *Effect of Cylinder Height.*—A standard test cylinder has a diameter one-half of its height. When the available specimens do not have these relative dimensions, the curve in figure 228 may be used to correct indicated strengths so that they will be comparable with those obtained from standard specimens. Since the curve is quite flat for ratios of 1.5 and over, small variations in height of specimen do not greatly affect strength.

TURBIDITY [28]

Designation 34

1. General.—The standard method for the determination of turbidity has been the Jackson candle turbidimeter. However, because the lowest turbidity value which can be measured on this instrument is 25 units and because the method is dependent upon the operator's ability to visually judge the point at which the candle light is indiscernible, the candle turbidimeter is generally being discarded in favor of photoelectric methods.

---

[28] Based on "Standard Methods for Examination of Water and Wastewater," part 163, 13th edition, 1971, American Public Health Association, American Water Works Association, Water Pollution Control Federation, Washington, D.C.

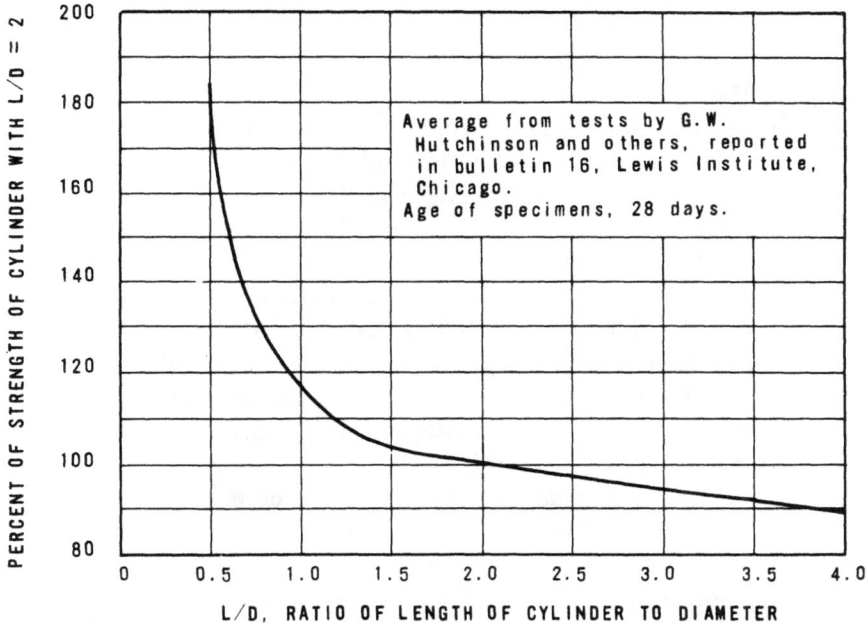

Figure 228.—Relation of length and diameter of specimen to compressive strength. 288–D–871.

Turbidity should preferably be determined on the same day the sample is taken. When longer storage is unavoidable, samples may be stored in the dark up to 24 hours. For even longer storage, treat each liter of sample with 1 gram mercuric chloride; however, prolonged storage is not recommended. All samples should be vigorously shaken before examination.

2. Photoelectric Method.—(a) *Apparatus.*—The nephelometer consists of a tungsten lamp light source, one or more photosensitive detectors, colorless glass sample tubes, and instrumentation necessary to provide direct turbidity readings.

(b) *Reagent Preparation.*—The following reagents should be prepared before beginning the test:

(1) *Turbidity-free water.*—Pass distilled water through a membrane filter having a pore size no greater than 100 millimicrometers, discarding the first 200 milliliters collected. Use the filtered water if it shows a lower turbidity than the distilled water. Otherwise, use the distilled water.

(2) *Stock turbidity suspension.*—Prepare the following solutions and suspension monthly:

Solution I—Dissolve 1.00 gram hydrazine sulfate, $(NH_2)_2 \cdot H_2SO_4$, in distilled water and dilute to 100 milliliters in a volumetric flask.

Solution II—Dissolve 10.00 grams hyexamethylenetetramine, $(CH_2)_6N_4$, in distilled water and dilute to 100 milliliters in a volumetric flask.

Stock Turbidity Suspension—In a 100-milliliter volumetric flask, mix 5.0 milliliters of solution I with 5.0 milliliters of solution II. Allow to stand 24 hours at $25° \pm 3°$ C, then dilute to the mark and mix. The turbidity of this suspension is 400 units.

(3) *Standard turbidity suspension.*—Dilute 10.00 milliliters of the stock turbidity suspension to 100 milliliters with turbidity-free water. The turbidity of this suspension is defined as 40 units. Prepare new suspension weekly.

(4) *Dilute turbidity standards.*—Dilute portions of the standard turbidity suspension with turbidity-free water as required. Prepare new suspension weekly.

(c) *Procedure.*—Calibrate the instrument in accordance with the manufacturer's instructions, running at least one standard in each range to be used. Turbidities less than 40 units are read directly from the instrument scale or from the appropriate calibration scale. For measurement of turbidities over 40 units the sample is diluted with turbidity-free water until the turbidity falls between 30 and 40 units. The turbidity is then calculated as follows:

$$\text{Turbidity units} = \frac{A \times (B + C)}{C}$$

where:

$A = $ Turbidity units found in diluted sample
$B = $ Volume of dilution water used
$C = $ Sample volume taken for dilution.

The sample tubes used must be of clear, colorless glass. They must be kept scrupulously clean, both inside and out. They should be discarded when they become scratched or etched. They must not be handled at all where the light strikes them but should be provided with sufficient extra length or with a protective case so that they may be handled properly.

3. Jackson Turbidimeter.—(a) *Apparatus.*—The Jackson candle turbidimeter consists of a graduated glass tube, a standard candle, and a support for the candle and tube. The glass tube is calibrated to read directly in parts per million and has a flat, polished bottom. Most of the tube should be enclosed in a metal case or other suitable case when observations are being made. The candle support has a spring or other device to keep the top rim of the candle pressed against the top plate of the candle support. A 400-milliliter beaker and a glass stirring rod may be conveniently used for transferring the water to the graduated tube.

(b) *Procedure.*—To secure uniform results, it is desirable that the flame be kept as nearly constant in size and in distance below the glass tube as is possible. This will require frequent trimming of the charred portion of the wick and frequent observations to see that the candle is in contact with the top plate of its supporting frame. Each time before lighting the candle, that portion of the charred wick which may be easily broken off with the fingers should be removed. The candle should not be kept lighted for more than a few minutes at a time as the flame has a tendency to increase in size. Observations should be made in a darkened room or with a black cloth over the head. The candle flame must be protected from drafts.

The turbidity of the sample is measured in terms of the amount of sample required in the calibrated tube above the candle to cause the flame to be invisible. The sample of water is shaken thoroughly and about 200 milliliters poured into the 400-milliliter beaker, stirring vigorously with a glass rod to prevent settling of the suspended solids. One-half inch or so of the water to be tested is placed in the glass tube, the candle lighted, and water from the beaker poured slowly into the tube. The water in the beaker should be stirred frequently during this process. Pouring should be very slow when the flame image becomes only faintly visible and should be stopped when it disappears. The removal of 1 percent of the suspension (water and suspended solids) should make the image visible again.

The glass tube is removed from the base and the turbidity read. The reading should be carefully checked, remembering that turbidity figures *increase* toward the bottom of the glass.

Samples having a turbidity of more than 1,000 parts per million should be diluted with one or more equal amounts of clear water until the turbidity falls below 1,000 parts per million. The turbidity of the original sample is then computed from the reading made on the diluted sample. For example, if the reading on the diluted sample is 500 parts per million and the amount of the original sample in the dilute sample is 1 part in 6, the turbidity of the original sample is 3,000.

All utensils should be carefully rinsed in clear water after using, and the bottom of the glass tube must be kept clear and clean, outside and inside. An occasional rinsing with weak hydrochloric acid is advisable. The candle and holder must be kept clean.

TEMPERATURE OF CONCRETE

Designation 35

The approximate temperature of concrete can be calculated from the temperatures of its ingredients by use of the following formula:

$$T = \frac{s(T_a W_a + T_c W_c) + T_f W_f + T_m W_m}{s(W_a + W_c) + W_f + W_m}$$

When ice is used as part of the mixing water, the following term should be included in the numerator: "minus *(heat of fusion)* (W_{ice})." In British units the heat of fusion of ice is 144 Btu per pound.

Weight symbol		*Temperature symbol*
W_a	Aggregates (surface dry)	T_a
W_c	Cement	T_c
W_f	Free moisture in aggregates	T_f
W_m	Mixing water	T_m

$s = 0.20$, assumed specific heat of dry materials.

$T =$ Temperature of concrete.

The formula may be used with any system of units and determines the temperature that would result from the blend of concrete materials if the concrete were not influenced by other heat losses or gains. Friction during the mixing process generates heat sufficient to raise the temperature of concrete about 1 °F. When cement first comes in contact with water, there is heat generated by heat of solution and heat of hydration which can raise the temperature of the concrete during the first 10 to 15 minutes by approximately 1 °F for each bag of cement per cubic yard of concrete. Therefore, when concreting in hot weather, the placing temperature of the concrete will usually be several degrees higher than the temperature determined by the formula unless correction factors are applied.

SAMPLING AIR-ENTRAINING AGENTS

Designation 36

1. **General.**—Air-entraining agents may be furnished either as liquid or powder. A 1-quart sample should be taken from a container selected at random in each shipment of liquid agent. The liquid sample is obtained by inserting a sampling tube or "thief" the full depth of the container. When the liquid has been stored for more than 6 months, the container should be agitated before the sample is taken.

A 1-pound sample should be taken from a sack selected at random in each shipment of powdered agent.

2. **Samples.**—Liquid samples should be placed in 1-quart cans fitted with friction-top lids. Sample cans should be closed tightly and the lids

spot soldered. Powder samples may be placed in metal cans or paper cartons. Samples should be carefully packed to avoid damage and spillage during shipment and should be shipped to the Chief, Division of Research, Bureau of Reclamation, Attn: D-1510, P O Box 25007, Denver, Colorado 80225. Each shipment should be properly identified as to manufacturer, quantity represented, construction contractor, and specification number. Form No. 7–1417, shown in figure 230, may be used for transmitting samples.

TESTING-MACHINE MAINTENANCE

Designation 37

1. **General.**—Testing machines require very little maintenance but should be checked periodically. To keep the machine functioning properly, check:

(1) Hydraulic oil in the reservoir often leaks past the packing between the piston and cylinder wall and sometimes drips from the high-pressure connections and the control valve. A reservoir low in oil will cause the load-indicating hand to fluctuate when a load is being applied with the piston extended. Insufficient oil in the reservoir will allow air to enter the system. The reservoir should be checked three or four times a year and, if low, should be replenished with a good grade of extra-heavy lubricating oil of approximately 80 to 100 viscosity.

(2) Air in the pressure gage line is indicated when the load-indicating hand fluctuates along any part of the dial when a specimen is being stressed and the piston is not extended. The air may be bled from this line by stressing a specimen to 10,000 or 20,000 pounds and loosening the connection to the Bourdon tube, allowing the oil to flow out into a can until evidence of air bubbles no longer appears in the oil stream. When the pressure gage line is opened, the load-indicating hand will fall back. This should not be taken as an indication that the load has been released from the specimen. Increasing the load will cause the oil to squirt forcibly from the end of the pressure line if the union has been disconnected. When bleeding the pressure line, it is better practice merely to loosen the connection to the Bourdon tube. This will allow the air and oil to emerge and the connection to be tightened when air bubbles cease to appear. If air is in the Bourdon tube, the Allen end plug in the extreme end of the tube is loosened and the oil and air released. The load should not be released from the machine until the union or plug has been tightened.

(3) Dust and dirt must be kept from getting between the piston and cylinder walls. Small particles of sand may become embedded in the piston wall, when the piston is left extended for any great length of time, and be carried down between the piston and cylinder

walls, thereby causing a binding action and a false load indication on the dial. This condition is indicated if the piston fails to return to its normal unloaded position when the bypass valve is open.

2. Servicing of Testing Machines.—When a testing machine is not functioning properly or when a calibration of the machine is due, the Chief, Division of Research, should be advised.

3. Calibration of Testing Machines.—Testing machines should be calibrated at intervals not greater than 2 years. Also, they should be calibrated immediately after being moved, after repairs or adjustments have been made to the weighing system, and whenever there is reason to doubt the accuracy of the results. This work must be done by one familiar with the proving ring and the required procedure if accurate readings of the applied loads are to be obtained.

4. Proving Ring.—The proving ring used in the calibration of testing machines is made of high-grade special alloy steel and is calibrated by the National Bureau of Standards, Washington, D.C. It must show certain characteristics under stress to be approved for the verification of testing machines. As careless handling may cause the ring to become useless as a verifying instrument, it is essential that it be handled with extra care. A clean cloth or gloves should be used in all handling operations to avert the possibility of corrosion or other damage to the surface of the ring. In no case should it be grabbed with a metal hook or handled in any way which might scratch the surface or change its characteristics.

5. Procedure.—(a) *Preparations for Testing.*—The spherical head of the testing machine is removed and replaced by the solid head. If a solid head is not available, the spherical head should be made as rigid as possible, with its bearing surface parallel with the upper bearing surface of the proving ring. The proving ring is installed by placing it directly upon a small, hardened bearing block placed on the compression table of the machine (fig. 229). Another hardened bearing block is placed on the top bearing surface of the ring. (The hardened blocks are supplied with the proving ring.) After the entire assembly is centered under the top bearing surface, the ring is preloaded to the capacity of the machine, the capacity of the dial, or that of the proving ring, whichever is least. Final calibrating runs should not be started until the ring has come to room temperature and has been stressed several times, as above noted. The best way to determine when the ring has reached room temperature is to take zero readings before and after the ring has been stressed. These readings should come within one division of each other on the micrometer wheel. When the zero readings are approximately equal after the ring has been stressed several times, it is conditioned and ready for the calibrating run. If the laboratory is not held at a fairly uniform temperature, the temper-

Figure 229.—Testing machine with proving ring in place prior to calibration of the machine. P382–706–11158.

ature of the ring will change, and accuracy of the readings will be impaired.

(b) *Reading the Proving Ring.*—A load applied by the testing machine to the proving ring causes a deformation of the ring. This deformation, from which the load can be accurately determined, is measured by means of a micrometer. The operation is as follows:

(1) Set the vibrating reed in motion.

(2) Move the micrometer wheel up slowly until the reed vibration is damped to about half frequency.

(3) Back the micrometer wheel off until the vibrating reed moves freely.

A point will be found between the two positions of the micrometer at which the vibrating reed slows down slightly and the humming sound changes tone. This is the point of contact that is desired and at which all readings should be taken. Speed and accuracy in reading the micrometer wheel are essential and develop with practice.

(c) *Testing.*—Preload the ring five or six times to its capacity until a constant micrometer reading at no load is obtained. This is called the zero reading. (*Caution: Do not overload the proving ring.*) A precautionary step is to set the micrometer wheel for the capacity of the ring as determined by the ratio $\left[\dfrac{maximum\ dial\ load}{maximum\ ring\ factor} \right]$. (The ring factor is obtained from the printed forms supplied with the proving ring.) This will prevent overloading the ring should the dial of the machine unknowingly read too low. After preloading, read and record the proving ring micrometer at no load or zero and again at a predetermined increment indicated on the calibration form supplied with the ring and shown by the load-indicating hand on the dial face. The load is then released slowly, and another zero reading is taken and recorded before loading the proving ring to the next higher predetermined load. This procedure is repeated until the entire dial has been encompassed. The zero readings before and after each new incremental load are averaged for each load.

As the load slowly approaches an increment to be read, the machine operator indicates this fact by warning, "Get ready." The observer moves the micrometer wheel of the proving ring into reading position and follows the vibrating reed at the proper reading clearance. When the load-indicating hand coincides with the increment line on the dial, the machine operator says "Read." The load should be applied very slowly from the time the warning is given until the reading is made, being careful not to exceed the increment of load. This allows the observer time to manipulate the micrometer wheel.

A minimum of three calibrations should be made and the data averaged for determining the accuracy of the machine.

6. Points to Observe for an Accurate Reading.—(a) Allow the ring to come to room temperature and flex it slowly five or six times before beginning the test. (See sec. 5(a).)

(b) Obtain the same tone of the vibrating reed at each reading of the micrometer. The reed should vibrate directly above the center of the micrometer wheel.

(c) Do not jam the vibrating reed by running the micrometer wheel up too far, and be sure to back the micrometer wheel off after each load is read.

(d) Shut off the battery after each reading. Under continuous use the battery will lose its charge quickly, thereby changing the tone of the vibrating reed.[29]

(e) Keep the hands away from the ring during a calibrating run; heat from the body changes the temperature of the ring. The micrometer wheel is the only part that need be touched during a calibration.

(f) Keep the load increasing during the calibration process.

(g) Handle the ring with care; it is a precision instrument.[30]

7. Calculations.—The average zero reading for each incremental load is subtracted from the reading of the proving ring for that load, and the results are entered in the "Ring deflection" column of the calibration form. These are the ring deflections at room temperature. They are corrected to 70° F or 23° C by one of the following equations:

$$d_{70} = d_t \ (1\text{-}k \ (t\text{-}70))$$
$$d_{23} = d_t \ (1\text{-}k \ (t\text{-}23))$$

---

[29] Remove the dry cell from the battery housing before placing the ring back in its box. Batteries corrode, then drip and stain the ring.

[30] Put a thin coat of oil on the ring before placing it in the box for reshipment.

where:

d_{70} = ring deflection at 70° F,
d_{23} = ring deflection at 23° C,
d_t = ring deflection at room temperature,
t = average room temperature, and
k = temperature correction factor.

The corrected deflections are multiplied by the ring factor, a constant obtained from the Bureau of Standards, to give the load indicated by the proving ring. The error is obtained by subtracting the load indicated by the ring from the load indicated by the dial. The percent error is derived by dividing the error by the dial load and multiplying by 100. Values of average, maximum, and minimum error are then evident.

8. Tolerance.—The limits of accuracy for either the 200,000-pound or the 300,000-pound dial are as follows: not more than 1 percent for all loads above 50,000 pounds, and not more than 500 pounds (one division on the testing machine dial) for loads below 50,000 pounds.

When these limits of error are exceeded, subsequent calibrations should be made. If no improvement is found, the calibration data together with a description of any improper behavior of the machine should be submitted to the Chief, Division of General Research, so that corrective measures may be recommended by the Denver laboratories.

SAMPLING CURING COMPOUNDS FOR CONCRETE

Designation 38

1. Sampling Methods.—Prior to withdrawing the sample, the contents of the drum should be thoroughly mixed. The method used will vary with the type of container and the facilities available. As 55-gallon drums are used in nearly all shipments, the following procedures are based on their use, but the same principles apply to other types of containers.

(a) *Bunghole-Type Drum—Compressed Air Available.*—This is the quickest and surest method for thoroughly mixing the liquid. A convenient length of ½- or ¾-inch pipe is connected with an air line which has been properly trapped to eliminate oil and moisture. The pipe is inserted through the bunghole to the bottom of the drum. Enough air pressure (about 20 lb/in²) is applied to cause a vigorous churning action throughout the contents but not so violent as to blow liquid out of the bunghole. The bottom of the drum is swept with the outlet end of the pipe until the material is completely mixed.

(b) *Bunghole-Type Drum—Compressed Air Not Available.*—This method should be used only when it is impracticable to use compressed air. The drum is turned on its side and rolled back and forth a total distance of about 200 feet. The drum is up-ended and a long wooden or metal slat-shaped paddle inserted through the bunghole and the contents stirred until the operator is satisfied that they are completely mixed. The drum is again turned on its side and rolled back and forth about 100 feet.

(c) *Open-Head Drum.*—The head of the drum is removed and a portion of the liquid poured off into a clean container. The remainder is stirred thoroughly with a power stirrer; all settled material should be uniformly distributed throughout the liquid. Hand stirring is inefficient and laborious and should be used only when a power stirrer is unobtainable. After the liquid in the drum has been thoroughly mixed, and while stirring is continued, the portion initially removed should be returned gradually to the drum. The entire contents should be stirred until complete blending is attained.

Immediately after the mixing is completed, the sample is dipped out and poured into the sample container. Several gallons should be withdrawn from a bunghole-type drum before the sample container is filled from the drum outlet.

2. Samples.—One sample should be taken of each batch or shipment of curing compound. Each sample should be placed in a 1-quart can fitted with a double friction-top lid. The cans should be filled to approximately 90 percent capacity, closed tightly, and the lids spot soldered. They should be carefully packed to avoid damage and spillage, and shipped to the Chief, Division of Research, Bureau of Reclamation, Attn: D-1510, P O Box 25007, Denver Colorado 80225. Each sample should be properly labeled and accompanied with a copy of filled-in form 7–1417. This form (fig. 230) may be obtained from the Denver office.

SAMPLING MASTIC JOINT SEALER

Designation 39

1. General.—Mastic joint sealer is supplied as a single-component, ready-mixed, homogeneous material composed of asphalt, rubber, inert filler, and suitable solvents. It is commonly shipped in 55-gallon drums or 5-gallon cans.

2. Sampling.—This material may separate slightly after long storage, and it may be necessary to remix the sealer before sampling. Since the compound is supplied in open-head drums, a mechanical mixer with a paddle of sufficient length to reach the bottom should in a short time, with thorough mixing, restore the material to its original condition.

7-1417 (8-70)
Bureau of Reclamation

ACCEPTANCE TESTS
For paint, concrete curing compound, joint filler,
roofing materials, dampproofing compound, etc.

TO: Denver Engineering and Research Center ATTN: 1520

	Date
	Project

FROM (Original to be returned to:)   DISTRIBUTION

		Feature
☐ Regional Director		Constr. Spec. No.
☐ Project Manager		
☐ E & RC Files - D-930		Item No.
☐ Laboratory Files - Attn. 1521		Ship Via:
		B/L No.

	SAMPLE 1	SAMPLE 2	SAMPLE 3
Purchase Order No.			
Paint Material			
Material Specification			
Manufacturer			
Quantity Represented			
Batch or Lot No.			
Item to be Painted			
Construction Contractor			
Date Received at Project			
Date Sampled			
Date Shipped			

REMARKS:

_____
Signature of Submitting Officer   Date

Instructions: Use this form to submit samples to Denver Laboratories for Acceptance Tests. Forward all copies (including two for Chief, Division of General Research's files) to Denver at the time samples are submitted. After testing, appropriate data on results of tests will be inserted on all copies and distribution made to designated offices from Denver.

THIS SPACE RESERVED FOR LABORATORY REPORT

Date Sample Received			
Date Tests Completed			
Acceptable			

REMARKS:

_____
Chief, Division of General Research   Date

GPO 834-432

Figure 230.—Sample transmittal form and test report. (Form 7–1417)

3. Samples.—A sample representing each lot of the sealer should be taken and placed in a 1-quart can. The can, fitted with a friction-top lid, should be filled to approximately 90 percent capacity. Sealing, labeling data, and shipping instructions for samples as given in designation 38 should be used.

METHOD OF TEST AND PERFORMANCE REQUIREMENTS FOR WATER-REDUCING, SET-CONTROLLING ADMIXTURES

Designation 40

1. Definition.—A water-reducing, set-controlling admixture (hereinafter referred to as WRA) is a material that, when used as an admixture in concrete, will reduce the quality of water and cementitious materials otherwise required in a similar concrete mix without the WRA to produce concrete of a given consistency and compressive strength and control the time of setting of the concrete.

2. Materials.—WRA's need not be limited to specific chemical compositions, but are commonly of two types: (a) lignosulfonic acids and their salts which, if in liquid form, should contain not less than 30 percent nor more than 50 percent solids; or (b) hydroxylated carboxylic acids and their salts which should be in the form of a water solution containing not less than 30 percent total dissolved solids consisting of a minimum of 99 percent organic carboxylic acids or their salts.

3. Performance Requirements.—Until 1977, Bureau of Reclamation construction specifications cited a Bureau standard for performance testing WRA's proposed for use. However, the Bureau now cites ASTM (American Society for Testing Materials) Designation: C 494, Standard Specification for Chemical Admixtures for Concrete. This ASTM standard includes all applicable methods of tests for chemical admixtures, including WRA's.

TEST FOR LIGHTWEIGHT PARTICLES IN AGGREGATE

Designation 41

1. General.—This test, graphically outlined in figure 231, is a procedure for determining the percentage of lightweight material in aggregate by means of sink-float separation in a heavy liquid of suitable specific gravity. This test is designed for 1,000-gram samples of sand passing the No. 4 screen and retained on the No. 8 screen and 5,000-gram samples of No. 4 to 3/4-inch and 3/4- to 1½-inch coarse aggregate size fractions. It is essentially adapted to control testing of wet aggregate produced by heavy-media processing. Samples should be representative of aggregate to be tested.

2. Chemicals.—(a) *Heavy Liquid.*—The heavy liquid should consist of a mixture of acetylene tetrabromide ($\approx$ s.g. 2.97) and petroleum naptha ($\approx$ s.g. 0.76) in such proportions that the desired specific gravity may be obtained. Other liquids such as perchloroethylene ($\approx$ s.g. 1.63) are suitable for use in this test, but they are more toxic, more volatile, and less stable.

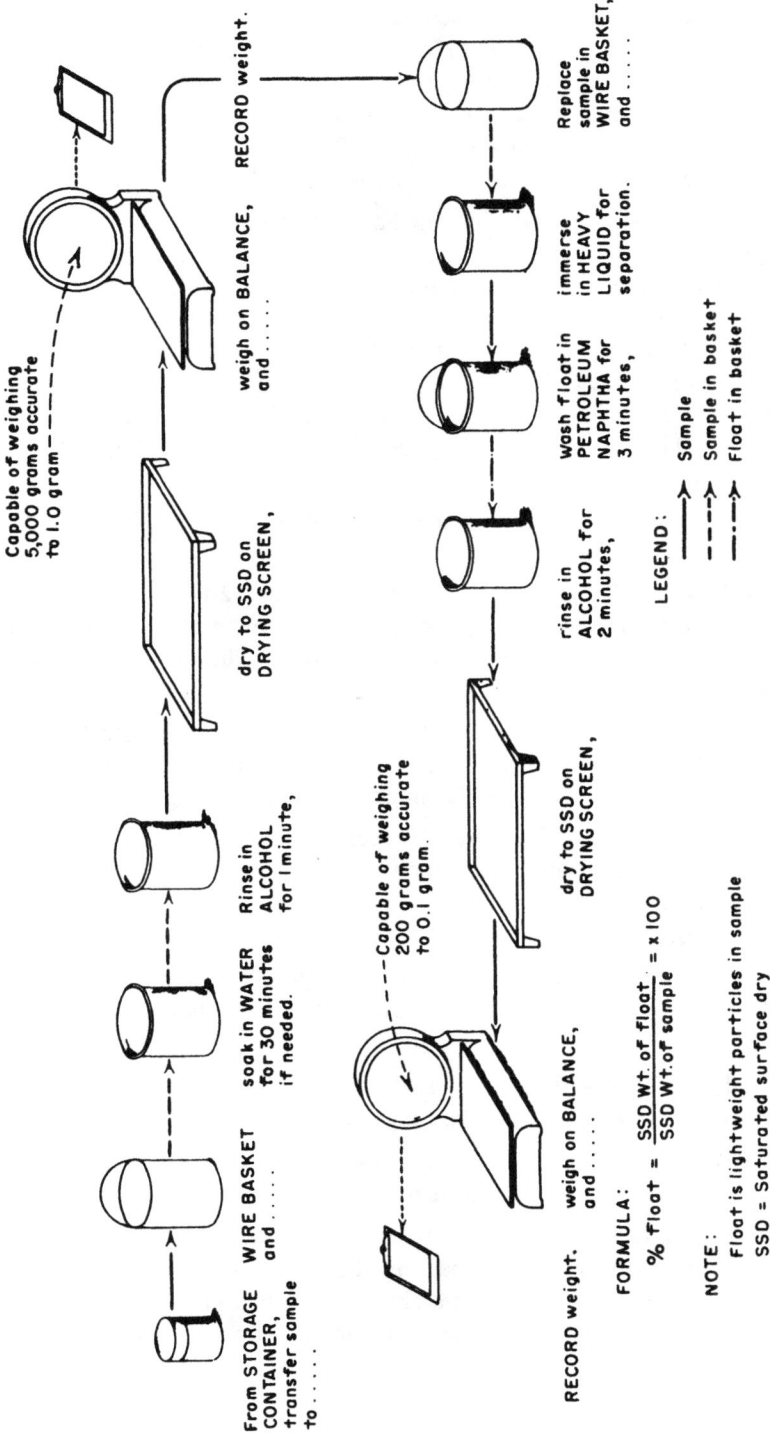

Figure 231.—Flow diagram for separation of lightweight pieces of aggregate. 288–D–2677.

(**Note**—*Acetylene tetrabromide and other liquids sometimes used for this test are highly toxic, both by absorption through the skin and by inhalation. Tests should be performed in a well-ventilated area, preferably under a hood, and care should be taken to avoid contact of the liquids with the skin or inhalation of the fumes. Use of protective clothing such as rubber gloves, rubber aprons, and goggles is advised.*)

The liquids employed in this test have differential rates of evaporation, causing a progressive change in the specific gravity of the liquid mixture; hence, periodic checking and adjustment of the specific gravity of the mixture are necessary.

(b) *Alcohol.*—The alcohol used for rinsing the lightweight aggregate fractions should be denatured alcohol conforming to formula 1, 12A, 28A, or 30, as defined in Title 26, Code of Federal Regulations, Part 212. These documents may be obtained from the Superintendent of Documents, Washington, D.C.

3. Apparatus.—The following equipment is necessary to conduct this test:

(a) *Test Containers.*—(1) *Sand.*—Five 2-liter stainless steel beakers, each approximately 5 inches in diameter, and two No. 10 mesh wire baskets with bail which will fit snugly, without binding, in the beakers.

(2) *Coarse aggregate.*—Five 12-quart stainless steel beakers, each approximately 10 inches in diameter, and two No. 10 mesh wire baskets with bail which will fit snugly, without binding, in the beakers.

(b) *Skimmers.*—Two No. 10 mesh saucer-shaped wire skimmers, 2-inch diameter for sand and 4-inch diameter for coarse aggregate, both with handles.

(c) *Hydrometers.*—Hydrometers accurate to 0.01 and suitable for determining specific gravities of heavy liquid, alcohol, and petroleum naphtha and hydrometer flasks of sufficient size for hydrometer measurements.

(d) *Balance.*—Balance or scale accurate within 0.1 percent of the test load at any point within the range of use.

(e) *Drying Screen.*—One 10-mesh wire screen tray, 18 inches by 36 inches by 1 inch deep, with legs.

4. Preparation of Samples.—Obtain representative samples of approximately 1,000 grams of sand passing the No. 4 screen and retained on the No. 8 screen and approximately 5,000 grams each of No. 4 to ¾-inch and ¾- to 1½-inch coarse aggregate size fractions. Wash as necessary to remove contaminating materials.

5. Test Procedure.—(a) *Separation.*—Place sample in appropriate wire basket and immerse basket with sample in water and soak for 30 minutes. (If the sample is already saturated, the soaking period will not be required.) Remove basket with sample and drain for 1 minute. Place basket with sample in alcohol.[31] Agitate sample by vertical reciprocating motion of the basket for 1 minute. Remove basket with sample and drain excess alcohol. Transfer sample from basket onto drying screen, spreading sample one particle in thickness. Dry sample to saturated surface-dry (SSD) condition (designation 9, sec. 3 or designation 10, sec. 4).

A saturated surface-dry condition exists when no free liquid remains on the surface of the particles.[32] Drying time to reach this condition may be reduced by use of a fan. Weigh saturated surface-dry sample and record weight. Place sample in wire basket and immerse in heavy liquid.[31]

Stir sample vigorously for 5 seconds, allow 15 seconds for turbulence to subside, and immediately skim off the floating particles. Skimming depth should be just sufficient to remove aggregate particles floating on the heavy liquid surface. Care should be taken while skimming not to create undue currents in the liquid which would disturb the settled material. After skimming, allow excess heavy liquid to drain from the skimmer back into the beaker.

Transfer skimmed float material to appropriate wire basket. Repeat stirring and skimming sequence until all lightweight material has been removed and transferred to the basket. Skimming should be completed within 5 minutes from the time the sample is immersed in heavy liquid.

(b) *Cleanup.*—Immerse the basket containing the float material in a beaker containing petroleum naphtha.[31] Agitate float material in the beaker of naphtha by vertical motion of basket for 3 minutes. Remove basket from beaker and drain excess naphtha for 1 minute. Remove remaining naphtha by agitating basket in alcohol for 2 minutes.[31] Remove basket and drain excess alcohol for 1 minute. The petroleum naphtha and alcohol rinse liquids should be replaced periodically to prevent excessive contamination (after 12 to 15 tests, depending on test conditions). Transfer rinsed material to drying screen and dry to saturated surface-dry condition as before, weigh, and record weight.

6. Calculations.—The percentage of float in the sample is determined by the following formula:

---

[31] Place basket of sand in a 2-liter beaker containing approximately 1.5 liters of the liquid, and coarse aggregate in a 12-quart beaker containing approximately 2 gallons of the liquid.

[32] *Caution:* Care must be exercised to ensure that the sand samples do not contain any free liquid when placed into the heavy liquids, as a film of liquid surrounding a heavyweight particle may cause it to float.

$$\text{Float, } \% = \frac{W_f}{W} \times 100$$

where:

W = total saturated surface-dry weight of sample, and
W_f = saturated surface-dry weight of float.

RELATIVE HUMIDITY TESTS FOR MASONRY UNITS

Designation 42

1. General.—Cracking in concrete masonry walls can be reduced by minimizing moisture shrinkage in the wall. To do this, it is necessary to preshrink the masonry units by adequate drying before they are placed in the wall. The degree of dryness needed will vary, but for any given locality the concrete should be in an air-dry condition corresponding to the average annual relative humidity of the outside air for that locality. This principle applies to all types of concrete masonry units regardless of type of aggregate or method of curing used.

The test that was used for many years for determining the permissible moisture content of concrete blocks failed to give a significant and realistic indication of the moisture condition of the concrete relative to the local air. The relative humidity method [33] overcomes this deficiency in that it provides a reliable indication, in about 20 to 60 minutes, of the state of dryness and potentiality for drying shrinkage of concrete blocks in subsequent exposure to atmospheric conditions.

In addition to its simplicity of operation, the apparatus can be easily calibrated or checked for accuracy and can be used under a variety of conditions. For instance, it can be used to determine the moisture condition of a block soon after sample blocks have been selected, or if desired, the test can be made at any convenient later time.

All concrete masonry units should be in an air-dry condition when placed in the structure. Stockpiles should be carefully protected against rain by storage under adequate cover and upon sufficient dunnage to prevent exposure to moisture until used. The air-dry condition is intended to be the moisture condition of concrete masonry units in a state of equilibrium with the average annual relative humidity at the project site, as established at the nearest U.S. Weather Station.

Upon delivery of concrete masonry units to the project site, representative samples should be selected from the stockpiles and tested for air-dry condition at Government expense. A tolerance of 5 percent relative

---

[33] "Moisture Condition of Hardened Concrete by the Relative Humidity Method," ASTM Designation C 427.

humidity above average annual relative humidity should be allowed when the ambient relative humidity of the air (before delivery) is above the average annual relative humidity for the locality. No concrete masonry units should be used when the relative humidity test shows them to contain more moisture than this limit.

Test for air-dry condition is described in ASTM Designation C 427, "Moisture Condition of Hardened Concrete by the Relative Humidity Method."

2. Hygrometer Calibration.—Before the apparatus is used for testing concrete blocks for the first time, the hygrometer must be calibrated. It must also be calibrated periodically later. (By calibration is meant testing for accuracy and establishing any corrections to be applied to the scale readings.) Calibration equipment consists of special saturated salt solutions supplied by the Denver laboratories if facilities are available for precision temperature control and complete calibration, or the instrument can be initially calibrated by the Denver laboratories.

3. Procedure.—The tests (whether on blocks or for calibration) should be conducted in a room which can be maintained at a fairly constant temperature during each test, that is, within 5° F of the average room temperature. Preferably, the average room temperature should not be less than 65° F nor more than 85° F.

Both the apparatus and concrete block specimens should be at the average room temperature before the beginning of the test. Each block specimen should be stored in a vaportight container (or wrapped in a plastic sheet) to protect it from moisture change prior to test and during the period required to attain room temperature.

The room should be provided with a table about 30 inches high, 24 inches wide, and 48 inches long, on which the apparatus is placed during test to facilitate reading and recording relative humidity and temperature within the container. The top of the table should be of nonabsorptive material and level in all directions.

A timer should be available that tilts on its base for easy viewing and that has a large readable dial, sweep second hand, and minute hand to make readings at desired time intervals up to 60 minutes.

Data should be recorded and plotted in a systematic manner as illustrated in sample data sheets, ASTM Designation C 427.

Low slump concrete consolidated by vibration is necessary for durable structures in severe climates. Photo P1236-600-66A

Restricted working area on a thin arch dam requires close coordination of construction operations and materials handling. Photo P459–640–2654NA

REFERENCE LIST OF INSPECTION ITEMS

The following inspection check list is adapted from the ACI Manual of Concrete Inspection published by Committee 311, Inspection of Concrete. The list is not intended to be used as an outline for daily check of work items but rather as a means to develop an individual inspector's reference list of items for an assignment in conjunction with particular specifications requirements.

GENERAL CONCRETING WORK

Preliminary Preparation.

Study of plans and specifications:

Requirements and permissible tolerances.

Items of payment.

Rights-of-way where involved.

Review of contractor's plans, including equipment, methods, and organization.

Establishment of method of keeping records and making reports.

Materials.

Aggregates:

Acceptance and control test evaluating grading, specific gravity, absorption, organic impurities, silt, percent lighter than a certain specific gravity, clay lumps, sodium sulfate soundness, Los Angeles abrasion test, freezing and thawing resistance, alkali-aggregate reactivity, and petrographic description.

Processing, including washing and screening, stockpiling and handling.

Cement and/or pozzolan:

Inspection reports, mill and acceptance tests.

Storage facilities and protection.

Provision for identification.

Schedule for use.

Water:

Source and supply, chemical suitability, turbidity.

Admixtures:

Type, use, tests and/or certification.

Reinforcement steel:

Certification of tests.

Size, identification, and tagging.

Bending.

Surface condition.

Other embedded material or equipment including anchors, waterstops, conduits, cooling pipes, drains, etc.

Design of concrete mix.

Trial mix and adjustment including computations of batch quantities and yield.

Requirements of materials control including quantities, identification, acceptability, uniformity, storage, handling, waste, scheduling of testing.

Preparation for Concrete Placement.

Lines and grades.

Excavation and foundations:

Class of excavation, location, dimensions, shape, surface preparation and drainage.

Forms:

Location.

Preparation of surface.

Alinement; allowances for finish tolerances; movement or settlement.

Stability, bearing and support adequacy, ties, and spacers.

Inspection openings, size, spacing, and location.

Final cleanup.

Reinforcement:

Size, diameter, length, bending, splicing, anchorage.

Location, number, minimum clear spacing, minimum coverage.

Stability, wiring, chair supports, and spacers.

Cleanness.

Fixtures location, stability, cleanness.

Openings.

Calibration of concrete control test equipment.

Calibration and checking of scales and batching equipment.

Mixer efficiency tests.

Provision for continuous operation.

Available protection against sun, rain, cold weather.

Adequate vibrating, finishing, and curing equipment including standby equipment as needed.

During Concreting Operations.

Working conditions:

Preparations for weather conditions completed; adequate lighting for night work; provisions for adequate accessibility to batching, mixing, and placing operations; safety requirements complied with; sufficient personnel.

Batching:

Cement, aggregate, water, admixtures.

Control of variations of free moisture in aggregate.
Check of batching devices as operating.
Mixer charging operations and sequence.
Check yield of concrete batch.

Mixing:
Minimum time, overloading.
Schedule of mixing, batches delayed in mixer.

Control of air-entrained concrete:
Accurate measurement of air-entraining agent, tests for air content of concrete, regulation of air content, adjustment of mix to compensate for air content, avoidance of excessive mixing or vibration, avoidance of wet consistency.

Compensation for free moisture in the aggregates.

Control of consistency, observation, concrete control tests.

Adjustments as required.

Transporting and handling:
Methods, no segregation, no excessive slump loss, time limits.

Placing:
Uniformity, continuous operation, preparation of contact surfaces, mortar bedding, vertical drop, means of introducing into forms, little or no flow after depositing, depth of layers, bleeding, systematic and uniform vibration, removal of temporary ties and spacers, deposition of rejected batches, any unusual requirements of deposition such as under water.

Consolidation:
Adequate vibrator equipment.
Specified vibrator frequency, sufficient size.
Adequate crew.
No overworking.

Construction joints:
Location, preparation of surface.

Expansion joints:
Joint material, location, alinement, stability, freedom from interference of subsequent movement.

Finishing of unformed surfaces:
Shallow surface layer of mortar, bleeding, no overworking, first floating, alinement of surface, final hard troweling, hair checks, protection from rain.

Finishing of formed surfaces:
Condition on removal of forms (honeycomb, peeling, ragged tie holes, ragged form lines, sand streaks).

Schedule of acceptance test when required.

After Concreting.

Protection from damage:

Impact, overloading, marring, traffic, spilling of oil on surface.
Excessive high and low temperatures, drying.

Time of removal of forms.

Minor repairs after form removal.

Curing:

Surfaces continuously moist, time of beginning curing, length of curing period, protection from heat and freezing.

Tests of Concrete.

Consistency tests.

Percent of entrained air.

Unit of weight test of fresh concrete.

Temperature.

Test of mixer efficiency.

Strength tests:

Molding, storing, handling specimens (standard conditions, field conditions), field tests, packing and shipping of specimens when required.

Specimens of hardened concrete (cores, beams).

Other tests.

Records and Reports.

Records, materials (quantity, quality, and source), mix computations, batching and mixing methods and equipment, placing and curing, weather, other special items.

Reports:

Daily, summary.

Diary.

Photographs.

Special Work

Cold-Weather Concreting.

Limiting temperatures and times:

Outdoor air, enclosure, materials.

Heating materials, contact surfaces, and enclosure; protection from drying ventilation; safety.

Removal of forms, protection from too rapid cooling.

Hot-Weather Concreting.

Available methods for lowering concrete temperature:

Cooling materials, prewetting aggregate, addition of ice, covering of work, evaporative cooling.

Scheduling of work.

Ready-Mixed Concrete.
Plant:
> Materials, batching, mixer capacity and condition, mixer efficiency tests.

Construction site:
> Elapsed time, mixer drum revolutions, tests for consistency and air content, strength test, temperature, delivery schedule, short waits, wash water, control, uniform discharge, no segregation, conveying to form.

Shotcrete.
> Materials (acceptability, quantities), conditions of equipment, preliminary mixing, air volume and pressures, water pressure, preparation of surfaces, application (thickness, bond, no sagging, construction joints cleanup), surface finish, curing, tests.

Pumped Concrete.
Application to construction:
> Methods, mechanical condition of equipment, cleanliness, capacity, arrangement, agitation, pipeline.

Mix design:
> Maximum size aggregate, percent sand, entrained air, slump.

Placing:
> Temperature control methods, slump loss control, no segregation, consolidation, scheduling delivery of concrete, curing.

Grouting Under Base Plates.
> Preparation of base, stiff mix, pack tightly, clearance, forms.

Pressure Grouting.
> Holes (depth, spacing, freedom from clogging).
> Materials (acceptability, quantities used), cement storage and orderly use.
> Injection (sequence, pressures, times, completeness of penetration, no damage to structure), leaks, calking material, equipment condition.
> Reports.

Two-Course Floors.
> Preparation of surface of base course, materials for top course, proportions and consistency, uniform screeding, rolling or tamping, first floating, final troweling, curing.

Terrazzo.
> Thickness of layers, uniformity, curing.

Stucco.
> Mortar, preparation of backing surface, bonding to backing surface, uniform finish, curing each layer.

Masonry.

Units:

Laboratory tests for strength and absorption.

Field inspection for size, shape, and soundness.

Construction:

Moisture content of units, completeness of bedding in mortar, alinement compliance with building code (mortar, minimum wall thickness, lateral support, bonding courses, supports for beams, openings in walls).

Pipe.

Method of manufacture.

Laboratory control tests of materials curing.

Units:

Laboratory tests for strength, absorption, and watertightness.

Field inspection for size, shape, soundness, interior surface smoothness; reinforcement (selected specimens); repairs.

Inspection reports.

Installation.

Alinement, bedding, joints (calking, filling, curing), backfilling, protection from damage.

Field tests of completed pipeline.

Cast Stone.

Laboratory tests for strength and absorption.

Field inspection for soundness and uniformity (match sample).

Ornamental Concrete.

Location and neat joining of molds, surface coating to avoid sticking or staining, curing.

Colored Concrete.

Pigments (integral or dust-on), matching of colors, thorough and intimate mixing of integral color with cement, uniform application and incorporation of dust-on color, troweling and finishing, curing.

Painting.

Cleaning surface, neutralizing surface, uniform application, curing portland cement paints.

Lightweight Concrete.

Lightweight aggregates (acceptability, prewetting, prevention of segregation).

Gas concrete (admixtures, timing of operations).

Test for unit weight.

Heavyweight Concrete.

Heavyweight aggregates (suitability, specific gravity, type, acceptance tests).

Mixing and handling same as conventional concrete; placing requirements and conditions specialized; tests.

Mass Concrete.
Times and rates of placement, avoidance of high or nonuniform temperatures, bonding of lifts, prevention of aggregate breakage and separation.

Preplaced Aggregate Concrete.
Placement of coarse aggregate, void content, composition and control of consistency of grout, sequence and pressures of grouting, completeness of filling of voids, condition of equipment, standby equipment, handling and measurement of materials for work.

Tiltup Construction.
Surface of casting platform, joints in sheet bondbreakers, timing and uniformity of liquid bondbreakers, alinement of edge forms, compaction of concrete at bottom corners, connections to columns, provision for expansion (if specified), strength of concrete at time of lifting, pickup points; avoidance of excessive pulling, jerking, or jarring.

Underwater Construction.
Avoidance of flowing water, temperatures, continuous placement, operation of tremie or bucket, minimizing of wash, protection from flowing water for several days.

Vacuum Concrete.
Allowance for reduction in thickness of slabs, timing and duration of application of vacuum, uniformity of processing, condition of mats.

Prestressed Concrete.
Strength of concrete at time of prestressing sheathing of reinforment (if specified), accurate placing of reinforcement, avoidance of obstruction or excessive friction, measurement of tension by means of jack pressure and/or lengthening of steel, thoroughness of grouting (if specified).

SELECTED REFERENCES PERTAINING TO CONCRETE

Materials

Backstrom, J. E., Burrows, R. W., Mielenz, R. C., and Wolkodoff, V. E., "Origin, Evolution, and Effects of the Air Void System in Concrete— Part 2, Influence of Type and Amount of Air-Entraining Agent," Proc. ACI, vol. 55, 1958, pp. 261–272.

Blanks, R. F., "Fly Ash as a Pozzolan," Proc. ACI, vol. 46, 1950, pp. 701–707.

Blanks, R. F., "The Use of Portlant-Pozzolan Cement by the Bureau of Reclamation," Proc. ACI, vol. 46, 1950, pp. 89–108.

Blanks, R. F., and Cordon, W. A., "Practices, Experiences, and Tests with Air-Entrainment Agents in Making Durable Concrete," Proc. ACI, vol. 45, 1949, pp. 469–487.

Blanks, R. F., and Gilliland, J. L., "False Set in Portland Cement," Proc. ACI, vol. 22, 1951, pp. 517–532.

Blanks, R. F., and Meissner, H. S., "The Expansion Test as a Measure of Alkali-Aggregate Reaction," Proc. ACI, vol. 42, 1946, pp. 517–540.

Brewer, H. W., and Burrows, R. W., "Coarse-Ground Cement Makes More Durable Concrete," Proc. ACI, vol. 47, 1951, pp. 353–360.

Dunstan, Jr., E. R., "Performance of Lignite and Subbituminous Fly Ash in Concrete–A Progress Report," Bureau of Reclamation REC-ERC-76-1, January 1976.

Higginson, E. C., and Wallace, G. B., "Control Testing for Separation of Lightweight Material from Aggregate," ASTM Bulletin No. 243, January 1960.

Kalousek, G. L., "Abnormal Set of Portland Cement—Causes and Correctives," Bureau of Reclamation REC-OCE-69-2, November 1969.

Kalousek, G. L., "Investigation of Shrinkage Compensating Cements," Bureau of Reclamation REC-OCE-70-43, October 1970.

Kalousek, G. L., Porter, L. C., and Harboe, E. M., "Past, Present, and Potential Developments of Sulfate-Resisting Concretes," ASTM, Journal of Testing and Evaluation, vol. 4, No. 5, pp. 347-354, September 1976.

Mielenz, R. C., "Petrography and Engineering Properties of Igneous Rocks," Engineering Monograph No. 1, Bureau of Reclamation. Revised 1961.

Mielenz, R. C., Greene, K. T., and Benton, E. J., "Chemical Test for the Reactivity of Aggregates with Cement Alkalies; Chemical Processes in Cement Aggregate Reaction," Proc. ACI, vol. 44, 1948, pp. 193-221.

Mitchell, L. J., "Thermal Expansion Tests of Aggregates, Neat Cements, and Concretes," Proc. ASTM, vol. 53, 1953, pp. 963-977.

Price, W. H., "Fly Ash in Heavy Construction," Concrete Laboratory Report No. C-828.2, Bureau of Reclamation, July 31, 1956. (Synopsis in Electrical World, vol. 146, No. 9, p. 17, August 27, 1956.)

Price, W. H., and Cordon, W. A., "Tests of Lightweight Aggregate Concrete Designed for Monolithic Construction," Proc. ACI, vol. 45, 1949, pp. 581-660.

Price, W. H., and Kretsinger, D. G., "Aggregates Tested by Accelerated Freezing and Thawing of Concrete," Proc. ASTM, vol. 51, 1951, pp. 1108-1119.

Savage, J. L., "Special Cements for Mass Concrete," Bureau of Reclamation, 1936, pp. 230.

Tuthill, L. H., "Developments in Methods of Testing and Specifying Coarse Aggregates," Proc. ACI, vol. 39, 1943, pp. 21-32.

Van Alstine, C. B., "Mixing Water Control by Use of a Moisture Meter," Proc. ACI, vol. 52, 1955, pp. 341-348.

Witte, L. P., and Mielenz, R. C., "Tests Used by the Bureau of Reclamation for Identifying Reactive Concrete Aggregate," Proc. ASTM, 1948, vol. 48.

Witte, L. P., Mielenz, R. C., and Glantz, O. J., "Effect of Calcination on Natural Pozzolans," ASTM Special Technical Publication No. 99, 1949.

Mixing, Handling, and Placing

ACI Committee 604, "Recommended Practice for Winter Concreting Methods," Proc. ACI, vol. 52, 1956, pp. 1025-1048.

ASTM Designation: A 305, "Tentative Specifications for Minimum Requirements for the Deformations of Deformed Steel Bars for Concrete Reinforcement."

Backstrom, J. E., Burrows, R. W., Mielenz, R. C., and Wolkodoff, V. E., "Origin, Evolution, and Effects of the Aid Void System in Concrete—Part 3, Influence of Water-Cement Ratio and Compaction," Proc. ACI, vol. 55, 1958, pp. 359-376.

Burnett, G. E., and Fowler, A. L., "Painting Exterior Concrete Surfaces with Special Reference to Pretreatment," Proc. ACI, vol. 43, 1947, pp. 1077-1086.

Burnett, G. E., and Spindler, M. R., "Effect of Time of Application of Sealing Compounds on the Quality of Concrete," Proc. ACI, vol. 49, 1953, pp. 193-200.

Higginson, E. C., "Some Effects of Vibration and Handling of Concrete Containing Entrained Air," Proc. ACI, vol. 49, 1953, pp. 1-12.

Rippon, C. S., "Construction Joint Cleanup Method at Shasta Dam," Proc. ACI, vol. 40, 1944, pp. 293-304.

Rippon, C. S., "Methods of Handling and Placing Concrete at Shasta Dam," Proc. ACI, vol. 39, 1943, pp. 1-8.

Shideler, J. J., Brewer, H. W., and Chamberlain, W. H., "Entrained Air Simplifies Winter Curing," Proc. ACI, vol. 47, 1951, pp. 449-459.

Tessitor, F., and Rosewarne, P., "Economy Through Better Control of Reinforcing Steel," Proc. ACI, vol. 47, 1951, pp. 333-340.

Tuthill, L. H., "Concrete Operations in the Concrete Ship Program," Proc. ACI, vol. 41, 1945, pp. 137-180.

Tuthill, L. H., "Inspection of Mass and Related Concrete Construction," Proc. ACI, vol. 46, 1950, pp. 349-359.

Tuthill, L. H., "Tunnel Lining Methods for Concrete Compared," Proc. ACI, vol. 37, 1941, pp. 29-48.

Tuthill, L. H., Glover, R. E., Spencer, C. H., and Bierce, W. B., "Insulation for Protection of New Concrete in Winter," Proc. ACI, vol. 48, 1952, pp. 253-272.

Vidal, E. N., and Blanks, R. F., "Absorptive Form Lining," Proc. ACI, vol. 38, 1942, pp. 253-268.

Wallace, G. B., "Insulation Facilitates Winter Concreting," Engineering Monograph No. 22, Bureau of Reclamation, October 1955.

Properties

Blanks, R. F., and Meissner, H. S., "Deterioration of Concrete Dams Due to Alkali-Aggregate Reaction," Proc. ASCE, vol. 71, No. 1, January 1945, pp. 3-18, 1089-1110.

Blanks, R. F., Meissner, H. S., and Rawhouser, C., "Cracking in Mass Concrete," Proc. ACI, vol. 34, 1938, pp. 477-495.

Blanks, R. F., Meissner, H. S., Tuthill, L. H., "Curing Concrete with Sealing Compounds," Proc. ACI, vol. 42, 1946, pp. 493-512.

Blanks, R. F., Vidal, E. N., Price, W. H., and Russel, F. M., "The Properties of Concrete Mixes," Proc. ACI, vol. 36, 1940, pp. 433-475.

DePuy, G. W., Kukacka, L. E., Auskern, A., Colombo, P., Romano, A., Steinberg, M., Causey, F. E., Cowan, W. C., Lockman, W. T., Smoak, W. G., "Concrete-Polymer Materials, Fifth Topical Report," Bureau of Reclamation REC-ERC-73-12, and Brookhaven National Laboratory BNL 50390 (TID-4500), December 1973.

Glover, R. E., "Calculation of Temperature Distribution in a Succession of Lifts Due to Release of Chemical Heat," Proc. ACI, vol. 34, 1938, pp. 105-116.

Glover, R. E., "Flow of Heat in Dams," Proc. ACI, vol. 31, 1935, pp. 113-124.

Graham, J. R., Backstrom, J. E., Redmond, M. C., Backstrom, T. E., Rubenstein, S. R., "Evaluation of Concrete for Desalination Plants," Bureau of Reclamation REC-ERC-71-15, March 1971.

Higginson, E. C., and Glantz, O. J., "The Significance of Tests for Sulfate Resistance of Concrete," Proc. ASTM, vol. 53, 1953, p. 1102.

Lorman, W. R., "The Theory of Concrete Creep," Proc. ASTM, vol. 40, 1940, pp. 1082-1102.

McConnell, D., Mielenz, R. C., Holland, W. Y., and Greene, K. T., "Cement-Aggregate Reaction in Concrete," Proc. ACI, vol. 44, 1948, pp. 93-128.

McHenry, Douglas, "A New Aspect of Creep in Concrete and Its Application to Design," Proc. ASTM, vol. 43, 1943, pp. 1069-1084.

Meissner, H. S., "Cracking in Concrete Due to Expansive Reaction Between Aggregate and High-Alkali Cement as Evidenced in Parker Dam," Proc. ACI, vol. 37, 1941, pp. 549-568.

Mielenz, R. C., Wolkodoff, V. E., Backstrom, J. E., and Burrows, R. W., "Origin, Evolution, and Effects of the Air Void System in Concrete—Part 4, The Air Void System in Job Concrete," Proc. ACI, vol. 55, 1958, pp. 507-517.

Mielenz, R. C., Wolkodoff, V. E., Backstrom J. E., and Flack, H. L., "Origin, Evolution, and Effects of the Air Void System in Concrete—Part 1, Entrained Air in Unhardened Concrete," Proc. ACI, vol. 55, 1958, pp. 95-122.

Ore, E. L., and Wallace, G. B., "Structural and Lean Mass Concrete as Affected by Water-Reducing, Set-Retarding Agents," ASTM Special Technical Publication No. 266.

Price, W. H., "Erosion of Concrete by Cavitation and Solids in Flowing Water, " Proc. ACI, vol. 43, 1947, p. 1109.

Price, W. H., "Factors Influencing Concrete Strength," Proc. ACI, vol. 47, 1951, pp. 417-432.

Price, W. H., and Wallace, G. B., "Resistance of Concrete and Protective Coatings to Forces of Cavitation," Proc. ACI, vol. 46, 1950, p. 109.

Raphael, J. M., "The Development of Stresses in Shasta Dam," Trans. ASCE, vol. 118, 1953, p. 53.

Rawhouser, C., "Cracking and Temperature Control of Mass Concrete," Proc. ACI, vol. 41, 1945, pp. 305-348.

Shideler, J. J., "Calcium Chloride in Concrete," Proc. ACI, vol. 48, 1952, pp. 537-559.

Shideler, J. J., and Chamberlin, W. H., "Early Strength of Concrete as Affected by Steam Curing Temperatures," Proc. ACI, vol. 46, 1950, pp. 273-283.

Tuthill, L. H., "Resistance of Cement to the Corrosive Action of Sodium Sulphate Solutions, " Proc. ACI, vol. 33, 1937, pp. 83-106.

Tuthill, L. H., "Resistance to Chemical Attack," ASTM Special Technical Publication No. 169, 1955, p. 188.

Witte, L. P., and Backstrom, J. E., "Properties of Heavy Concrete Made with Barite Aggregates," Proc. ACI, vol. 51, 1955, pp. 65-88.

Witte, L. P., and Backstrom, J. E., "Some Properties Affecting the Abrasion Resistance of Air-Entrained Concrete, " Proc. ASTM, vol. 51, 1951, pp. 1141-1155.

INDEX

specific gravity, 57, 121
stockpiling, 192
stream, 57, 87
strength, 56
surface moisture
  control, 206
  effect on slump, 246
test holes, pits, and trenches, 98, 105
tests, 121, 202 (*see also* specific test
  concerned)
tunnel lining, 62
undersize, 63, 192
unit weights of, 125
volume batching, 212
volume change, 56
weathering effects, 54, 92
weighing equipment (field operations), 229
wood fragments in, 202
Coatings, cement-mortar, for steel pipe,
  478
  inplace, 478
  shop-applied, 480
Cobbles, 62
Coefficient of expansion and contraction
  aggregates, 56
  cement paste, 18, 56
  concrete, 18, 56
  mortar, 56
  rocks, 56
Coefficient of variation
  controllable factors, 173, 176
  test data, 172
  uncontrollable factors, 176
Cold joints, 72, 301, 303
Cold weather precautions
  accelerator, use of, 70
  curing, 383
  form insulation, 385
  mixing, 256
  placing, 71, 383
Colored floor finishes, 465
Colorimetric test, 122
  sand, 531
Compaction by vibration, 290, 308,
  335, 348, 350, 456
  effect on cold joints, 303
Compaction, precast concrete pipe, 343
Compounds, sealing, 299, 376
  application, 376
Compressive strength (*see* Strength,
  compressive)

Concrete (*see also* more specialized
  subjects)
construction data, 165
control
  definition, 147
  importance, 78, 245
  of placing, 280
  report forms and narrative, 165,
    167
  temperature, 79
definition, 1
face, 300
harsh, 5
impact hammer for testing, 253
interior, 300
materials, 85, 181
mixes, 5, 131
precooling, 81
prestressed, 449
properties, 3, 33
  effect of various factors, 33
  quality control, 245, 253
  segregation, detrimental effect, 78
  selected references pertaining to, 600
Conductivity, thermal, 32, 40
Conglomerates, 91
Consistency, 5, 132, 246 (*see also*
  Slump)
  meters, 248
Consolidation (*see also* Compaction and
  Vibration)
  canal lining, 350
  during placement, 292
  importance, 36, 79
  mass concrete, 301
  precast concrete pipe, 343
  tunnel invert, 309
  vibrators, 350
Construction joints
  bond and requirements, 262
  cleanup operations, 262
  definition, 262
  formed, treatment, 270
  grooves, 277
  longitudinal, 309
  mortar layer, 292
  tunnel inverts, 309
  watertightness, 262, 292
Construction progress reports, 165
Contaminating substances
  in aggregate, 54, 93, 122, 202
  in water, 68

Surface preparation for cement-mortar coatings, 479, 480
Suspended solids, water, 70
Sustained load, effect, 28
Swiss hammer (*see* Impact test hammer)

Talus deposits, 87
Tanks
  curing, 161
  mixing water, 233, 243
Tannic acid, 9
Tar distillate, effect on hardened concrete, 9
Temperature
  cement, effect on mass concrete, 210
  concrete, computation, 578
  curing, effect on concrete, 255, 381
  control, 161, 381, 568
  drop, 385
  placing, control, 79, 81, 243, 255
  variations, stress, 81
  wet-bulb, 384
Tensile stress, 27
Terrazzo floors, 466
Test (*see also* Testing methods)
  compression, f a c t o r s affecting strength, 572
  cylinder
    capping, 569
    curing, 19, 161, 252, 568, 572
    molds
      cans, 252, 567
      cast-iron, 565
      pneumatically applied mortar, 274
    number required, 252
    packing and shipping, 569
    significance, 252
    storage and handling, 567
  data, 172
    criteria for strength, 173
    statistical analysis of, 172
  holes (aggregate deposits), 98, 102, 116
    cased, 99
    uncased, 102
  pits (aggregate deposits), 98, 105, 116
  pozzolanic materials, 126
  trenches (aggregate deposits), 98, 113
  unit weight, 254, 528

Testing machine maintenance and calibration, 579
Testing methods (designations 1 through 43)
  abrasion (Los Angeles test) of coarse aggregate, 545
  absorption
    coarse aggregate, 521
    sand, 518
  acceptance test for water-reducing, set-retarding agents, 586
  aggregate (*see also* specific test for)
    content of fresh concrete, 550
    fine (sand), mortar-making properties of, 541
  air content of fresh concrete, 550, 554
  air-entraining agent, sampling, 578
  batching for machine-mixed laboratory concrete, 562
  bleeding of concrete, 556
  capping, cylinders, 569
  cement content of fresh concrete, 550
  clay and silt in sand, 532
    friable particles in aggregate, 529
  colorimetric test (sand), 531
  compressive strength of concrete (cylinder tests), 572
  concrete (fresh and hardened) (*see* specialized subjects)
  consistency and slump, fresh concrete, 548
  cylinders, casting in
    cans, 567
    cast-iron molds, 565
  false set in cement, 512
  gravel (*see* Testing methods, aggregate)
  Jackson turbidimeter, turbidity of water, 576
  lightweight material in coarse aggregate, 536
  lightweight material in sand, 535
  lightweight particles in aggregate (heavy media separation), 586
  Los Angeles (abrasion) test, 545
  mastic joint filler, sampling, 584
  mixer performance test, 558
  mixing in laboratory, 562
  mortar-making properties of fine aggregate, 541
  organic impurities in sand, 541

✿U.S. GOVERNMENT PRINTING OFFICE: 1989 676-106/05103